Mechanical, Corrosive and Tribological Degradation of Metal Coatings and Modified Metallic Surfaces

Mechanical, Corrosive and Tribological Degradation of Metal Coatings and Modified Metallic Surfaces

Editors

Matic Jovičević-Klug
Patricia Jovičević-Klug
László Tóth

Basel • Beijing • Wuhan • Barcelona • Belgrade • Novi Sad • Cluj • Manchester

Editors

Matic Jovičević-Klug
Max-Planck-Institut für
Eisenforschug
Düsseldorf
Germany

Patricia Jovičević-Klug
Max-Planck-Institut für
Eisenforschug
Düsseldorf
Germany

László Tóth
Óbuda University, Bánki
Donát Faculty of Mechanical
and Safety Engineering,
Institute of Mechanical
Engineering and Technology
Budapest
Hungary

Editorial Office
MDPI
St. Alban-Anlage 66
4052 Basel, Switzerland

This is a reprint of articles from the Special Issue published online in the open access journal *Coatings* (ISSN 2079-6412) (available at: https://www.mdpi.com/journal/coatings/special_issues/Modified_Metallic_Surfaces).

For citation purposes, cite each article independently as indicated on the article page online and as indicated below:

Lastname, A.A.; Lastname, B.B. Article Title. *Journal Name* **Year**, *Volume Number*, Page Range.

ISBN 978-3-7258-0275-3 (Hbk)
ISBN 978-3-7258-0276-0 (PDF)
doi.org/10.3390/books978-3-7258-0276-0

© 2024 by the authors. Articles in this book are Open Access and distributed under the Creative Commons Attribution (CC BY) license. The book as a whole is distributed by MDPI under the terms and conditions of the Creative Commons Attribution-NonCommercial-NoDerivs (CC BY-NC-ND) license.

Contents

Patricia Jovičević-Klug, Matic Jovičević-Klug and László Tóth
Mechanical, Corrosive, and Tribological Degradation of Metal Coatings and Modified Metallic Surfaces
Reprinted from: *Coatings* **2022**, *12*, 886, doi:10.3390/coatings12070886 . . . . **1**

Van Cao Long, Umut Saraç, Mevlana Celalettin Baykul, Luong Duong Trong, Ştefan Ţălu and Dung Nguyen Trong
Electrochemical Deposition of Fe–Co–Ni Samples with Different Co Contents and Characterization of Their Microstructural and Magnetic Properties
Reprinted from: *Coatings* **2022**, *12*, 346, doi:10.3390/coatings12030346 . . . . **4**

Zhao Wang, Zhaohui Cheng, Yong Zhang, Xiaoqian Shi, Mosong Rao and Shangkun Wu
Effect of Voltage on the Microstructure and High-Temperature Oxidation Resistance of Micro-Arc Oxidation Coatings on AlTiCrVZr Refractory High-Entropy Alloy
Reprinted from: *Coatings* **2023**, *13*, 14, doi:10.3390/coatings13010014 . . . . **17**

Patricia Jovičević-Klug, László Tóth and Bojan Podgornik
Comparison of K340 Steel Microstructure and Mechanical Properties Using Shallow and Deep Cryogenic Treatment
Reprinted from: *Coatings* **2022**, *12*, 1296, doi:10.3390/coatings12091296 . . . . **28**

Patricia Jovičević-Klug, Matic Jovičević-Klug and Bojan Podgornik
Unravelling the Role of Nitrogen in Surface Chemistry and Oxidation Evolution of Deep Cryogenic Treated High-Alloyed Ferrous Alloy
Reprinted from: *Coatings* **2022**, *12*, 213, doi:10.3390/coatings12020213 . . . . **39**

Zhongyu Dou, Haili Jiang, Rongfei Ao, Tianye Luo and Dianxi Zhang
Improving the Surface Friction and Corrosion Resistance of Magnesium Alloy AZ31 by Ion Implantation and Ultrasonic Rolling
Reprinted from: *Coatings* **2022**, *12*, 899, doi:10.3390/coatings12070899 . . . . **54**

Joanna Sypniewska and Marek Szkodo
Influence of Laser Modification on the Surface Character of Biomaterials: Titanium and Its Alloys—A Review
Reprinted from: *Coatings* **2022**, *12*, 1371, doi:10.3390/coatings12101371 . . . . **66**

Rijun Wang, Fulong Liang, Xiangwei Mou, Lintao Chen, Xinye Yu, Zhujing Peng and Hongyang Chen
Development of an Improved YOLOv7-Based Model for Detecting Defects on Strip Steel Surfaces
Reprinted from: *Coatings* **2023**, *13*, 536, doi:10.3390/coatings13030536 . . . . **90**

Zhanyuan Yang, Hong Li, Xingqiang Cui, Jinke Zhu, Yanhui Li, Pengfei Zhang and Junru Li
Highly Efficient CuInSe$_2$ Sensitized TiO$_2$ Nanotube Films for Photocathodic Protection of 316 Stainless Steel
Reprinted from: *Coatings* **2022**, *12*, 1448, doi:10.3390/coatings12101448 . . . . **107**

Juanjuan Han, Wei Zheng, Qingqiang Chen, Jie Sun and Shubo Xu
Study on Size Effect of Surface Roughness Based on the 3D Voronoi Model and Establishment of Roughness Prediction Model in Micro-Metal Forming
Reprinted from: *Coatings* **2022**, *12*, 1659, doi:10.3390/coatings12111659 . . . . **123**

Yuriy Stulov, Vladimir Dolmatov, Anton Dubrovskiy and Sergey Kuznetsov
Electrochemical Synthesis of Functional Coatings and Nanomaterials in Molten Salts and Their Application
Reprinted from: *Coatings* 2023, 13, 352, doi:10.3390/coatings13020352 141

Mingyang Chen, Guangming Lu and Gang Wang
Discrimination of Steel Coatings with Different Degradation Levels by Near-Infrared (NIR) Spectroscopy and Deep Learning
Reprinted from: *Coatings* 2022, 12, 1721, doi:10.3390/coatings12111721 167

Xiaohong Wang, Tingjun Hu, Tengfei Ma, Xing Yang, Dongdong Zhu, Duo Dong, Junjian Xiao, and et al.
Mechanical, Corrosion, and Wear Properties of TiZrTaNbSn Biomedical High-Entropy Alloys
Reprinted from: *Coatings* 2022, 12, 1795, doi:10.3390/coatings12121795 178

Germain Boissonnet, Ewa Rzad, Romain Troncy, Tomasz Dudziak and Fernando Pedraza
High Temperature Oxidation of Enamel Coated Low-Alloyed Steel 16Mo3 in Water Vapor
Reprinted from: *Coatings* 2023, 13, 342, doi:10.3390/coatings13020342 194

Yutong Duan, Honggang Lei and Shihong Jin
Experimental Study on Fatigue Performance of Welded Hollow Spherical Joints Reinforced by CFRP
Reprinted from: *Coatings* 2022, 12, 1585, doi:10.3390/coatings12101585 206

Hongjun Ni, Chunyu Lu, Yu Zhang, Xingxing Wang, Yu Zhu, Shuaishuai Lv and Jiaqiao Zhang
Effects of Sodium Carbonate and Calcium Oxide on Roasting Denitrification of Recycled Aluminum Dross with High Nitrogen Content
Reprinted from: *Coatings* 2022, 12, 922, doi:10.3390/coatings12070922 221

Shengfang Zhang, Guangming Lv, Fujian Ma, Ziguang Wang and Yu Liu
Influence of Contact Interface Friction on Plastic Deformation of Stretch-Bend Forming
Reprinted from: *Coatings* 2022, 12, 1043, doi:10.3390/coatings12081043 234

Maxim Sergeevich Vorobyov, Elizaveta Alekseevna Petrikova, Vladislav Igorevich Shin, Pavel Vladimirovich Moskvin, Yurii Fedorovich Ivanov, Nikolay Nikolaevich Koval, Tamara Vasil'evna Koval, and et al.
Steel Surface Doped with Nb via Modulated Electron-Beam Irradiation: Structure and Properties
Reprinted from: *Coatings* 2023, 13, 1131, doi:10.3390/ coatings13061131 258

Wei Wang, Yuan Wang, Jingqi Huang and Lunbo Luo
Mathematical Model of Surface Topography of Corroded Steel Foundation in Submarine Soil Environment
Reprinted from: *Coatings* 2022, 12, 1078, doi:10.3390/coatings12081078 275

Shixue Zhang, Yunlong Ding, Zhiguo Zhuang and Dongying Ju
Corrosion Resistance of Mg/Al Vacuum Diffusion Layers
Reprinted from: *Coatings* 2022, 12, 1439, doi:10.3390/coatings12101439 291

Patricia Jovičević-Klug and Michael Rohwerder
Sustainable New Technology for the Improvement of Metallic Materials for Future Energy Applications
Reprinted from: *Coatings* 2023, 13, 1822, doi:10.3390/coatings13111822. 310

Editorial

Mechanical, Corrosive, and Tribological Degradation of Metal Coatings and Modified Metallic Surfaces

Patricia Jovičević-Klug [1,2,*], Matic Jovičević-Klug [3,*] and László Tóth [4,*]

1 Department of Metallic Materials and Technology, Institute of Metals and Technology, Lepi Pot 11, 1000 Ljubljana, Slovenia
2 Jožef Stefan International Postgraduate School, Jamova Cesta 39, 1000 Ljubljana, Slovenia
3 Department of Microstructure Physics and Alloy Design, Max Planck Institute for Iron Research, Max-Planck-Straße 1, 40237 Düsseldorf, Germany
4 Bánki Donát Faculty of Mechanical and Safety Engineering, Óbuda University, Népszínház U. 8., 1081 Budapest, Hungary
* Correspondence: p.jovicevic-klug@mpie.de (P.J.-K.); m.jovicevic-klug@mpie.de (M.J.-K.); toth.laszlo@bgk.uni-obuda.hu (T.L.)

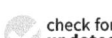

Citation: Jovičević-Klug, P.; Jovičević-Klug, M.; Tóth, L. Mechanical, Corrosive, and Tribological Degradation of Metal Coatings and Modified Metallic Surfaces. *Coatings* 2022, 12, 886. https://doi.org/10.3390/coatings12070886

Received: 13 June 2022
Accepted: 21 June 2022
Published: 22 June 2022

Publisher's Note: MDPI stays neutral with regard to jurisdictional claims in published maps and institutional affiliations.

Copyright: © 2022 by the authors. Licensee MDPI, Basel, Switzerland. This article is an open access article distributed under the terms and conditions of the Creative Commons Attribution (CC BY) license (https://creativecommons.org/licenses/by/4.0/).

Mechanical, corrosive, and tribological degradation of metal and metal coatings is just one of the challenges faced by numerous industries. The industries which are most commonly affected by the degradation of the material are the tool industry, medicine, the electronic industry, the transport industry, aeronautics, the mining industry, and the energy sector. For this reason, it is important to constantly adapt and search for the further improvement of metallic surfaces and metal coatings.

Mechanical degradation of metals normally occurs under the influence of various external forces such as tension, compression, and shear that are commonly present during various material processing and applications such as grinding, agitation, and extrusion [1].

Corrosive degradation of the metal involves redox reactions, under which the metal ions are lost by the dissolution at the anode (oxidation) in a corrosive environment [2]. Despite the fact that different types of coatings are used for improvement of corrosion resistance for metallic surfaces, there is still room for improvement that is needed for specific application that require various mechanical and corrosive properties simultaneously. Furthermore, the development of new coatings is ongoing, due to the continuous learning of corrosion–coating interactions and the necessity of optimizing the coating technology for the development of more affordable and sustainable coatings.

Tribological degradation, which is mostly bound to the contact interaction of two materials and their loss in mass due to this interaction, is another form of material degradation that is the most commonly considered aspect in applications with moveable parts. In this case, wear (abrasion and adhesion) and galling of components are the major forms of mechanically based degradation, which induce the deformation and removal of interacting materials [3]. Additionally, when oxidative or corrosive conditions are present, the mechanical wear is enhanced by the corrosion reactions, leading to tribocorrosion that usually causes rapid degradation of the material [4].

To counter the above degradation effects, coatings are most commonly applied as a laminar layer of certain material that covers the base material with aim of protecting it in order to sustain its properties, with the added properties of the coating layer under specific conditions. When choosing the right coating for the selected material, several parameters have to be considered: thickness, roughness, defects on the coating, tensile strength, elongation, resistivity, structure, residual stresses, adhesion, hardness, ductility, electrical properties, magnetic properties, and anti-corrosion resistivity. All of these coating properties are important as they have impact on its final properties and applicability under the targeted environment and conditions [5,6]. Metallic coatings can consist of metals (Ni, Zn, Cr etc.), metal composites (usually deposited on expensive materials), or metal alloys

(Pb-Sn-Cu, Ni-Cr-Al-Y etc.). Coatings can be produced by immersion, spray, crystallization, cladding, and galvanization [5,6].

Modification of the metallic surface is another possibility to enhance the surface of the material to different degradation conditions, at which the modified base material is used as the surface protector. In contrast to coatings, this technique allows higher compatibility and integrability of the surface to the bulk, reducing the possibility of delamination, decohesion, and interface-weakening between the surface and bulk of the material [5,6].

The surface modification and coating development can be done by laser technology [7], electron beam technology [8], ion implantation [9,10], glow discharge technology [11], chemical vapor deposition (CVD) [11,12], and vacuum deposition by physical vapor deposition (PVD) [13]. Laser technique can also be used for formation of thin and hard coatings, where coatings can be formed by the fusion of alloying elements with gas method (CVD and PVD), by the pure vapor deposition method (PVD), or by pyrolytic and photochemical formation, and even by chemical methods (LCVD) [13]. Electron beam deposition (also known as EBPVD) is a method of using electron beams in a vacuum to irradiate evaporated material, so that it can form as a thin film on the base material. Ion implantation technique is used for the modification of the surface structure in the superficial layer, which can change the surface in a physical as well as chemical fashion through direct ion implantation or plasma ion integration [14]. Glow discharge technology covers methods such as carburizing, carbonitriding, nitriding, and sulfonitriding, as well as boriding and siliciding [11]. Glow discharge methods are characterized by the use of non-equilibrium, low temperature, and non-isothermal plasma, which is formed as a result of the continuous drawing of energy from the electric field [11]. CVD and plasma-assisted CVD (PACVD) are treatments whereby the deposition of a layer from the gas phase is formed through additional participation of a chemical reaction or series of them. In the majority of cases, these techniques are applied for the development of anti-abrasion and anti-corrosion layers [13]. PVD is a technique of surface engineering during which the coating material is evaporated and condensed on the target material in a vacuum. With prior treatment of the substrate material, the fusion and compatibility of the coating can be significantly enhanced [13].

In the last few years, with the increased demands on reduction of material consumption, recycling of the material, environmental safety, improvement of the material performance and life cycle, and last but not least, cost saving of the production, it has become a further necessity to develop new alternatives to enhance both the surface as well as the bulk of materials in a single processing procedure [15]. As such, development and optimization of heat treatment procedures has obtained increasing emphasis in recent research. One of the possible alternatives, which is also a green technology, is cryogenic treatment, during which the material is exposed to cryogenic temperatures (sub-zero temperatures) [16,17]. The advantage of cryogenic treatment in comparison to coatings and surface modification techniques is that it can change the properties of the material from surface to core, whereas coatings and modification of metallic material are confined only to the properties of the surface layer. As a result, the latter only works as long as the coating/modified surface remains stable in protecting the material core, whereas heat treatment-based techniques can omit this dependency.

To summarize, combining different heat treatment methods and surface engineering methods (coatings and modification of the surface layer) is the best way to improve materials' properties and provide the most optimal method for improving materials' resistance and performance.

Funding: This research received no external funding.

Conflicts of Interest: The authors declare no conflict of interest.

References

1. Niaounakis, M. *Biopolymers: Processing and Products*; Elsevier: Waltham, MA, USA, 2015.
2. LeBozec, N.; Thierry, D.; Peltola, A.; Luxem, L.; Luckeneder, G.; Marchiaro, G.; Rohwerder, M. Corrosion Performance of Zn–Mg–Al Coated Steel in Accelerated Corrosion Tests Used in the Automotive Industry and Field Exposures. *Mater. Corros.* **2013**, *64*, 969–978. [CrossRef]
3. Hutchings, I.; Shipway, P. *Tribology: Friction and Wear of Engineering Materials: Second Edition*; Elsevier Inc.: Amsterdam, The Netherlands, 2017.
4. Munoz, A.I.; Espallargas, N.; Mischler, S. *Tribocorrosion*; SpringerBriefs in Applied Sciences and Technology; Springer International Publishing: Cham, Germany, 2020.
5. Burakowski, T. Metal Surface Engineering-Status and Perspectives of Development. *Int. Cent. Sci. Tech. Inf.* **1990**, *6*, 166–178.
6. Burakowski, T. A Word about Surface Engineering. *Metall. Heat Treat. Surf. Eng.* **1993**, *121–123*, 16–31.
7. Dubik, A. Laser Application. *WNT Wars.* **1991**, *1*, 1–22.
8. Oczoce, K. The Shaping of Materials by Concentrated Fluxes of Energy. *Publ. Rzesz. Tech. Univ. Rzesz.* **1988**.
9. Burakowski, T.; Wierchon, T. *Surface Engineering of Metals, Priciples, Equipment, Technologies*; CRC Press LLC.: Boca Raton, FL, USA, 1999.
10. Podgórski, A.; Jagielski, J.; Gawlik, G. Perfection of Ion Implantation Method in Order to Obtain Optimum Properties of the Technical Surface Layer. *Mechanik* **1990**, *11*, 393–396.
11. Weston, G.F. *Cold Cathode Glow Discharge Tubes*; Academic Press: London, UK, 1968.
12. Marciniak, A. Processes of Heating and Nitriding of a Cathode in Conditions of Glow Discharge. Ph.D. Thesis, Warsaw University of Technology, Warsaw, Poland, 1983.
13. Mattox, D.M. *Handbook of Physical Vapor Deposition (PVD) Processing: Film Formation, Adhesion, Surface Preparation and Contamination Control*; Noyes Publications: Park Ridge, NJ, USA, 1998.
14. van Dorp, W.F.; Hagen, C.W. A Critical Literature Review of Focused Electron Beam Induced Deposition. *J. Appl. Phys.* **2008**, *104*, 81301. [CrossRef]
15. Vogl, V.; Åhman, M.; Nilsson, L.J. The Making of Green Steel in the EU: A Policy Evaluation for the Early Commercialization Phase. *Clim. Policy* **2020**, *21*, 78–92. [CrossRef]
16. Jovičević-Klug, P.; Podgornik, B. Review on the Effect of Deep Cryogenic Treatment of Metallic Materials in Automotive Applications. *Metals* **2020**, *10*, 434. [CrossRef]
17. Toth, L. Cryogenic Treatment against Retained Austenite. In Proceedings of the Mérnöki Szimpózium a Bánkiban, Budapest, Hungary, 3–5 May 2021; pp. 181–186.

Article

Electrochemical Deposition of Fe–Co–Ni Samples with Different Co Contents and Characterization of Their Microstructural and Magnetic Properties

Van Cao Long [1], Umut Saraç [2,*], Mevlana Celalettin Baykul [3], Luong Duong Trong [4], Ştefan Ţălu [5,*] and Dung Nguyen Trong [1,6,*]

1. Institute of Physics, University of Zielona Góra, Prof. Szafrana 4a, 65-516 Zielona Góra, Poland; vancaolong2020@gmail.com
2. Department of Science Education, Bartın University, Bartın 74100, Turkey
3. Department of Metallurgical and Materials Engineering, Eskişehir Osmangazi University, Eskişehir 26480, Turkey; cbaykul@ogu.edu.tr
4. Department of Electronic Technology and Biomedical Engineering, Hanoi University of Science and Technology, Hanoi 100000, Vietnam; luong.duongtrong@hust.edu.vn
5. The Directorate of Research, Development and Innovation Management (DMCDI), Technical University of Cluj-Napoca, 15 Constantin Daicoviciu St., 400020 Cluj-Napoca, Romania
6. Faculty of Physics, Hanoi National University of Education, 136 Xuan Thuy, Cau Giay, Hanoi 100000, Vietnam
* Correspondence: usarac@bartin.edu.tr (U.S.); stefan.talu@auto.utcluj.ro (T.Ş.); dungntsphn@hnue.edu.vn (D.N.T.)

Citation: Cao Long, V.; Saraç, U.; Baykul, M.C.; Trong, L.D.; Ţălu, Ş.; Nguyen Trong, D. Electrochemical Deposition of Fe–Co–Ni Samples with Different Co Contents and Characterization of Their Microstructural and Magnetic Properties. *Coatings* **2022**, *12*, 346. https://doi.org/10.3390/coatings12030346

Academic Editors: Matic Jovičević-Klug, Patricia Jovičević-Klug and László Tóth

Received: 1 February 2022
Accepted: 3 March 2022
Published: 6 March 2022

Publisher's Note: MDPI stays neutral with regard to jurisdictional claims in published maps and institutional affiliations.

Copyright: © 2022 by the authors. Licensee MDPI, Basel, Switzerland. This article is an open access article distributed under the terms and conditions of the Creative Commons Attribution (CC BY) license (https://creativecommons.org/licenses/by/4.0/).

Abstract: In this study, to explore the effect of Co contents on the electroplated Fe–Co–Ni samples, three different Fe–Co_{33}–Ni_{62}, Fe–Co_{43}–Ni_{53}, and Fe–Co_{61}–Ni_{36} samples were electrochemically grown from Plating Solutions (PSs) containing different amounts of Co ions on indium tin oxide substrates. Compositional analysis showed that an increase in the Co ion concentration in the PS gives rise to an increment in the weight fraction of Co in the sample. In all samples, the co-deposition characteristic was described as anomalous. The samples exhibited a predominant reflection from the (111) plane of the face-centered cubic structure. However, the Fe–Co_{61}–Ni_{36} sample also had a weak reflection from the (100) plane of the hexagonal close-packed structure of Co. An enhancement in the Co contents caused a strong decrement in the crystallinity, resulting in a decrease in the size of the crystallites. The Fe–Co_{33}–Ni_{62} sample exhibited a more compact surface structure comprising only cauliflower-like agglomerates, while the Fe–Co_{43}–Ni_{53} and Fe–Co_{61}–Ni_{36} samples had a surface structure consisting of both pyramidal particles and cauliflower-like agglomerates. The results also revealed that different Co contents play an important role in the surface roughness parameters. From the magnetic analysis of the samples, it was understood that the Fe–Co_{61}–Ni_{36} sample has a higher coercive field and magnetic squareness ratio than the Fe–Co_{43}–Ni_{53} and Fe–Co_{33}–Ni_{62} samples. The differences observed in the magnetic characteristics of the samples were attributed to the changes revealed in their phase structure and surface roughness parameters. The obtained results are the basis for the fabrication of future magnetic devices.

Keywords: cauliflower-like agglomerates; Co contents; crystallinity; Fe–Co–Ni thin film samples; magnetic properties; phase structure; pyramidal particles; roughness parameters

1. Introduction

Nanostructured ferromagnetic materials in the form of thin films are widely used in many technological applications and attract great attention because of their good physical and magnetic features [1–5]. To date, many physical and chemical growth techniques have been developed that are utilized in the production process of magnetic thin film samples. Among the growth techniques developed, the electrochemical deposition technique has been successfully used in computer read/write heads and Micro–ElectroMechanical Systems (MEMS) applications due to its unique features [1,3,6–12]. It is well known that ternary ferromagnetic alloy films are interesting soft magnetic materials due to their high saturation magnetization and low coercive field [4,6]. The conducted studies showed that the Fe, Ni, and Co components in binary Ni–Co, Ni–Fe and Fe–Cu and ternary Ni–Co–Cu, Ni–Fe–Cu, Co–Fe–Cu and Fe–Co–Ni magnetic materials grew by the electrochemical deposition technique on Indium Tin Oxide (ITO) covered glass substrates which can be tuned by controlling the Fe, Ni and Co ion concentrations in the Plating Solutions (PSs), respectively [13–21]. However, the relative compositions of Co and Fe components in the samples were found be higher than those in the PSs for different electroplating parameters [6,13,14,16–24], which is in good agreement with the definition defined by Brenner [25]. In recent years, scientists have flexibly used a variety of methods to study the structural, mechanical, and magnetic properties of nanomaterials and thin films. With the simulation method, the influence of size, heating rate, temperature and annealing time has been successfully studied on the structure, electronic structure, phase transition and mechanical properties of metal Ni [26–28], Fe [29], Al [30,31], Alloy AuCu [32,33], CuNi [34–36], NiAu [37], FeC [38], FeNi [39,40], AgAu [41], AlNi [42], and NiCu [43].

In addition, with the magnetism of nanomaterials and thin films, the influence of nanoparticle size and shell thickness has been successfully studied on the Curie Tc phase transition temperature of Fe nanoparticles [44], the influence of external magnetic field and size on the temperature Neel TN phase transition of Fe_2O_3 thin films [45]. The obtained results show that the Neel TN transition temperature is always smaller than the Curie Tc phase transition temperature, the cause of this phenomenon is due to the Topo effect. With the experimental method, the authors have successfully studied the effects of Fe ion concentration in the PS. Furthermore, the deposition potential applied during electroplating process on the chemical composition, some physical properties, and magnetic characteristics of ternary ferromagnetic thin film samples were investigated [21,24]. These experimental studies clearly demonstrated that the surface performance, magnetic and structural characteristics were affected significantly by the deposit composition caused by the variation of the electroplating parameters [21,24]. On the other hand, in a former study, Fe–Co–Ni deposits were electrochemically manufactured on titanium sheets from a chloride–sulfate–tartaric acid medium at different Co^{2+}/Ni^{2+} ion ratios [46]. In another study, Fe–Co–Ni films were electrochemically fabricated on copper substrates from an ammonium–chloride–based PS at different Co^{2+}/Ni^{2+} ion ratios [47]. In addition, in a very recent study, nanocrystalline Ni–Co–Fe coatings were electroplated on copper plates from a sulfate–citrate PS at different Co ion concentrations [3]. To make magnetic Fe–Co–Ni thin films, researchers can use many methods such as evaporation and electrochemical deposition [48]. Among these, thin films obtained by the electrochemical deposition method have very high uniformity. However, in this work, Fe–Co–Ni samples were electroplated on ITO substrates and the Co contents in the samples were tuned by the amount of Co ion concentration in the sulfate-based PS. The structure, morphology, and magnetic characteristics of the resultant samples were discussed with respect to their Co contents. The results showed that the crystallinity, crystallite size, phase structure, particle shape, particle size, magnetic properties, and roughness parameters of the samples are strongly dependent on their Co contents. The obtained results will be applied in different fields of science and technology.

2. Materials and Method

The Fe–Co–Ni thin film samples were produced from PSs composed of Ni sulfate (0.07 M), Fe sulfate (0.0020 M), boric acid (0.1 M), and various Co sulfate concentrations (0.016 M, 0.024 M and 0.040 M). The samples were deposited galvanostatically at the same current density of -10 mA/cm^2 from freshly prepared PSs (pH value was 5.2 ± 0.1 and temperature was 22 ± 1 °C) without stirring. The electroplating processes were performed by employing a three-electrode system. A platinum sheet was utilized as a counter electrode, whereas a Saturated Calomel Electrode (SCE) was served as a reference electrode. The samples were grown on ITO coated glass substrates used as a working electrode. Before the plating process, the substrates were first rinsed in an acetone solution and then in ethanol solution. After that, the substrates were cleaned by an ultrasonic bath using deionized water. The crystal structure was defined by a Rigaku SmartLab X–Ray Diffractometer (XRD) (Rigaku Cooperation—Tokyo, Japan). The XRD measurements were carried out in the 2θ range between 40° and 54° at a scanning step of 0.01° using CuK$_\alpha$ radiation source. The compositional analysis was performed by an Energy Dispersive X–ray (EDX) spectroscopy. The X-ray diffraction beam is shined at the sample with a very narrow angle of incidence to increase the length of the X-ray beam that interacts with the thin film, keeping the sample stationary and rotating the receiver. Then the resulting diffraction beam appears on a concentric circle, recording the reflected beam intensity and first-order diffraction spectrum. An Oxford X–max 50 detector (Oxford Instruments, High Wycombe, UK) was used for the EDX measurements under an operating voltage of 20.00 kV. To study the surface structure, a Tescan MAIA3 Scanning Electron Microscopy (SEM) (TESCAN, Brno, Czech Republic) was used. The SEM measurements were done under the same operating voltage of 5.00 kV at room temperature. The particle sizes were determined from the SEM images using a freely available image processing and analysis software (ImageJ) (Software version for imageJ is 1.8.0). The roughness parameters were determined using a Veeco Multimode V Atomic Force Microscopy (AFM) (Veeco Instruments İnc., Santa Barbara, CA, USA) and evaluated using a WSxM 5.0 develop 9.4 software package [49]. To reveal the effect of the Co contents on the coercive field and squareness ratio, magnetic measurements were carried out by means of a JDAW–2000D model Vibrating Sample Magnetometer (VSM) (Xiamen Dexing Magnet Tech. Co., Ltd., City-Country: Xiamen, China) at ambient temperature and pressure by applying the external magnetic field parallel to the sample plane.

3. Results and Discussion

This paper aimed to study the impact of the Co contents on the structure, morphology, and magnetic characteristics of the Fe–Co–Ni deposits. To obtain the samples with various Co contents, the samples were grown onto ITO–coated glass substrates from PSs comprising different concentrations of Co ions using the electrochemical deposition technique. The potential–time transient curves are given in Figure 1.

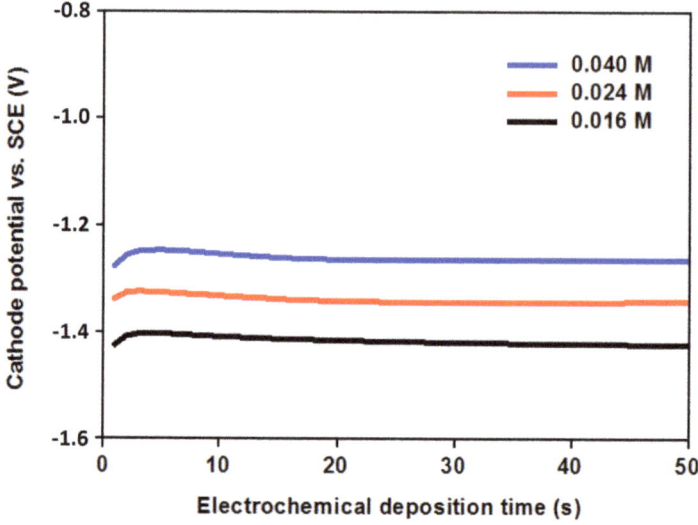

Figure 1. Potential–time transient curves of the samples electroplated at different ion concentrations of Co.

From Figure 1, it was understood that the samples can be grown properly from PSs containing different concentrations of Co ions owing to their stable cathode potentials. On the other hand, the cathode potential was detected to be higher for the sample electrochemically deposited from the PS with 0.016 M Co ion concentration compared to those determined for the samples deposited from PSs with 0.024 M and 0.040 M Co xml: confirmned with AE 1. keep email based on word 2. keep the ORCID 3. keep the format of not English for City in affs 4. keep the hyphen in figures 5. keep the two "Figure 6 indicates the AFM images of the samples. The samples possessed globular particles of various sizes" in textion concentrations.

EDX analyses showed that the samples electrochemically grown from PSs with different concentrations of Co ions have different Fe, Co, and Ni compositions. The compositional differences are shown in Table 1.

Table 1. The EDX data, phase structure, mean crystallite size, roughness parameters, coercive field, and squareness ratio of the samples.

	Co Ion Concentration (M)		
	0.016	0.024	0.040
Co (wt.%)	32.9	43.2	60.9
Ni (wt.%)	62.4	52.7	35.7
Fe (wt.%)	4.7	4.1	3.4
Resultant sample	Fe–Co_{33}–Ni_{62}	Fe–Co_{43}–Ni_{53}	Fe–Co_{61}–Ni_{36}
Phase structure	fcc	fcc	Fcc + hcp
Mean crystallite size (nm)	21.6	20.2	15.6
RMS roughness (nm)	14.4	17.8	28.4
Average roughness (nm)	11.0	14.0	21.8
Average particle size (nm)	~150	~14.0	~250
Coercive field (Oe)	36	51	121
Squareness ratio (%)	9.2	17.6	23.6

The sample grown from the PS with the lowest Co ion concentration of 0.016 M contained the lowest Co contents (32.9 wt.%), but the highest Ni (62.4 wt.%) and Fe (4.7 wt.%) compositions. In contrast to that, the sample electrochemically deposited from the PS with the highest Co ion concentration of 0.040 M included the highest Co contents (60.9 wt.%), but the lowest Ni (35.7 wt.%) and Fe (3.4 wt.%) compositions. The Co, Ni, and Fe compositions of the sample fabricated from the PS with an intermediate Co ion concentration of 0.024 M were 43.2, 52.7 and 4.1 wt.%, respectively. In summary, as the Co ion concentration in the PS was increased, the weight proportions of Ni and Fe components decreased, while the weight proportion of the Co component in the Fe–Co–Ni samples increased. Thus, three different ternary Fe–Co_{33}–Ni_{62}, Fe–Co_{43}–Ni_{53} and Fe–Co_{61}–Ni_{36} samples with different Co contents were fabricated. In recent studies [3,50], it was reported that the Co contents in electrochemically manufactured Ni–Co–Fe coatings and Co–Fe–Ni alloying micropillars increased but the Fe and Ni compositions decreased as the Co^{2+} ion concentration in the PS increased. Similar results were also found in Fe–Co–Ni deposits electroplated in a former study [46], which was consistent with our results. On the other hand, in this work, the presence or absence of Anomalous Co–Deposition (ACD) was also explored. In this context, the relative Co, Ni, and Fe ion percentages in the PSs were compared to the relative Co, Ni, and Fe compositions in the samples.

As seen in Figure 2, the relative Co (Fe and Ni) composition in the sample increased with increasing relative Co (Fe and Ni) ion percentage in the PS. However, in all cases, the Co contents in the samples were determined to be higher than the Co ion percentage in the PSs (Figure 2a). The same phenomenon found for the Co component was also detected for the Fe component (Figure 2b). This revealed the preferential electrochemical deposition for the Co and Fe components. However, the relative Ni composition in the samples was found to be lower than the relative Ni ion percentage in the PSs (Figure 2c). This indicated that the reduction of Ni components was inhibited. Thus, it was understood that the ACD behavior took place for all Co ions in the PS. The order of ACD was also revealed via composition ratio value (CRV).

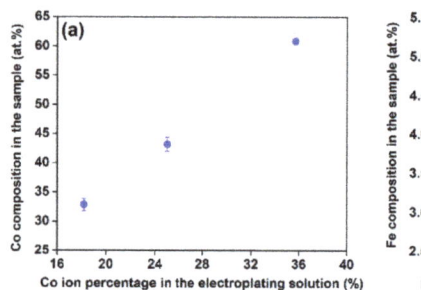 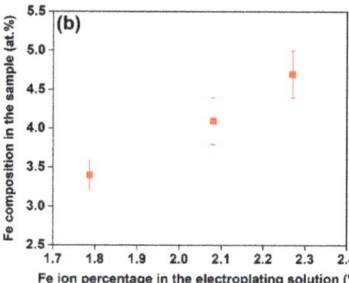 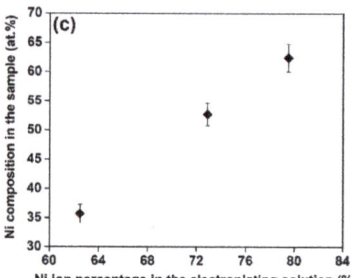

Figure 2. The relative Co (**a**), Fe (**b**), and Ni (**c**) composition in the samples against the relative Co (Fe and Ni) ion percentage in the PSs.

The CRV for a Co component is described by the following expression [46].

$$\text{CRV for Co} = \frac{\text{relative composition of Co in the sample}}{\text{relative ion concentration of Co in the PS}} \quad (1)$$

The relative concentration of Co ions in the PS is given by the following expression:

$$\text{Relative concentration of Co ions} = \frac{[\text{CoSO}_4]}{[\text{CoSO}_4 + \text{NiSO}_4 + \text{FeSO}_4]} \times 100 \quad (2)$$

The above procedure was also applied to calculate the CRV_{Fe} and CRV_{Ni}. From the results of the analysis depicted in Figure 3, it was understood that the CRV_{Fe} and CRV_{Co} were higher than one, while the CRV_{Ni} was lower than one, revealing that the reduction rate of Ni^{2+} was significantly lower than the reduction rates of Fe^{2+} and Co^{2+} during the deposition process. This phenomenon confirmed the creation of ACD behavior, which is the characteristic feature for the electrochemical deposition of iron–group alloys. Furthermore, the reduction rate of Fe^{2+} was higher compared to the reduction rate of Co^{2+} as the CRV_{Co} was lower than CRV_{Fe} [3,6,19,24,46,51–53]. Therefore, the degree of ACD characteristics of Co–Ni was lower than Fe–Ni. In addition, the CRV_{Fe}/CRV_{Co} ratio was determined to be lower than the CRV_{Co}/CRV_{Ni} ratio, revealing that the order of ACD was Fe–Ni > Co–Ni > Fe–Co for all Co ion concentrations in the PS, which is in good agreement with the findings of conducted studies [3,6,24,46].

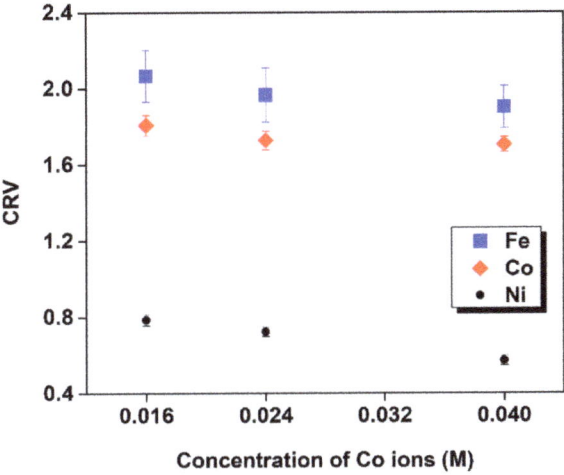

Figure 3. The CRV$_{Fe}$, CRV$_{Co}$ and CRV$_{Ni}$ as a function of the Co ion concentration in the PS.

The phase structure of the samples produced in this work was investigated via XRD analysis. The resulting XRD patterns are shown in Figure 4.

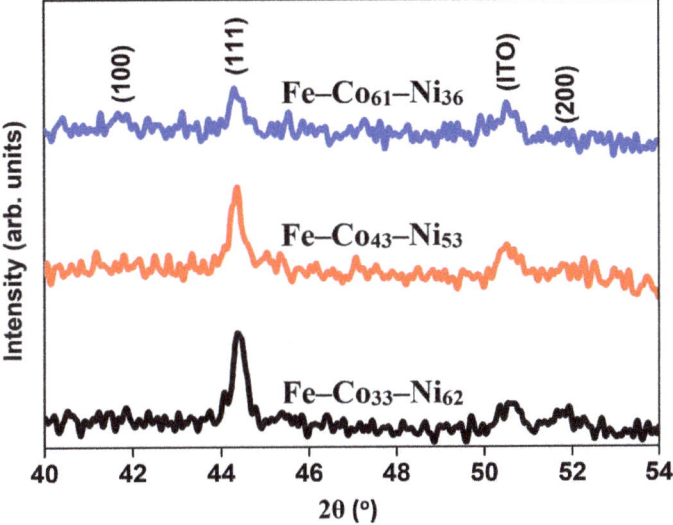

Figure 4. XRD patterns of the samples.

Obviously, in all samples, the (111) diffraction peak of the face–centered cubic (fcc) crystal structure observed at the angular position of about $2\theta = 44.4°$ was the most intense irrespective of the Co contents. The single phase structure (fcc) obtained for the Fe–Co$_{33}$–Ni$_{62}$ and Fe–Co$_{43}$–Ni$_{53}$ samples shows a good agreement with the XRD patterns of the ternary ferromagnetic materials with similar compositions produced in previous studies [6,21,24,54,55]. In addition to that, compared to the Fe–Co$_{33}$–Ni$_{62}$ and Fe–Co$_{43}$–Ni$_{53}$ samples, the XRD pattern of the Fe–Co$_{61}$–Ni$_{36}$ sample also revealed the presence of the (100) diffraction peak with low intensity related to the hexagonal close–packed (hcp) phase structure which occurred at about $2\theta = 41.7°$. At high Co contents, a transition from single phase structure (fcc) to dual phase structure (fcc + hcp or fcc + bcc) was also reported in electrochemically grown binary Ni–Co films, Co–Ni–Al$_2$O$_3$ composite coatings, and ternary ferromagnetic films of Fe, Co, and Ni [12,13,17,47,56]. Alongside the phase transition, an increment in the Co contents resulted in a significant decrement in the intensities of both (111) and (200) diffraction peaks, reflecting a strong reduction in the crystallization (Figure 4). This also caused a change in the crystallite size of the samples. The crystallite size (D) of the produced samples was determined by Scherrer's equation [57]:

$$D = [0.9\lambda / B\cos\theta] \times [180°/\pi] \qquad (3)$$

where λ, B and θ represent the wavelength of CuKα radiation, Full–Width at Half Maximum (FWHM) value, and Bragg diffraction angle, respectively. To estimate the B and θ values, XRD patterns were fitted by Lorentzian curves. It was revealed that mean crystallite size decreases from 21.6 to 15.6 nm as the Co contents in the samples increases from 33 to 61 wt.%, indicating that the crystallite size of the Co–rich samples is smaller compared to the Co–poor samples. The decrease in the crystallite size with the Co contents was also reported in Fe–Co–Ni films electroplated on copper substrates from ammonium–chloride-based PSs [47]. The cause of the increase (decrease) in the size of Fe–Co–Ni is due to the lattice constant of Fe, Co, and Ni atoms and the interaction between electrons leads to the appearance of size effect. The 3-D surface microtexture can be characterized for a deeper understanding of the nano-scale patterns by stereometric [58–60] and fractal/multifractal analyses [61–64].

The surface topography was studied by means of SEM images analysis of the samples. The SEM device we used has a resolution of 100 KX, the crystals can be observed in Figure 5.

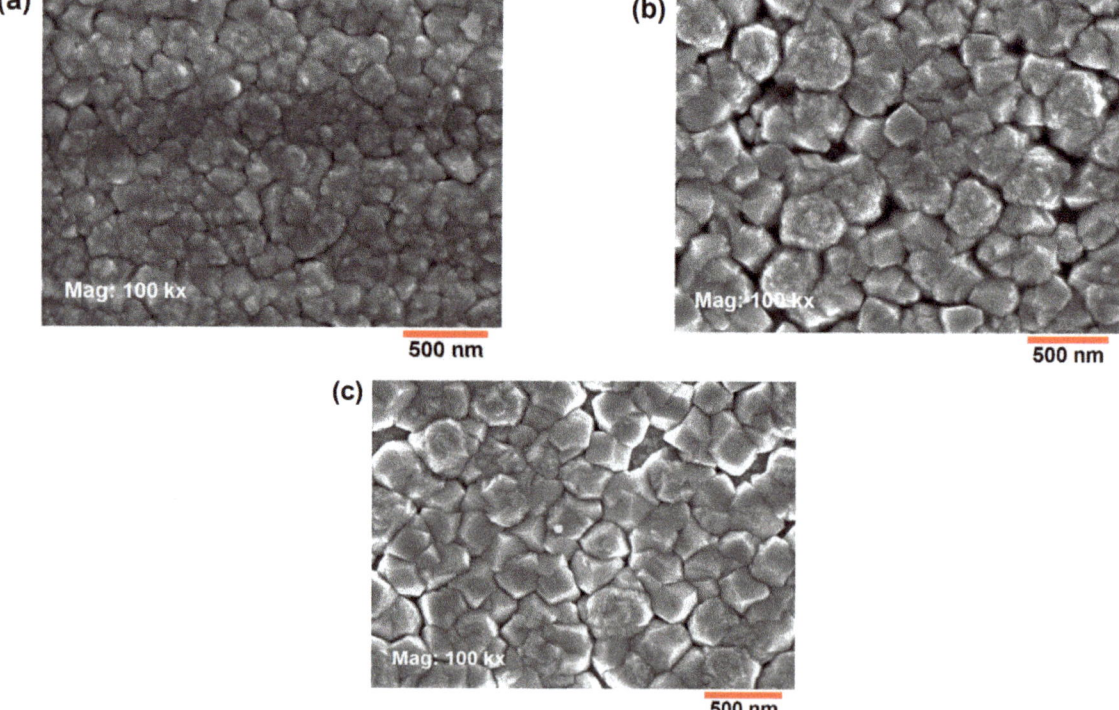

Figure 5. SEM images of the samples Fe–Co$_{33}$–Ni$_{62}$ (**a**), Fe–Co$_{43}$–Ni$_{53}$ (**b**), and Fe–Co$_{61}$–Ni$_{36}$ (**c**), respectively.

Figure 5a showed that the Fe–Co$_{33}$–Ni$_{62}$ sample exhibits a surface topography comprising only cauliflower–like agglomerates.

However, as seen from Figure 5b,c, the Fe–Co$_{43}$–Ni$_{53}$ and Fe–Co$_{61}$–Ni$_{36}$ samples had a surface topography covered with a mixture of pyramidal particles and cauliflower–like agglomerates. The surface morphology of a material is due to the lattice constant, the electronic interactions between different atoms caused.

In all samples, one cauliflower–like agglomerate was composed of grains. An increment in the Co contents caused an enhancement in the size of the cauliflower–like agglomerates and a decrease in their number. The average width of the cauliflower–like agglomerates was found to be about 150 nm for the Fe–Co$_{33}$–Ni$_{62}$ sample. However, the Fe–Co$_{43}$–Ni$_{53}$ and Fe–Co$_{61}$–Ni$_{36}$ samples had larger cauliflower–like agglomerates with an average width of about 270 and 350 nm, respectively. The size of the crystal nuclei is larger than the average size of the particles. The cause of this phenomenon is due to the difference between the lattice constants and the interactions between the electronic structures which leads to the formation of nuclei crystallization in the form of cauliflower.

On the other hand, the average width of the pyramidal particles was determined to be about 190 and 220 nm for the Fe–Co$_{43}$–Ni$_{53}$ and Fe–Co$_{61}$–Ni$_{36}$ samples, respectively. Thus, the Fe–Co$_{61}$–Ni$_{36}$ sample exhibited larger pyramidal particles than the Fe–Co$_{43}$–Ni$_{53}$ sample. Consequently, as also listed in Table 1, the Fe–Co$_{33}$–Ni$_{62}$, Fe–Co$_{43}$–Ni$_{53}$ and Fe–Co$_{61}$–Ni$_{36}$ samples had particles with an average width of approximately 150, 210, and 250 nm, respectively, which was in good agreement with the findings reported in the literature [47]. Further studies on the morphological characteristics were also carried out using an AFM. Figure 6 indicates the AFM images of the samples. The samples possessed globular particles of various sizes. The size of globular particles increased and their number decreased when the Co content increased, which is consistent with the findings of the SEM analysis. The surface morphology of the material is caused by the lattice constant, the electronic force of interaction between the atoms of the material. On the other hand, the influence of the Co content on the particle size can be ascribed to different cathode potentials caused by the concentration of Co ions in the PS. Conducted studies showed that the particle size decreased when the cathode potential was increased [24].

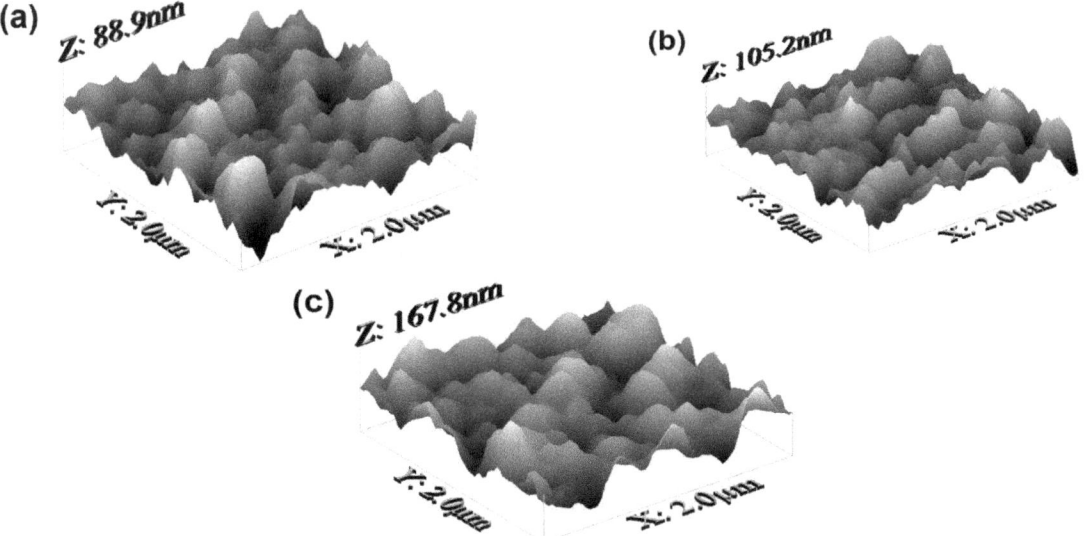

Figure 6. AFM images of the samples (a) Fe–Co$_{33}$–Ni$_{62}$, (b) Fe–Co$_{43}$–Ni$_{53}$, and (c) Fe–Co$_{61}$–Ni$_{36}$, respectively.

Therefore, the particle size increased with the Co content since the cathode potential decreased with increasing Co ion concentration in the PS (Figure 1). The roughness parameters were determined from the AFM images. As distinctly noticed in Table 1, the Co content had a significant effect on the surface roughness parameters. When the Co content in the samples increases, the electronic interaction of Co with Fe and Ni increases, so the size of crystal nuclei increases, and the number of spheres is reduced leading to the increased surface roughness of the material.

Figure 6 indicates the AFM images of the samples. The samples possessed globular particles of various sizes.

The obtained results revealed that the surface roughness increased the cause is due to when Co content increased, leading to the electronic interaction of Co atoms with Fe, Ni increases, and the size of crystal nuclei increases and the number of spheres reduces to increased surface roughness of the material.

In this paper, the thickness and size of the considered thin films are not investigated especially, but the thickness of the thin film has an average size equal to the size of the crystal nuclei such as: 150 nm with Fe–Co_{33}–Ni_{62}, 270 nm with Fe–Co_{43}–Ni_{53}, and 350 nm with Fe–Co_{61}–Ni_{36}.

Figure 7 exhibits the normalized in–plane magnetic hysteresis loops measured to determine the magnetic characteristics with respect to their Co contents. The results obtained from the magnetic analysis are collected in Table 1. Although all samples exhibited a ferromagnetic behavior with a magnetic hardness being between soft and hard [3,17,21,22,24,55,65], the Co contents played a significant role in the coercive field. The coercive field increased considerably from 36 to 121 Oe as the Co contents in the samples increased from 33 to 61 wt.%. Increasing the content of Co leads to increased crystallization, increasing the size of the crystal nuclei (magnetic domain) leading to an increase in the coercive field of the material, which is shown in Figure 7. The increase in the coercive field with the Co content may also be attributed to an increment in the surface roughness.

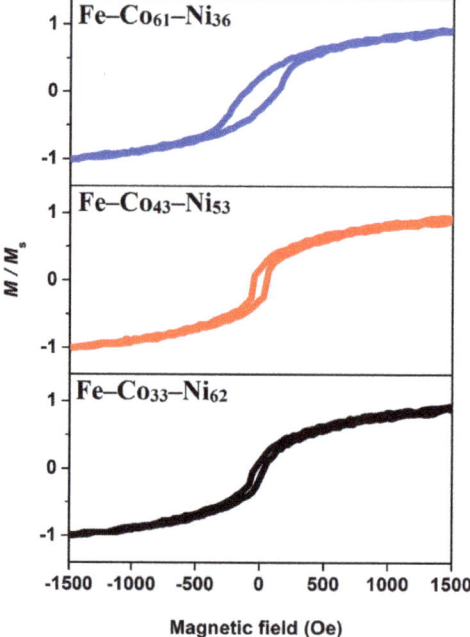

Figure 7. Hysteresis loops of the samples with different compositions.

The variations observed in the average surface roughness and coercive field with respect to the Co contents are shown in Figure 8.

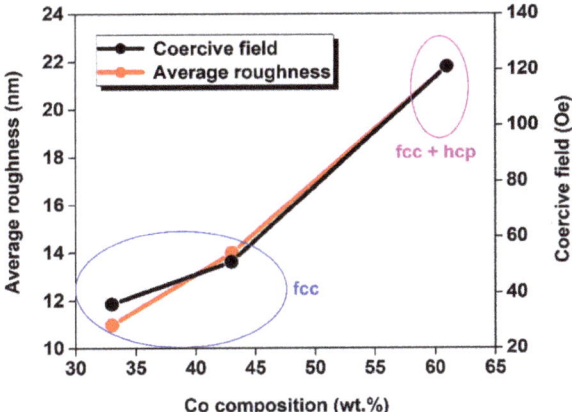

Figure 8. The correlation between the coercive field and the surface roughness.

As clearly evidenced in Figure 8, there was a direct correlation between the surface roughness and the coercive field as also reported in the electrochemically deposited ternary ferromagnetic films of Fe, Co, and Ni [21,24,65,66]. In addiiton, the Fe–Co_{61}–Ni_{36} sample had a much higher coercive field than the Fe–Co_{33}–Ni_{62} and Fe–Co_{43}–Ni_{53} samples. This abrupt increase in the coercive field with increasing Co contents from 43 to 61 wt.% may also be ascribed to the appearance of the (100) diffraction peak of the hcp phase structure of Co. In previous studies [13,17,47], it was shown that the Ni–Co and Fe–Co–Ni films with single phase structure at low Co contents exhibit a much lower coercive field than the Ni–Co and Fe–Co–Ni films with dual phase structure at high Co contents. On the other hand, the produced samples were found to have very low squareness ratios ranging from 9.2 to 23.6%. Such low squareness ratios correspond to the formation of an in–plane hysteresis loop with a vertical magnetization component as also observed in electroplated Ni–Co/ITO and Fe–Co–Ni/ITO thin film samples [21,65]. Furthermore, as clearly seen from Table 1, a gradual increment in the magnetic squareness ratio was detected with the Co contents, revealing a decrement in the vertical component of magnetization.

4. Conclusions

This work aimed to obtain the ternary Fe–Co–Ni samples with various Co contents and reveal the differences in the morphological, magnetic, and structural properties with respect to their Co contents. According to the compositional analysis, a change in the Co ion concentration of the PS significantly affected the deposit composition. It was understood that the Fe–Co–Ni sample with higher Co contents could be obtained when the sample was electrochemically deposited from the PS including higher Co ion concentration. It was also revealed that the co–deposition characteristic (anomalous) and its order (Fe–Ni > Co–Ni > Fe–Co) were not affected by the amount of Co ions in the PS. The resultant samples exhibited the predominant reflection from the (111) plane of the fcc crystal structure. Unlike the Fe–Co_{33}–Ni_{62} and Fe–Co_{43}–Ni_{53} samples, the Fe–Co_{61}–Ni_{36} sample with the highest Co contents exhibited a weak reflection from the (100) plane of the hcp crystal structure of Co. Compared to the Fe–Co_{61}–Ni_{36} and Fe–Co_{43}–Ni_{53} samples, the crystallinity was found to be stronger for the Fe–Co_{33}–Ni_{62} sample. The size of the crystallites decreased from 21.6 to 15.6 nm as the Co contents in the sample increased from 33 to 61 wt.%. A surface structure covered with a mixture of pyramidal particles and cauliflower–like agglomerates was detected for the Fe–Co_{43}–Ni_{53} and Fe–Co_{61}–Ni_{36} samples, whereas the Fe–Co_{33}–Ni_{62} sample had a surface structure consisting of only cauliflower–like agglomerates.

Moreover, compared to others, the Fe–Co_{33}–Ni_{62} sample exhibited a more compact surface morphology consisting of smaller cauliflower–like agglomerates with an average width of 150 nm. As the Co contents enhanced, the average and RMS surface roughness parameters increased significantly from 11.0 to 21.8 nm and from 14.4 to 28.4 nm, respectively. The samples produced were magnetically semi-hard. However, the Fe–Co_{61}–Ni_{36} (121 Oe) thin film sample exhibited a noticeably higher coercive field compared to the Fe–Co_{43}–Ni_{53} (51 Oe) and Fe–Co_{33}–Ni_{62} (36 Oe) thin film samples, which was attributed to the phase transition from single phase structure (fcc) to dual phase structure (fcc + hcp) and an abrupt enhancement in the surface roughness parameters. An increase in the Co contents from 33 to 61 wt.% also induced an enhancement in the magnetic squareness ratio from 9.2 to 23.6%, reflecting a decrement in the vertical component of magnetization.

Author Contributions: V.C.L.: Validation, writing and editing. U.S.: conceptualization, methodology, investigation. M.C.B.: validation, writing and editing. L.D.T.: writing and editing. Ş.T.: writing-editing. D.N.T.: conceptualization, methodology, investigation, resources, supervision, writing—original draft—review and editing. All authors have read and agreed to the published version of the manuscript.

Funding: This research was financially supported by the Scientific Research Projects Commission of Bartın University under the project number 2018–FEN–A–021.

Institutional Review Board Statement: Not applicable.

Informed Consent Statement: Not applicable.

Data Availability Statement: The data that support the findings of this study are available from the corresponding author upon reasonable request.

Acknowledgments: The authors would like to thank Çağdaş Denizli for taking AFM images and Malik Kaya for his assistance in the electrodeposition process of the samples.

Conflicts of Interest: The authors declare no conflict of interest. The funders had no role in the design of the study; in the collection, analyses, or interpretation of data; in the writing of the manuscript, or in the decision to publish the results. Neither author has a financial or proprietary interest in any material or method mentioned.

References

1. Hemeda, O.M.; Tawfik, A.; El–Sayed, A.H.; Hamad, M.A. Synthesis and characterization of semi–crystalline NiCoP film. *J. Supercond. Nov. Magn.* **2015**, *28*, 3629–3632. [CrossRef]
2. Boubatra, M.; Azizi, A.; Schmerber, G.; Dinia, A. The influence of pH electrolyte on the electrochemical deposition and properties of nickel thin films. *Ionics* **2012**, *18*, 425–432. [CrossRef]
3. Ledwig, P.; Kac, M.; Kopia, A.; Falkus, J.; Dubiel, B. Microstructure and properties of electrodeposited nanocrystalline Ni–Co–Fe coatings. *Materials* **2021**, *14*, 3886. [CrossRef]
4. Betancourt-Cantera, L.G.; Sánchez-de Jesús, F.; Bolarín–Miró, A.M.; Gallegos–Melgar, A.; Mayen, J.; Betancourt-Cantera, J.A. Structural analysis and magnetic characterization of ternary alloys (Co–Fe–Ni) synthesized by mechanical alloying. *J. Mater. Res. Technol.* **2020**, *9*, 14969–14978. [CrossRef]
5. Milyaev, M.A.; Bannikova, N.S.; Naumova, L.I.; Proglyado, V.V.; Patrakov, E.I.; Glazunov, N.P.; Ustinov, V.V. Effective Co–rich ternary CoFeNi alloys for spintronics application. *J. Alloys Compd.* **2021**, *854*, 157171. [CrossRef]
6. Budi, S.; Manaf, A. The effects of saccharin on the electrodeposition of NiCoFe films on a flexible substrate. *Mater. Res. Express* **2021**, *8*, 086513. [CrossRef]
7. Kayani, Z.N.; Riaz, S.; Naseem, S. Structural and magnetic properties of FeCoNi thin films. *Indian J. Phys.* **2014**, *88*, 165–169. [CrossRef]
8. Safavi, M.S.; Tanhaei, M.; Ahmadipour, M.F.; Adli, R.G.; Mahdavi, S.; Walsh, F.C. Electrodeposited Ni–Co alloy–particle composite coatings: A comprehensive review. *Surf. Coat. Technol.* **2020**, *382*, 125153. [CrossRef]
9. Song, J.; Yoon, D.; Han, C.; Kim, D.; Park, D.; Myung, N. Electrodeposition characteristics and magnetic properties of CoFeNi thin film alloys. *J. Korean Electrochem. Soc.* **2002**, *5*, 17–20. [CrossRef]
10. Pingale, A.D.; Owhal, A.; Katarkar, A.S.; Belgamwar, S.U.; Rathore, J.S. Recent researches on Cu–Ni alloy matrix composites through electrodeposition and powder metallurgy methods: A review. *Mater. Today Proc.* **2021**, *47*, 3301–3308. [CrossRef]
11. Alizadeh, M.; Safaei, H. Characterization of Ni–Cu matrix, Al2O3 reinforced nano–composite coatings prepared by electrodeposition. *Appl. Surf. Sci.* **2018**, *456*, 195–203. [CrossRef]

12. Sattawitchayapit, S.; Yordsri, V.; Panyavan, P.; Chookajorn, T. Solute and grain boundary strengthening effects in nanostructured Ni–Co alloys. *Surf. Coat. Technol.* **2021**, *428*, 127902. [CrossRef]
13. Tian, L.; Xu, J.; Qiang, C. The electrodeposition behaviors and magnetic properties of Ni–Co films. *Appl. Surf. Sci.* **2011**, *257*, 4689–4694. [CrossRef]
14. Su, X.; Qiang, C. Influence of pH and bath composition on properties of Ni–Fe alloy films synthesized by electrodeposition. *Bull. Mater. Sci.* **2012**, *35*, 183–189. [CrossRef]
15. Saraç, U.; Baykul, M.C. Comparison of microstructural and morphological properties of electrodeposited Fe–Cu thin films with low and high Fe: Cu ratio. *Adv. Mater. Sci. Eng.* **2013**, *2013*, 971790. [CrossRef]
16. Saraç, U.; Baykul, M.C. Microstructural and morphological characterizations of nanocrystalline Ni–Cu–Fe thin films electrodeposited from electrolytes with different Fe ion concentrations. *J. Mater. Sci. Mater. Electron.* **2014**, *25*, 2554–2560. [CrossRef]
17. Saraç, U.; Baykul, M.C.; Uguz, Y. Differences observed in the phase structure, grain size–shape, and coercivity field of electrochemically deposited Ni–Co thin films with different Co contents. *J. Supercond. Nov. Magn.* **2015**, *28*, 3105–3110. [CrossRef]
18. Saraç, U.; Kaya, M.; Baykul, M.C. Synthesis of Ni–Fe thin films by electrochemical deposition technique and characterization of their microstructures and surface morphologies. *Turk. J. Phys.* **2017**, *41*, 536–544. [CrossRef]
19. Saraç, U.; Kaya, M.; Baykul, M.C. The influence of deposit composition controlled by changing the relative Fe ion concentration on properties of electroplated nanocrystalline Co–Fe–Cu ternary thin films. *Turk. J. Phys.* **2018**, *42*, 136–145.
20. Saraç, U.; Baykul, M.C. An investigation of structural properties and surface morphologies of electrochemically fabricated nanocrystalline Ni–Co–Cu/ITO deposits with different compositions. *Turk. J. Phys.* **2019**, *43*, 372–382.
21. Saraç, U.; Baykul, M.C. Studying structural, magnetic and morphological features of electrochemically fabricated thin films of Co–Ni–Fe with different Fe compositions. *Thin Solid Films* **2021**, *736*, 138901. [CrossRef]
22. Saraç, U.; Baykul, M.C.; Uguz, Y. The influence of applied current density on microstructural, magnetic, and morphological properties of electrodeposited nanocrystalline Ni–Co thin films. *J. Supercond. Nov. Magn.* **2015**, *28*, 1041–1045. [CrossRef]
23. Hanafi, I.; Daud, A.R.; Radiman, S. Potentiostatic electrodeposition of Co–Ni–Fe thin films from sulfate medium. *J. Chem. Technol. Metall.* **2016**, *51*, 547–555.
24. Saraç, U.; Kaya, M.; Baykul, M.C. A comparative study on microstructures, magnetic features and morphologies of ternary Fe–Co–Ni alloy thin films electrochemically fabricated at different deposition potentials. *J. Supercond. Nov. Magn.* **2019**, *32*, 917–923. [CrossRef]
25. Brenner, A. *Electrodeposition of Alloys Principles and Practice*; Academic Press: New York, NY, USA, 1963.
26. Dung, N.T.; Cuong, C.N.; Hung, V.T. Molecular dynamics study of microscopic structures, phase transitions and dynamic crystallization in Ni nanoparticles. *RSC Adv.* **2017**, *7*, 25406–25413.
27. Hue, D.T.M.; Coman, G.; Hoc, N.Q.; Dung, N.T. Influence of heating rate, temperature, pressure on the structure, and phase transition of amorphous Ni material: A molecular dynamics study. *Heliyon* **2020**, *6*, e05548.
28. Dung, N.T. Z-AXIS deformation method to investigate the influence of system size, structure phase transition on mechanical properties of bulk nickel. *Mater. Chem. Phys.* **2020**, *252*, 123275.
29. Dung, N.T.; van, C.L. Effects of number of atoms, shell thickness, and temperature on the structure of Fe nanoparticles amorphous by molecular dynamics method. *Adv. Civ. Eng.* **2021**, *2021*, 9976633.
30. Dung, N.T.; Phuong, N.T. Understanding the heterogeneous kinetics of Al nanoparticles by simulations method. *J. Mol. Struct.* **2020**, *1218*, 128498.
31. Quoc, T.T.; Trong, D.N. Molecular dynamics factors affecting on the structure, phase transition of Al bulk. *Phys. B Condens. Matter* **2019**, *570*, 116–121. [CrossRef]
32. Dung, N.T.; van, C.L.; van, D.Q.; Ţălu, Ş. First-principles calculations of structural and electronic structural properties of Au–Cu alloys. *J. Compos. Sci.* **2022**; in press.
33. Dung, N.T.; Cuong, N.C.; van, D.Q. Study on the effect of doping on lattice constant and electronic structure of bulk AuCu by the density functional theory. *J. Multiscale Model.* **2020**, *11*, 2030001.
34. Dung, N.T.; van, C.L. Factors affecting the depth of the Earth's surface on the heterogeneous dynamics of $Cu_{1-x}Ni_x$ alloy, x = 0.1, 0.3, 0.5, 0.7, 0.9 by molecular dynamics simulation method. *Mater. Today Commun.* **2021**, *29*, 102812.
35. Dung, N.T. Some factors affecting structure, transition phase and crystallized of CuNi nanoparticles. *Am. J. Mod. Phys.* **2017**, *6*, 66–75. [CrossRef]
36. Dung, N.T.; Phuong, N.T. Molecular dynamic study on factors influencing the structure, phase transition and crystallization process of $NiCu_{6912}$ nanoparticle. *Mater. Chem. Phys.* **2020**, *250*, 123075.
37. Dung, N.T.; van, C.L.; Ţălu, Ş. The structure and crystallizing process of NiAu alloy: A molecular dynamics simulation method. *J. Compos. Sci.* **2021**, *5*, 18.
38. Hoc, N.Q.; Viet, L.H.; Dung, N.T. On the melting of defective FCC interstitial alloy γ-FeC under pressure up to 100 Gpa. *J. Electron. Mater.* **2020**, *49*, 910–916. [CrossRef]
39. Dung, N.T.; Kien, P.H.; Phuong, N.T. Simulation on the factors affecting the crystallization process of FeNi alloy by molecular dynamics. *ACS Omega* **2019**, *4*, 14605–14612.
40. Dung, N.T. Influence of impurity concentration, atomic number, temperature and tempering time on microstructure and phase transformation of $Ni_{1-x}Fe_x$ (x = 0:1, 0.3, 0.5) nanoparticles. *Mod. Phys. Lett. B* **2018**, *14*, 1850204. [CrossRef]

41. Van, C.L.; van, D.Q.; Dung, N.T. Ab initio calculations on the structural and electronic properties of AgAu alloys. *ACS Omega* **2020**, *5*, 31391–31397.
42. Dung, N.T.; Phuong, N.T. Factors affecting the structure, phase transition and crystallization process of AlNi nanoparticles. *J. Alloys Compd.* **2020**, *812*, 152133.
43. Quoc, T.T.; Trong, D.N. Effect of heating rate, impurity concentration of Cu, atomic number, temperatures, time annealing temperature on the structure, crystallization temperature and crystallization process of $Ni_{1-x}Cu_x$ bulk; x = 0.1, 0.3, 0.5, 0.7. *Int. J. Mod. Phys. B* **2018**, *32*, 1830009.
44. Dung, T.N.; Cuong, C.N.; Toan, T.N.; Hung, K.P. Factors on the magnetic properties of the iron nanoparticles by classical Heisenberg model. *Phys. B* **2018**, *532*, 144–148.
45. Dung, N.T.; van, C.L.; Țălu, Ș. the study of the influence of matrix, size, rotation angle, and magnetic field on the isothermal entropy, and the néel phase transition temperature of Fe_2O_3 nanocomposite thin films by the Monte-Carlo simulation method. *Coatings* **2021**, *11*, 1209–1216.
46. Yang, Y. Preparation of Fe–Co–Ni ternary alloys with electrodeposition. *Int. J. Electrochem. Sci.* **2015**, *10*, 5164–5175.
47. Yanai, T.; Koda, K.; Kaji, J.; Aramaki, H.; Eguchi, K.; Takashima, K.; Nakano, M.; Fukunaga, H. Electroplated Fe–Co–Ni films prepared in ammonium-chloride-based plating baths. *AIP Adv.* **2018**, *8*, 056127. [CrossRef]
48. Omata, F. Magnetic properties of Fe–Co–Ni films with high saturation magnetization prepared by evaporation and electrodeposition. *IEEE Transl. J. Magn. Jpn.* **1990**, *5*, 17–28. [CrossRef]
49. Horcas, I.; Fernández, R.; Gómez-Rodríguez, J.M.; Colchero, J.; Gómez-Herrero, J.; Baro, A.M. WSXM: A software for scanning probe microscopy and a tool for nanotechnology. *Rev. Sci. Instrum.* **2007**, *78*, 013705. [CrossRef]
50. Tseng, Y.T.; Wu, G.X.; Lin, J.C.; Hwang, Y.R.; Wei, D.H.; Chang, S.Y.; Peng, K.C. Preparation of Co–Fe–Ni alloy micropillar by microanode-guided electroplating. *J. Alloys Compd.* **2021**, *885*, 160873. [CrossRef]
51. Zhang, Y.; Ivey, D.G. Electrodeposition of nanocrystalline CoFe soft magnetic thin films from citrate–stabilized baths. *Mater. Chem. Phys.* **2018**, *204*, 171–178. [CrossRef]
52. Qiang, C.; Xu, J.; Xiao, S.; Jiao, Y.; Zhang, Z.; Liu, Y.; Tian, L.; Zhou, Z. The influence of pH and bath composition on the properties of Fe–Co alloy film electrodeposition. *Appl. Surf. Sci.* **2010**, *257*, 1371–1376. [CrossRef]
53. Ramazani, A.; Almasi-Kashi, M.; Golafshan, E.; Arefpour, M. Magnetic behavior of as-deposited and annealed CoFe and CoFeCu nanowire arrays by ac–pulse electrodeposition. *J. Cryst. Growth* **2014**, *402*, 42–47. [CrossRef]
54. Budi, S.; Kurniawan, B.; Mott, D.M.; Maenosono, S.; Umar, A.A.; Manaf, A. Comparative trial of saccharin–added electrolyte for improving the structure of an electrodeposited magnetic FeCoNi thin film. *Thin Solid Films* **2017**, *642*, 51–57. [CrossRef]
55. Kockar, H.; Demirbas, O.; Kuru, H.; Alper, M.; Karaagac, O.; Haciismailoglu, M.; Ozergin, E. Electrodeposited NiCoFe films from electrolytes with different Fe ion concentrations. *J. Magn. Magn. Mater.* **2014**, *360*, 148–151. [CrossRef]
56. Wu, G.; Li, N.; Zhou, D.; Mitsuo, K. Electrodeposited Co–Ni–Al_2O_3 composite coatings. *Surf. Coat. Technol.* **2004**, *176*, 157–164. [CrossRef]
57. Wang, J.; Lei, W.; Deng, Y.; Xue, Z.; Qian, H.; Liu, W.; Li, X. Effect of current density on microstructure and corrosion resistance of Ni graphene oxide composite coating electrodeposited under supercritical carbon dioxide. *Surf. Coat. Technol.* **2019**, *358*, 765–774. [CrossRef]
58. Țălu, Ș. *Micro and Nanoscale Characterization of Three Dimensional Surfaces. Basics and Applications*; Napoca Star Publishing House: Cluj-Napoca, Romania, 2015.
59. Méndez, A.; Reyes, Y.; Trejo, G.; Stępień, K.; Țălu, Ș. Micromorphological characterization of zinc/silver particle composite coatings. *Microsc. Res. Tech.* **2015**, *78*, 1082–1089. [CrossRef] [PubMed]
60. Grayeli-Korpi, A.R.; Luna, C.; Arman, A.; Țălu, Ș. Influence of the oxygen partial pressure on the growth and optical properties of RF-sputtered anatase TiO_2 thin films. *Results Phys.* **2017**, *7*, 3349–3352. [CrossRef]
61. Mwema, F.M.; Akinlabi, E.T.; Oladijo, O.P.; Fatoba, O.S.; Akinlabi, S.A.; Țălu, Ș. Advances in manufacturing analysis: Fractal theory in modern manufacturing. In *Modern Manufacturing Processes*, 1st ed.; Kumar, K., Davim, J.P., Eds.; Woodhead Publishing Reviews: Mechanical Engineering Series; Elsevier: Amsterdam, The Netherlands, 2020; pp. 13–39. [CrossRef]
62. Țălu, Ș.; Stach, S.; Raoufi, D.; Hosseinpanahi, F. Film thickness efect on fractality of tin-doped In_2O_3 thin films. *Electron. Mater. Lett.* **2015**, *11*, 749–757. [CrossRef]
63. Țălu, Ș.; Morozov, I.A.; Yadav, R.P. Multifractal analysis of sputtered indium tin oxide thin film surfaces. *Appl. Surf. Sci.* **2019**, *484*, 892–898. [CrossRef]
64. Țălu, Ș.; Stach, S.; Ramazanov, S.; Sobola, D.; Ramazanov, G. Multifractal characterization of epitaxial silicon carbide on silicon. *Mater. Sci.-Pol.* **2017**, *35*, 539–547. [CrossRef]
65. Denizli, Ç.; Sarac, U.; Baykul, M.C. Controlling the surface morphologies, structural and magnetic properties of electrochemically fabricated Ni–Co thin film samples via seed layer deposition. *J. Mater. Sci. Mater. Electron.* **2020**, *31*, 4279–4286. [CrossRef]
66. Rhen, F.M.F.; Roy, S. Dependence of magnetic properties on micro–to nanostructure of CoNiFe films. *J. Appl. Phys.* **2008**, *103*, 103901. [CrossRef]

Article

Effect of Voltage on the Microstructure and High-Temperature Oxidation Resistance of Micro-Arc Oxidation Coatings on AlTiCrVZr Refractory High-Entropy Alloy

Zhao Wang [1], Zhaohui Cheng [2], Yong Zhang [2], Xiaoqian Shi [2,*], Mosong Rao [2] and Shangkun Wu [2]

[1] Shaanxi Provincial Office of Defense Science, Technology and Industry, Xi'an 710021, China
[2] School of Materials Science and Chemical Engineering, Xi'an Technological University, Xi'an 710021, China
* Correspondence: shixiaoqian0220@163.com

Abstract: In order to improve the high-temperature oxidation resistance of refractory high-entropy alloys (RHEAs), we used micro-arc oxidation (MAO) technology to prepare ceramic coatings on AlTiCrVZr alloy, and the effects of voltage on the microstructure and high-temperature oxidation resistance of the coatings were studied. In this paper, the MAO voltage was adjusted to 360 V, 390 V, 420 V, and 450 V. The microstructure, elements distribution, chemical composition, and surface roughness of the coatings were studied by scanning electron microscopy (SEM), energy dispersive (EDS), X-ray photoelectron spectroscopy (XPS), and white-light interferometry. The matrix alloy and MAO-coated samples were oxidized at 800 °C for 5 h and 20 h to study their high-temperature oxidation resistance. The results showed that as the voltage increased, the MAO coating gradually became smooth and dense, the surface roughness decreased, and the coating thickness increased. The substrate elements and solute ions in the electrolyte participated in the coating formation reaction, and the coating composition was dominated by Al_2O_3, TiO_2, Cr_2O_3, V_2O_5, ZrO_2, and SiO_2. Compared with the substrate alloy, the high-temperature oxidation resistance of the MAO-coated samples prepared at different voltages was improved after oxidation at 800 °C, and the coating prepared at 420 V showed the best high-temperature oxidation resistance after oxidation for 20 h. In short, MAO coatings can prevent the diffusion of O elements into the substrate and the volatilization of V_2O_5, which improves the high-temperature oxidation resistance of AlTiCrVZr RHEAs.

Keywords: refractory high-entropy alloys; micro-arc oxidation; voltage; microstructure; high-temperature oxidation resistance

1. Introduction

With the development of aerospace technology, the requirement for the high-temperature performance of components is increasing. At present, most of the materials used in the high-temperature field are nickel-based superalloys, but the application temperature is only between 116 °C and 1277 °C, which have been unable to meet the demand [1,2]. Refractory high-entropy alloys (RHEAs) composed of three or more refractory elements (Ti, V, Cr, Zr, Nb, Mo, Hf, Ta, and W), and other elements in approximate molar ratios, have the potential to be applied to high-temperature structural parts at higher temperatures because of their excellent room-temperature and high-temperature mechanical properties [3–5]. For example, the yield strength of NbMoTaW and VNbMoTaW RHEAs at 1600 °C exceeds 400 MPa, far exceeding the yield strength of Inconel 718 nickel-based superalloy at 1000 °C of 200 MPa. However, refractory high-entropy alloys tend to rapidly form volatile oxides (MoO_3, WO_3, V_2O_5, etc.) and easily spalling oxides (Nb_2O_5 and Ta_2O_5) that do not have high-temperature protection capabilities at high temperatures, resulting in high levels of oxidation in high-temperature aerobic environments [6–9]. Poor high-temperature oxidation resistance has become one of the key factors restricting the application of refractory high-entropy alloys in high-temperature fields, and it has become particularly urgent to improve the high-temperature oxidation resistance of RHEAs.

Micro-arc oxidation (MAO) is a surface modification technology with simple operation, environmental protection, and high efficiency. The ceramic coating formed has the characteristics of strong adhesion to the matrix and uniform growth of the coating, along with the ability to improve the high-temperature oxidation resistance of the matrix metal [10–12]. It was found that after the preparation of the MAO coating dominated by Al_2O_3 on the surface of the Ti-45Al-8.5Nb alloy and oxidation at 900 °C for 100 h, the minimum weight gain was only 0.396 g/m^2, which significantly improved the high-temperature oxidation resistance [13]. Of course, some progress has also been made in the study of MAO technology to improve the high-temperature oxidation resistance of $AlTiNbMo_{0.5}Ta_{0.5}Zr$ and AlTiCrVZr refractory high-entropy alloys [10,14]. However, the current research on the use of MAO technology to improve the high-temperature oxidation resistance of RHEAs is still in its initial stage. Voltage as one of the key influencing factors of MAO technology plays a decisive role in the electric field strength in the MAO process, affects the migration rate of anions and ions, and then affects the growth rate, density, and bonding strength of the coating [15–18]. The thickness, density, and bonding strength of the coating are closely related to the high-temperature oxidation resistance of the coating. At present, the effect of MAO voltage on the high-temperature oxidation resistance of RHEAs has not been reported.

In this paper, by adjusting the voltage, the MAO coating was prepared in situ on the AlTiCrVZr RHEA. The effects of voltage on the micromorphology, roughness, chemical composition, and high-temperature oxidation resistance of the surface and cross-section of the MAO coating were studied. These research results can promote the development of MAO technology and RHEAs, and provide a reference for the surface modification of RHEAs applied in high-temperature environments.

2. Experimental Process

2.1. Sample Pretreatment

The purchased commercial AlTiCrVZr RHEA was processed into blocks with a size of 7 mm × 7 mm × 5 mm, and all samples were ground step by step with 240#~2000# sand paper, and then sonicated in anhydrous ethanol for 20 min, dried in air, and set aside.

2.2. Preparation of the MAO Coating

Using the MAO-20H power supply independently developed by Xi'an Technological University, the working voltages were 360 V, 390 V, 420 V, and 450 V, and the MAO coating was prepared on the surface of the AlTiCrVZr high-entropy alloy. The electrolyte was 50 g/L Na_2SiO_3 + 25 g/L $(NaPO_3)_6$ + 5 g/L NaOH, the frequency was 600 Hz, the duty cycle was 8%, the MAO treatment lasted 5 min, and the electrolyte temperature remained between 30 °C and 35 °C. The samples treated by MAO were sonicated in anhydrous ethanol and dried for further study.

2.3. Microstructure Analysis and Performance Characterization

Scanning electron microscopy (SEM, TESCAN VEGA 3-SBH, Taseken Trading Company, Shanghai, China) was used to observe the surface and cross-sectional morphology of the coating, and the elemental content of the coating was studied with an equipped energy-dispersive X-ray spectrometer (EDS, TESCAN VEGA 3-SBH, Taseken Trading Company, Shanghai, China). The white-light interferometer (ZeGage, ZYGO, Connecticut, Middlefield, CT, USA) was used to obtain the surface 3D morphology of the coating, and the surface roughness analysis was performed. An Axis Ultra DLD X-ray Photoelectron Spectrometer (XPS) (Shimadzu Corporation, Hadano, Japan) was used to analyze the bond composition. The experimental temperature of the high-temperature oxidation resistance test was 800 °C, and the experimental time was 5 h and 20 h. The matrix and the samples prepared using different voltages were placed in an alumina crucible and then placed in a high-temperature chamber furnace where the temperature had risen to 800 °C for the oxidation test. After oxidation at 800 °C, the cross-sectional morphology of the coating was observed by SEM, and the diffusion of the elements was analyzed by EDS.

3. Results and Discussion

3.1. Microstructure of the Coating

Figure 1 shows the surface morphologies of the MAO coatings prepared using different voltages, and there are obvious pores on the coating, showing a typical "volcano crater" morphology. During the growth of the coating, the presence of holes as discharge channels is unavoidable, and the formation of holes is mainly attributed to the outward escape of internal molten oxides and gases [10]. As the voltage increased, less particulate matter accumulated on the coating. This is because the higher the voltage, the more intense the reaction; the generated molten particles will be connected to each other, and it is difficult to observe the obvious particle aggregation phenomenon. In addition, the higher the voltage, the more energy and heat are released during the MAO process, and the more melt is generated, which will cover the uneven area formed in the early stage, making the coating smoother and smoother. Of course, higher voltages can also cause greater thermal stress in the melt during the electrolyte cooling, resulting in more microcracks being observed on the coatings [19].

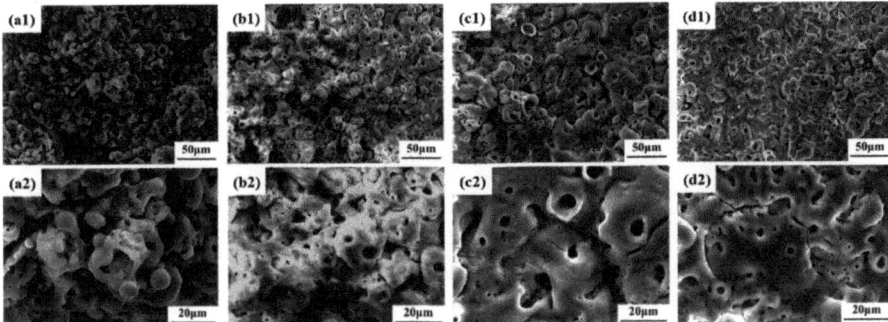

Figure 1. Surface morphologies of the coatings prepared using different voltages: (**a1,a2**) 360 V, (**b1,b2**) 390 V, (**c1,c2**) 420 V, and (**d1,d2**) 450 V.

The surface element content of the MAO coatings prepared using different voltages is shown in Table 1. The coating was dominated by O and Si elements and contained a small amount of P elements, all of which were derived from the electrolyte, and the content was related to their concentration. The matrix elements were all involved in the coating formation reaction, of which the content of Al, Ti, and Zr was relatively high, while the content of V and Cr was relatively small. The difference in elemental content is most likely related to the size of the oxygen affinity, and the greater the oxygen affinity, the more likely the element is to oxidize, and the higher the content. The less electronegative the element, the greater the affinity for oxygen [20,21]. The electronegative size of the five elements of Al, Ti, Cr, V, and Zr is ordered as: Zr < Ti < Al < V < Cr, so the content of Al, Ti, and Zr in the coating will be higher than the content of V and Cr [21]. With the increase in the MAO voltage, the content of O, Al, Ti, and Zr elements on the coating gradually decreased, while the content of Si and P elements increased. This may be because the probability of the matrix elements reaching the surface of the coating through the discharge channel decreases with the increase in the thickness of the coating; this is mainly applicable to the solute ions participating in the coating forming reaction.

Table 1. Surface element content of coatings prepared using different voltages (at.%).

Voltages	O	Al	Si	P	Ti	Cr	V	Zr
360 V	71.9	3.0	11.9	1.1	3.1	2.3	1.3	5.4
390 V	70.4	3.0	16.0	0.2	3.0	1.1	1.4	4.9
420 V	70.6	2.9	16.3	0.4	2.7	1.0	1.4	4.7
450 V	69.8	2.6	15.6	1.2	2.8	1.9	1.5	4.5

Figure 2 presents the 3D morphologies and surface roughness of the coatings prepared using different voltages, and the coating area selected for different test samples was 834.370 μm × 834.370 μm. There was no obvious trend observed regarding the influence of voltage on the surface roughness, but it is undeniable that the coating prepared at a low voltage was coarsest, while the surface roughness of the coating prepared at a high voltage was smallest. The surface roughness of the coating prepared at 360 V voltage was 10.299 μm, and the surface roughness of the coating prepared at 450 V voltage was 7.343 μm, which represents a large difference in roughness. The measurement results of the coatings surface roughness were consistent with those observed in Figure 1. The temperature of the micro-arc discharge area rose rapidly at higher working voltages, resulting in the formation of more molten oxides, which plays a role in repairing and leveling the coating.

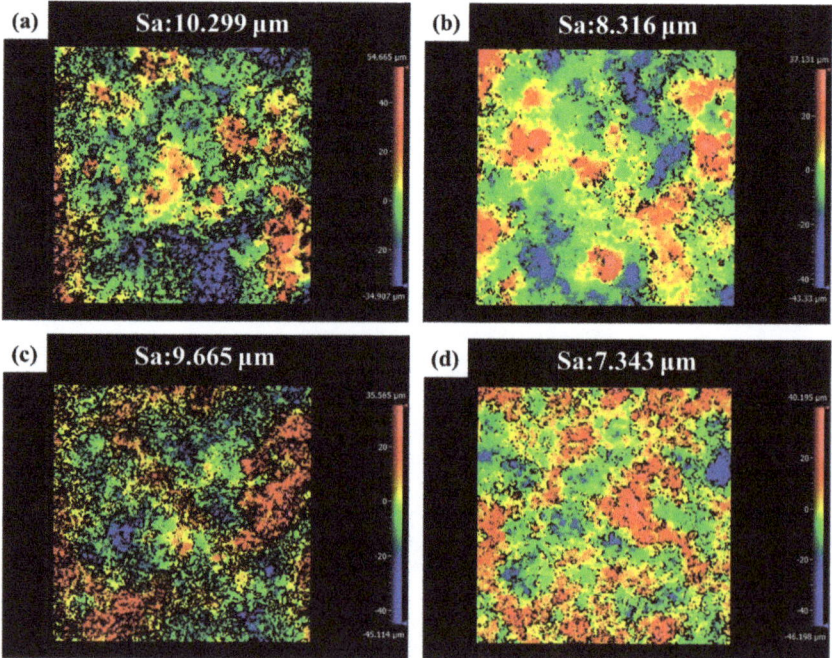

Figure 2. Two–dimensional morphologies of coatings prepared using different voltages: (**a**) 360 V, (**b**) 390 V, (**c**) 420 V, and (**d**) 450 V.

Cross-sectional morphologies of the coatings prepared using different voltages are shown in Figure 3. It can be seen that the coating and the matrix were well bonded, there was no obvious gap, and the interface junction was uneven. There were pores of different sizes at the cross-section of the coating, but these pores did not penetrate the entire coating. The thickness of the coating prepared at 360 V, 390 V, and 420 V did not differ much, all of which were around 40 μm, and the thickest coating prepared at 450 V was approximately

50 µm. In general, the higher the voltage, the thicker the MAO coating. Of course, this is not absolute, because the junction between the MAO coating and the matrix interface is uneven, which may lead to an inconsistent coating thickness between the two microregions. The higher the working voltage, the more intense the MAO reaction and the higher the growth rate of the coating, resulting in an increase in the thickness of the coating. Although the thicker coatings prepared at 450 V provide better protection for the matrix, larger cracks can be clearly observed in the cross-sectional topography that could cause the coating to peel off during high-temperature oxidation, which is extremely detrimental to improving high-temperature oxidation resistance.

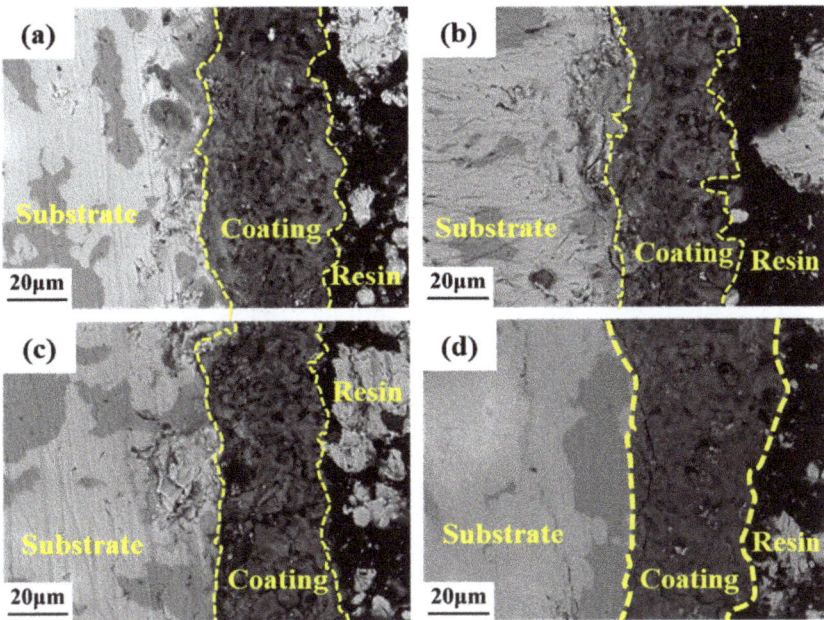

Figure 3. Cross-sectional morphologies of coatings prepared using different voltages: (**a**) 360 V, (**b**) 390 V, (**c**) 420 V, and (**d**) 450 V.

3.2. Chemical Composition of the Coating

In order to determine the presence of elements in the MAO coating on the AlTiCrVZr RHEA, XPS was used for analysis. The XPS full spectrum of the MAO coating prepared at 420 V is shown in Figure 4, and the eight peaks of Al2p, Si2p, Zr3d, C1s, Ti2p, V2p, O1s, and Cr2p were detected in the coating; of these, C1s was the source of contamination. It can be seen from the XPS full spectrum that the coating was mainly composed of O, Si, Al, Ti, Cr, V, and Zr elements, indicating that the solute elements and matrix elements in the electrolyte were involved in the coating-forming reaction.

The XPS high-resolution spectrum of each element on the coating prepared at 420 V is shown in Figure 5. There was only one peak in the Al2p spectrum, corresponding to a binding energy of 75.51 eV, indicating that the predominant form of Al present was Al_2O_3 [22]. The binding energies corresponding to the two peaks of Ti2p were 459.46 eV and 465.22 eV, indicating that Ti was present in the coating in the form of TiO_2 [23]. From the high-resolution spectrum of Cr2p, it can be seen that the binding of Cr2p had two peaks at 577.44 eV and 587.23 eV, corresponding to the Cr–O bond, representing Cr_2O_3 [10]. The peaks in the V2p spectrum were 517.99 eV and 523.87 eV, corresponding to the V–O bond, representing V_2O_5 [24]. The high-resolution spectrum of Zr3d showed that it had two peaks, corresponding to binding energies of 183.32 eV and 185.73 eV, and the surface

Zr present in the coating was ZrO$_2$ [25]. The peak of the Si2p spectrum was 103.10 eV, indicating that the Si present in the coating was SiO$_2$ [26].

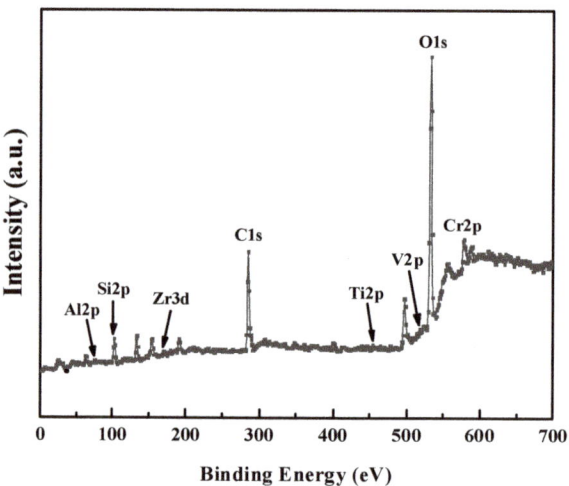

Figure 4. XPS full spectrum of the coating prepared at 420 V.

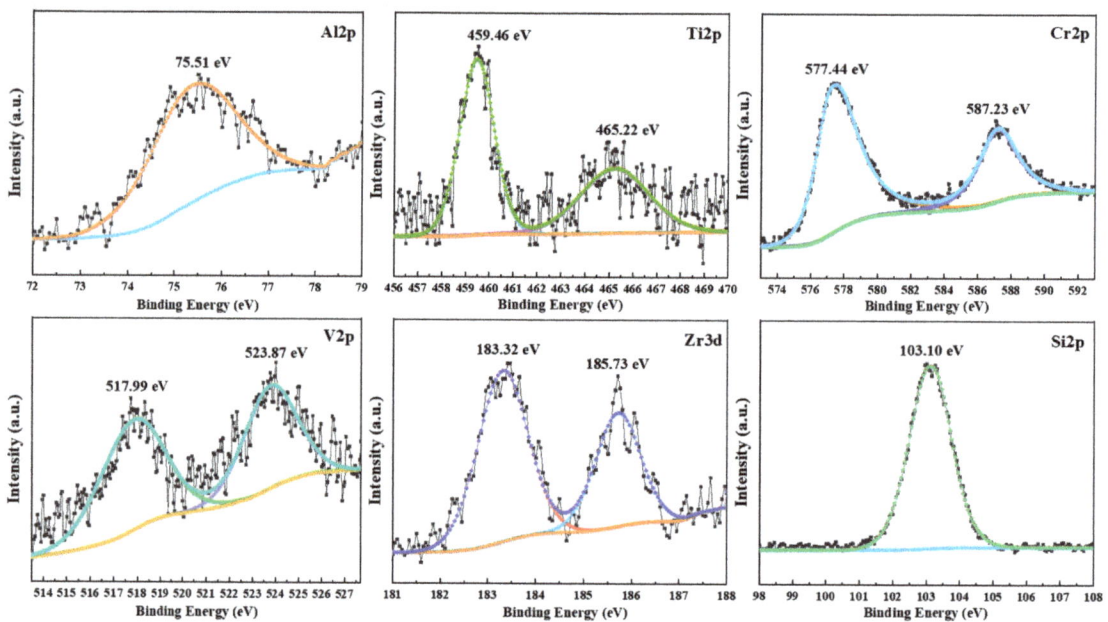

Figure 5. High-resolution spectrum of each element of the coating prepared at 420 V.

The matrix elements and solute ions in the electrolyte were involved in the coating-forming reaction, and the coating composition was dominated by Al$_2$O$_3$, TiO$_2$, Cr$_2$O$_3$, V$_2$O$_5$, ZrO$_2$, and SiO$_2$. Many studies have shown that Al$_2$O$_3$, TiO$_2$, Cr$_2$O$_3$, ZrO$_2$, and SiO$_2$ formed in the coating can prevent oxygen from diffusing directly into the matrix, thereby improving the high-temperature oxidation resistance of RHEAs [14,27]. We believe that the MAO coating prepared on AlTiCrVZr RHEA has the potential to improve the high-temperature oxidation resistance of matrix alloy.

3.3. High-Temperature Oxidation Resistance

Figure 6 shows the cross-sectional morphologies and elemental distribution of the MAO samples prepared using different voltages at 800 °C for 5 h. It can be seen that, after oxidation at 800 °C for 5 h, the thickness of the matrix diffusion layer exceeded 700 μm, which was significantly greater than the thickness of the diffusion layer of samples after MAO. In addition, we observed the presence of significant cracks and a small number of holes in the diffusion layer of the matrix. The generation of cracks and holes is attributed to thermal stress and the volatilization of oxides, respectively. The MAO samples prepared at 360 V, 420 V, and 450 V showed little difference in the thickness of the diffusion layer after high-temperature oxidation, of approximately 500 μm. The thickness of the diffusion layer of the MAO sample prepared at 390 V after high-temperature oxidation was only approximately 400 μm. Combined with the microstructure of the coating, it was found that the MAO sample prepared at 390 V exhibited good high-temperature oxidation resistance, which is closely related to the compactness of the coating and fewer microscopic defects.

The EDS results showed that the O element was enriched in the diffusion layer, and uncoated samples had a high content in the diffusion layer and a large amount inside the matrix, which indicates that the MAO coating hinders the diffusion of O elements to the inside of the matrix to a certain extent. In addition, the content of Ti, Al, Cr, and Zr elements in the diffusion layer after high-temperature oxidation of the MAO sample was lower than that of the matrix. The content of element V in the diffusion layer is always very low, because the V_2O_5 formed after oxidation is extremely volatile at 800 °C, resulting in a very low content of V elements and creating holes in the diffusion layer.

Figure 7 shows the cross-sectional morphologies and elemental distribution of the MAO samples prepared using different voltages and the matrix after oxidation at 800 °C for 20 h. After oxidation at 800 °C for 20 h, the thickness of the matrix diffusion layer exceeded 1500 μm, which was greater than the thickness of the diffusion layer of the MAO-coated samples. Although the MAO samples prepared at 390 V at 800 °C oxidation for 5 h showed good high-temperature oxidation resistance, the protection effect of the coating was poor when the high-temperature oxidation time increased to 20 h. The MAO sample prepared at 420 V had the smallest diffusion layer thickness, of approximately 800 μm, and the best high-temperature oxidation resistance. The high-temperature oxidation resistance of the coating prepared at 450 V is known to be poor as the micro-arc discharge reaction at higher voltages is violent, and the heat released results in additional cracks in the coating; O_2 very easily invades the inside of the matrix through cracks; therefore, the coating loses its protective effect on the matrix.

From the results of cross-sectional EDS after high-temperature oxidation, it can be seen that O elements were enriched in the diffusion layer, while the matrix elements were less abundant in the diffusion layer; in particular, the content of V elements was extremely low. With the increase in high-temperature oxidation time, O_2 gradually invaded the interior of the matrix and combined with V elements to form V_2O_5 at high temperatures. V_2O_5 is extremely volatile at 800 °C, resulting in an extremely low V content in the diffusion layer. Combined with the cross-sectional morphology and element distribution after 5 h of high-temperature oxidation, it can be seen that the MAO coating can hinder the diffusion of O elements to the interior of the matrix and the volatilization of V_2O_5 to a certain extent. The obstruction effect is not only related to the compactness of the coating, but also to the thickness and internal defects of the coating. The key to improving the high-temperature oxidation resistance of the MAO coating is to prepare a uniform and dense coating with good bonding with the matrix and to reduce cracks in the coating as much as possible. Increasing the thickness of the coating when the above conditions are met can more effectively hinder the contact of O_2 with the matrix metal, thereby significantly improving the high-temperature oxidation resistance of RHEAs.

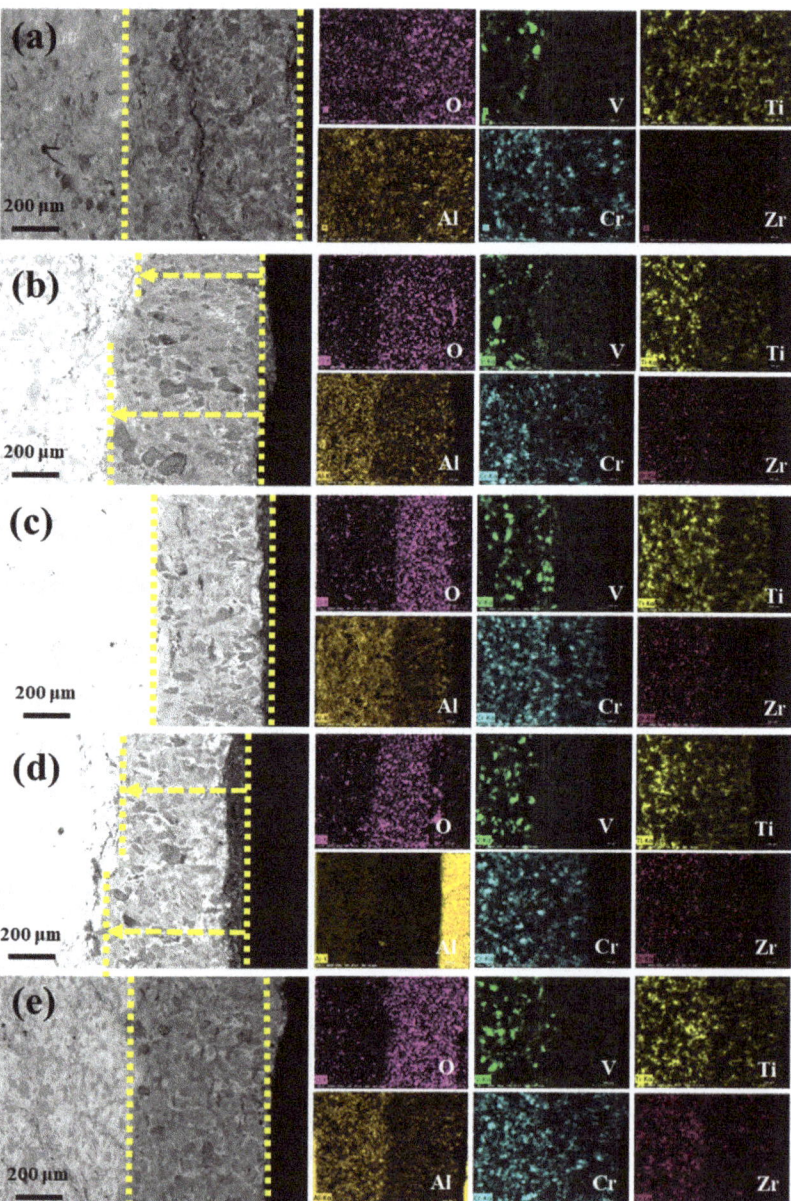

Figure 6. Cross-sectional morphologies and elements distribution of samples prepared using different voltages after oxidation at 800 °C for 5 h: (**a**) matrix, (**b**) 360 V, (**c**) 390 V, (**d**) 420 V, and (**e**) 450 V.

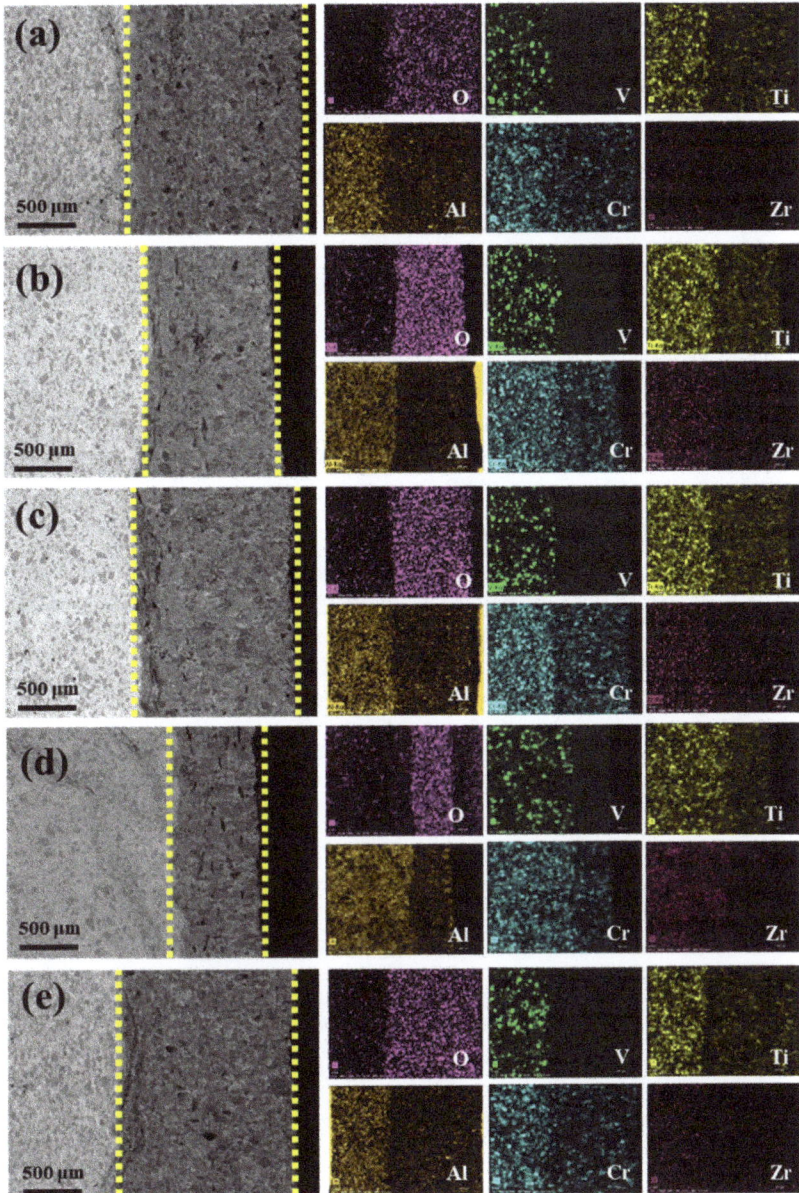

Figure 7. Cross-sectional morphologies and elements distribution of samples prepared using different voltages after oxidation at 800 °C for 20 h: (**a**) matrix, (**b**) 360 V, (**c**) 390 V, (**d**) 420 V, and (**e**) 450 V.

4. Conclusions

(1) Under different MAO voltages, ceramic coatings with a thickness of 40–50 μm were prepared on the AlTiCrVZr RHEA. With the increase in voltage, the surface of the coating was smoother and denser, but the internal defects of the ceramic coating increased, especially the obvious microcracks in the coating prepared at 450 V.

(2) The solute ions and matrix elements contained in the electrolyte during the MAO process were involved in the coating-forming reaction, and the coating composition was mainly Al_2O_3, TiO_2, Cr_2O_3, V_2O_5, ZrO_2, and SiO_2.

(3) Compared with the matrix alloy, the high-temperature oxidation resistance of MAO-coated samples prepared using different voltages was improved after 5 h and 20 h of oxidation at 800 °C. Among them, the coating prepared at 420 V exhibited better high-temperature oxidation resistance after long-term oxidation for 20 h.

Author Contributions: Conceptualization, Z.W. and X.S.; methodology, Z.C.; validation, Z.W. and Y.Z.; formal analysis, M.R.; investigation, S.W.; resources, Z.C.; data curation, Z.C.; writing—original draft preparation, Z.W.; writing—review and editing, X.S.; visualization, Y.Z.; supervision, M.R.; project administration, Y.Z.; funding acquisition, X.S. All authors have read and agreed to the published version of the manuscript.

Funding: This work was financially supported by the National Natural Science Foundation of China (No. 52071252) and the Key research and development plan of Shaanxi province industrial project (2021GY-208 and 2021ZDLSF03-11).

Institutional Review Board Statement: Not applicable.

Informed Consent Statement: Not applicable.

Data Availability Statement: Not applicable.

Conflicts of Interest: The authors declare that they have no known competing financial interests or personal relationships that could have appeared to influence the work reported in this paper.

References

1. Chen, J.; Zhou, X.; Wang, W.; Liu, B.; Lv, Y.; Yang, W.; Xu, D.; Liu, Y. A review on fundamental of high entropy alloys with promising high–temperature properties. *J. Alloys Compd.* **2018**, *760*, 15–30. [CrossRef]
2. Ma, M.; Han, A.; Zhang, Z.; Lian, Y.; Zhao, C.; Zhang, J. The role of Si on microstructure and high-temperature oxidation of $CoCr_2FeNb_{0.5}Ni$ high-entropy alloy coating. *Corros. Sci.* **2021**, *185*, 109417. [CrossRef]
3. Stepanov, N.D.; Yurchenko, N.Y.; Panina, E.S.; Tikhonovsky, M.A.; Zherebtsov, S.V. Precipitation-strengthened refractory $Al_{0.5}CrNbTi_2V_{0.5}$ high entropy alloy. *Mater. Lett.* **2017**, *188*, 162–164. [CrossRef]
4. Kumari, P.; Gupta, A.K.; Mishra, R.K.; Ahmad, M.; Shahi, R.R. A Comprehensive Review: Recent Progress on Magnetic High Entropy Alloys and Oxides. *J. Magn. Magn. Mater.* **2022**, *554*, 169142. [CrossRef]
5. Liu, S.S.; Zhang, M.; Zhao, G.L.; Wang, X.H.; Wang, J.F. Microstructure and properties of ceramic particle reinforced FeCoNiCrMnTi high entropy alloy laser cladding coating. *Intermetallics* **2022**, *140*, 107402. [CrossRef]
6. Kuang, J.; Zhang, P.; Wang, Q.; Hu, Z.; Liang, X.; Shen, B. Formation and oxidation behavior of refractory high-entropy silicide (NbMoTaW)Si_2 coating. *Corros. Sci.* **2022**, *198*, 110134. [CrossRef]
7. Lu, S.; Li, X.; Liang, X.; Shao, W.; Yang, W.; Chen, J. Effect of Al content on the oxidation behavior of refractory high-entropy alloy $AlMo_{0.5}NbTa_{0.5}TiZr$ at elevated temperatures. *Int. J. Refract. Met. Hard Mater.* **2022**, *105*, 105812. [CrossRef]
8. Shao, L.; Yang, M.; Ma, L.; Tang, B.-Y. Effect of Ta and Ti content on high temperature elasticity of HfNbZrTa1−xTix refractory high-entropy alloys. *Int. J. Refract. Met. Hard Mater.* **2021**, *95*, 105451. [CrossRef]
9. Kao, P.-C.; Hsu, C.-J.; Chen, Z.-H.; Chen, S.-H. Highly transparent and conductive $MoO_3/Ag/MoO_3$ multilayer films via air annealing of the MoO_3 layer for ITO-free organic solar cells. *J. Alloys Compd.* **2022**, *906*, 164387. [CrossRef]
10. Shi, X.; Yang, W.; Cheng, Z.; Shao, W.; Xu, D.; Zhang, Y.; Chen, J. Influence of micro arc oxidation on high temperature oxidation resistance of AlTiCrVZr RHEA. *Int. J. Refract. Met. Hard Mater.* **2021**, *98*, 105562. [CrossRef]
11. Jia, X.; Song, J.; Zhao, L.; Jiang, B.; Yang, H.; Zhao, T.; Zhao, H.; Liu, Q.; Pan, F. The corrosion behaviour of MAO coated AZ80 magnesium alloy with surface indentation of different sizes. *Eng. Fail. Anal.* **2022**, *136*, 106185. [CrossRef]
12. Yang, W.; Xu, D.; Guo, Q.; Chen, T.; Chen, J. Influence of electrolyte composition on microstructure and properties of coatings formed on pure Ti substrate by micro arc oxidation. *Surf. Coat. Technol.* **2018**, *349*, 522–528. [CrossRef]
13. Cheng, F.; Li, S.; Gui, W.; Lin, J. Surface modification of Ti-45Al-8.5 Nb alloys by microarc oxidation to improve high-temperature oxidation resistance. *Prog. Nat. Sci. Mater. Int.* **2018**, *28*, 386–390. [CrossRef]
14. Cheng, Z.; Yang, W.; Xu, D.; Wu, S.; Yao, X.; Lv, Y.; Chen, J. Improvement of high temperature oxidation resistance of micro arc oxidation coated $AlTiNbMo_{0.5}Ta_{0.5}Zr$ high entropy alloy. *Mater. Lett.* **2020**, *262*, 127192. [CrossRef]
15. Yilmaz, M.S.; Sahin, O. Applying high voltage cathodic pulse with various pulse durations on aluminium via micro-arc oxidation (MAO). *Surf. Coat. Technol.* **2018**, *347*, 278–285. [CrossRef]
16. Du, Q.; Wei, D.; Wang, Y.; Cheng, S.; Liu, S.; Zhou, Y.; Jia, D. The effect of applied voltages on the structure, apatite-inducing ability and antibacterial ability of micro arc oxidation coating formed on titanium surface. *Bioact. Mater.* **2018**, *3*, 426–433. [CrossRef]

17. Wang, J.; Wang, J.; Lu, Y.; Du, M.H.; Han, F.Z. Effects of single pulse energy on the properties of ceramic coating prepared by micro-arc oxidation on Ti alloy. *Appl. Surf. Sci.* **2015**, *324*, 405–413. [CrossRef]
18. Zhang, J.; Fan, Y.; Zhao, X.; Ma, R.; Du, A.; Cao, X. Influence of duty cycle on the growth behavior and wear resistance of micro-arc oxidation coatings on hot dip aluminized cast iron. *Surf. Coat. Technol.* **2018**, *337*, 141–149. [CrossRef]
19. Bai, L.; Dong, B.; Chen, G.; Xin, T.; Wu, J.; Sun, X. Effect of positive pulse voltage on color value and corrosion property of magnesium alloy black MAO ceramic coating. *Surf. Coat. Technol.* **2019**, *374*, 402–408. [CrossRef]
20. Douillard, J.-M.; Salles, F.; Henry, M.; Malandrini, H.; Clauss, F. Surface energy of talc and chlorite: Comparison between electronegativity calculation and immersion results. *J. Colloid Interface Sci.* **2007**, *305*, 352–360. [CrossRef]
21. Qteish, A. Electronegativity scales and electronegativity-bond ionicity relations: A comparative study. *J. Phys. Chem. Solids* **2019**, *124*, 186–191. [CrossRef]
22. Guan, S.; Qi, M.; Wang, C.; Wang, S.; Wang, W. Enhanced cytocompatibility of Ti6Al4V alloy through selective removal of Al and V from the hierarchical MAO coating. *Appl. Surf. Sci.* **2021**, *541*, 148547. [CrossRef]
23. Lan, N.; Yang, W.; Gao, W.; Guo, P.; Zhao, C.; Chen, J. Characterization of ta-C film on micro arc oxidation coated titanium alloy in simulated seawater. *Diam. Relat. Mater.* **2021**, *117*, 108483. [CrossRef]
24. Rajesh, V.; Veeramuthu, K.; Shiyamala, C. Investigation of the morphological, optical and electrochemical capabilities of V_2O_5/MWCNT nanoparticles synthesized using a microwave autoclave technique. *JCIS Open* **2021**, *4*, 100032. [CrossRef]
25. Barreca, D.; Battiston, G.A.; Gerbasi, R.; Tondello, E.; Zanella, P. Zirconium Dioxide Thin Films Characterized by XPS. *Surf. Sci. Spectra* **2000**, *7*, 303–309. [CrossRef]
26. Xu, G.; Shen, X. Fabrication of SiO_2 nanoparticles incorporated coating onto titanium substrates by the micro arc oxidation to improve the wear resistance. *Surf. Coat. Technol.* **2019**, *364*, 180–186. [CrossRef]
27. Cao, Y.; Liu, Y.; Liu, B.; Zhang, W.D.; Wang, J.W.; Meng, D.U. Effects of Al and Mo on high temperature oxidation behavior of RHEAs. *Trans. Nonferrous Met. Soc. China* **2019**, *29*, 1476–1483. [CrossRef]

Disclaimer/Publisher's Note: The statements, opinions and data contained in all publications are solely those of the individual author(s) and contributor(s) and not of MDPI and/or the editor(s). MDPI and/or the editor(s) disclaim responsibility for any injury to people or property resulting from any ideas, methods, instructions or products referred to in the content.

Article

Comparison of K340 Steel Microstructure and Mechanical Properties Using Shallow and Deep Cryogenic Treatment

Patricia Jovičević-Klug [1,2,*], László Tóth [3,*] and Bojan Podgornik [1,2]

1. Department of Metallic Materials and Technology, Institute of Metals and Technology, Lepi pot 11, 1000 Ljubljana, Slovenia
2. Jožef Stefan International Postgraduate School, Jamova c. 39, 1000 Ljubljana, Slovenia
3. Bánki Donát Faculty of Mechanical and Safety Engineering, Óbuda University, Népszínház u. 8., 1081 Budapest, Hungary
* Correspondence: patricia.jovicevicklug@imt.si (P.J.-K.); toth.laszlo@bgk.uni-obuda.hu (L.T.); Tel.: +386-1470-1990 (P.J.-K.); +36-1-66-5342 (L.T.)

Citation: Jovičević-Klug, P.; Tóth, L.; Podgornik, B. Comparison of K340 Steel Microstructure and Mechanical Properties Using Shallow and Deep Cryogenic Treatment. *Coatings* **2022**, *12*, 1296. https://doi.org/10.3390/coatings12091296

Academic Editor: Diego Martinez-Martinez

Received: 20 July 2022
Accepted: 31 August 2022
Published: 2 September 2022

Publisher's Note: MDPI stays neutral with regard to jurisdictional claims in published maps and institutional affiliations.

Copyright: © 2022 by the authors. Licensee MDPI, Basel, Switzerland. This article is an open access article distributed under the terms and conditions of the Creative Commons Attribution (CC BY) license (https://creativecommons.org/licenses/by/4.0/).

Abstract: In this research, Böhler K340 cold work tool steel was subjected to three different heat treatment protocols, conventional heat treatment (CHT), shallow cryogenic treatment (SCT), and deep cryogenic treatment (DCT). The study compares the effect of SCT and DCT on the microstructure and consequently on the selected mechanical properties (micro- and macroscale hardness and impact toughness). The study shows no significant difference in macroscale hardness after the different heat treatments. However, the microhardness values indicate a slightly lower hardness in the case of SCT and DCT. Microstructure analysis with light (LM) and scanning electron microscopy (SEM) indicated a finer and more homogenous microstructure with smaller lath size and preferential orientation of the martensitic matrix in SCT and DCT samples compared to CHT. In addition, the uniform precipitation of more spherical and finer carbides is determined for both cryogenic treatments. Moreover, the precipitation of small dispersed secondary carbides is observed in SCT and DCT, whereas in the CHT counterparts, these carbide types were not detected. X-ray diffraction (XRD) and electron backscatter diffraction (EBSD) confirms that SCT and DCT are very effective in minimizing the amount of retained austenite down to 1.8 vol.% for SCT and even below 1 vol.% for the DCT variant.

Keywords: shallow cryogenic treatment (SCT); deep cryogenic treatment (DCT); cold work tool steel; microstructure; mechanical properties

1. Introduction

Cold work tool steels are used for producing tools for cutting and forming processes such as punching, blanking, cold rolling, extruding, deep drawing, bending, and pressing. Therefore, these steels need high hardness, high abrasive and adhesive wear resistance, high compressive strength, high toughness, and improved dimensional stability [1]. One of these steels is also Böhler K340 cold work tool steel, which has a higher Cr (around 8 wt.%) and C (around 1.1%) alloying content and is produced through electroslag remelting (ESR) [2–4]. The ESR technology is mainly used to reduce inhomogeneity, such as segregations and shrink voids, and to improve the cleanness level, especially in regard to larger non-metallic inclusions [1].

Due to its unique chemical composition and properties, Böhler K340 has been highly demanded in the industry for high-performance applications [2,5]. However, in order to achieve these favorable properties also, the applied heat treatment needs to be well defined and optimized. The most commonly used heat treatment in the tool industry is conventional heat treatment (CHT), which consists of slow heating to the hardening temperature, holding at that temperature for a certain amount of time to obtain homogenous austenite grains and tailor their size to adapt the final steel properties. This part of the process can be considerably segmented into different steps, i.e., preheating, during which the heated steel is held at intermediate temperatures to thermally equalize the surface and the core. After the hardening process, the material is quenched (rapidly cooled) at high cooling rates to temperatures below 100 °C and finally double or triple tempered at a lower temperature than the hardening temperature [1]. As a result, the microstructure of such steels normally consists of tempered martensite, retained austenite (RA), and precipitated carbides (MxC and MxCy) [5].

Another option is also to expose the material to cryogenic temperatures performed within the heat treatment path. During cryogenic treatment/technology (CT), steel is exposed to subzero temperatures in order to come close to the martensite finish temperature (M_f). This enables us to obtain the highest possible martensite fraction in the steel in order to gain the preferable higher hardness and strength properties [6–8]. However, this may cause micro-cracking and cracking of the steel if it is not followed by a tempering step [9,10]. Nevertheless, cold work tool steels are generally triple tempered at higher tempering temperatures without CT [2,11–14], which normally results in a low RA content and sufficient dimensional stability [15]. Part of CT is also shallow cryogenic treatment (SCT) (temperature to −160 °C) and deep cryogenic treatment (DCT) (temperatures below −160 °C) [16]. Both treatments are an effective method to reduce the amount of RA [14,17] in order to improve materials properties and, with it, increase the tool service life [18]. As such, CT can improve the mechanical properties (hardness, impact toughness) [8,17], resulting from the highest martensite fraction as well as due to the increased precipitation and formation of very small carbides between or within the tempered martensitic grains [12,19,20]. The newly formed carbides reduce the internal stress of the martensite and also act as buffers for the microcrack propagations [21]. SCT and DCT can both induce precipitation of finer and more spherical carbides [22], resulting in a more homogenous microstructure [22,23]. However, SCT and DCT have, in the end, different levels of effect on the microstructure due to the different cooling temperatures and, with it, different effects on the final properties [7].

The aim of this research study was to compare the effect of shallow (SCT) and deep cryogenic treatments (DCT) to conventional heat treatment (CHT) in terms of changes induced in microstructure and selected mechanical properties (hardness, microhardness, and impact toughness) of cold work tool steel K340. The study also aims to provide new insight into basic research of rival cryogenic treatments.

2. Materials and Methods

2.1. Material and Heat Treatment

The selected experimental material was Böhler K340 Isodur cold work tool steel produced by electroslag remelting (ESR) technology. The samples were cut from a rectangular raw bar with charmless wire electrical discharge machining (WEDM) using Fi 240 S2P machine from GF Machines and Technologies, Biel, Switzerland. Material was cut into small pieces of 55 × 10 × 10 mm that were afterward fine-ground and polished. The chemical composition of steel was analyzed with a Hitachi PMP2 type instrument (Hitachi, Uedem, Germany). The given chemical composition in wt.% is: 1.12% C, 0.92% Si, 0.39% Mn, 8.25% Cr, 2.19% Mo, 0.27% Ni, 0.15% Nb, 0.40% V, 0.14% W, and Fe as base.

Specimens were processed using three different heat treatments: the first was conventional heat treatment (CHT), the second was heat treatment involving subzero cooling after quenching to −150 °C (123 K), referred to as shallow cryogenic temperature or treatment (SCT), and the third heat treatment involving deep cryogenic treatment (DCT), where the samples were cooled down to −196 °C (77 K).

The selected heat treatment for each group is provided in Figure 1. All samples were first austenitized and quenched in a single step, either in a horizontal vacuum furnace IPSEN VTTC-324R, Ipsen, Kleve, Germany (DCT) or in a vacuum furnace IVA Schemetz IU72, Menden, Germany (CHT and SCT), with uniform high-pressure gas quenching using N_2 at the pressure of 8 bars (average quenching rate was approximately 7–8 °C s^{-1}). After quenching, the CHT group went directly for triple tempering at an average temperature of 555 °C for 2 h, whereas SCT and DCT groups went first for cryogenic treatment (CT) and then followed by triple or single tempering under the same conditions (555 °C/2 h), Figure 1. The second group was subjected to SCT, performed immediately after quenching by cooling the specimens in the IVA Schmetz IU72 vacuum furnace, using the cool plus cryogenic process in liquid nitrogen at −150 °C for 50 min, and finalized by triple tempering. The third group, DCT, was gradually immersed in liquid nitrogen for 24 h (1 day) at −196 °C after quenching, followed by only a single tempering cycle, which is reported to be adequate enough to provide the nearly complete RA transformation into martensite, when combined with DCT [24].

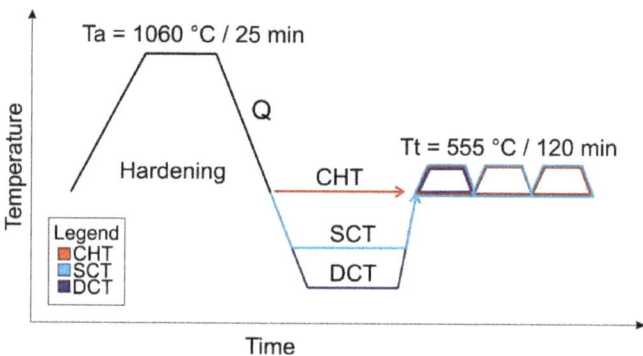

Figure 1. Scheme of all three heat treatments (CHT = conventional heat treatment, SCT = shallow cryogenic treatment and DCT = deep cryogenic treatment) with selected heat treatment temperature (Ta = austenitization temperature and time and Tt = tempering temperature and time).

2.2. Methods

2.2.1. Mechanical Testing

The effectiveness of the heat treatments was checked by evaluating three mechanical properties: hardness, microhardness, and impact toughness. Hardness was measured by Rockwell C hardness measurement with Instron B2000, Instron; Norwood, MA, USA, according to SIST EN ISO 6508-1:2016 standard. Microhardness was measured with Instron Tukon 2100B, Instron, Norwood, MA, USA, according to standard SIST EN ISO 6507-1:2016. For both hardness measurements, 10 samples were measured. For impact testing, a Charpy impact test machine of type RM 201 by VEB WPM, Leipzig, Germany, was used and performed on 10 samples.

2.2.2. Microstructure and Phase Analysis

Basic microstructural analysis was performed on polished and etched (etched by Nital for a few seconds, depending on each sample [25]) samples by ZEISS Axio Imager, Carl Zeiss, Oberkochen, Germany. Detailed microstructure characterization was performed with SEM, SM-6500 F, Jeol, Tokyo, Japan, where secondary electrons (SE), backscattered electrons (BSE), energy dispersive X-ray spectroscopy (EDX) with 15 keV electron beam Oxford EDX INCA Energy 450, detector type INCA X-SIGHT LN_2, Oxford Instruments, UK and electron backscatter diffraction (EBSD) with 15 keV electron beam and current of around 5 nA were used. For EBSD data analysis, OIM Analysis software, EDAX, Ametek Inc., Warrendale, PA, USA was employed. Phase and phase fraction determination of samples via X-ray diffraction (XRD) was carried out on PANalytical 3040/60, Almelo, The Netherlands. The phase identification analysis and interpretation were performed using COD database references, and their fractions were evaluated using a combination of Rietveld refinement and Toraya method [26]. The standard evaluation error for each phase was determined to be 1–2 vol.%.

3. Results

3.1. Microstructure and Phase Analysis

Microstructural analysis was first conducted with light (optical) microscopy in order to obtain an overview of the final microstructure of all three heat treatments, CHT, SCT, and DCT (Figure 2a–c).

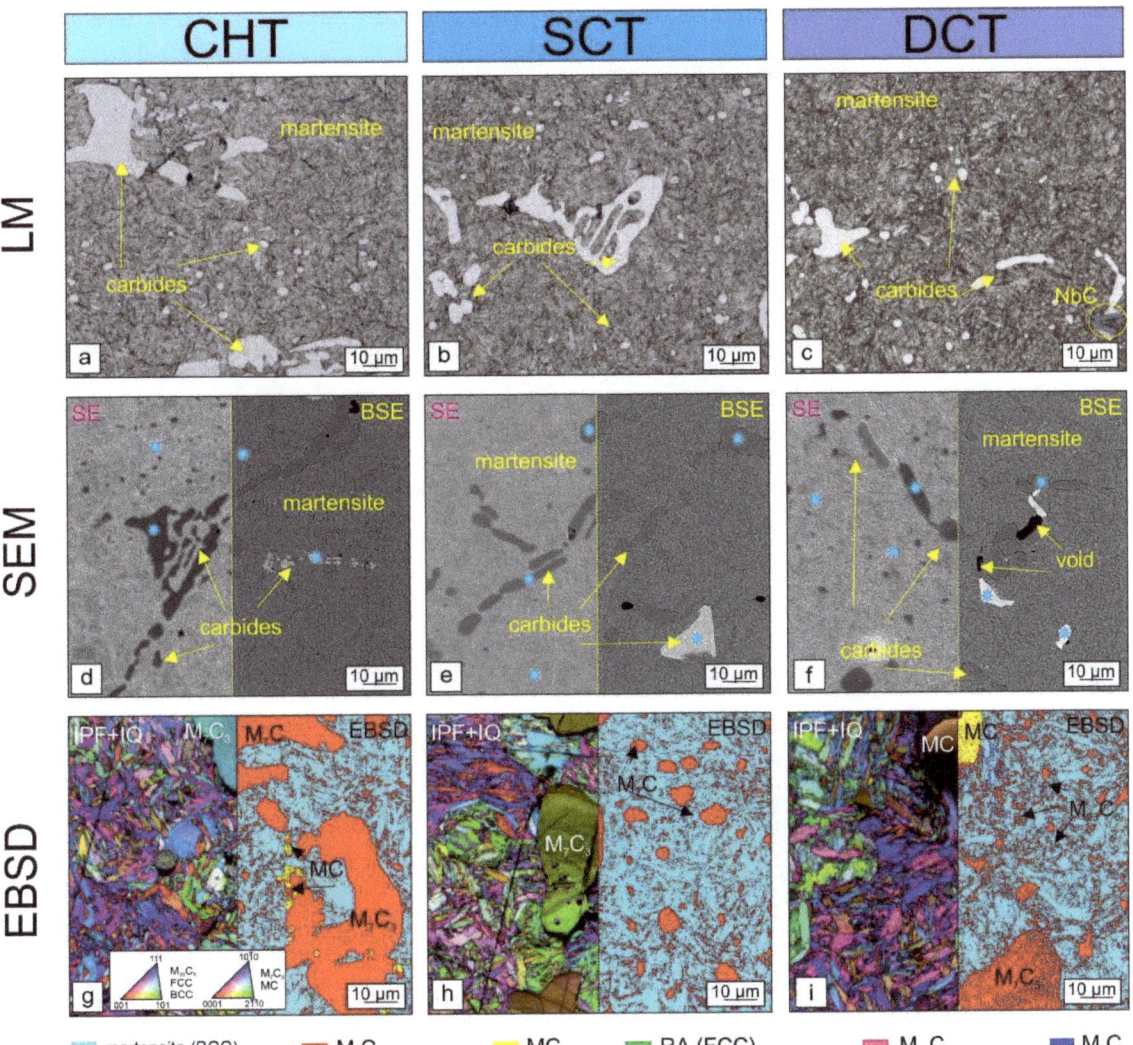

Figure 2. The microstructure of all three heat treatment groups CHT-conventional heat treatment (**a,d,g**), SCT-shallow cryogenic treatment (**b,e,h**) and DCT-deep cryogenic treatment (**c,f,i**) observed with light microscopy (LM), scanning electron microscopy (SEM) under secondary electrons (SE) and backscatter electrons (BSE) and with electron backscatter diffraction (EBSD) represented through inverse pole figures (IPF-+IQ). The phases, which are presented are martensite (BCC), retained austenite (RA) and different carbides (MC, M_7C_3, $M_{23}C_6$ and M_3C_2).

In-detail analysis of the microstructure, such as type, size, number, volume fraction, and distribution of phases conducted by SEM, EDS, and EBSD, is presented in

Figures 2, 3 and 4a and by XRD in Figure 4b. For all three groups, the matrix consists of lath martensite with some amount of RA (see Table 1). On average, the martensitic laths were finer with each cryogenic treatment by 21% and 33% with SCT and DCT, respectively (Figures 2a–f and 3). Accordingly, the size of martensitic laths is roughly 22% smaller with DCT compared to SCT. The martensitic laths of both SCT and DCT samples also display an overall preferential orientation along [101] and [001] directions compared to the laths of the CHT sample. Furthermore, in DCT samples, the martensitic laths displayed even more distinctively set orientation towards the [101] and [001] directions compared to the laths of the SCT sample (Figure 2g–i).

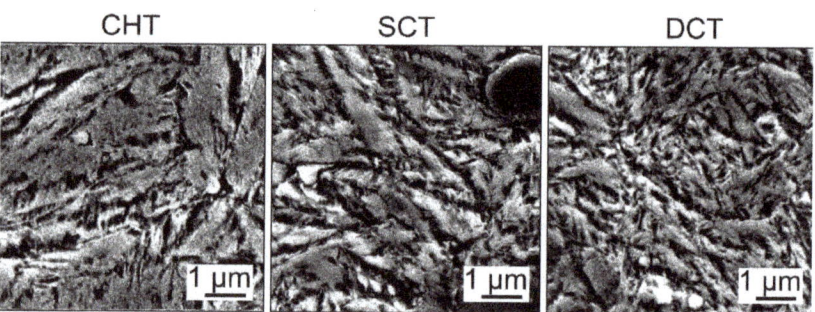

Figure 3. Enlarged SEM micrographs of the etched martensitic matrix presenting the martensitic lath refinement with shallow cryogenic treatment (SCT) and deep cryogenic treatment (DCT) compared to conventional heat treatment (CHT).

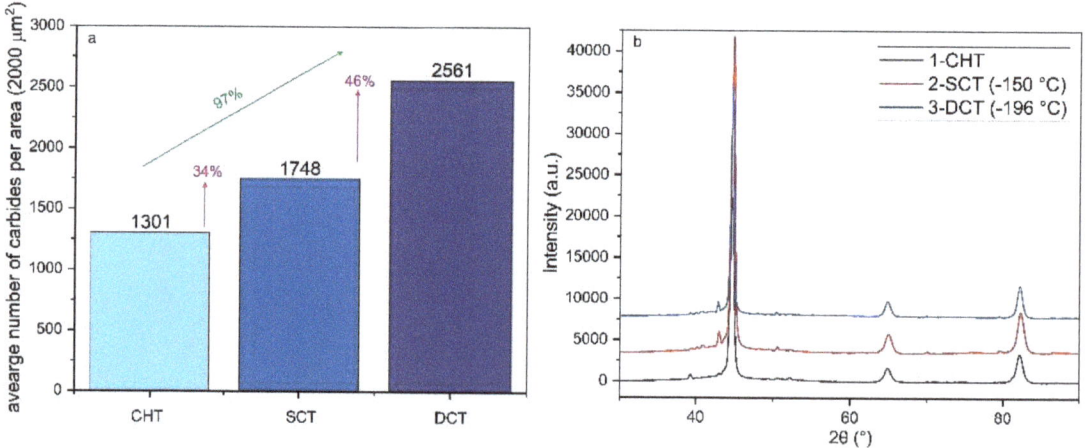

Figure 4. (a) the graph represents the average number of carbides per area (2000 μm^2), where at least 5 locations were measured per sample. (b) X-ray diffraction (XRD) of CHT (black line), SCT (red line) and DCT (blue line) samples.

In regards to the precipitated carbides, the majority of them are formed in-between martensitic laths. The analysis of the number of precipitated carbides revealed that both SCT and DCT increased the precipitation of carbides by 34% and 97%, respectively (Figure 4a), equating to a 46% increase in precipitation with DCT compared to SCT. Furthermore, the SCT and DCT have been shown to influence the shape and average size of the carbides, at which more spherical-shaped carbides are present, with a reduced chance of agglomerations for the cryogenic variants compared to the CHT samples. Carbides distribution and size analysis, determined with EBSD, also reveals that SCT and DCT promote precipitation of finer carbides (≤0.5 μm) and induce a more homogenous microstructure compared to CHT counterparts, especially DCT (Figure 2d–f).

Table 1. Mean volumetric fractions of phases present in the samples; CHT—conventionally heat-treated; SCT—shallow cryogenically heat-treated; DCT—deep cryogenically heat-treated.

Phase (vol.%)	Heat Treatment		
	CHT	SCT (−150 °C)	DCT (−196 °C)
Martensite	53.0	55.0	50.5
RA	6.2	1.8	0.9
MC (V, Nb)	5.2	4.8	4.9
M_7C_3 (Cr, Fe)	16.7	24.6	28.3
$M_{23}C_6$ (Cr, Fe)	5.8	6.1	8.1
M_3C_2 (Cr, Fe)	13.1	7.7	7.3

XRD data (Figure 4b, Table 1) confirm that the matrix of all three samples consists mainly of martensite, with the highest vol.% for SCT (55.0). In all three groups, RA is still present. However, the presence of RA is strongly decreased by both cryogenic treatments, from 6.2 vol.% in CHT to 1.8 vol.% by SCT (reduction by 71%) and even below 1 vol.% by DCT (reduction higher than 85%), being below the detection limit. In all three heat treatment groups, the presence of M_2C, M_7C_3, M_3C_2 and $M_{23}C_6$ carbides is determined.

The crystal structure and chemical composition of carbides are determined with a combination of XRD and EBSD (Figures 2g–i and 4b and Table 1) and EDS (Figure 2d–f, blue marking where the analysis was performed) and EDS mapping (Figure 5). Carbides MC are shown to be enriched by V and Nb, M_7C_6, $M_{23}C_6$, and M_3C_2 with Cr and Fe. Further analysis showed the difference in carbide precipitation between SCT and DCT compared to CHT. The greatest difference is observed in relation to M_7C_3 precipitation, where SCT precipitation is increased by roughly 50% and with DCT by around 70%, compared to the CHT group. The $M_{23}C_6$ carbide group is predominant in the DCT group (8.1 vol.%), whereas in SCT (6.1 vol.%) and in CHT (5.8 vol.%), the values are similar. DCT also yields an increase in $M_{23}C_6$ precipitation compared to the other two groups by approximately 35%. The general tendency of M_3C_2 reflects a decrease in their formation for both CTs, equating to a 40% decrease.

3.2. Mechanical Properties

3.2.1. Hardness and Microhardness

The results of hardness measurements (Figure 6a) show a slight decrease of less than 5% in hardness from CHT to SCT/DCT and from 60 HRC to 59 HRC. For both cryogenic treatments, the values of hardness are the same. However, for the microhardness (Figure 6b), values drop from 810 HV 0.1 for CHT to about 760 HV 0.1 for SCT and DCT (around 7% decrease compared to CHT).

3.2.2. Impact Toughness

Impact toughness results presented in energy required to break CVN samples of tested cold work tool steel Böhler K340 Isodur (Figure 6c) show significant improvement for both cryogenic treatments. Impact toughness increased from 4 J for CHT to 8.5 J for SCT (increase by roughly 113%) and 8 J for DCT (increase by 100%), accordingly. Considering measurement uncertainty and scatter, both cryogenic treatments show similar results. These results are also complementary to the results of decreased hardness after both cryogenic treatments, SCT and DCT, accordingly.

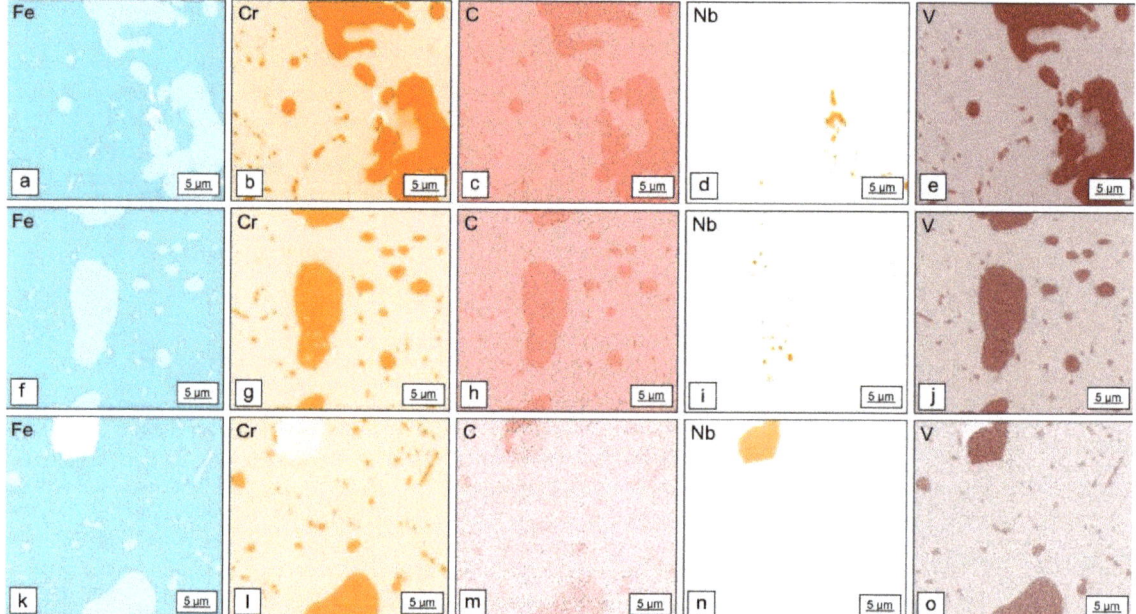

Figure 5. EDS-mapping of all three samples: CHT (**a–e**), SCT (**f–j**) and DCT (**k–o**) for Fe, Cr, C, Nb and V.

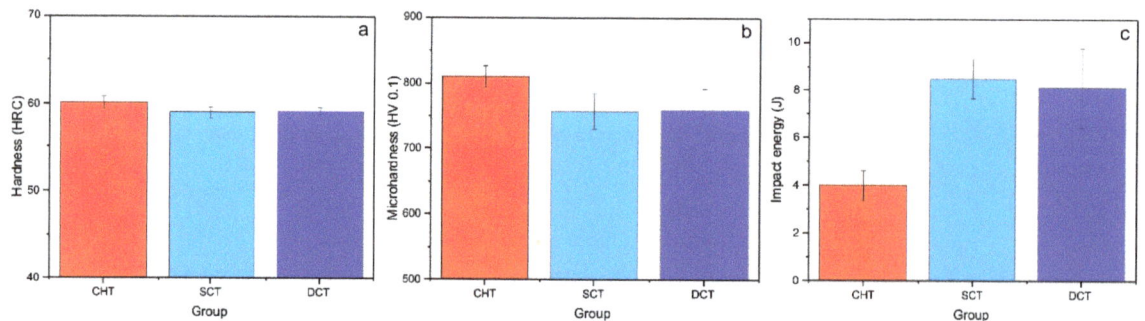

Figure 6. Tested mechanical properties: (**a**) hardness in HRC, (**b**) microhardness in HV 0.1 and (**c**) impact toughness in J.

4. Discussion

4.1. Influence of Different Cryogenic Treatments on Microstructure

Microstructural changes were strongly linked to the selected heat treatment. Both samples, SCT and DCT, after tempering, show a reduced amount of RA compared to the CHT sample, which confirms that CT reduces the amount of RA [27,28]. Furthermore, with DCT, additional conversion of RA into martensite by roughly 50% compared to SCT is achieved due to the lower treatment temperature (−196 °C). Additionally, the CTs also modified the martensitic matrix, which was more refined and preferably oriented along [101] and [001], which was also observed by Jovičević-Klug et al. 2021 [29] for a different type of cold work tool steel and other steels. The reasoning for such a change is related to the preferential formation of martensite laths from austenite during direct non-diffusional conversion under cryogenic temperature (case of SCT and DCT) rather than through the decomposition of RA with tempering (case of CHT). As a result, the martensitic laths for the

SCT case do show a slightly lower orientational preference compared to the DCT sample due to the small amount of RA not transformed during SCT that still partially decomposes during the subsequent tempering stage.

The observation of carbide precipitation showed that both SCT and DCT induce precipitation of carbides. In this study, the precipitation of the $M_{23}C_6$ carbide group is the highest for the DCT group. However, for SCT, no significant difference compared to CHT was determined. The reason for induced overall carbide precipitation in the DCT group is due to the very low temperature, which causes carbon redistribution that influences the carbide formation during the later tempering stage [30]. For the DCT group also, the highest amount of M_7C_3 carbide group precipitation was observed, which could be linked to the suppression of the precipitation of the stable M_3C_2 during tempering, which was also confirmed by Jurči et al. 2021 for another cold work tool steel [31]. Furthermore, this dynamic can also be linked to the C and Cr content and their distribution in the matrix, as observed by Jovičević-Klug et al. 2021 [23]. The additional observation of the precipitated carbide groups also shows that both SCT and DCT groups have equally lower M_3C_2 carbides compared to CHT. The possible explanation is that already cryogenic temperature below $-150\ °C$ (both SCT and DCT) induces the precipitation of transition carbides [31] that effectively modify the carbide evolution pathway during tempering.

4.2. Hardness and Microhardness

In selected cold work tool steel Böhler K340 Isodur a decrease in hardness and microhardness after application of cryogenic treatment (SCT and DCT) was observed.

The decrease in (micro)hardness after the application of cryogenic treatment can be correlated to the decrease in carbon content in the martensitic matrix after SCT/DCT, which then leads to reduced solid solution strengthening. A similar effect was observed by Li et al. 2018 [32] and Jovičević-Klug et al. 2021 for other steels [21]. In addition, both SCT and DCT alter the carbide precipitation, as the RA decomposition does not occur during tempering in contrast to the CHT group. Subsequently, this causes an increase in the number and density of homogeneously distributed carbides (see Section 3.1. Microstructure and phase analysis), but with finer sizes. For this reason, the matrix has a dominant effect during the indentation measurement. The overall drop in the hardness of the CTs compared to CHT can also be related to the change in the carbide precipitation that increases the occurrence of M_7C_3 compared to M_3C_2. As the M_3C_2 are generally harder compared to the M_7C_3, the hardness and microhardness should also generally drop for both CTs.

4.3. Impact Toughness

The changes in impact toughness of different heat treatments are strongly linked to the matrix and its cohesion. The increase in impact toughness for both cryogenic treatments (SCT and DCT) is correlated to the finer martensitic laths compared to the CHT group. The increase in impact toughness can also be correlated to a more homogeneous distribution of the carbides and the cohesion between carbides and matrix. In both SCT and DCT, the additional increased M_7C_3 carbide precipitation was present compared to CHT. As a result, these carbides form a more spherically-shaped form that reduces the tearing of the matrix due to reduced local stress concentration that normally forms due to oblique-shaped edges of other nano-sized carbides, such as M_3C_2. Moreover, the slight change in the impact energy for SCT can be explained by the higher presence of RA compared to DCT. In spite of the higher RA in CHT, the presence of M_3C_2 carbides effectively reduce the impact energy due to their higher brittleness and lower cohesion with the matrix. Furthermore, the increase in impact energy after SCT and DCT can be correlated to the decrease in carbon presence in the martensitic solid solution during cryogenic treatment, which increases the toughness of the martensitic matrix. A similar observation was also observed in high-alloyed steels by Li et al. 2016 [33,34].

4.4. Comparison of SCT vs. DCT

When comparing SCT and DCT cryogenic treatments, some distinctive differences can be observed favoring one or another type. The more effective treatment for minimizing the amount of RA within the matrix is DCT, where the RA vol.% is below 1%. In DCT also, finer martensitic laths and increased $M_{23}C_6$ precipitation are present compared to the SCT, which may favor an increase in (micro)hardness for specific heat treatment parameters (austenitizing, tempering). In both treatments, the increase in carbide precipitation is observed with a more homogenous distribution.

However, there are some important changes in the precipitated types of carbides. In both SCT and DCT, the transient carbide type M_3C_2 is reduced, which additionally influences the overall toughness of the steel. All these differences add up to the change in mechanical properties. As a result, the usage of the different CTs depends on the targeted properties. For an increase in hardness of the material, DCT might be preferable, whereas, for an increase in toughness, SCT would be preferable for cold work tool steel Böhler K340 Isodur.

5. Conclusions

The study investigated the influence of different cryogenic treatments, shallow (SCT) and deep cryogenic treatment (DCT), on cold work tool steel Böhler K340 Isodur in correlation to changes in microstructure and mechanical properties, namely hardness and impact toughness. The following conclusions have been established:

I. SCT and DCT are effective methods in lowering RA presence within the matrix by 71% and 82%, accordingly. SCT and DCT groups have finer martensitic laths, which are oriented along [101] and [001]. The martensitic laths are finer by 21% and 33% with SCT and DCT, respectively.

II. SCT and DCT influence the carbide precipitation of $M_{23}C_6$ (5% by SCT and 35% by DCT) and M_7C_3 (50% by SCT and 70% by DCT) carbide groups and also reduce the formation of transient carbide group (M_3C_2), which is directly linked to the cryogenic temperatures.

III. Hardness and microhardness were not significantly influenced (increase by roughly 5% for both groups) by SCT and DCT and are thus considered to not be a consistent indicator for comparing cryogenic treatments in relation to the microstructural changes.

IV. Impact toughness was increased by both cryogenic treatments by more than 100% (by SCT 113% and by DCT 100%).

V. For an increase in hardness of the cold work tool steel Böhler K340 Isodur, DCT is recommended, whereas, for an increase in toughness, SCT is preferable.

Author Contributions: Conceptualization, P.J.-K., L.T. and B.P.; methodology, P.J.-K., L.T. and B.P.; validation, P.J.-K., L.T. and B.P.; investigation, P.J.-K., L.T. and B.P.; resources, L.T. and B.P.; writing—original draft preparation, P.J.-K., L.T. and B.P.; visualization, P.J.-K.; writing-and editing P.J.-K., L.T. and B.P., supervision, P.J.-K., L.T. and B.P. All authors have read and agreed to the published version of the manuscript.

Funding: This research was funded by Slovenian Research Agency (ARRS), grant number P2-0050.

Institutional Review Board Statement: Not applicable.

Informed Consent Statement: Not applicable.

Data Availability Statement: Not applicable.

Acknowledgments: Authors would like to thank vacuum-heat treatment, mechanical and metallographic labs at IMT, Barbara Šetina Batič for the help with XRD measurements and for fruitful discussion and help with data to Matic Jovičević-Klug.

Conflicts of Interest: The authors declare no conflict of interest.

References

1. Totten, G.E. *Steel Heat Treatment Handbook*, 2nd ed.; CRC Press: Boca Raton, FL, USA, 2007.
2. Tóth, L.; Réka, F. The Effects of Quenching and Tempering Treatment on the Hardness and Microstructures of a Cold Work Steel. *Int. J. Eng. Manag. Sci.* **2019**, *4*, 286–294. [CrossRef]
3. Toth, L. Cryogenic Treatment against Retained Austenite. In Proceedings of the Mérnöki Szimpózium a Bánkiban, Budapest, Hungary, 18 November 2021; pp. 181–186.
4. Chumanov, I.V.; Chumanov, V.I. Technology for Electroslag Remelting with Rotation of the Consumable Electrode. *Metallurgist* **2001**, *45*, 125–128. [CrossRef]
5. Zamborsky, D.S. Control of Distortion in Tool Steels. In *The Heat Treating Source Book*; ASM International: Materials Park, OH, USA, 1998; pp. 73–79.
6. Baldissera, P.; Delprete, C. Deep Cryogenic Treatment: A Bibliographic Review. *Open Mech. Eng. J.* **2008**, *2*, 1–11. [CrossRef]
7. Senthilkumar, D. Cryogenic Treatment:Shallow and Deep. In *Encyclopedia of Iron, Steel, and Their Alloys*; Totten, G.E., Colas, R., Eds.; Taylor and Francis: New York, NY, USA, 2016; pp. 995–1007. ISBN 9781351254496.
8. Pellizzari, M.; Molinari, A. Deep Cryogenic Treatment of Cold Work Tool Steel. In Proceedings of the 6th International Tooling Conference, Stockholm, Sweden, 10–13 September 2002; pp. 657–669.
9. Kalsi, N.S.; Sehgal, R.; Sharma, V.S. Cryogenic Treatment of Tool Materials: A Review. *Mater. Manuf. Processes* **2010**, *25*, 1077–1100. [CrossRef]
10. Sonar, T.; Lomte, S.; Gogte, C. Cryogenic Treatment of Metal–A Review. In *Materials Today: Proceedings*; Elsevier: Amsterdam, The Netherlands, 2018; Volume 5, pp. 25219–25228.
11. Amini, K.; Nategh, S.; Shafyei, A. Influence of Different Cryotreatments on Tribological Behavior of 80CrMo12 5 Cold Work Tool Steel. *Mater. Des.* **2010**, *31*, 4666–4675. [CrossRef]
12. Podgornik, B.; Uršič, D.; Paulin, I. Effectiveness of Deep Cryogenic Treatment in Improving Mechanical and Wear Properties of Cold Work Tool Steels. *Int. J. Microstruct. Mater. Prop.* **2017**, *12*, 216. [CrossRef]
13. Li, J.; Cai, X.; Wang, Y.; Wu, X. Multiscale Analysis of the Microstructure and Stress Evolution in Cold Work Die Steel during Deep Cryogenic Treatment. *Materials* **2018**, *11*, 2122. [CrossRef]
14. Kus, M.; Jurci, P.; Durica, J. Microstructure and Hardness of Cold Work Vanadis 6 Steel after Subzero Treatment at $-140\ ^\circ\mathrm{C}$. *Adv. Mater. Sci. Eng.* **2018**, *2018*, 1–7.
15. Su, Y.Y.; Chiu, L.H.; Chen, F.S.; Lin, S.C.; Pan, Y.T. Residual Stresses and Dimensional Changes Related to the Lattice Parameter Changes of Heat- Treated JIS SKD 11 Tool Steels. *Mater. Trans.* **2014**, *55*, 831–837. [CrossRef]
16. Jovičević-Klug, P.; Podgornik, B. Review on the Effect of Deep Cryogenic Treatment of Metallic Materials in Automotive Applications. *Metals* **2020**, *10*, 434. [CrossRef]
17. Jurči, P.; Ptačinová, J.; Sahul, M.; Dománková, M.; Dlouhy, I. Metallurgical Principles of Microstructure Formation in Sub-Zero Treated Cold-Work Tool Steels–a Review. *Matériaux Tech.* **2018**, *106*, 104–113. [CrossRef]
18. Zhou, Z.C.; Du, J.; Yan, Y.J.; Shen, C.L. The Recent Development of Study on H13 Hot-Work Die Steel. *Solid State Phenom.* **2018**, *279*, 55–59. [CrossRef]
19. Villa, M.; Somers, M.A.J. Cryogenic Treatment of Steel: From Concept to Metallurgical Understanding. In Proceedings of the 24th International Feration for Heat Treatment and Surface Engineering Congress, Nice, France, 26–29 June 2017.
20. Min, N.; Li, H.M.; Xie, C.; Wu, X.C. Experimental Investigation of Segregation of Carbon Atoms Due to Sub-Zero Cryogenic Treatment in Cold Work Tool Steel by Mechanical Spectroscopy and Atom Probe Tomography. *Arch. Metall. Mater.* **2015**, *60*, 1110–1113. [CrossRef]
21. Jovičević-Klug, P.; Puš, G.; Jovičević-Klug, M.; Žužek, B.; Podgornik, B. Influence of Heat Treatment Parameters on Effectiveness of Deep Cryogenic Treatment on Properties of High-Speed Steels. *Mater. Sci. Eng. A* **2022**, *829*, 142157. [CrossRef]
22. Jovičević-Klug, P.; Jovičević-Klug, M.; Podgornik, B. Effectiveness of Deep Cryogenic Treatment on Carbide Precipitation. *J. Mater. Res. Technol.* **2020**, *9*, 13014–13026. [CrossRef]
23. Jovičević-Klug, P.; Jovičević-Klug, M.; Sever, T.; Feizpour, D.; Podgornik, B. Impact of Steel Type, Composition and Heat Treatment Parameters on Effectiveness of Deep Cryogenic Treatment. *J. Mater. Res. Technol.* **2021**, *14*, 1007–1020. [CrossRef]
24. Pellizzari, M. Influence of Deep Cryogenic Treatment on the Properties of Conventional and PM High Speed Steels. *Metall. Ital.* **2008**, *100*, 17–22.
25. Jovičević-Klug, P.; Lipovšek, N.; Jovičević-Klug, M.; Podgornik, B. Optimized Preparation of Deep Cryogenic Treated Steel and Al-Alloy Samples for Optimal Microstructure Imaging Results. *Mater. Today Commun.* **2021**, *27*, 102211. [CrossRef]
26. Toraya, H. A New Method for Quantitative Phase Analysis Using X-Ray Powder Diffraction: Direct Derivation of Weight Fractions from Observed Integrated Intensities and Chemical Compositions of Individual Phases. *J. Appl. Crystallogr.* **2016**, *49*, 1508–1516. [CrossRef]
27. Oppenkowski, A.; Weber, S.; Theisen, W. Evaluation of Factors Influencing Deep Cryogenic Treatment That Affect the Properties of Tool Steels. *J. Mater. Processing Technol.* **2010**, *210*, 1949–1955. [CrossRef]
28. Tyshchenko, A.I.; Theisen, W.; Oppenkowski, A.; Siebert, S.; Razumov, O.N.; Skoblik, A.P.; Sirosh, V.A.; Petrov, Y.N.; Gavriljuk, V.G. Low-Temperature Martensitic Transformation and Deep Cryogenic Treatment of a Tool Steel. *Mater. Sci. Eng. A* **2010**, *527*, 7027–7039. [CrossRef]

29. Jovičević-Klug, P.; Jenko, M.; Jovičević-Klug, M.; Šetina Batič, B.; Kovač, J.; Podgornik, B. Effect of Deep Cryogenic Treatment on Surface Chemistry and Microstructure of Selected High-Speed Steels. *Appl. Surf. Sci.* **2021**, *548*, 1–11. [CrossRef]
30. Zhirafar, S.; Rezaeian, A.; Pugh, M. Effect of Cryogenic Treatment on the Mechanical Properties of 4340 Steel. *J. Mater. Processing Technol.* **2007**, *186*, 298–303. [CrossRef]
31. Jurči, P.; Bartkowska, A.; Hudáková, M.; Dománková, M.; Čaplovičová, M.; Bartkowski, D. Effect of Sub-Zero Treatments and Tempering on Corrosion Behaviour of Vanadis 6 Tool Steel. *Materials* **2021**, *14*, 3759. [CrossRef]
32. Li, B.; Li, C.; Wang, Y.; Jin, X. Effect of Cryogenic Treatment on Microstructure and Wear Resistance of Carburized 20CrNi2MoV Steel. *Metals* **2018**, *8*, 808. [CrossRef]
33. Yan, Y.; Luo, Z.; Liu, K.; Zhang, C.; Wang, M.; Wang, X. Effect of Cryogenic Treatment on the Microstructure and Wear Resistance of 17Cr2Ni2MoVNb Carburizing Gear Steel. *Coatings* **2022**, *12*, 281. [CrossRef]
34. Li, H.; Tong, W.; Cui, J.; Zhang, H.; Chen, L.; Zuo, L. The Influence of Deep Cryogenic Treatment on the Properties of High-Vanadium Alloy Steel. *Mater. Sci. Eng. A* **2016**, *662*, 356–362. [CrossRef]

Article

Unravelling the Role of Nitrogen in Surface Chemistry and Oxidation Evolution of Deep Cryogenic Treated High-Alloyed Ferrous Alloy

Patricia Jovičević-Klug [1,2,*], Matic Jovičević-Klug [3] and Bojan Podgornik [1,2]

1 Department of Metallic Materials and Technology, Institute of Metals and Technology, Lepi pot 11, 1000 Ljubljana, Slovenia; bojan.podgornik@imt.si
2 Jožef Stefan International Postgraduate School, Jamova cesta 39, 1000 Ljubljana, Slovenia
3 Department of Microstructure Physics and Alloy Design, Max Planck Institute for Iron Research, Max-Planck-Straße 1, 40237 Düsseldorf, Germany; m.jovicevic-klug@mpie.de
* Correspondence: patricia.jovicevicklug@imt.si; Tel.: +386-14701-990

Citation: Jovičević-Klug, P.; Jovičević-Klug, M.; Podgornik, B. Unravelling the Role of Nitrogen in Surface Chemistry and Oxidation Evolution of Deep Cryogenic Treated High-Alloyed Ferrous Alloy. *Coatings* **2022**, *12*, 213. https://doi.org/10.3390/coatings12020213

Academic Editor: Francesco Di Quarto

Received: 12 January 2022
Accepted: 4 February 2022
Published: 6 February 2022

Publisher's Note: MDPI stays neutral with regard to jurisdictional claims in published maps and institutional affiliations.

Copyright: © 2022 by the authors. Licensee MDPI, Basel, Switzerland. This article is an open access article distributed under the terms and conditions of the Creative Commons Attribution (CC BY) license (https://creativecommons.org/licenses/by/4.0/).

Abstract: The role of nitrogen, introduced by deep cryogenic treatment (DCT), has been investigated and unraveled in relation to induced surface chemistry changes and improved corrosion resistance of high-alloyed ferrous alloy AISI M35. The assumptions and observations of the role of nitrogen were investigated and confirmed by using a multitude of complementary investigation techniques with a strong emphasis on ToF-SIMS. DCT samples display modified thickness, composition and layering structure of the corrosion products and passive film compared to a conventionally heat-treated sample under the same environmental conditions. The changes in the passive film composition of a DCT sample is correlated to the presence of the so-called ghost layer, which has higher concentration of nitrogen. This layer acts as a precursor for the formation of green rust on which magnetite is formed. This specific layer combination acts as an effective protective barrier against material degradation. The dynamics of oxide layer build-up is also changed by DCT, which is elucidated by the detection of different metallic ions and their modified distribution over surface thickness compared to its CHT counterpart. Newly observed passive film induced by DCT successfully overcomes the testing conditions in more extreme environments such as high temperature and vibrations, which additionally confirms the improved corrosion resistance of DCT treated high-alloyed ferrous alloys.

Keywords: deep cryogenic treatment; nitrogen; passive film; ToF-SIMS; surface chemistry; corrosion resistance

1. Introduction

One of the challenges in numerous industries is corrosion of the material (degradation of material), especially in industries involved with the application of metallic materials or their production (such as steel and tool industry [1], oil and gas industry [2], medicine [3], electronic industry [4], automotive industry [5], aerospace industry [6] and nuclear industry [7]). Corrosion resistance of metallic surfaces can be enhanced by various methods. Heat treatment, as one such method, is used to tailor the microstructure of metallic material and with it the surface characteristics and behavior, which in turn influences the corrosion resistance of the treated metals [8]. One of the potentially developing heat treatment methods is deep cryogenic treatment (DCT), in which on one hand the impact on the climate is minimal and the costs of production are low [9], and on the other hand, the modification of metallic material is induced from the surface down to the bulk core [10]. During DCT the material is exposed to cryogenic temperatures (under −160 °C and usually in liquid nitrogen (N_2)) in order to improve mechanical properties such as hardness [11] and fracture toughness [12]. However, DCT can also improve the oxidation behavior of a metallic surface and improve corrosion resistance [1,13]. The effect of DCT on corrosion resistance of ferrous alloys has been associated with the altercation of their microstructure [1,13,14], oxidation behavior of the surface layer [15] and modification of the development dynamics of corrosion products [13]. However, the corrosion resistance of ferrous alloys and its modification through heat treatment is also influenced by three main factors: the chemical composition of ferrous alloys [16], the corrosion environment [17] and biophysical factors [17]. The corrosion resistance and its modification with previously mentioned factors is often discussed in connection to the development of a passivation layer and its degradation and reformation with temporal corrosion progress. The corrosion dynamics of passive film composition and breakdown of passive film of ferrous alloys have been widely studied solely in connection to stainless steels. Furthermore, the research on the role of alloying elements in steel materials has been mainly focused on Cr and its common corrosion products, such as Cr(III) oxide and hydroxide, which are believed to play the main role in passive film [18]. The research on the effect of other alloying elements on corrosion resistance is seldom researched and is thus highly dubious.

The beneficial effect of N on passive film and corrosion resistance, has been studied in connection with stainless steels, for which N has been shown to play an important role in decreasing weight loss of material and increasing the pitting potential [18]. Additionally, N is believed to accelerate repassivation of stainless steels, especially in the presence of dissolved chloride [13,18]. N presence in passive film in stainless steels was determined to be in elemental form incorporated within the metallic matrix (solid solution) or chemically bound in various compounds (nitrides, nitrates, nitrites). N can participate and influence corrosion processes during exposure of material to corrosive environments when N is part of the metallic surface (in the matrix as alloying element [19] or later added during the process of nitriding [20]) or as part of a passive film in the form of dissolved species (such as NHX). The interfacial N was proven to be negatively charged, which can influence the passivation film dynamics [18]. There are some studies performed in correlation to N presence in metallic matrix, which indicate that N improves corrosion resistance. However, there is a lack of studies correlating the presence of N with dynamics of the passive film [18]. There are three possible mechanisms responsible for the influence of N on the corrosion properties, which are merely based on observation of stainless steels:

(1) Anodic segregation of N during dissolution, which can act as a barrier in active dissolution and consequently changes the passivation process, which then influences the corrosion resistance [21].
(2) The increase of pH in pits is a consequence of the formation of ammonium ions, which buffer the solution of passive film regions [22].
(3) Negatively charged N ions and uncharged N enrichment beneath the passive film reduce the electric potential difference, which effectively acts as a protective layer of the passivation layer [21,22]. The interphase region directly between the metallic surface and passive film (active dissolution and active passivation) has not been researched in detail.

The literature review showed that there is a lack of studies on the DCT effect in correlation to the role of N on corrosion behavior and surface modification as well as passive film build-up in high-alloyed ferrous alloys. Previous studies mostly concentrated on improving the corrosion resistance of predominantly low-alloyed ferrous alloys [23–25], with a handful of research studies performed on high-alloyed steels such as stainless steels and maraging steels [26–29]. However, in addition to these two classes there is an extensive group of other high-alloyed ferrous alloys, such as high-strength steels, heat-resistant ferrous alloys, low-density ferrous alloys, etc., which are used in corrosive environments and require sufficient corrosion resistance for their application. This indicates that there is a considerable research gap in this field with great potential to understand and explain the effect of DCT on the corrosion behavior of high-alloyed ferrous alloys. Furthermore, this study is a follow-up study of previous surface and corrosion investigations of high-alloyed ferrous alloys from our group. Jovičević-Klug et al. 2021 [13,15] focused on the effect of DCT in relation to alloying elements (Co, Cr, Fe, V, W), surface modification and the N dynamic in DCT-treated tool steels. It was determined that these different factors result in improved corrosion properties with DCT, which was also observed by our previous study, Voglar et al. 2021 [1]. The changes to high-alloyed ferrous alloy AISI M35 were observed after DCT in correlation with surface chemistry and surface modification, which indicated that DCT samples display a different evolution of corrosion products compared to conventionally heat-treated ones. It was discovered that green rust (GR) plays an important role in increased corrosion resistance of DCT samples through preferential build-up of magnetite over the pre-existing GR layer. The main open question is the exact understanding of how GR is incorporated into the corrosion development and how the nitrogen is influencing such development. For these reasons, this study continues to further elucidate the corrosion and oxidative behavior of the surface of AISI M35 in more detail with correlative investigation with optical microscopy (OM), scanning electron microscopy (SEM), energy-dispersive X-ray spectroscopy (EDS), time-of-flight secondary ion mass spectroscopy (ToF-SIMS) and X-ray diffraction (XRD) methods. Furthermore, this paper also aims to demonstrate the advantages of DCT in regards to corrosion behavior of high-alloyed ferrous alloys by revealing surface-sensitive details related to N dynamics and GR.

2. Materials and Methods
2.1. Materials and Experimental Details

The selected high-alloyed ferrous alloy was high-speed steel AISI M35, provided by SIJ Metal, Ravne (Slovenia). This alloy was selected based on our previous research and due to its lower corrosion resistance, making it easier to follow, and explain in detail, dynamic surface chemistry and corrosion behavior in relation to DCT-induced changes, when the alloy is exposed to chloride-ions-enriched medium. The chemical composition of the investigated material is in wt.%, 0.90 C, 0.34 Mn, 0.004 S, 4.10 Cr, 5.20 Mo, 6.22 W, 2.01 V, 4.52 Co and 76.70 Fe. For the experiments cube samples (10 mm × 10 mm × 10 mm) were used. Samples were prepared with two different heat treatments: the first was conventional heat treatment (CHT), performed according to the steel producer recommendations with quenching in nitrogen gas at 5 bar, and the second was DCT, performed by controlled gradual immersion of material into liquid nitrogen (austenitization temperature 1160/2 min and tempering temperature 620/2 h (3× cycles for CHT and 1× cycle for DCT after DCT)). The soaking time of DCT was 24 h, the soaking temperature was −196 °C and the warming/cooling rate was approximately 10 °C·min^{-1}, with DCT placed after quenching and before a single step of tempering. The samples were mechanically grounded with silicon carbide (SiC) emery paper down to 1000 grit and then polished with diamond paste down to 1 µm, and as the last step they were ultrasonically rinsed in alcohol. The polished samples (Ra = 0.05 µm) were then immersed in chloride-ions-enriched medium seawater (pH 8.22 ± 0.08, salinity 37.31‰ ± 1.50‰; all other water-soluble ions are in µmol·L^{-1}: NO_2^- 0.09 ± 0.12, NO_3^- 0.38 ± 0.27, PO_4^{3-} 0.05 ± 0.01, NH_4^+ 0.45 ± 0.36, SO_4^{2-} 1.48 ± 1.70, Na^+ 1.96 ± 1.19, Mg^{2+} 0.42 ± 0.27, Ca^{2+} 1.49 ± 0.96, Cl^- 1.15 ± 0.68, K^+ 0.30 ± 0.35) for 1 day in and 7 days.

In order to enhance the role of nitrogen and alloying elements (Co, Cr, Fe, V, W) as potential corrosion inhibitors in DCT-treated high-alloyed ferrous alloy, two additional 1 h tests with different corrosion environments were performed, one with elevated temperature, where the chloride-ions-enriched medium was heated up to 100 °C (CHT+T/DCT+T) and another with elevated temperature and vibration, with chloride-ions enriched heated up to 100 °C and specimens subjected to vibrations of 25 Hz (CHT+T+V/DCT+T+V). Elevated temperatures of corrosion environment lead to higher corrosion rates of ferrous alloys because of the increase in electrochemical reactions at higher temperatures, being also strongly affected by chemical compositions and microstructure of the alloy [30]. Vibrations on the other hand can have diverse effect on corrosion resistance of ferrous alloys, either improving or deteriorating it due to the acoustic cavitation and delamination of corrosion products [31]. Afterwards, for the study of pitting corrosion, the corrosion products were removed from the sample surfaces according to C.3.1. G1-90 ASTM standard [32].

2.2. Microstructure and Oxide Characterization

The surface of samples was firstly investigated with light optical microscope (LM), Zeiss Axio Vario, Carl Zeiss, Oberkochen, Germany and then with a scanning electron microscope (SEM), JEOL JSM-6500F, Jeol, Tokyo, Japan and locally chemically analyzed using energy-dispersive spectroscopy (EDX), Oxford EDS INCA Energy 450, Oxford Instruments, Abingdon, UK. The detailed microstructure and corrosion products analyses for high-alloyed ferrous alloy AISI M35 is based on our previous studies by Jovičević-Klug et al. 2020 [33] and Jovičević-Klug et al. 2021 [13]. The phase/oxide characterization and identification were performed by XRD, PANalytical 3040/60, Almelo, The Netherlands. The XRD data were measured from 15° to 90° of the 2θ angle. The phase identification was performed using COD database references on a selected location.

2.3. Surface Characterization

Electrochemical results are based on results of the study of Voglar et al. 2021 [1], whereas the main values are provided in Supplementary Material 1. After corrosion products removal, the topography analysis of samples and evaluation of pitting depth was performed by focus variation microscopy with Alicona InfiniteFocus, Bruker, Graz, Austria. The local chemistry of the surface and in-depth chemical profiles were performed by time-of-flight secondary ion mass spectroscopy (ToF-SIMS), IONTOF GmbH, Münster, Germany. The primary beam was Bi_3^+ with energy of 30 keV. The analytical velocity of the depth profile was 0.2 nm/s and the detection limit of the oxide/metal interface species was around 1 ppm. The areas of 250 × 250 μm² (512 × 512 px) and 50 × 50 μm² (256 × 256 px) were measured with positive and negative surface spectra (secondary ions in the range m/z of 0-875), with 180 nm lateral resolution. The angle of ions was 45° and etching in-profile was performed by Cs (2 keV). The correction of results was performed according to Poisson statistics [34]. The statistics were performed in SPSS (PASWStatistics18). The influence of topography is also considered in the interpretation of the individual 2D ToF-SIMS images through the total-ion signal images. The total-ion signal images for all the investigated sample surface are provided for reference in Supplementary Material 2.

3. Results and Discussion

3.1. Corrosion Products

To confirm the testing conditions also control groups with synthetic chloride-ions-enriched medium-synthetic saltwater (3.5% NaCl) were performed, which provided similar results of evolution and presence of corrosion products, for both CHT and DCT sample groups. However, the major difference compared to natural seawater was that less intense signals of corrosion products were observed for the synthetic one. In order to enhance the corrosion and to emphasize the differences between the two treatment groups (CHT and DCT) natural seawater was selected. The first set of specimens (control group) was exposed to chloride-ions-enriched medium at room temperature (CHT/DCT), aimed at verifying the observations from our previous studies. The XRD data showed that generally the corrosion products after 1 and 7 days exposure (Figure 1a) are similar for both heat treatments (CHT and DCT). The main identified phases on the surface of all tested samples are mainly calcium carbonate ($CaCO_3$), halite (NaCl), iron as matrix (Fe), carbides, goethite (α-FeO(OH); Figure 1b), lepidocrocite (γ-FeO(OH); Figure 1c), magnetite (Fe_3O_4; Figure 1d) and green rust (GR I and GR II; Figure 1e). In our previous study [13], the presence of other oxides was also confirmed, but they are not relevant to this study due to their lower volumetric presence within the surface. The observation of GR I+II also confirmed our previous findings that GR I+II is predominantly present on DCT samples, on which, later in time, magnetite (Fe_3O_4) is formed. It is suggested that GR is formed due to the presence of newly formed anions (NO_3^-, SO_4^{2-} and Cl^+) that compensate the pre-existing ammonium cations (NH_4^+). The main sources of nitrogen for GR for both heat treatment groups are suggested to be ions presence in chloride-ions-enriched medium and the metallic surface of AISI M35. In order to observe dynamics of nitrogen and other alloying elements (Co, Cr, Fe, V, W) in correlation with GR I+II and magnetite, which could act as the base underlying corrosion products (denoted as passive film), the samples with 1 day exposure in chloride-ions-enriched medium were selected for further study. Samples exposed for 7 days develop extended presence of other corrosion products that obscure the underlying initial corrosion layers relevant for this research.

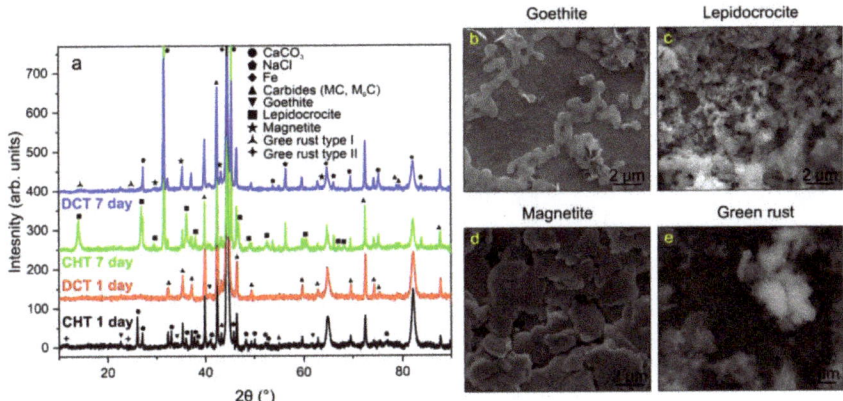

Figure 1. (a) The XRD data of corrosion products for 1 day and 7 days exposure in chloride-ions-enriched medium. (b–e) Observed corrosion products.

3.2. Characterization of Layers

The characterization of layers (oxide layer, passive film and modified layer (ghost layer)) for both heat-treated samples (CHT and DCT) after 1 day exposure was investigated by SEM/EDX and SEM-FIB techniques in combination with XRD data and supplementary ToF-SIMS technique. The SEM/EDX-FIB analysis provided insight into the dynamics of selected alloying elements (Cr, Fe, V, W, Mo), whereas Co is excluded due to its homogenous distribution in both samples. The cross-section of CHT sample revealed that beside O, also Cr, Fe, Mo, W and V act as the main elements within the surface interphase between the metallic surface and corrosive environment, presented in Supplementary Material 3. In the DCT sample, similar results were obtained (Supplementary Material 3) with one major difference. In the DCT sample, a layer enriched with N and depleted of other elements, previously dubbed the "ghost layer" [35] is formed between the metallic surface and the oxide layer. The passive film, which is correlated with the improved corrosion resistance (Supplementary Material 1), is suggested to be enriched with N-dissolved species and acts as the inhibitor for the GR I+II growth in DCT samples and, later, magnetite (Figure 2a,d). This phenomenon (passive film induced by DCT) is observed in correlation to DCT and corrosion properties of high-alloyed ferrous alloys for the first time. In addition to this layer, the specific development of overlaying oxide layer (partly part of passive film) is also found, at which the first inner layer is determined to be magnetite and afterwards the outer layer is composed of other oxide/hydroxide species that are also present for CHT sample. In addition, in the DCT sample minor depletion of alloying elements is observed, which is explained to originate from the protection barrier formed by the passive film as presented in Figure 2c,d.

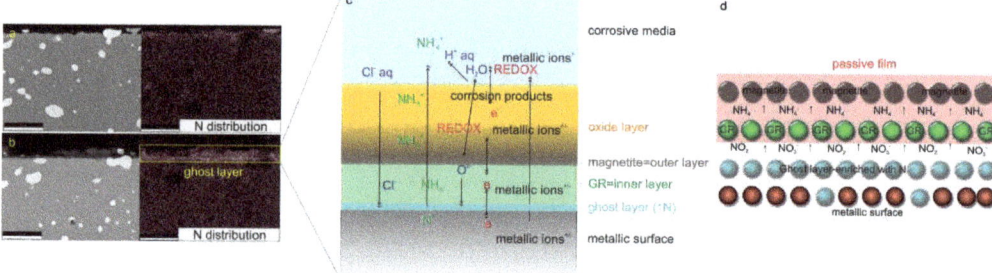

Figure 2. (**a**) SEM-mapping results of CHT sample, (**b**) SEM-mapping results of DCT sample, (**c**) schematic representation of layers and (**d**) the schematic representation of passive film and ghost layer in DCT sample.

3.3. ToF-SIMS Surface Analyses

To provide a deeper insight into the newly observed surface chemistry dynamic of the passive film and ghost layer in correlation to GR in a DCT sample, ToF-SIMS surface analysis and in-depth profiling (Chapter 3.4) were carried out on CHT and DCT specimens after 1 day exposure in chloride-ions-enriched medium at room temperature (21 °C). The ToF-SIMS images were used to identify differences in the surface distribution of ionic species related to corrosion product and their development on the samples' surface. The samples are assumed to be chemically and microstructurally homogenous and the corrosion products (oxide layer) and passive layer are homogeneously distributed on the sample surface. Nitrogen content (N) was measured with CN^- signal, due to the small ionization yield of N with ToF-SIMS. To indirectly indicate the corrosion propagation of samples, the isotope behavior of Cl was tracked, as Cl is the element with highest electron affinity that can be present within the corrosion products. For this reason, the mapping of Cl^- ion and its isotope variants $^{35}Cl/^{37}Cl$ were measured. Furthermore, to track the dynamics of other main alloying elements (C, Co, Cr, Fe, V and W), both negative and positive polarity measurements were performed. The species selected for each selected alloying element are: C (C^-, CH^-, C_2^-, C_2H^-, CHO_2^-, $C_2H_2O_2^-$), Cr (Cr^+, $CrOH^-$), Co (Co^+, CoO^-), Cl (Cl^-), Fe (Fe^+, $^{54}Fe^+$, FeH^+), N (CN^-), V (V^+, VO^-), W ($^{186}WO_4H^-$, WO_4H^-, $^{183}WO_4H^-$, $^{182}WO_4H^-$, $^{186}WO_3^-$, $^{183}WO_3^-$, $^{182}WO_3^-$, WO_3^-). Unfortunately, no Mo species could be detected for the description of Mo dynamics.

Figures 3–5 show the results of low-magnification ToF-SIMS analysis of both polarities, presenting cumulative ion images of selected ion species for individual elements. The first image (Figures 3a, 4a and 5a,b) presents the distribution of Cl^- ions, because it can indicate the passive film structure through its incorporation within the passive film and can possibly explain the breakdown dynamics [36] of the "ghost layer" formed by DCT. The comparison of CHT (Figure 3a) and DCT (Figure 4a) samples clearly shows higher abundance of Cl^- ions for Cl of nitrides in CHT sample in the martensitic matrix of the sample. This provides one of the first pieces of direct evidence that DCT affects the passive film properties. The source of nitrogen for DCT is the immersion media from which the nitrogen adsorbs onto the surface and later diffuses during high-temperature treatment into the outermost layers of the material as interstitially dissolved. The nitrogen incorporation is considered to occur through vacancies and intergranular spacings that are present in the martensitic lattice. Moreover, this incorporation could be the primary mechanism, which is latter accompanied in different environments by the secondary mechanism of formation of NH_4^+ ions. This results in electro-potential discharge [37] of the newly formed "ghost layer" (passive layer) and reduced degradation of the material.

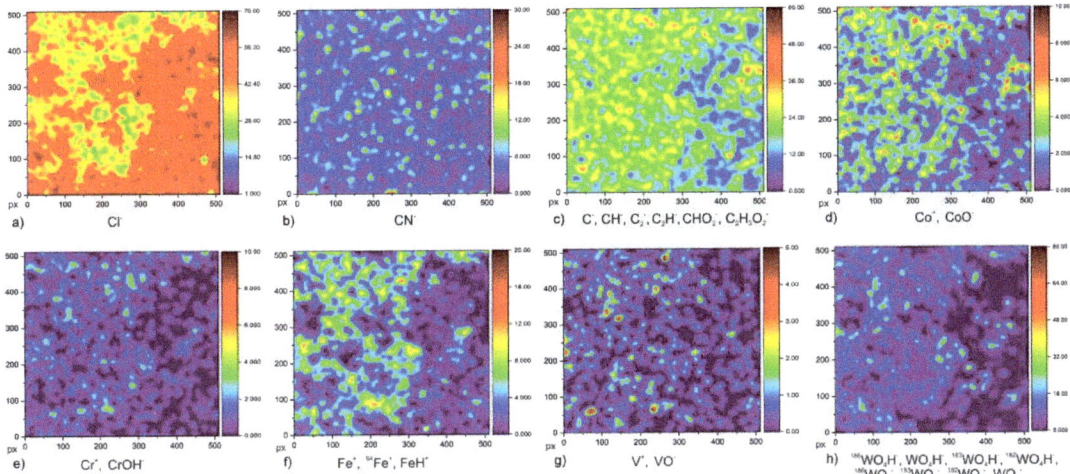

Figure 3. ToF-SIMS analysis of conventionally heat-treated sample (CHT) after 1 day of exposure in chloride-ions-enriched medium at 21 °C. (**a**) Cl^- ions; (**b**) CN^- ions; (**c**) C^-, CH^-, C_2^-, C_2H^-, CHO_2^-, $C_2H_2O_2^-$ ions; (**d**) Co^+, CoO^- ions; (**e**) Cr^+, $CrOH^-$ ions; (**f**) Fe^+, $^{54}Fe^+$, FeH^+ ions; (**g**) V^+, VO^- ions; (**h**) $^{186}WO_4H^-$, WO_4H^-, $^{183}WO_4H^-$, $^{182}WO_4H^-$, $^{186}WO_3^-$, $^{183}WO_3^-$, $^{182}WO_3^-$, WO_3^- ions.

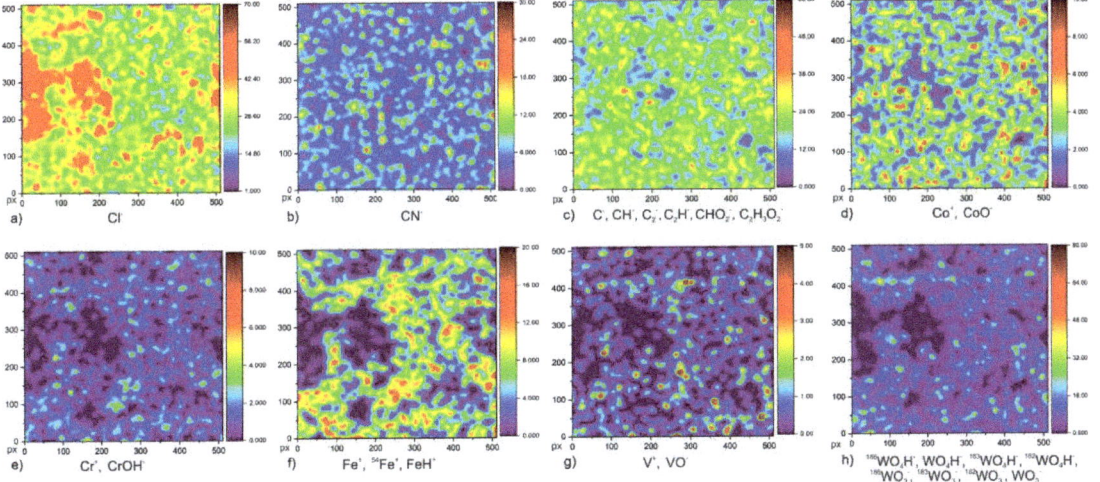

Figure 4. ToF-SIMS analysis of deep cryogenic heat-treated sample (DCT) after 1 day of exposure in chloride-ions-enriched medium at 21 °C. (**a**) Cl^- ions; (**b**) CN^- ions; (**c**) C^-, CH^-, C_2^-, C_2H^-, CHO_2^-, $C_2H_2O_2^-$ ions; (**d**) Co^+, CoO^- ions; (**e**) Cr^+, $CrOH^-$ ions; (**f**) Fe^+, $^{54}Fe^+$, FeH^+ ions; (**g**) V^+, VO^- ions; (**h**) $^{186}WO_4H^-$, WO_4H^-, $^{183}WO_4H^-$, $^{182}WO_4H^-$, $^{186}WO_3^-$, $^{183}WO_3^-$, $^{182}WO_3^-$, WO_3^- ions.

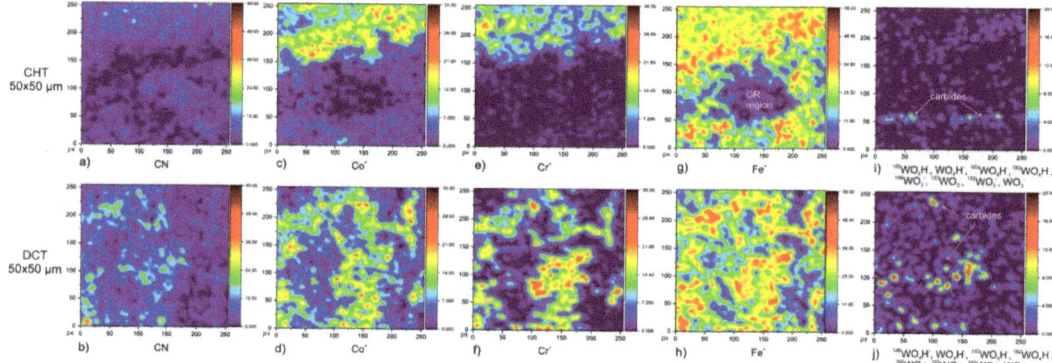

Figure 5. Higher magnification ToF-SIMS analysis of both conventionally (CHT) and deep cryogenic heat-treated samples (DCT) after 1 day of exposure in chloride-ions-enriched medium at 21 °C for selected ion groups. (**a,b**) CN$^-$ ions; (**c,d**) Co$^+$ ions; (**e,f**) Cr$^+$ ions; (**g,h**) Fe$^+$ ions; (**i,j**) ^{186}WO$_4$H$^-$, WO$_4$H$^-$, ^{183}WO$_4$H$^-$, ^{182}WO$_4$H$^-$, ^{186}WO$_3$$^-$, ^{183}WO$_3$$^-$, ^{182}WO$_3$$^-$, WO$_3$$^-$ ions.

The next alloying element analyzed was cobalt (ion forms of Co$^-$ and CoO$^-$), Figures 3d, 4d and 5b–g. Compared to other alloying elements of the investigated alloy, cobalt is mainly present in the martensitic matrix in the form of a solid solution. When comparing both samples, CHT (Figures 3d and 5c) and DCT (Figures 4d and 5d), no significant difference in distribution or abundance could be observed (the spatial distribution of corrosion crust was taken into account). Chromium distribution (Cr$^+$ and CrOH$^-$ forms); Figures 3e, 4e and 5e,f, clearly indicates regions of carbides, which are the main source of chromium. However, from the maps and the corrosion crust analysis chromium was determined to not play a primary role in the corrosion resistance mechanism of the investigated alloy, neither for CHT nor DCT samples. This can be explained by the lower amount (4.10 wt.%) of chromium compared to stainless steels (>10 wt.%), for which Cr is the main contributor to the corrosion resistance. In the current case (AISI M35) cobalt in high content (4.52 wt.%) has a prevalent role in corrosion resistance over chromium [38]. From this, and in general even distribution of Co, it is postulated that the underlying passive film is partially formed from cobalt. Nevertheless, the higher local presence of chromium in Figure 5e,f indicates local pitting of the samples (evaluated in the next section), if not correlated to carbides. Similar relation of Cr with local pitting has been also observed in previous study with cross-sectional SEM investigation [13].

Iron distribution was measured by Fe$^+$, ^{54}Fe$^+$, FeH$^+$ ions (Figures 3f, 4f and 5g,h). On one hand, the distribution of iron indicates the location of specific carbides when combined with other elemental maps (C, Cr, W). On the other hand, the absence of iron species in the corrosion crust indicates the presence and position of GR I+II (Figure 5g,h and Figure 6), when associated with the characteristic elements (Cl and S) of GR for CHT and DCT samples. The map distribution of vanadium (V$^+$, VO$^-$ ions) in CHT (Figure 3g) and DCT (Figure 4g) samples show the peaks corresponding to the distribution of carbides in the matrix. Due to its concentration in the investigated alloy (2.01 wt.%), vanadium is not supposed to play an important role in the corrosion resistance of the samples. However, vanadium could theoretically help improve corrosion resistance by reducing both the formation of salt film and the formation rate of pits, as suggested by Ras et al. [39]. The distribution of the most abundant (6.22 wt.%) alloying element, tungsten (^{186}WO$_4$H$^-$, WO$_4$H$^-$, ^{183}WO$_4$H$^-$, ^{182}WO$_4$H$^-$, ^{186}WO$_3$$^-$, ^{183}WO$_3$$^-$, ^{182}WO$_3$$^-$, WO$_3$$^-$ ions), shows localized concentrations, which correspond to carbides enriched by tungsten (M$_6$C [33]), present in both CHT (Figures 3h and 5i) and DCT (Figures 4h and 5j) samples. No other significant information was obtained from tungsten distribution and its influence on corrosion resistance.

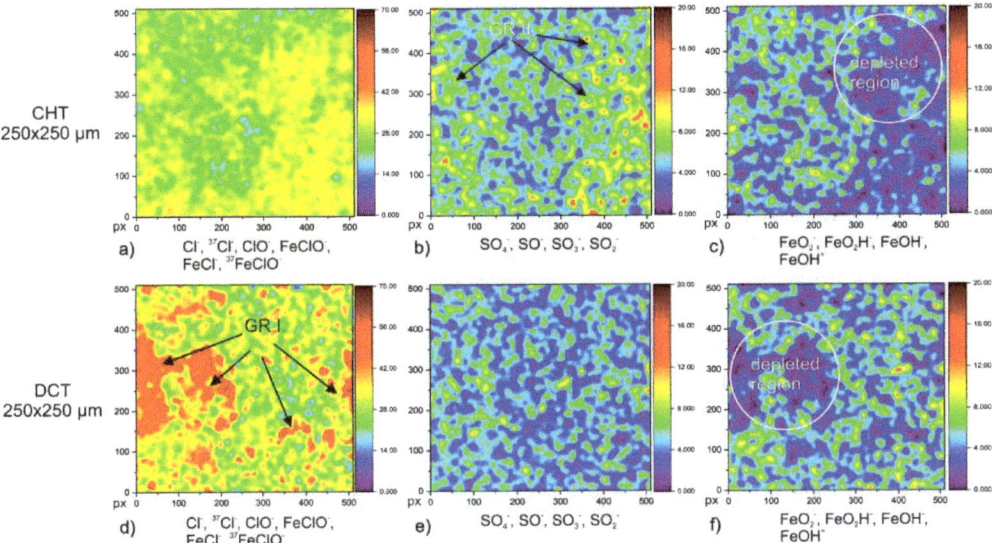

Figure 6. ToF-SIMS analysis of both conventionally (CHT) and deep cryogenic heat-treated samples (DCT) after 1 day of exposure in chloride-ions-enriched medium at 21 °C for (**a**,**d**) green rust type I, (**b**,**e**) green rust type II and (**c**,**f**) Fe-oxides.

Figure 6 shows the distribution of green rust, type I (a,b) and type II (c,d), and Fe-oxides/hydroxides (e,f) on the samples surface. The ToF-SIMS analysis clearly shows the predominant formation of GR type I (combination of ions Cl^-, $^{37}Cl^-$, ClO^-, $FeClO^-$, $FeCl^-$ and $^{37}FeClO^-$) in the DCT sample, whereas GR type II (combination of ions SO_4^-, SO^-, SO_3^- and SO_2^-) has higher abundance in the CHT sample, corresponding well with XRD results (Figure 1a). Absence or reduction of GR type II could not be linked to any improvement in corrosion resistance. Additionally, excessive presence of green rust was observed for the DCT sample compared to the CHT sample. The higher presence of GR for the DCT sample is associated with the increased concentration of nitrogen on the surface, which together with chlorine is the basis for the development of GR (excess amount of NO_3^-, Cl^- and NH_4^+ ions) [40]. This confirms our previous observations [13] and proposed theory that GR I plays an important role in the improvement of the corrosion resistance of the DCT sample. The proposed mechanism for improved corrosion resistance lies in the formation of GR I, as an inner part of the oxide layer formed over the passive layer. The GR I then acts as a buffer layer for the preferential development of magnetite, which is later covered by other Fe oxides/hydroxides as part of the oxide layer. Such a mechanism of GR preferential growth of magnetite was confirmed by Sumoondur et al. 2008 [41]. The distribution of GR was additionally correlated and confirmed with Fe-oxides/hydroxides, as indicated in Figure 6c,f.

3.4. ToF-SIMS Depth Profiling

ToF-SIMS depth profiling was performed with a negative ion charge. The ToF-SIMS depth profile analysis includes all corrosion products of the crust. The oxide layer is identified as the layer with an initial increase in oxides and hydroxides, which is defined as 0–2000 s of sputter time for the CHT sample and 0–1000 s for the DCT sample, respectively. The passive film is the region between the oxide layer and the metallic substrate, expressed as a second increase in the intensity of alloying elements (Figures 7 and 8). The passive film for CHT is 2000–3200 s and for DCT, 1000–3800 s of sputtering time, respectively. The region of metallic substrate is defined as the region of rapid decrease in oxides/hydroxides presence and constant value of alloying elements, which starts for the CHT sample after 3200 s and for the DCT sample after 3800 s of sputtering (Figures 7 and 8).

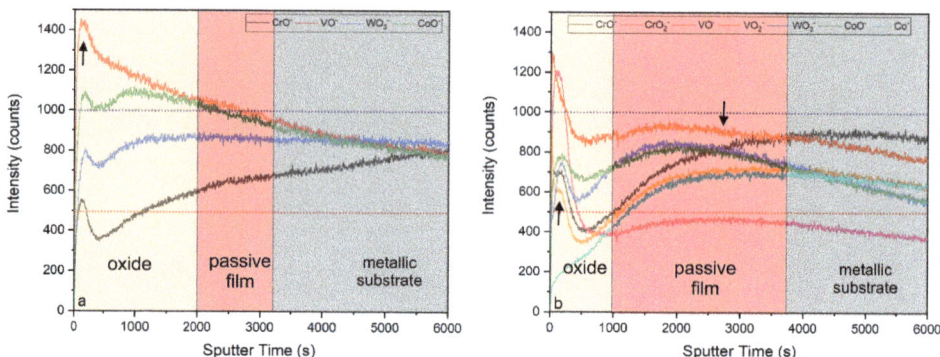

Figure 7. ToF-SIMS depth profiles (negative polarity) after 1 day exposure in chloride-ions-enriched medium at room conditions for selected alloying elements Cr, V, W and Co. (**a**) CHT sample and (**b**) DCT sample.

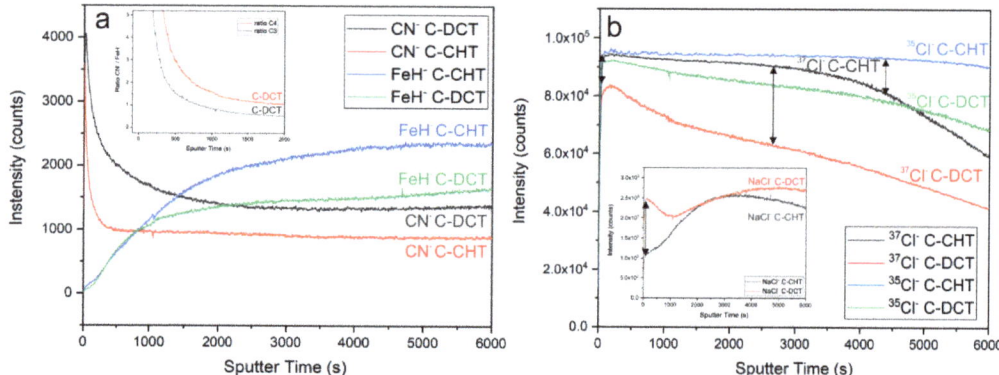

Figure 8. ToF-SIMS depth profiles (negative polarity) after 1 day exposure in chloride-ions-enriched medium at room conditions. (**a**) depth profile of nitrogen and iron and (**b**) depth profile of $^{35}Cl^-$, $^{37}Cl^-$ and $NaCl^-$.

When comparing the ions CrO^-, VO^-, WO_3^- and CoO^- in CHT (Figure 7a) and DCT samples (Figure 7b), a more constant distribution of alloying elements throughout the passive film is observed for DCT compared to CHT. Additionally, the ions in CHT sample (Figure 7a) have higher intensity compared to DCT (Figure 7b), in both the oxide layer and passive film. CrO^-, as a representative for chromium distribution, has its main peak at ~130 s for the CHT sample (Figure 7a) and at ~100 s of sputtering for the DCT sample (Figure 7b), followed by a continuous increase in intensity towards the metallic substrate. In addition to CrO^-, CrO_2^- was also detected in the DCT sample, with its main peak at ~125 s and then a significant drop occurring within the oxide layer. This significant drop indicates the inner layer of the oxide film, which is the transition layer between the outer oxide layer and the passive film. A similar structure of the oxide film has been reported by other researchers [42]. VO^- results show main peaks at ~125 s of sputtering for CHT (Figure 7a) and ~25 s for the DCT (Figure 7b) sample. Afterwards, the intensity drops towards the value of the metallic substrate. However, for the DCT sample a peak of VO_2^- ions is also observed at ~20 s. Tungsten ion WO_3^- signal shows a shift in the main peak of DCT to ~1750 s (Figure 7b) compared to the CHT peak at ~1180 s (Figure 7a). The results of CoO$^-$, provided main peaks at ~1050 s for CHT and at ~2010 s for DCT. In the DCT sample species, Co- was also observed with a main peak at ~3125 s. These results and the initial peaks, which form due to the matrix effect [43] caused by oxide proximity, provide the correlation to the oxide thickness. The oxides for CHT are generally thicker and more enriched with the alloying elements Cr and V compared to DCT samples, which also correlates well with the ToF-SIMS spatial images and findings (Figures 3 and 4). The distribution and ratio of ions is also different, especially within the passive film, which clearly indicates different dynamics of the passive film development for both samples.

Considering the role of nitrogen in the DCT sample, the depth profile of CN$^-$ and FeH$^-$ was also analyzed. The CN$^-$ profile (Figure 8a) clearly shows a higher intensity of nitrogen in all surface layers (outer and inner layer of oxide film, and passive film) for the DCT sample compared to the CHT sample. In both cases the main peak is located at the beginning of the measurements. However, for the CHT sample a significantly stronger drop is present compared to DCT. The source of the nitrogen peak on the surface is related to the absorbed nitrogen and short-range diffusion from the quenching in nitrogen gas during heat treatment. However, for the DCT sample higher values of N below the material surface are a result of the DCT contribution, which is a consequence of the incorporation of nitrogen into the material during immersion into the liquid nitrogen. It is postulated that through boiling and bubbling and local high-pressure variations on the materials surface, the nitrogen is incorporated into the deeper portions of the material as well as adsorbed on the surface. Afterwards, during the heat treatment the surface-bounded nitrogen can diffuse deeper into the material due to the elevated temperatures of the tempering procedure (>600 °C). FeH$^-$ was measured in order to determine the relative nitrogen presence within the iron matrix of both samples (Figure 8a). The ratio of CN$^-$/FeH$^-$ confirms the increased presence of nitrogen in the DCT sample. This clearly indicates that DCT has an important role in the surface chemistry of the selected alloy, which can be also implicated on other similar alloys, in bulk or thin film form, when treated with DCT. Furthermore, in order to observe the surface chemistry dynamics of oxide and passive film formation in correlation to GR, the spectra of ^{35}Cl$^-$, ^{37}Cl$^-$ and NaCl$^-$ were investigated. The depth profile shows lower signal of both isotopes ^{35}/^{37}Cl in the DCT sample, as in the CHT sample. The main peaks of ^{35}Cl$^-$ and ^{37}Cl$^-$ are at ~315 and at ~245 s of sputtering for DCT and at ~285 and at ~235 s for the CHT sample, respectively. Another interesting dynamic was observed for NaCl$^-$. Beside the signal of NaCl$^-$ for DCT being higher compared to the CHT sample, it drops after the initial peak and then slowly increases to its main peak at ~4810 s. This indicates the change in surface chemistry. For CHT a lower initial signal is observed, followed by an increase, with the main peak at ~2855 s, and then a drop in the signal, indicating less stable passive film formation of the CHT sample. The ratio between isotopes for each sample could possibly indicate the preferable GR formation in each sample. However, to obtain clear and reliable results, an isotope tracking method by introduction of stable oxygen isotopes should be applied and tested in future studies.

The results show that the reason behind the higher corrosion resistance of the DCT sample can be attributed to the protective nature of the DCT-induced passive film, which is composed of nitrogen and oxides/hydroxides of alloying elements (Cr, Co, V, W), on which the GR type I grows and acts as a precursor for magnetite and, later, other corrosion products. The results also indicate that the oxide layer and passive film for the DCT sample are thicker compared to the CHT sample, which additionally indicates that a more physically stable corrosion layer forms, and with it increased corrosion resistance when applying DCT.

3.5. Testing Effectiveness of Newly Characterized Passive Film Behavior for DCT-Treated High-Alloyed Ferrous Alloy

In order to test the stability of the newly observed passive film induced by DCT, three different environments were chosen for testing the hypothesis of higher stability of passive film. CHT and DCT samples were exposed to different environments (C-control environment, T-increased temperature of chloride-ions-enriched medium (100 °C) and T+V-increased temperature of chloride-ions-enriched medium (100 °C) and vibrations (25 Hz, 1 h). In the first step, the composition of corrosion products in all three environments was measured. The XRD data, presented in Figure 9a, show the presence of different corrosion products. It is confirmed that for DCT, GR I forms after just 1 h of exposure, and this is observed for all three testing conditions, whereas GR II is mainly present in CHT samples and also forms after 1 h exposure in all conditions. The increased formation of magnetite can also be observed for DCT samples, compared to the CHT counterparts. Other corrosion products present in samples are calcium carbonate, halite, iron as matrix and goethite (Figure 9a). In the second step, the weight loss of each sample group (10 samples per group) was measured in order to observe the difference in corrosion performance between DCT and CHT samples. The average weight loss (mg) after 1 h of exposure for all six testing groups is shown in Figure 9b. In the control group (room temperature, standard conditions) the weight loss and corrosion rate of CHT and DCT was 2.6 ± 0.1 mg and 0.03 ± 0.001 mm/y, and 0.6 ± 0.01 mg and 0.006 ± 0.0005 mm/y, respectively. Under the second condition (elevated temperature; T), the weight loss increased due to the thermal influence on the corrosion propagation. Nevertheless, for the second condition the weight loss and corrosion rate were lower for DCT in comparison with CHT (CHT+T: 382.7 ± 5 mg and 4.11 ± 0.1 mm/y; DCT+T: 302.4 ± 3 mg and 0.84 ± 0.05 mm/y). Under the last condition with higher temperature and vibrations (T+V), the weight loss and corrosion rate for CHT and DCT samples are 77.7 ± 2 mg and 3.25 ± 0.1 mm/y, and 36 ± 2 mg and 0.39 ± 0.01 mm/y, respectively. In all three environments, the DCT samples display improved corrosion properties, which is clearly emphasized in conditions with a higher temperature. The data suggest that the DCT-induced passive film actively increases material corrosion resistance. The improvement in control environment is 75% in terms of weight loss and 80% in corrosion rate. In a high-temperature environment improvement is 20% and 79%, and in a high-temperature environment combined with vibrations it is 53% and 88%, respectively.

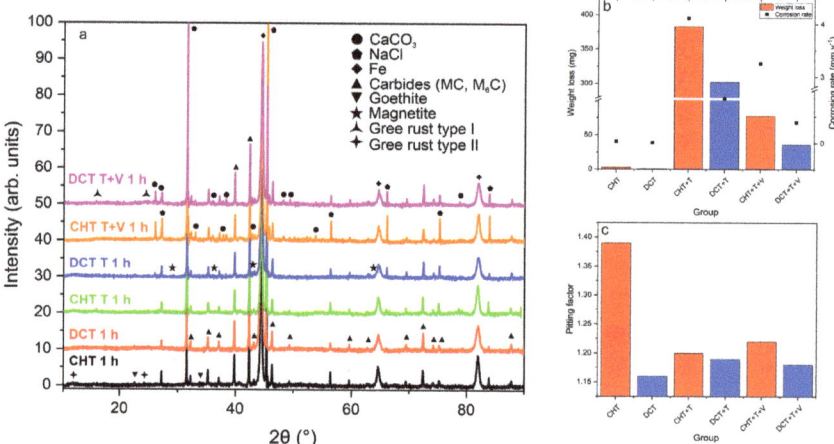

Figure 9. (a) XRD data for control samples (CHT, DCT) and samples (CHT+T, DCT+T, CHT+T+V, DCT+T+V) tested in two different environments; T-higher temperature and T+V-higher temperature and vibrations. (b) Weight loss (columns) and corrosion rate (black dots); (c) pitting factor of each sample group.

In order to test the stability of the newly observed passive film induced by DCT, three different environments were chosen to test the hypothesis of the higher stability of passive film. CHT and DCT samples were exposed to different environments (C-control environment, T-increased temperature of chloride-ions-enriched medium (100 °C) and T+V-increased temperature of chloride-ions-enriched medium (100 °C) and vibrations (25 Hz, 1 h). In the first step, the composition of corrosion products in all three environments was measured. The XRD data, presented in Figure 9a, shows presence of different corrosion products. It is confirmed that for DCT GR I forms after just 1 h of exposure, observed for all three testing conditions. Whereas GR II is mainly present in CHT samples and also forms already after 1 h exposure in all conditions. The increased formation of magnetite can also be observed for DCT samples, compared to its CHT counterparts. Other corrosion products present in samples are calcium carbonate, halite, iron as matrix and goethite (Figure 9a). In the second step the weight loss of each sample group (10 samples per group) was measured in order to observe the difference in corrosion performance between DCT and CHT samples. The average weight loss (mg) after 1 h of exposure for all six testing groups is shown in Figure 9b. In the control group (room temperature, standard conditions) the weight loss and corrosion rate of CHT and DCT was 2.6 ± 0.1 mg and 0.03 ± 0.001 mm/y, and 0.6 ± 0.01 mg and 0.006 ± 0.0005 mm/y, respectively. Under the second condition (elevated temperature; T), the weight loss increased due to the thermal influence on the corrosion propagation. Nevertheless, for the second condition the weight loss and corrosion rate were lower for DCT in comparison to CHT (CHT+T: 382.7 ± 5 mg and 4.11 ± 0.1 mm/y; DCT+T: 302.4 ± 3 mg and 0.84 ± 0.05 mm/y). Under the last condition with higher temperature and vibrations (T+V), the weight loss and corrosion rate for CHT and DCT samples are 77.7 ± 2 mg and 3.25 ± 0.1 mm/y, and 36 ± 2 mg and 0.39 ± 0.01 mm/y, respectively. In all three environments, the DCT samples display improved corrosion properties, which is clearly emphasized in conditions with higher temperature. The data suggest that the DCT-induced passive film actively increases material corrosion resistance. The improvement in the control environment is 75% in terms of weight loss and 80% in terms of the corrosion rate. In a high-temperature environment the improvement is 20 and 79%, and in a high-temperature environment combined with vibrations it is 53 and 88%, respectively.

4. Conclusions

In this work the role of nitrogen, introduced by deep cryogenic treatment, was investigated in relation to the surface modification of high-alloyed ferrous alloy. Using time-of-flight secondary ion mass spectroscopy (ToF-SIMS), the nitrogen was confirmed to be present in a larger quantity in DCT samples compared to their conventionally heat-treated (CHT) counterparts. The nitrogen acts as a building block for the formation of a thin corrosion buffer layer dubbed as a ghost layer, which facilitates the preferential formation of green rust type I. The modification is considered to result from the formation of additional ion species (NO_3^- and NH_4^+) that modify the local environment and ionic exchange between the alloy surface and corrosive medium. In turn, the green rust layer acts as a precursor for the formation of magnetite, which reduces the corrosion propagation due to its high density. As a result, the DCT samples exhibit lower corrosion rates and wear loss, which was also confirmed in more extreme environments that involved elevated temperatures and vibrations. From these experiments it was confirmed that the DCT-/Induced passive film is more stable than the passive film of the CHT counterpart. Furthermore, the modified passivation of the material with DCT leads to different corrosion product development to when the material is conventionally heat-treated. DCT also induces changes in the inclusion of the different alloying elements in the formation of the passivation layer as well as the formation of different ionic species that were detected and monitored with ToF-SIMS. The presence of different ions indicates a change in the oxidation behavior of the metallic surface as well as the formation of different oxides, which goes hand in hand with our previous findings on the alloy's oxidation dynamics in air. With these results, this study provides the first proof of the influence of DCT on surface behavior through incorporation of nitrogen into a sample surface, and with it an answer to the improved corrosion response of the ferrous alloys in a chloride-ions-enriched environment.

Supplementary Materials: The following supporting information can be downloaded at: https://www.mdpi.com/article/10.3390/coatings12020213/s1.

Author Contributions: Conceptualization, P.J.-K., M.J.-K. and B.P.; methodology, P.J.-K. and M.J.-K.; investigation, P.J.-K. and M.J.-K.; resources, P.J.-K.; writing—original draft preparation, P.J.-K. and M.J.-K.; writing—review and editing, P.J.-K., M.J.-K. and B.P.; visualization, P.J.-K. and M.J.-K.; supervision, B.P. All authors have read and agreed to the published version of the manuscript.

Funding: This research was funded by Slovenian Research Agency (ARRS), Ljubljana, Slovenia, Grant Nos. P2-0050 and J2-9211.

Institutional Review Board Statement: Not applicable.

Informed Consent Statement: Not applicable.

Data Availability Statement: The raw processed data required to reproduce these findings cannot be shared at this time as the data also form part of an ongoing study.

Acknowledgments: Acknowledgement goes to T. Kranjec (IMT, Slovenia) for the help in the metallographic lab. For seawater data, authors thank K. Klun, Marine Biology Station, National Institute of Biology, Slovenia. For the help with ToF-SIMS spectra authors would like to thank N. Valle from LIST, Luxembourg; and J. Kovač and J. Ekar from Jožef Stefan Institute, Slovenia.

Conflicts of Interest: The authors declare no competing interests.

References

1. Voglar, J.; Novak, Ž.; Jovičević-Klug, P.; Podgornik, B.; Kosec, T. Effect of deep cryogenic treatment on corrosion properties of various high-speed steels. *Metals* **2020**, *11*, 14. [CrossRef]
2. Popoola, L.T.; Grema, A.S.; Latinwo, G.K.; Gutti, B.; Balogun, A.S. Corrosion problems during oil and gas production and its mitigation. *Int. J. Ind. Chem.* **2013**, *4*, 1–15. [CrossRef]
3. Eliaz, N. Corrosion of metallic biomaterials: A review. *Materials* **2019**, *12*, 407. [CrossRef]
4. Comizzoli, R.B.; Frankenthal, R.P.; Milner, P.C.; Sinclair, J.D. Corrosion of electronic materials and devices. *JSTOR* **1986**, *234*, 340–345. [CrossRef]
5. LeBozec, N.; Blandin, N.; Thierry, D. Accelerated corrosion tests in the automotive industry: A comparison of the performance towards cosmetic corrosion. *Mater. Corros.* **2008**, *59*, 889–894. [CrossRef]
6. Benavides, S. Corrosion in the aerospace industry. *Corros. Control Aerosp. Ind.* **2009**, 1–14. [CrossRef]
7. Cattant, F.; Crusset, D.; Féron, D. Corrosion issues in nuclear industry today. *Mater. Today* **2008**, *11*, 32–37. [CrossRef]
8. Park, J.Y.; Park, Y.S. The effects of heat-treatment parameters on corrosion resistance and phase transformations of 14Cr-3Mo martensitic stainless steel. *Mater. Sci. Eng. A* **2007**, *448–451*, 1131–1134. [CrossRef]
9. Fundazioa, E. Cryogenic Treatment Improves the Characteristics of Materials and Cuts Their Costs—ScienceDaily. Available online: https://www.sciencedaily.com/releases/2014/03/140319085424.htm (accessed on 19 July 2021).
10. Jovičević-Klug, P.; Podgornik, B. Review on the effect of deep cryogenic treatment of metallic materials in automotive applications. *Metals* **2020**, *10*, 434. [CrossRef]
11. Tóth, L. Examination of the properties and structure of tool steel EN 1.2379 due to different heat treatments. *Eur. J. Mater. Sci. Eng.* **2018**, *3*, 1–7.
12. Gogte, C.L.; Peshwe, D.R.; Paretkar, R.K. Influence of Cobalt on the Cryogenically Treated W-Mo-V High Speed Steel. In Proceedings of the Advances in Cryogenic Engineering, Spokane, WA, USA, 12–14 June 2012; pp. 1175–1182.
13. Jovičević-Klug, M.; Jovičević-Klug, P.; Kranjec, T.; Podgornik, B. Cross-effect of surface finishing and deep cryogenic treatment on corrosion resistance of AISI M35 steel. *J. Mater. Res. Technol.* **2021**, *14*, 2365–2381. [CrossRef]
14. Hemath Kumar, G.; Mohit, H.; Purohit, R. Effect of deep cryogenic treatment on composite material for automotive Ac system. *Mater. Today Proc.* **2017**, *4*, 3501–3505. [CrossRef]
15. Jovičević-Klug, P.; Jenko, M.; Jovičević-Klug, M.; Šetina Batič, B.; Kovač, J.; Podgornik, B. Effect of deep cryogenic treatment on surface chemistry and microstructure of selected high-speed steels. *Appl. Surf. Sci.* **2021**, *548*, 1–11. [CrossRef]
16. Kim, B.; Kim, S.; Kim, H. Effects of alloying elements (Cr, Mn) on corrosion properties of the high-strength steel in 3.5% NaCl solution. *Adv. Mater. Sci. Eng.* **2018**, *2018*, 1–13. [CrossRef]
17. Hou, X.; Gao, L.; Cui, Z.; Yin, J. Corrosion and protection of metal in the seawater desalination. *IOP Conf. Ser. Earth Environ. Sci.* **2018**, *108*, 022037. [CrossRef]
18. Fu, Y.; Wu, X.; Han, E.H.; Ke, W.; Yang, K.; Jiang, Z. Effects of nitrogen on the passivation of nickel-free high nitrogen and manganese stainless steels in acidic chloride solutions. *Electrochim. Acta* **2009**, *54*, 4005–4014. [CrossRef]

19. Grabke, H.J. The role of nitrogen in the corrosion of iron and steels. *ISIJ Int.* **1996**, *36*, 777–786. [CrossRef]
20. Leskovšek, V.; Podgornik, B. Simultaneous ion nitriding and tempering after deep cryogenic treatment of PM S390MC HSS. *Int. Heat Treat. Surf. Eng.* **2013**, *7*, 115–119. [CrossRef]
21. Olefjord, I.; Clayton, C.R. Surface composition of stainless steel during active dissolution and passivation. *ISIJ Int.* **1991**, *31*, 134–141. [CrossRef]
22. Olefjord, I.; Wegrelius, L. The influence of nitrogen on the passivation of stainless steels. *Corros. Sci.* **1996**, *38*, 1203–1220. [CrossRef]
23. Ramesh, S.; Bhuvaneshwari, B.; Palani, G.S.; Mohan Lal, D.; Mondal, K.; Gupta, R.K. Enhancing the corrosion resistance performance of structural steel via a novel deep cryogenic treatment process. *Vacuum* **2019**, *159*, 468–475. [CrossRef]
24. Ramesh, S.; Bhuvaneswari, B.; Palani, G.S.; Lal, D.M.; Iyer, N.R. Effects on corrosion resistance of rebar subjected to deep cryogenic treatment. *J. Mech. Sci. Technol.* **2017**, *31*, 123–132. [CrossRef]
25. Bensely, A.; Shyamala, L.; Harish, S.; Mohan Lal, D.; Nagarajan, G.; Junik, K.; Rajadurai, A. Fatigue behaviour and fracture mechanism of cryogenically treated En 353 steel. *Mater. Des.* **2009**, *30*, 2955–2962. [CrossRef]
26. Cai, Y.; Luo, Z.; Zeng, Y. Influence of deep cryogenic treatment on the microstructure and properties of AISI304 austenitic stainless steel A-TIG weld. *Sci. Technol. Weld. Join.* **2017**, *22*, 236–243. [CrossRef]
27. Baldissera, P.; Delprete, C. Deep cryogenic treatment of AISI 302 stainless steel: Part II-fatigue and corrosion. *Mater. Des.* **2010**, *31*, 4731–4737. [CrossRef]
28. Tian, J.; Wang, W.; Shahzad, M.B.; Yan, W.; Shan, Y.; Jiang, Z.; Yang, K. A new maraging stainless steel with excellent strength–toughness–corrosion synergy. *Materials* **2017**, *10*, 1293. [CrossRef]
29. Tian, J.L.; Wang, W.; Shahzad, M.B.; Yan, W.; Shan, Y.Y.; Jiang, Z.H.; Yang, K. Corrosion resistance of Co-containing maraging stainless steel. *Acta Metall. Sin. (Engl. Lett.)* **2018**, *31*, 785–797. [CrossRef]
30. Hsu, H.W.; Tsai, W.T. High temperature corrosion behavior of siliconized 310 stainless steel. *Mater. Chem. Phys.* **2000**, *64*, 147–155. [CrossRef]
31. Vasyliev, G.S.; Novosad, A.A.; Pidburtnyi, M.O.; Chyhryn, O.M. Influence of ultrasound vibrations on the corrosion resistance of heat-exchange plates made of AISI 316 steel. *Mater. Sci.* **2019**, *54*, 913–919. [CrossRef]
32. *ASTM Standard Practice for Preparing, Cleaning, and Evaluating Corrosion Test Specimens*; ASTM International: West Conshohocken, PA, USA, 1999; pp. 1–8.
33. Jovičević-Klug, P.; Jovičević-Klug, M.; Podgornik, B. Effectiveness of deep cryogenic treatment on carbide precipitation. *J. Mater. Res. Technol.* **2020**, *9*, 13014–13026. [CrossRef]
34. Stephan, T. TOF-SIMS in Cosmochemistry. *Planet. Space Sci.* **2001**, *49*, 859–906. [CrossRef]
35. Jovičević-Klug, P.; Kranjec, T.; Jovičević-Klug, M.; Podgornik, B. Modification of Steel Corrosion Resistance in Seawater with Deep Cryogenic Treatment. In Proceedings of the Proceedings of the 60th Conference of Metallurgists, COM 2021: The Canadian Institute of Mining, Metallurgy and Petroleum, Toronto, ON, Canada, 16–19 August 2021; pp. 1–4.
36. Esmaily, M.; Malmberg, P.; Shahabi-Navid, M.; Svensson, J.E.; Johansson, L.G. A ToF-SIMS Investigation of the Corrosion Behavior of Mg Alloy AM50 in Atmospheric Environments. *Appl. Surf. Sci.* **2016**, *360*, 98–106. [CrossRef]
37. Ghanem, W.A. Effect of Nitrogen on the Corrosion Behavior of Austenitic Stainless Steel in Chloride Solutions. In Proceedings of the EUROCORR 2004-European Corrosion Conference: Long Term Prediction and Modelling of Corrosion, Toronto, ON, Canada, 12–14 July 2004; pp. 1–8. [CrossRef]
38. Metikoš-Huković, M.; Babić, R. Passivation and corrosion behaviours of cobalt and cobalt–chromium–molybdenum alloy. *Corros. Sci.* **2007**, *49*, 3570–3579. [CrossRef]
39. Ras, M.H.; Pistorius, P.C. Possible mechanisms for the improvement by vanadium of the pitting corrosion resistance of 18% chromium ferritic stainless steel. *Corros. Sci.* **2002**, *44*, 2479–2490. [CrossRef]
40. Alcántara, J.; de la Fuente, D.; Chico, B.; Simancas, J.; Díaz, I.; Morcillo, M. Marine atmospheric corrosion of carbon steel: A review. *Materials* **2017**, *10*, 406. [CrossRef]
41. Sumoondur, A.; Shaw, S.; Ahmed, I.; Benning, L.G. Green rust as a precursor for magnetite: An in situ synchrotron based study. *Mineral. Mag.* **2008**, *72*, 201–204. [CrossRef]
42. Wang, Z.; Paschalidou, E.M.; Seyeux, A.; Zanna, S.; Maurice, V.; Marcus, P. Mechanisms of Cr and Mo enrichments in the passive oxide film on 316L austenitic stainless steel. *Front. Mater.* **2019**, *6*, 1–12. [CrossRef]
43. Priebe, A.; Xie, T.; Bürki, G.; Pethö, L.; Michler, J. The matrix effect in TOF-SIMS analysis of two-element inorganic thin films. *J. Anal. At. Spectrom.* **2020**, *35*, 1156–1166. [CrossRef]

Article

Improving the Surface Friction and Corrosion Resistance of Magnesium Alloy AZ31 by Ion Implantation and Ultrasonic Rolling

Zhongyu Dou *, Haili Jiang, Rongfei Ao, Tianye Luo and Dianxi Zhang

College of Physics and Electronic Science, Anshun University, Anshun 561000, China; j200109012022@163.com (H.J.); arf1028@163.com (R.A.); rotiondemon@126.com (T.L.); xiwa_315@163.com (D.Z.)
* Correspondence: 2012620@asu.edu.cn

Abstract: The use of the magnesium alloy AZ31 is common in aviation and biomedicine; however, this alloy has poor friction and corrosion resistance. Here, mechanical grinding, ultrasonic rolling, and ultrasonic rolling + ion implantation were performed on the magnesium alloy surface to study the effect of the treatment process on the friction and corrosion resistance of the magnesium alloy surface. The results show that the surface roughness of the magnesium alloy treated by ultrasonic rolling + ion injection is reduced more than mechanical grinding and ultrasonic rolling. The friction coefficient is the lowest, the wear resistance is the best, and new phase nitrogen compounds appear on the surface. The results of SBF (simulated body fluid) solution immersion showed that the sample treated via this composite process had the lowest corrosion rate, which was 62.45% and 58.47% lower than that of the mechanically ground samples. The surface was relatively intact after the corrosion test, and the corrosion resistance was the best. These results can provide a new strategy for magnesium alloy surface protection.

Keywords: magnesium alloy; ultrasonic rolling; ion implantation; friction and wear; corrosion

Citation: Dou, Z.; Jiang, H.; Ao, R.; Luo, T.; Zhang, D. Improving the Surface Friction and Corrosion Resistance of Magnesium Alloy AZ31 by Ion Implantation and Ultrasonic Rolling. *Coatings* **2022**, *12*, 899. https://doi.org/10.3390/coatings12070899

Academic Editors: Matic Jovičević-Klug, Alina Vladescu, Patricia Jovičević Klug and László Tóth

Received: 26 May 2022
Accepted: 22 June 2022
Published: 25 June 2022

Publisher's Note: MDPI stays neutral with regard to jurisdictional claims in published maps and institutional affiliations.

Copyright: © 2022 by the authors. Licensee MDPI, Basel, Switzerland. This article is an open access article distributed under the terms and conditions of the Creative Commons Attribution (CC BY) license (https://creativecommons.org/licenses/by/4.0/).

1. Introduction

Magnesium alloys are a green and lightweight material and have a high specific strength, high specific stiffness, and low density. They can be used in new energy vehicles, biomedicine, aviation, and other fields [1,2], but their wear and corrosion resistance are severely restricted. Alloy composition deployment, alloy-processing technology, and alloy surface-coating technologies have all been proposed [3–7] and have enhanced the development of magnesium, its alloys, and alloy applications. However, these traditional protective-layer methods can only effectively protect the surface of the magnesium alloy when the protective layer exists. The magnesium alloy will still quickly corrode in a corrosive medium when the protective layer on the surface is damaged by corrosion. Thus, a protective coating is used. It is important to study whether the corrosion products are harmful to the body. The wear of ultra-high-molecular-weight polyethylene (UHMWPE) releases polyethylene wear particles, which can trigger a negative reaction of the body and promote osteolysis [8]. The biocompatibility of the protective layer needs to be considered.

Surface nanoscale treatment processes are a common and effective method. Researchers have performed different treatment processes such as surface mechanical grinding [9,10], shot peening [11,12], and laser treatment [13]. Although these have improved the mechanical properties, the surface quality of the material is reduced, which affects the friction and corrosion resistance. Liu et al. [14] found that the surface nanostructured layer of GW63K magnesium alloy after SMAT treatment had poor plasticity and toughness, thus resulting in worse wear resistance versus untreated alloys. Liu Mengen et al. [15] found that the corrosion resistance of high-energy shot peening on AZ31 magnesium alloy in 5% (mass fraction) NaCl solution is lower than that of untreated samples due to the formation of a large number of cracks on the surface during shot peening. As a result, the corrosion contact surface increases, thus resulting in a significantly higher corrosion rate of the shot-peened sample than the unpeened sample. The ultrasonic surface-rolling process (USRP) combines traditional rolling processes and ultrasonic technology to refine grains and improve performance. The surface quality of the workpiece is significantly improved [16]. Zhang Haiquan et al. [17] strengthened ZK60 magnesium alloys via a surface-rolling strengthening process. The results showed that rolling strengthening can significantly reduce the surface roughness of the material and introduce residual compressive stress to the surface of the material. Yang et al. [18] treated the magnesium alloy AZ31 via ultrasonic rolling and found that the surface grains of magnesium alloys were refined after rolling strengthening; the roughness was reduced and the friction properties were improved.

High-energy ion implantation technology (HEII) utilizes high-speed ion bombardment of pre-infiltrated elements from the target to achieve metallurgical bonding with the plated metal, thus improving the friction and corrosion resistance of the material. The material itself does not deform [19]. Lei et al. implanted Al ions into the surface of AZ31 magnesium alloy, and the friction and wear results showed that the wear rate of magnesium alloy was reduced by 30% [20]. Zhou et al. [21] used Zr to implant magnesium alloy ZK60 and found that the friction and corrosion resistance of magnesium alloy was effectively improved after implantation. Other studies [22,23] reported that N/Ti ion implantation improved the friction and corrosion resistance of magnesium alloy AZ31.

Dingshun [24] carried out USRP + PN (plasma nitriding) composite treatment on pure titanium TA2, and the infiltrated layer showed the best wear resistance and friction reduction performance. Dawen et al. [25] used USRP + HEII for composite treatment of 316 L. The surface hardness was increased by 57.8% versus single-HEII-treated samples; the thickness of the infiltration layer was nearly double that of a single-HEII-treated sample. In conclusion, ultrasonic rolling and ion implantation are effective means of improving the surface properties of magnesium alloys. Surprisingly, the influence of the composite treatment technology on the properties of the magnesium alloys is rarely reported.

Therefore, the effects of mechanical grinding, ultrasonic rolling, and USRP + HEII on the surface structure, friction resistance, and corrosion resistance of magnesium alloys are reported in this study. The results offer a reference for the development of treatment technologies for magnesium alloy surface protection.

2. Materials and Methods

2.1. Material and Sample Preparation

The test material was rolled AZ31 magnesium alloy purchased from a domestic factory; the chemical composition is shown in Table 1. Magnesium alloy sheets were rolled and strengthened using ultrasonic-rolling equipment (customized). The size was $15 \times 15 \times 5$ mm as cut by a wire electric discharge machine. High-energy N ion implantation was performed using ion implantation equipment (Southwestern Institute of Physics, Chengdu, China). The magnesium alloy was polished with 1000# and 2000# water-grinding sandpaper and then polished; this sample was marked as S1. The sample after ultrasonic rolling was designated as S2. The static pressure was 0.15 MPa, the feed speed was 4000 mm/min, and the rolling treatment was 1 pass. The ion implantation sample after rolling was marked as S3. It had an implantation energy of 40 keV. The implantation dose was 1×10^{18} icons/cm^2 and the vacuum was 3.9×10^{-3} Pa.

Table 1. Chemical composition of AZ31 magnesium alloy (wt%).

Al	Zn	Mn	Si	Ca	Cu	Fe	Ni	Mg
2.5–3.5	0.6–1.4	0.2–1.0	0.08	0.04	0.04	0.003	0.001	Remainder

2.2. Microstructure and Performance Characterization

Magnesium alloy samples with different treatments were analyzed with an X-ray diffractometer (PANalytical X Pert PRO, Almelo, Holland). The detection angle was 20–80°, the speed was 2°/min, and the step size was 0.013/s. The three-dimensional topography and surface roughness of the treated magnesium alloy surfaces were measured by atomic force microscopy (Bruker Dimension Icon AFM, Karlsruhe, Germany). The surface mechanical and friction properties were investigated using a microhardness tester (HMV-G, Kyoto, Japan) and a friction and wear-testing machine (Bruker UMT-2, Karlsruhe, Germany). A Phase Shift MicroXAM-3D (MicroXAM-3D, Milpitas, USA) measured the wear volume of the samples after wear to judge the extent of wear. The hardness test selected five points to obtain the average value. The load was 0.98 N and the duration was 10 s. The room-temperature dry-friction test was performed with 10 mm Al_2O_3 balls (HRC95). The circumferential speed was 100 rpm/min, the load was 10 N, and the duration was 20 min. The wear-scar radius was 12 mm.

The AZ31 magnesium alloy samples were ultrasonically cleaned in ethanol before soaking, and then dried in cold air. The samples with different treatments were encapsulated with oxidized resin with an exposed area of 1 cm². They were then weighed with an electronic balance. The encapsulated samples were soaked in SBF at 37 ± 0.5 °C for 48 h, and 55% of the solution was renewed every 24 h to simulate natural human body-fluid renewal. The sample was soaked in a 5% NaCl solution for 72 h. Specimens were cleaned according to ASTM Standard G31-72 and then weighed. Each sample was placed in concentrated nitric acid for 5 min to remove corrosive products. The degradation performance of the samples under different treatment processes in SBF (simulated body fluid) and NaCl solution was evaluated via the weight-loss method. The corrosion rate **V** was calculated by the weight-loss method (G31-72 standard) using the following formula [26]: $v = (K \times \Delta m)/(A \times t)$, where K is the constant pole, Δm is the mass loss of the sample before and after soaking, A is the exposed area of the sample, and t is the soaking time. Scanning electron microscopy (VEGA3, Brno, Czech Republic) was used to observe the surface morphology of the magnesium alloy samples after immersion to remove corrosive products. Before observing the corrosion morphology, anhydrous ethanol was used for ultrasonic cleaning for 5 min and dried with cold air.

3. Results

3.1. Phase Analysis

Figure 1 shows XRD patterns of the AZ31 magnesium alloys treated under different processes. The XRD peaks of the compounds were found by comparing PDF cards (#35-0821, #45-0946, #01-1289). No new diffraction peaks appeared in the XRD patterns after ultrasonic-rolling treatment versus untreated samples. The untreated samples and the ultrasonic-rolling test show only Mg phases and MgO phases in the samples, thus indicating that the ultrasonic-rolling process did not lead to the formation of new phases in the samples. There were no obvious changes in the diffraction spectrum due to the low content of $Mg_{17}Al_{12}$ phase in the original material [27]. The XRD pattern of the ultrasonic-rolling sample after ion implantation had Mg phases and MgO phases. There was also an Mg_3N_2 phase formed after N ion implantation, thus indicating that the ions and the inherent elements in the matrix would combine with each other during the implantation process to form a new phase. The diffraction peak position and intensity of the surface phase of the sample changed before and after implantation due to the deformation of the surface-lattice structure caused by the internal stress generated upon bombardment of the ion implantation. Holes formed on the surface and generated dislocations and many defects. The formation of an amorphous structure affected the preferred orientation of the same phase in the grains, thus resulting in changes in the position and intensity of its diffraction peaks.

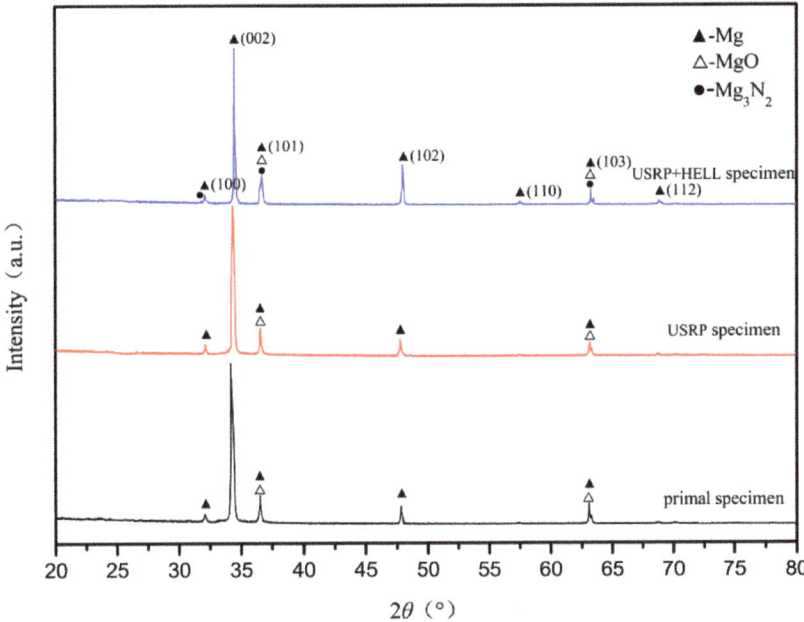

Figure 1. XRD patterns of magnesium alloy samples treated by different processes.

3.2. Surface Roughness and Microhardness

Surface roughness is an important indicator to measure the surface quality of materials and is an important factor affecting the performance of its mechanical parts. Figure 2 shows the three-dimensional surface morphologies of the magnesium alloy samples after different treatments. Surface roughness and microhardness information is shown in Table 2. The roughness of the AZ31 magnesium alloy after ultrasonic rolling was greatly reduced, and the surface quality was greatly improved. The surface quality of the magnesium alloy after ion implantation was further improved, and the average roughness value was reduced by 60% compared to the polished AZ31 magnesium alloy. The surface hardness increased by 23% after ultrasonic-rolling treatment, and the surface hardness was further improved after composite processing.

Table 2. Roughness and microhardness information.

Sample (#)	Primal Specimen	USRP Specimen	USRP + HELL Specimen
RMS roughness (nm)	87.4	42.7	35.0
Average roughness (nm)	64.2	34.3	26.0
Microhardness (HV)	60.2	73.3	81.6

Plastic flow occurred on the surface of the USRP specimen under the action of multi-directional force during ultrasonic rolling; thus, the "peaks" on the material surface flowed into the "valleys," significantly reducing the mechanical defects (scratches) of the original specimen. The addition of lubricating oil on the surface of the sample further reduced the friction between the ball of the processing head and the surface of the sample, and thus the ultrasonic surface-rolling treatment significantly reduced the surface roughness of the sample [28]. Further reduction of surface roughness after ion implantation may have been due to sputtering, etching, and diffusion processes under this implantation dose. The results of the microhardness showed an increase in microhardness and the formation of a hardened layer after ultrasonic surface-rolling treatment. These results are due to the fact that ultrasonic-rolling treatment can produce better deformation hardening effects and fineness in the surface layer within a certain depth of the material. The grain-strengthening effect is caused by grain refinement, strain strengthening, and residual compressive stress. Studies have shown [29,30] that a smaller grain leads to greater microhardness. Hard phases such as Mg_3N_2 in the modified layer after N ion implantation are the main reasons for the increased microhardness.

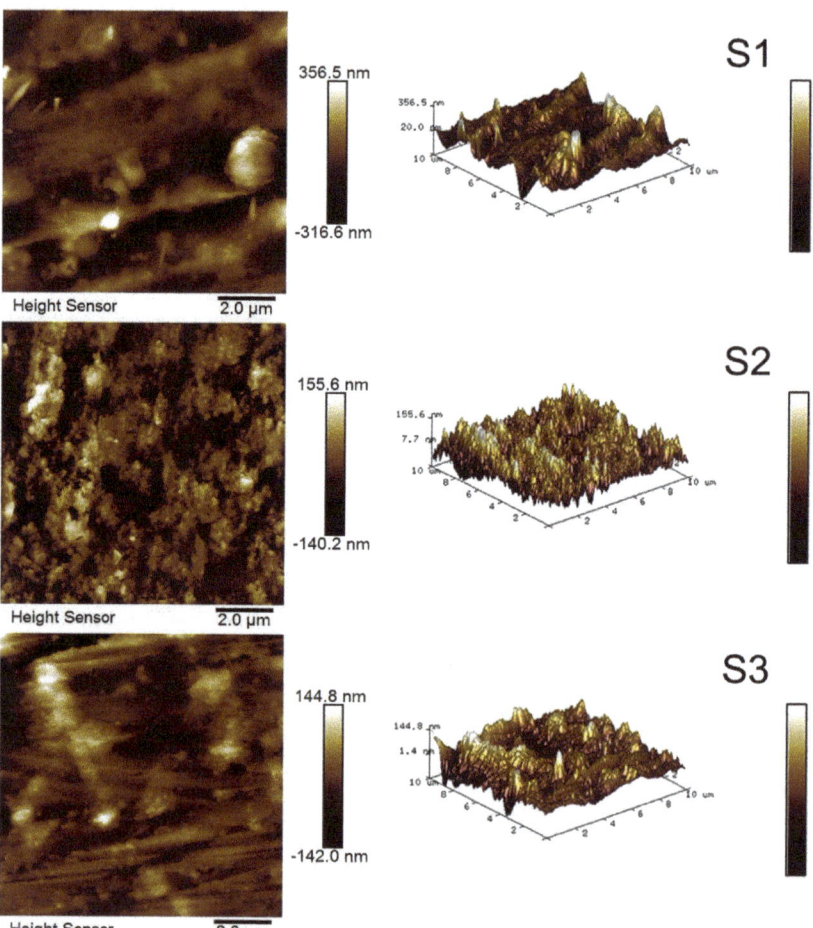

Figure 2. Three–dimensional AFM images of the samples: (**S1**) primal specimen, (**S2**) USRP specimen, (**S3**) USRP + HELL specimen.

3.3. Friction and Wear Performance

The coefficient of friction is the ratio of the frictional force between two surfaces to the vertical force acting on one surface. A smaller coefficient of friction leads to more wear resistance. The coefficient of friction is related to such factors as the surface roughness, hardness, and strength. The friction coefficient curves of AZ31 magnesium alloy samples treated with different processes are shown in Figure 3: At the beginning of friction, the friction coefficient of magnesium alloy samples treated with different processes increased with almost the same slope and then fluctuated within a certain interval. In the initial running-in wear stage, the softer substrate was first worn away by the harder surface of the friction pair, thus resulting in furrows, fractures, or chips; the friction factor was larger. The friction coefficient then stabilized. The average friction coefficient of the original AZ31 magnesium alloy was about 0.321 in the 1200 s test period. The friction coefficient of the samples in the ultrasonic rolling place were improved due to the improved surface quality. The average friction coefficient was about 0.29, and the friction coefficient of the USRP specimen was smaller than the ground specimen throughout the entire friction process. The friction coefficient of the samples after N ion implantation was further reduced, and the average friction coefficient was about 0.276 due to the combined effect of the emergence of hardened phase nitrides, hardness enhancement, and surface roughness after ion implantation.

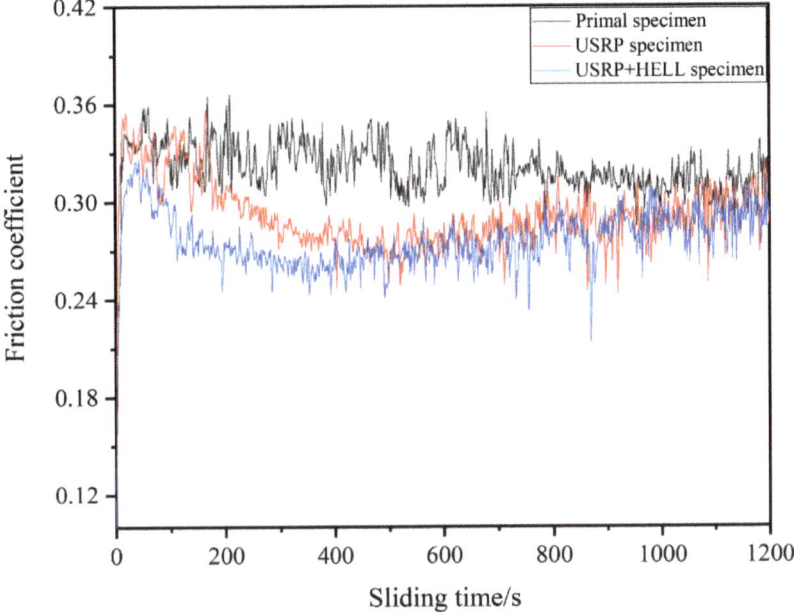

Figure 3. Variation curve of the friction coefficient of different samples.

The amount of wear directly reflects the wear resistance of the material. Thickness, mass, and volume are three ways to characterize the wear amount. The wear resistance was evaluated by measuring the volume wear of the samples treated with different processes. The volume wear of the samples with the three treatments is shown in Figure 4. The figure shows that the volume wear of the samples treated by ultrasonic rolling was significantly smaller than the original magnesium alloy under the same experimental conditions. The volume wear of the composite-treated samples was the smallest and was related to the friction coefficient. The performance was consistent, thus indicating that the ultrasonic-rolling process can significantly improve the wear resistance of magnesium alloy materials. The wear resistance of the materials treated by the USRP + HELL composite process was further improved, which proves that ion-implantation technology based on prefabricated nanostructured layers is an effective method to improve the friction and wear properties of magnesium alloys. Figure 5 shows the SEM morphology of the wear track. The morphology of sample S1 showed obvious grooves and pits, the plastic deformation was serious, and a large amount of wear-scar debris appeared on the surface of the wear scar, indicating that the polished magnesium alloy sample had adhesive wear and abrasive particles, which is the main wear mechanism of magnesium alloys. Compared with sample S1, the furrows caused by micro-cutting in the rolled samples were still more obvious, but the grooves were shallow and narrow, the plastic deformation was reduced, and the abrasive wear condition was improved. The surface of the sample S3 treated by the composite process was relatively flat, the plastic deformation was greatly improved, the adhesion of debris was greatly improved compared to the former two, and the wear debris was granular. The shape of the wear debris was proportional to the degree of wear [31], indicating that the load-bearing capacity of the specimen was improved after the composite treatment process, which is consistent with the performance of the irradiation strengthening study [32], and the results of the wear volume loss also illustrate this point. The improvement in friction and wear performance was due to the substantial reduction of surface roughness and the increased surface hardness; it may also be that high residual compressive stress was introduced into the surface layer, forming a gradient nanostructure that inhibited the initiation and expansion of microcracks in the surface layer and improved the friction and wear properties of the material.

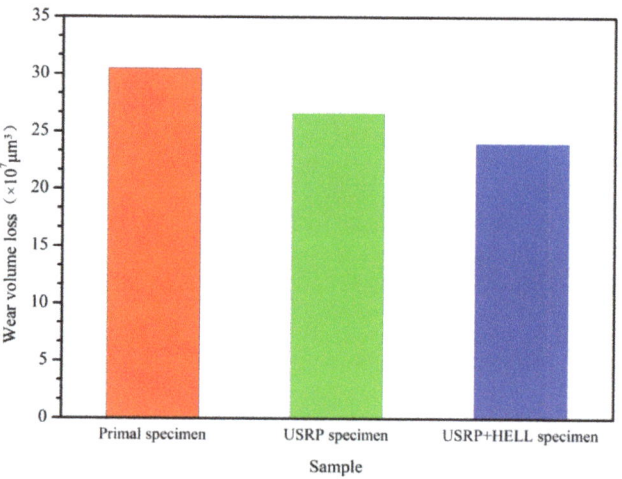

Figure 4. Wear volume loss of samples with different treatment processes.

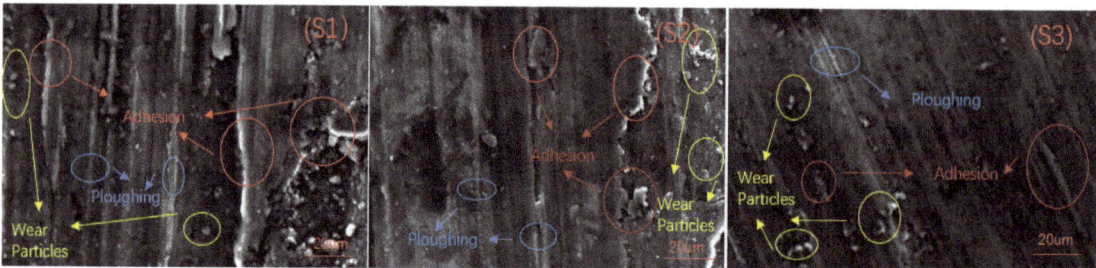

Figure 5. SEM images of worn morphologies of the (**S1**) primal sample; (**S2**) USRP sample; (**S3**) USRP + HELL sample.

3.4. Corrosion Performance

The corrosion morphology of the sample in Figure 6 shows that the surface of sample S1 (Figure 6a) had a large and deep corrosion area after corrosion in the SBF solution; there were corrosion impurities after the corrosion reaction. The residue covered most of the surface with obvious corrosion pits. The report by [33] pointed out that the corrosion of AZ31 after immersion in SBF solution appeared as voids, and the corrosion products included MgO/Mg(OH)$_2$ and Ca$_{10}$(PO$_4$)$_6$(OH)$_2$ phases. The corrosion pits of sample S2 (Figure 6b) were much smaller, and most of them were pitting pits. The surface of sample S3 (Figure 6c) was corroded in a semi-uniform way. The fine corrosion cracks were distributed along the grain boundaries in a network shape. After the immersion test in NaCl solution, the corrosion products were dominated by Mg(OH)$_2$ phase [34,35]. It can be seen that the corrosion conditions were greatly improved after different treatments (versus Figure 6c,d), and the surface of the sample treated via the composite process was much smoother than sample (S1) in terms of corrosion resistance.

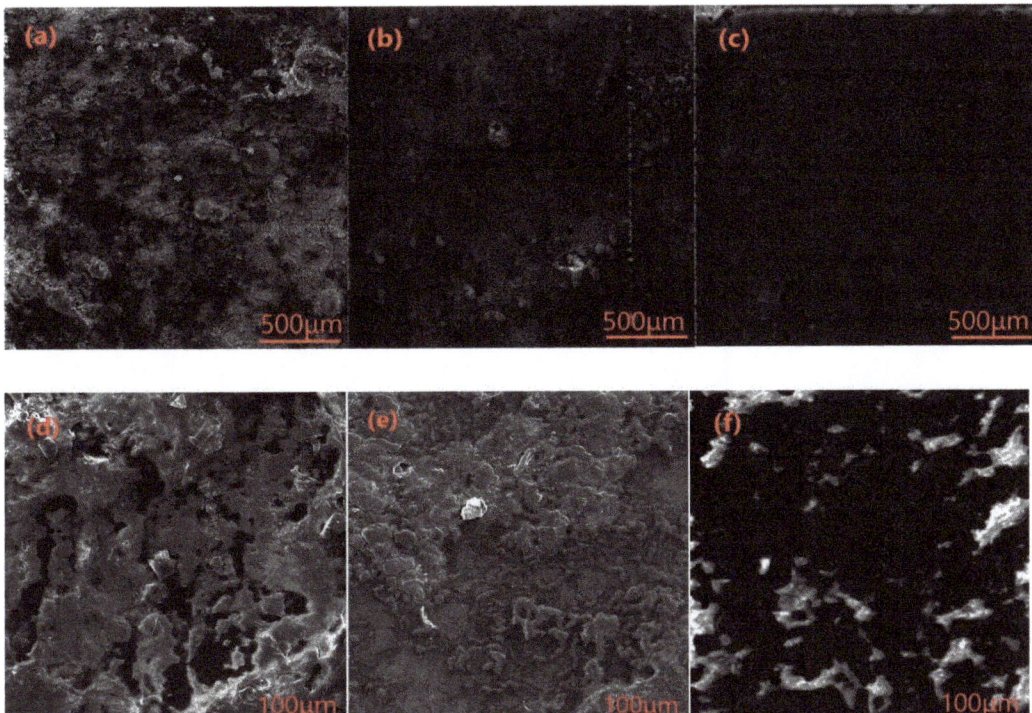

Figure 6. Typical SEM images of samples with different treatments immersed in SBF solution and 5 wt% NaCl aqueous solution: (**a,d**) primal sample; (**b,e**) USRP sample; (**c,f**) USRP + HELL sample.

Figure 7 shows the corrosion information of the magnesium alloy samples obtained after the immersion experiment. The performance was relatively consistent in the two corrosion solutions. Ref. [33] pointed out that magnesium alloy AZ31 had the highest degradation rate of Mg in SBF solution compared with other solutions, which is also the reason why AZ31 magnesium alloy degrades the fastest in SBF solution. The ultrasonic-rolling treatment reduced the corrosion rate of the surface of the magnesium alloy sample due to the ultrasonic-rolling process. The surface of the magnesium alloy then formed a high-density plastic deformation layer. The crystal size was refined versus the original magnesium alloy sample, and the grain boundary was significantly increased. Many grain boundaries blocked the continuous erosion of the sample and prevented the development trend of tiny cracks and the widening of the etched holes in the substrate [36]. Refs. [37–39] pointed out that the surface roughness and grain size of the alloy significantly affect the corrosion resistance, and the nanostructured surface grains enhance the formation of the surface passivation layer, thereby improving the corrosion resistance of the material. Therefore, the surface roughness and grain refinement of the samples after ultrasonic rolling improved the corrosion resistance of the samples. The ion-implantation process after ultrasonic rolling further delayed the corrosion reaction on the surface of the sample, and the degradation rates of the samples in SBF and NaCl solutions were 62.45% and 58.47% lower, respectively, than those of the mechanically ground samples. In addition to grain refinement and compressive stress after ultrasonic rolling, N ions, as an interstitial element, formed an interstitial solid solution after implantation, which made it easy to form an amorphous surface and improve the resistance to pitting corrosion. After ionization and acceleration of ion implantation, high-energy ions were implanted into the surface of the workpiece, and a series of collisions occurred with atoms and electrons near the surface, generating strong energy and resulting in changes in the structure and organization of the effective processing layer. The modified layer was composed of the compounds MgO and Mg_3N_2 with very good corrosion resistance. When the implantation dose reached a certain critical value, the implanted layer became disordered, and the structure had good anti-oxidation and anti-corrosion ability [40,41]. The residual pressure increased the strength of the $Mg(OH)_2$ protective film due to the large residual stress generated after rolling and ion implantation; thus, it improved the corrosion resistance [42].

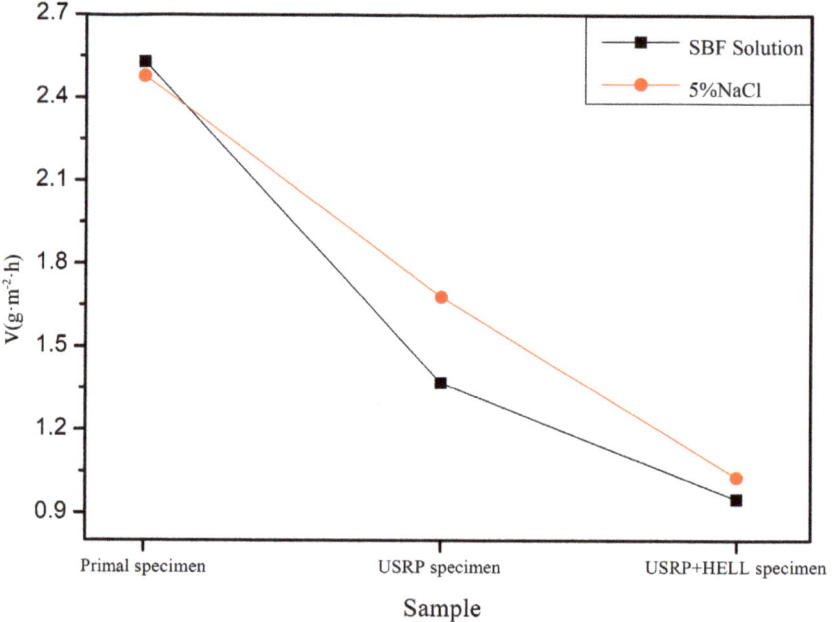

Figure 7. Corrosion information of magnesium alloy specimens after immersion experiments in two solutions.

In summary, the comprehensive effects of surface roughness, grain refinement, and residual compressive stress after USRP + HELL composite process further improved the corrosion resistance of magnesium alloy surfaces, thus indicating that the USRP + HELL composite is the best way to improve the friction and corrosion resistance of magnesium alloys. This is an effective method, and it is worth further studying the effect of composite treatment process parameters on the microstructure and properties of magnesium alloys.

4. Conclusions

In this study, magnesium alloys were processed via different treatment processes. The results showed that compared with mechanical grinding and ultrasonic rolling, the magnesium alloy treated by the USRP + HELL composite process had the lowest surface roughness, the highest hardness, the lowest friction coefficient, and the best wear resistance, and the adhesion wear on the surface was greatly improved. The reason for the improved friction and wear performance was due to the combined effect of the appearance of the hardened phase nitride after ion implantation, the surface roughness, and the residual compressive stress. The immersion experiment showed that the USRP treatment process could improve its corrosion resistance, but the corrosion rate was further reduced after the composite treatment process, which was 62.45% and 58.47% lower than that of the mechanically ground samples, and the larger residual stress further improved the corrosion resistance of magnesium alloys. The effects of three different treatment processes on the wear resistance and corrosion resistance of magnesium alloys are discussed, but the effects of USRP process and ion implantation process parameters on the microstructure and properties of magnesium alloys are not discussed. The effects of USRP + HELL process parameters on the surface structure and friction and corrosion resistance of magnesium alloys need to be further studied, and the research results can provide new references and ideas for surface-protection technology.

Author Contributions: Conceptualization, Z.D. and D.Z.; methodology, H.J.; software, H.J.; validation, Z.D., R.A. and T.L.; formal analysis, R.A.; investigation, H.J., R.A. and T.L.; resources, Z.D.; data curation, Z.D.; writing—original draft preparation, Z.D.; writing—review and editing, D.Z.; visualization, D.Z.; project administration, Z.D.; funding acquisition, Z.D. All authors have read and agreed to the published version of the manuscript.

Funding: This research was funded by the Youth Growth Project of The Education Department of Guizhou Province (2022040), the Key Laboratory of Materials Simulation and Computing of Anshun University (Asxyxkpt201803), the Key Supporting Discipline of Materials and Aviation of Anshun College (2020), and the National Undergraduate Innovation and Entrepreneurship Training Program of China (202110667005).

Institutional Review Board Statement: Not applicable.

Informed Consent Statement: Not applicable.

Data Availability Statement: No new data were created or analyzed in this study. Data sharing is not applicable to this article.

Acknowledgments: The author would like to give thanks for the equipment and technical support provided by Mechanical and Electrical Research and Design Institute of Guizhou Province, the Youth Growth Project of The Education Department of Guizhou Province (2022040), the Key Laboratory of Materials Simulation and Computing of Anshun University (Asxyxkpt201803), the Key Supporting Discipline of Materials and Aviation of Anshun College (2020), and the National Undergraduate Innovation and Entrepreneurship Training Program of China (202110667005).

Conflicts of Interest: The authors declare no conflict of interest.

References

1. Esmaily, M.; Svensson, J.E.; Fajardo, S.; Birbilis, N.; Frankel, G.S.; Virtanen, S.; Arrabal, R.; Thomas, S.; Johansson, L.G. Fundamentals and advances in magnesium alloy corrosion. *Prog. Mater. Sci.* **2017**, *89*, 92–193. [CrossRef]
2. Ding, W.; Wu, Y.; Peng, L.; Zeng, X.; Lin, D.; Chen, B. Research and application development of advanced magnesium alloys. *Mater. China* **2010**, *29*, 37–45.
3. Ma, R.; Lv, S.; Xie, Z.; Yang, Q.; Yan, Z.; Meng, F.; Qiu, X. Achieving high strength-ductility in a wrought Mg–9Gd–3Y–0.5Zr alloy by modifying with minor La addition. *J. Alloy. Compd.* **2021**, *884*, 1–12. [CrossRef]
4. Jana, A.; Das, M.; Balla, V.K. Effect of heat treatment on microstructure, mechanical, corrosion and biocompatibility of Mg-Zn-Zr-Gd-Nd alloy. *J. Alloy. Compd.* **2020**, *821*, 153462. [CrossRef]
5. Chen, B.; Wang, D.; Zhang, L.; Geng, G.; Yan, Z.; Eckert, J. Correlation between the crystallized structure of Mg67Zn28Ca5 amorphous alloy and the corrosion behavior in simulated body fluid. *J. Non-Cryst. Solids* **2020**, *553*, 120473. [CrossRef]
6. Zaludin, M.A.F.; Jamal, Z.A.Z.; Derman, M.N.; Kasmuin, M.Z. Fabrication of calcium phosphate coating on pure magnesium substrate via simple chemical conversion coating: Surface properties and corrosion performance evaluations. *J. Mater. Res. Technol.* **2019**, *8*, 981–987. [CrossRef]
7. Zhou, Z.; Zheng, B.; Lang, H.; Qin, A.; Ou, J. Corrosion resistance and biocompatibility of polydopamine/hyaluronic acid composite coating on AZ31 magnesium alloy. *Surf. Interfaces* **2020**, *20*, 100560. [CrossRef]

8. Jamari, J.; Ammarullah, M.I.; Santoso, G.; Sugiharto, S.; Supriyono, T.; Prakoso, A.T.; Basri, H.; van der Heide, E. Computational contact pressure prediction of CoCrMo, SS 316L and Ti6Al4V femoral head against UHMWPE acetabular cup under gait cycle. *J. Funct. Biomater.* **2022**, *13*, 64. [CrossRef]
9. Tao, N.R.; Wang, Z.B.; Tong, W.P.; Sui, M.L.; Lu, J.; Lu, K. An investigation of surface nanocrystallization mechanism in Fe induced by surface mechanical attrition treatment. *Acta Mater.* **2002**, *50*, 4603–4616. [CrossRef]
10. Xia, S.; Liu, Y.; Fu, D.; Jin, B.; Lu, J. Effect of surface mechanical attrition treatment on tribological behavior of the AZ31 alloy. *J. Mater. Sci. Technol.* **2016**, *32*, 1245–1252. [CrossRef]
11. Wu, S.X.; Wang, S.R.; Wang, G.Q.; Yu, X.C.; Liu, W.T.; Chang, Z.Q.; Wen, D.S. Microstructure, mechanical and corrosion properties of magnesium alloy bone plate treated by high-energy shot peening. *Trans. Nonferrous Met. Soc. China* **2019**, *29*, 1641–1652. [CrossRef]
12. Liu, H.; Jiang, C.; Chen, M.; Wang, L.; Ji, V. Surface layer microstructures and wear properties modifications of Mg-8Gd-3Y alloy treated by shot peening. *Mater. Charact.* **2019**, *158*, 109952. [CrossRef]
13. Guo, Y.; Wang, S.; Liu, W.; Sun, Z.; Zhu, G.; Xiao, T. Effect of laser shock peening on tribological properties of magnesium alloy ZK60. *Tribol. Int.* **2020**, *144*, 106138. [CrossRef]
14. Liu, Y.; Jin, B.; Li, D.J.; Zeng, X.Q.; Lu, J. Wear behavior of nanocrystalline structured magnesium alloy induced by surface mechanical attrition treatment. *Surf. Coat. Technol.* **2015**, *261*, 219–226. [CrossRef]
15. Liu, M.E.; Sheng, G.M.; Yin, L.J. Effects of high energy shot peening for magnesium alloy AZ31 on the corrosion properties and microhardness. *Funct. Mater.* **2012**, *43*, 2702–2704.
16. Fei, Z.; Xuchao, S.G. Effect of Surface Ultrasonic Rolling Treatment on Fatigue Properties of AISI304 Stainless Steel. *Chin. J. Hot. Manuf. Technol.* **2017**, *46*, 136–140.
17. Haiquan, Z.; Peiquan, G. Experimental Research on Surface Rolling Strengthening of ZK60 Magnesium Alloy. *Chin. J. Manuf. Technol. Mach. Tool* **2020**, *02*, 93–97.
18. Xilian, Y.; Yu, Z.; Wenting, H.; Meihong, H. Effect of ultrasonic surface rolling on friction and wear properties of AZ31B magnesium alloy. *Spec. Cast. Non-Ferr. Alloy.* **2020**, *40*, 1214–1218.
19. Yangyang, T.; Linbo, L.; Chao, W.; Zhao, F.; Weibo, M. Research Status of Ultrasonic Surface Rolling Nanotechnology. *Chin. J. Surf. Technol.* **2021**, *50*, 160–169.
20. Lei, M.K.; Li, P.; Yang, H.G.; Zhu, X.M. Wear and corrosion resistance of Al ion implanted AZ31 magnesium alloy. *Surf. Coat. Technol.* **2007**, *201*, 5182–5185. [CrossRef]
21. Ba, Z.; Jia, Y.; Dong, Q.; Li, Z.; Kuang, J. Effect of Zr ion implantation on wear and corrosion resistance of magnesium alloy. *J. Mater. Heat Treat.* **2019**, *40*, 135–142.
22. Fei, C.; Hai, Z.; Suo, C.; Fanxiu, L.; Chengming, L. Corrosion resistance properties of AZ31 magnesium alloy after Ti ion implantation. *Rare Met.* **2007**, *26*, 142–146.
23. Liu, H.; Xu, Q.; Jiang, Y.; Wang, C.; Zhang, X. Corrosion resistance and mechanical property of AZ31 magnesium alloy by N/Ti duplex ion implantation. *Surf. Coat. Technol.* **2013**, *228* (Suppl. 1), S538–S543. [CrossRef]
24. WenHua, X.; ShouZhou, W. Effect of surface nanocrystallization pretreatment on tribological properties of nitrogen layers of 316L stainless steel. *Mater. Prot.* **2017**, *50*, 23–27.
25. Zhao, X.H.; Nie, D.W.; Xu, D.S.; Liu, Y.; Hu, C.H. Effect of gradient nanostructures on tribological properties of 316L stainless steel with high energy ion implantation tungsten carbide. *Tribol. Trans.* **2019**, *62*, 189–197. [CrossRef]
26. ASTM G31-72. Standard Practice for Laboratory Immersion Corrosion Testing of Metals. ASTM: West Conshohocken, PA, USA, 1990.
27. Jialong, Z.; Liwei, L.; Wei, K.; Bo, C.; Minhao, L.; Yutian, F.; Min, M. Effect of cryogenic treatment on microstructure and mechanical properties of AZ31 magnesium alloy Sheet after Rolling. *J. Plast. Eng.* **2010**, *29*, 126–133.
28. Zhao, X.H.; Liu, K.C.; Xu, D.S.; Liu, Y.; Hu, C.H. Effects of ultrasonic surface rolling processing and subsequent recovery treatment on the wear resistance of AZ91D Mg alloy. *Materials* **2020**, *13*, 5705. [CrossRef]
29. Yu, H.; Xin, Y.; Wang, M.; Liu, Q. Hall-petch relationship in Mg alloys: A review. *J. Mater. Sci. Technol.* **2018**, *34*, 248–256. [CrossRef]
30. Andani, M.T.; Lakshmanan, A.; Sundararaghavan, V.; Allison, J.; Misra, A. Quantitative study of the effect of grain boundary parameters on the slip system level Hall-Petch slope for basal slip system in Mg-4Al. *Acta Mater.* **2020**, *200*, 148–161. [CrossRef]
31. Ye, H.; Sun, X.; Liu, Y.; Rao, X.; Gu, Q. Effect of ultrasonic surface rolling process on mechanical properties and corrosion resistance of AZ31B Mg alloy. *Surf. Coat. Technol.* **2019**, *372*, 288–298. [CrossRef]
32. Li, P.; Han, X.G.; Xin, J.P.; Zhu, X.P.; Lei, M.K. Wear and corrosion resistance of AZ31 magnesium alloy irradiated by high-intensity pulsed ion beam. *Nucl. Instrum. Methods Phys. Res.* **2008**, *266*, 3945–3952. [CrossRef]
33. Mena-Morcillo, E.; Veleva, L. Degradation of AZ31 and AZ91 magnesium alloys in different physiological media: Effect of surface layer stability on electrochemical behaviour. *J. Magnes. Alloy.* **2020**, *8*, 667–675. [CrossRef]
34. Wang, L.; Shinohara, T.; Zhang, B. Corrosion behavior of Mg, AZ31, and AZ91 alloys in dilute NaCl solutions. *J. Solid State Electrochem.* **2010**, *14*, 1897–1907. [CrossRef]
35. Xin, R.; Li, B.; Li, L.; Liu, Q. Influence of texture on corrosion rate of AZ31 Mg alloy in 3.5 wt.% NaCl. *Mater. Des.* **2011**, *32*, 4548–4552. [CrossRef]

36. Jinzhong, L.; Shijie, J.; Liujun, W.; Kaiyu, L. Effect of Laser Shock-Ultrasonic Rolling Composite Process on Mechanical Properties of AZ91D Magnesium Alloy. *Chin. J. Jilin Univ. Eng. Technol. Ed.* **2020**, *50*, 1301–1309.
37. Li, Y.; Zhang, T.; Wang, F.H. Effect of micro crystallization on corrosion resistance of AZ91D alloy. *Electrochim. Acta* **2006**, *51*, 2845. [CrossRef]
38. Pandey, V.; Singh, J.K.; Chattopadhyay, K.; Srinivas, N.C.S.; Singh, V. Influence of ultrasonic shot peening on corrosion behavior of 7075 aluminum alloy. *J. Alloy. Compd.* **2017**, *723*, 826–840. [CrossRef]
39. Ye, W.; Li, Y.; Wang, F. The improvement of the corrosion resistance of 309 stainless steel in the transpassive region by nano-crystallization. *Electrochim. Acta* **2009**, *54*, 1339–1349. [CrossRef]
40. Yanzhang, L.; Shaoyu, Q.; Xiaotao, Z.; Li, W.; Xinquan, H. A Study of Effect of Nitrogen Implantation on Corrosion Properties of Ti-Al-Zr Alloy. *Nucl. Phys. Rev.* **2006**, *23*, 202–206.
41. Xuewei, T.; Zhangzhong, W.; Zhixin, B.; Shixiao, K.; Qixiang, H. Research progress on surface modification technology of magnesium alloy ion implantation. *Mater. Rev.* **2014**, *28*, 112–115.
42. Heng, C.; Lin, L. Influence of residual stress on local corrosion behavior of metal materials. *Chin. J. Eng. Sci.* **2019**, *41*, 929–939.

Review

Influence of Laser Modification on the Surface Character of Biomaterials: Titanium and Its Alloys—A Review

Joanna Sypniewska * and Marek Szkodo

Faculty of Mechanical Engineering and Ship Technology, Gdansk University of Technology, Narutowicza, 11/12, 80-233 Gdansk, Poland
* Correspondence: joanna.sypniewska@pg.edu.pl; Tel.: +48-608-458-720

Abstract: Laser surface modification is a widely available and simple technique that can be applied to different types of materials. It has been shown that by using a laser heat source, reproducible surfaces can be obtained, which is particularly important when developing materials for medical applications. The laser modification of titanium and its alloys is advantageous due to the possibility of controlling selected parameters and properties of the material, which offers the prospect of obtaining a material with the characteristics required for biomedical applications. This paper analyzes the effect of laser modification without material growth on titanium and its alloys. It addresses issues related to the surface roughness parameters, wettability, and corrosion resistance, and discusses how laser modification changes the hardness and wear resistance of materials. A thorough review of the literature on the subject provides a basis for the scientific community to develop further experiments based on the already investigated relationships between the effects of the laser beam and the surface at the macro, micro, and nano level.

Keywords: laser modification; titanium; titanium alloys

Citation: Sypniewska, J.; Szkodo, M. Influence of Laser Modification on the Surface Character of Biomaterials: Titanium and Its Alloys—A Review. *Coatings* **2022**, *12*, 1371. https://doi.org/10.3390/coatings12101371

Academic Editor: Matic Jovičević-Klug

Received: 21 August 2022
Accepted: 8 September 2022
Published: 20 September 2022

Publisher's Note: MDPI stays neutral with regard to jurisdictional claims in published maps and institutional affiliations.

Copyright: © 2022 by the authors. Licensee MDPI, Basel, Switzerland. This article is an open access article distributed under the terms and conditions of the Creative Commons Attribution (CC BY) license (https://creativecommons.org/licenses/by/4.0/).

1. Introduction

Surface modifications of materials provide the base for achieving specific material properties for specific applications. The main advantage of laser surface modification is the ability to improve the properties of different materials [1,2]. In addition, the precision, elasticity, the possibility of parameters control, and the repeatability of this process are indicated as an advantage, as is the small heat-affected zone resulting from the modification [3,4]. Because of the high degree of process sophistication, it is also possible to carry out the process without direct human intervention, thus reducing the risk of negative effects on the body [5].

Researchers mainly use lasers such as a femtosecond laser [6], Nd: YAG laser [7–10], CO_2 laser [11], diode [12,13], and fiber laser [14]. Firstly, they provide the possibility to carry out modification processes by adding a coating of the same or another material. Moreover, they are used for modifications with no further addition of materials [1]. Figure 1 presents a classification of laser machining based on an increase or no increase in material. This article is primarily concerned with laser remelting and texturing, as these two techniques are the most commonly used in the modification of biomaterials. Laser remelting is a technique based on remelting the surface of a material to change its morphology and structure without a specific modification objective, to generally improve the properties affected by a laser beam or, for example, when there is talk of modifying the density of the material or the hardness [7,15–17]. Laser texturing involves melting the material and then cooling it to produce a specific pattern on the surface of the material [18]. A combination of laser hardening and laser texturing is also found in the literature [19]. Laser hardening is a technique aimed at improving the hardness of materials using a laser beam. Furthermore, it has the advantage of being able to increase the wear resistance and improve fatigue properties [20]. Surface modification with no addition of a material, which will be discussed in the following work, directly affects a change in wettability and roughness, corrosion resistance, hardness of the material, and wear resistance. Laser modification is closely related to a change in the surface microstructure and influences the roughness and wettability of materials [21–24]. The advantages of using lasers to modify biomedical materials are the ability to control the effect of heat on the surface structure of the materials due to the local action of a laser pulse on the surface, the small heat-affected zone, and the fact that the process is clean and does not cause material loss [25,26].

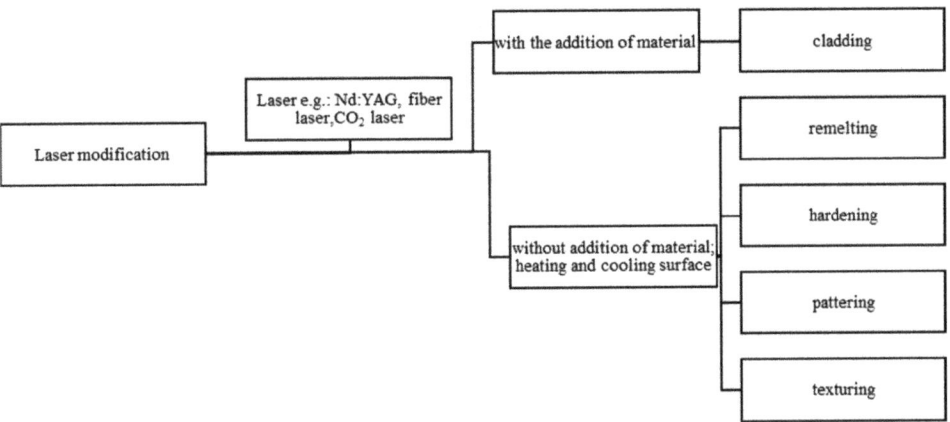

Figure 1. Graphic interpretation of laser modification possibilities based on [1,13,20–22].

The process of the heat treatment of materials by laser is mainly related to the treatment of metallic materials such as aluminum, steel, and its alloys, as well as titanium [2,3,27,28]. Many research efforts are currently focused on gaining a thorough understanding of the properties of titanium and its alloys. Further, efforts are being made to improve these properties to produce a material dedicated to applications.

Titanium and its alloys are widely used in various industries due to their low density. This family of materials has its uses in the manufacture of parts for motorcycles, and sports vehicles to reduce their weight [29,30]. In the aerospace industry, these materials are the third group of materials, after nickel-based materials, in terms of their frequency of use, not only because of their low density, but mainly because of their resistance to corrosion and high temperatures (e.g., alpha alloys and α/β alloys) [31–33]. Considerable attention is given to this material when it comes to medical applications [34,35]. In applications in the field of dental implantology, titanium alloys are used for the production of dental crowns and bridges primarily due to the reduced risk in an allergic reaction and they have a beneficial effect on the process of osseointegration [36,37]. The use of titanium and its alloys is based on its high corrosion resistance, better than steel and cobalt–chromium alloys, owing to the properties of self-assimilation, which are beneficial due to the environment of the body in which the implants are placed [38–41]. In the case of titanium and its alloys, it is said to have little effect on the human body. Pure titanium is a biocompatible material, as are all of its alloys. The alloying elements used to produce titanium alloys, however, do pose a problem, but it should be noted, that titanium and its alloys have hemolytic indices below a value that would indicate the possibility of the formation of embolisms or clots as a result of the presence of this material in the body [42]. Studies by Chen et al. indicate that the presence of vanadium and aluminum in the human body leads to disorders of the nervous system, brain diseases, and circulatory diseases, while also affecting the softening of bone tissue; however, it is indicated that alloys containing aluminum and vanadium ($\alpha + \beta$ alloys) have mechanical properties on the same level as β alloys. In implantology, aluminum–vanadium alloys are used primarily to produce fracture fixation plates, spinal column components, and connecting elements such as rods, wires, and screws [43]. Review papers over the years related to modifications using laser beams have reported on the benefits of modification for industrial applications (matrix-enhanced laser modification) [44], but they mainly confirm the benefits for biomedical applications [45,46]. Paper [45] focuses on the effect of laser surface texturing on the antimicrobial properties and biological activity of titanium, while a 2005 article [46] provides an overview of the possible types of surface modifications available for various biomaterials and discusses in detail laser modification in the context of the biocompatibility of the modified material. The present review focuses on the mechanical properties of the surface of titanium and its alloys after laser modification. Attention is paid to the properties directly related to the requirements for biomedical materials, such as an adequate material hardness, wear, corrosion resistance, and affinity of the modified surface to bone-forming cells.

A literature review on the effects of laser modification on titanium and its alloys was conducted to present the current state of the art in this field and to highlight the topicality of the problem of the laser modification of titanium and its alloys for biomedical applications due to the requirements that are placed on biomaterials.

2. Methods

This systematic review used the databases: ResearchGate, Science Direct, and Scopus. Google Scholar was also used to analyze the trends in the laser modification of titanium and its alloys. The searching strategy was based on the "laser modification", "laser treatment", "laser remelting", "laser surface modification of titanium and its alloys", and "modification of titanium and its alloys" terms. A bibliography of 175 literature references was collected and extracted from over 200 collected papers. Figure 2 shows the number of articles from each year. The literature review was based mainly on works from 2019 to 2022, which allowed us to discuss issues according to the current state of knowledge on the laser surface modification of titanium and its alloys for selected parameters and properties. The paper discusses the effect of the laser treatment of titanium and its alloys on the surface roughness and wettability, corrosion resistance, and hardness with the indications of micro and macro hardness and wear resistance.

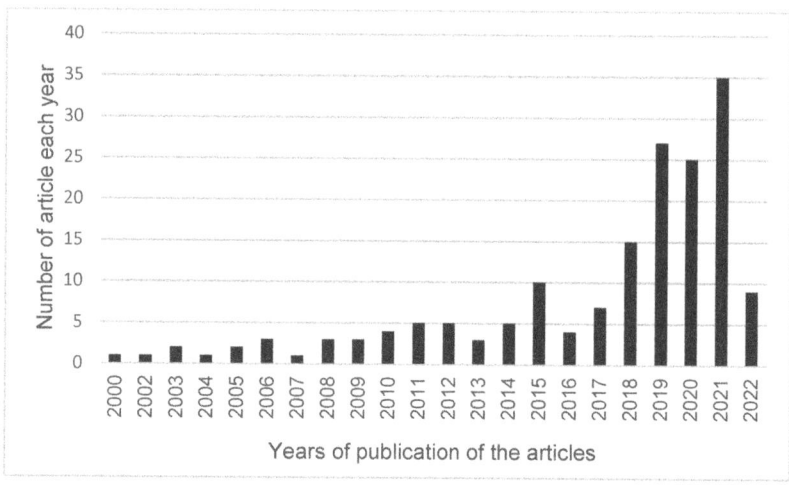

Figure 2. Distribution of the number of articles used per year.

3. Roughness and Wettability

Roughness and surface wettability are very often compared by authors to determine the relationship between these properties [47,48]. The study of these parameters in the case of titanium modifications is very important due to the use of titanium in biomedicine. The evaluation of the roughness parameter is important because an increase in this index has a positive effect on the cell adhesion during osseointegration and on the connection between the bone and an implant [49]. The use of titanium and its alloys in biomedicine is indicated for the manufacture of artificial hip joints, artificial knee joints, bone plates, fracture fixation screws, prosthetic heart valves, pacemakers, and the production of artificial hearts [50,51]. Depending on the area of aspiration of titanium and its alloys, a different surface roughness is required of a biomaterial. For example, heart valve implants, the artificial heart, and other components of the circulatory system must be made of a medium to low-roughness titanium to avoid the formation of areas where blood cells could agglomerate and form blockages, thus stopping the blood flow [52]. In addition, titanium itself possesses anti-thrombogenic properties, and for implant applications, it is necessary to improve the properties of the structure. In the case of osseous dental implants, it is said that components with a high degree of surface roughness are necessary due to the osseointegration process that occurs. An increase in the surface roughness is associated with an increase in the growth surface of osteoblasts and protein polyfusion [53]. Menci et al. [24] point out that laser modification with different types of lasers (e.g., a fiber laser or Nd: YAG laser) is necessary, as the hip implants in question are composed of several elements and each part must have different surface properties. The acetabulum and femoral stem need to be rougher to stimulate bone osseointegration, while the femoral head and distal part of the femoral stem need to be smoother to minimize wear.

In paper [49] it was shown that the roughness of the pore walls created by laser modification was characterized by the roughness parameters, Sa and Sq, with a higher value than on the inter-pore surfaces. The values of Sa and Sq for the inter-pore surfaces were 0.029 µm and 0.04 µm, respectively, while those for the pore walls were 0.126 and 0.149 µm, respectively. The authors concluded from their studies on bone deposition that a higher pore roughness improves the migration of bone-forming cells and also increases the possible surface area for osteoblast attachment to the implant. Furthermore, a high surface roughness is desirable due to increased biointegration and a decreased risk of implant rejection in the body [8].

In papers [8,54–60], an Nd: YAG laser was used to modify titanium alloys and pure titanium. In Table 1, the Nd: YAG laser operating parameters are presented for the literature reviewed subject. In the study, the Nd: YAG laser modification was performed for different laser operating parameters. The effects of varying laser operating parameters on the surface roughness and wettability, which are directly related to the phenomenon of bone cell adhesion, were investigated in a study from 2013, where Györgyey et al. showed [58] a decrease in the R_a roughness parameters using a frequency-doubled Q-switched Nd: YAG laser and a KrF excimer laser. In a 2020 editorial [8] it was observed that the modification of the titanium alloy Ti13Nb13Zr and pure titanium CP-Ti using an Nd: YAG laser caused an increase in the surface roughness. In addition, it was shown that the higher the laser power, the higher the roughness parameters. Additionally, in research work [8] it was pointed out that titanium alloys were rougher than pure titanium before and after laser treatment. Research of [57,61–63] also confirmed that laser modification, regardless of the type of laser, along with increasing the laser parameters, increased the roughness. In paper [57] it was shown that laser processing increased the roughness parameters by approximately 2.8 to 7.5 times at different frequencies relative to the native material. AFM surface texture tests at 10 Hz and 7 Hz successively yielded R_a parameters of 394.35 nm and 279.53 nm, respectively, while this parameter for the reference sample of the Ti6Al4V titanium alloy was 127.20 nm [57].

Table 1. Parameters of Nd: YAG laser modification to improve the roughness of titanium and its alloys.

Laser	Material	Environment	Laser Output Energy	Voltage	Laser Pulse Duration	Irradiation Time	Scanning Speed	Frequency	Average Expansive Area	Fluency/ Power	Shot of Pulses	Roughness Parameters	Reference
Nd: YAG	Ti13Nb13Zr and CP-Ti Ti15Mo and cpTi grade 2	air/argon	5; 30 mJ	-	150 ps	5s	-	10 Hz	-	-	50	(Table 2)	[8]
	technically pure VT-1-00 titanium	Air	-	-	-	-	0–200 mm/s	20–35 Hz	14 mm²	1.9 J/cm³	-	$R_a = 3.95$ μm $R_z = 21.2$ μm $R_{max} = 29.01$ μm $S_m = 91$ μm	[54]
		Air	-	250–400 V	3–10 ms	-	-	1–5 Hz	-	-	1–5		[55]
	Ti6Al4V	Argon	-	-	-	-	30 mm/min	-	-	200 W	-	$R_a = 0.45$ μm	[56]
	CP4) Ti rods	Air	-	-	7 ms	-	1 mm/s	1, 3, 5, 7, 10, 15, 20 Hz	-	average power 300 W; peak power 2.1 kW 1–1.5 J/cm³	-	R_a for 10 Hz = 0.394 μm R_a for 7 Hz = 0.127 R_a μm	[58]
	Ti6Al4V	Air	95 mJ	-	-	120 ns	-	10 Hz	-	0.95 W	-		[59]
	Ti45Nb	Air, argon, nitrogen	-	-	150 ps	5 ns; 15 s	-	10 Hz	7.1×10^{-4} cm²	0.13–0.38 J/cm²	50, 150	(Table 2)	[60]

Table 2. Surface roughness depended on the laser modification environment [8,60].

Sample [60]									
Laser modification environment	Argon				Air			Nitrogen	
Laser modifications parameters	5 mJ 50 pulses	5 mJ 150 pulses	15 mJ 150 pulses	5 mJ 50 pulses	5 mJ 150 pulses	15 mJ 150 pulses	5 mJ 50 pulses	5 mJ 150 pulses	15 mJ 150 pulses
Surface roughness [μm]	1.026	2.053	1.414	0.551	3.062	0.921	0.884	0.949	0.839	1.148	0.647	0.804
Sample [8]									
	Ti13Nb13Zr								
Laser modification environment	Argon		Air		Argon				
Laser modifications parameters	5 mJ 5 s	30 mJ 5 s	5 mJ 5 s	30 mJ 5 s	5 mJ 5 s	30 mJ 5 s			
Surface roughness [μm]	0.428	0.988	0.701	1.366					
Sample [8]									
	CP-Ti								
Laser modification environment	Argon		Air		Argon				
Laser modifications parameters	5 mJ 5 s	30 mJ 5 s	5 mJ 5 s	30 mJ 5 s	5 mJ 5 s	30 mJ 5 s			
Surface roughness [μm]	0.258	0.931	0.927	1.842					

The references state that the laser treatment of titanium surfaces allows for controlling the roughness parameters, R_a, R_z, and R_{max}, in a wide range [55]. The use of a modern fiber laser engraving method [64] allowed the achievement of a surface with a higher roughness than the Ti6Al4V native material. Meanwhile, comparing the results of the roughness measurements at different laser operating parameters, it was observed that with an increase in the laser operating parameters (e.g., groove distance and frequency), there was a decrease in the value of the roughness and an increase in the value of the wetting angle, which was associated with the determination of the surface of the samples as hydrophobic.

Studies [8,60,65] also indicate that the operating environment of the Nd: YAG laser affects the surface roughness and morphologies. The results of EDS [8] have shown that the presence of argon affects the reduction of oxygen molecules for both commercially pure titanium and Ti13Nb13Zr titanium alloy samples; however, the modification carried out in the presence of air caused a strong passivation of the coating and, thus, increased the amount of oxygen on the surface, while the surface roughness for the modified surfaces in the presence of air was lower. Conducting the modification in the presence of argon for each parameter variant results in the roughest surface (Table 2). The laser modification in a nitrogen environment for low power and number of pulses presented in work [60] showed a higher roughness than for the same modification in the presence of air, while in the case of a modification in the presence of argon the modified surfaces—for different laser powers of 5 mJ and 15 mJ and the number of pulses of 50 and 150—were characterized by the highest roughness among all the modified surfaces. The conclusions in [60] are confirmed in [65], where the reason for the lower surface roughness after modification in a nitrogen environment was due to the formation of titanium and nitrogen compounds and the formation of a smaller heat-affected zone. The authors of that article also indicated that for a smaller number of pulses performed, the surface roughness was lower.

Lawrence et al. [56] 2006, presented the idea that laser processing increases the roughness parameters and improves the surface wettability. In Table 3, the literature on the results of the influence of a laser treatment on the character of the surface wettability is summarized. Menci et al. [24] searched for a direct relationship between the contact angle and the Sa parameter for Ti11.5Mo6Zr4.5Sn titanium alloy samples modified by an Nd: YAG laser and fiber laser for different parameters. Wenzel's claim was referred to, which states that increasing the roughness parameter (r) increases the wettability of the surface (where θy is the contact angle of an ideal flat surface and θw is the contact angle of a rough surface) [66]:

$$cos\theta_w = rcos\theta_y$$

The realization of this equation in the work of [24] indicates hydrophilic surface properties because the initial properties of titanium were improved by changing the surface texture (increase in roughness) and there was an increase in droplet diffusion [67].

Table 3. Wettability results on titanium and its alloys observed in the article.

Surface Property	Author, Year, and Reference	Short Conclusion
	Hao et al., 2005 [68]	Observation and study of the surface of laser-modified Ti6Al4V alloy showed an increase in the surface wettability which is beneficial for medical applications of titanium alloys. The increase in the contact angle after laser modification is a result, according to the authors, of an increase in the surface energy of the modified material and an increase in roughness parameters.
	Lawrence et al., 2006 [56]	Lawrence et al. demonstrated that laser treatment improves the surface wettability as a result of a change in the surface energy, an increase in the oxygen content, and an increase in the surface roughness. The cell studies carried out revealed an increase in bone cell adhesion and proliferation for Ti6Al4V titanium alloy samples subjected to laser modification, compared to a titanium alloy without modification.

Table 3. Cont.

Surface Property	Author, Year, and Reference	Short Conclusion
Wettability	Cunha, A. et al., 2013 [69]	Obtaining an anisotropic surface by modification is beneficial for controlling the surface wettability and this property is also indicated to improve stem cell adhesion.
	May et al., 2015 [70]	The surface anisotropy of titanium alloy Ti6Al4V subjected to fiber-laser system modification was demonstrated. Wetting angles were smaller for the measurement performed perpendicularly for each laser operating frequency. The formation of contact anisotropy after laser modification was related to the frequency of the laser work, and increasing this parameter decreased the anisotropy.
	Raimbault O. et al., 2016 [71]	The paper focused on the bioactivity of cells towards a femtosecond laser-modified surface but also examined the wettability of the Ti6Al4V titanium alloy, which was determined by measuring the contact angle. It was shown that the storage medium had a great influence on the change of the wettability characteristics of the modified samples. Samples stored in boiling water were slower to change their character to hydrophobic ones due to the slowing down of the passivation process, and the atmospheric environment accelerated these changes.
	Rotella 2017 [72]	The authors used three methods for the surface modification of titanium alloy Ti6Al4V, one of them was a femtosecond laser treatment. The hydrophobic character of the laser-modified samples was observed, but at the same time, it was pointed out that this was not a disadvantage of such a surface because it gives, in a long-term context, a chance for a stronger bonding of the cells with the laser-modified implant.
	Lu et al., 2018 [73]	A laser treatment at different laser fluence values was applied to pure titanium samples and then they were chemically treated. For each of the modification combinations, it was shown that the femtosecond laser modifications decreased the contact angle immediately after the laser treatment, while the contact angle increased after the modification. The possibility to obtain a stable structure with hydrophobic properties was pointed out in the paper as the most important advantage of such a modification.
	Pires et al., 2017 [54]	The Nd: YAG laser treatment produced a superhydrophilic surface. The laser-modified surface consisted of more oxygen, which was one of the factors influencing the change in surface wettability. It was indicated that the use of this type of laser allows for the control of parameters important for bone cells.
	Menci et al., 2019 [24]	In this study, a laser beam modification was performed using two different types of Nd: YAG laser and fiber laser, for different laser wavelengths. It was shown that a fiber laser processing ns 1064 nm produces the highest surface roughness with the greatest reduction in wetting angle. The paper also presents the possibility of using individual lasers with specific parameters to process specific implant components because of the roughness that can be achieved with them.
	Murillo et al., 2019 [74]	It was observed that immediately after the modification of a Ti6Al4V titanium alloy with a UV ns laser and IR-fs pulsed lasers, the surface exhibited hydrophilic properties. In this study, the effect of the sample holding environment of titanium on material aging was investigated.

Table 3. Cont.

Surface Property	Author, Year, and Reference	Short Conclusion
	Shaikh et al., 2019 [75]	In this study, a decrease in the contact angle was observed for Ti6Al4V titanium alloy samples which underwent laser modification. It was also observed that the surface of the samples after laser treatment became hydrophilic immediately after the modification; however, during the storage of the material, the contact angle was tested again, and the results showed a change in the surface character toward a hydrophobic one. The authors suggested that this could be due to the oxidation of the modified film as well as contaminants deposited on the sample (the samples were stored in an atmospheric environment).
	Dou et al., 2020 [76]	An increase in hydrophilicity with an increasing laser fluence was observed. The surface hydrophilicity was not stable, and the wettability of the surface changed to hydrophobic properties with time. The need for research on the stability of surface hydrophilicity was indicated.
	Wang et al., 2021 [77]	A 355 mm UV laser modification of commercially pure titanium was carried out. In this study, the possibility of controlling the wettability by light and sample heating was demonstrated. The samples showed a superhydrophilic surface immediately after laser modification.
	Mukherjee et al., 2021 [78]	In this paper, the laser modification of titanium alloy Ti6Al4V using a Yb-doped fiber laser was carried out. It was shown that the surface produced by the laser was anisotropic, which revealed that the contact angle for water was different for a parallel and perpendicular incidence of a drop on the surface. It was shown that in the direction parallel to the laser beam direction, the wetting of the surface was higher as a result of droplet propagation along the corrugation grooves.
	Wang et al., 2021 [79]	A Ti6Al4V titanium alloy was modified with a UV laser at a wavelength of 355 nm. The results of the contact angle measurements were presented for three conditions: for the untreated sample (hydrophilic surface), the sample after laser treatment (superhydrophilic surface), and the sample modified with a laser and additionally subjected to a heat treatment (hydrophobic surface). For the same samples, an erosion test was performed and it was observed that the fastest erosion process occurred for the laser-modified samples and the slowest for those with a hydrophobic surface.
	Singh et al., 2021 [80]	In this study, a CO_2 laser modification was carried out on titanium Ti6Al4V alloys. After the modification, the values of the contact angle and surface energy were investigated. It was found that for the laser-modified surface, the contact angle was higher than for the unmodified samples, and the surface energy also increased. It was also found that a decrease in the surface energy resulted in a decrease in the affinity of the modified surface for bacteria, which is beneficial for the potential use of the material in implant production.
	Li et al., 2022 [81]	Pure titanium samples were subjected to laser surface texturing. Wetting angle studies were carried out using distilled water and modified-simulated body fluid (m-SBF). For both fluids, the laser-modified surface showed an increased wettability. It was indicated that the drops on the structure with a higher roughness realized Wenzel's law, which explained the decrease in the contact angle. The wetting angle for the water was higher than for the m-SBF, which gave information that cells would grow better on such a substrate.

The literature identifies four terms for surface wettability: superhydrophilic, hydrophilic, hydrophobic, and superhydrophobic, concerning a drop of water falling on a surface Figure 3 [59].

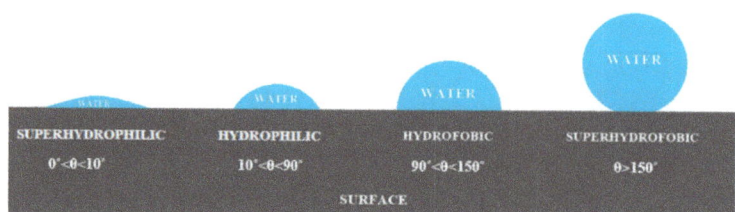

Figure 3. Categories of wettability based on [59,66].

The researchers, Lawrence et al. [56], Menci et al. [24] and Pries et al. [54], treated titanium alloys successively with an Nd: YAG and Yd: YAG laser, and based on contact angle studies, they showed that a laser modification increases the wettability of the sample surface successively for fluids such as human blood, human blood plasma, glycerol and 4-acetanol [56] and water [24]. A decrease in the wetting angle is also shown in Figure 4 for the modification made with the Nd: YAG laser [82]. Pries et al. [54] revealed that the contact angle measured after a laser treatment was 0°. Singh et al. [80] on the other hand, observed that the laser modification of a titanium alloy resulted in a decreased surface roughness, resulting in coatings with a lower wettability than unmodified samples.

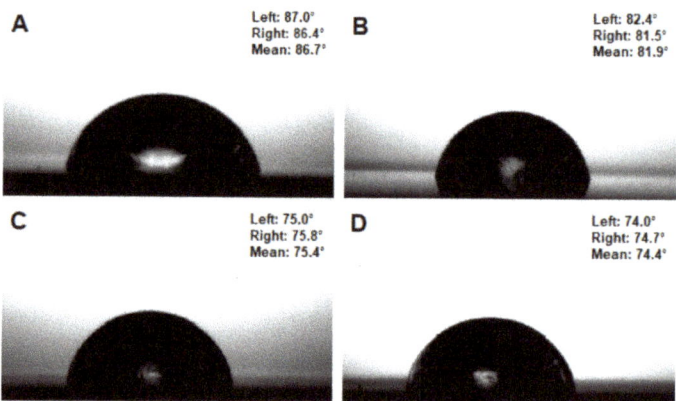

Figure 4. Wettability of material for (**A**) base material Ti13Nb13Zr, (**B**) laser modified sample 700 W, (**C**) laser modified sample 1000 W, and (**D**) laser modified sample 700 + 1000 W [82].

Researchers *Lawrence et al.* [56] *Menci et al.* [24] and *Pries et al.* [54] treated titanium alloys successively with Nd: YAG and Yd: YAG laser, based on contact angle studies they showed that the laser modification increases the wettability of the sample surface successively for fluids such as human blood, human blood plasma, glycerol and 4-acetanol [56] and water [24]. The decrease in wetting angle is also shown in Figure 4 for the modification made with the Nd: YAG laser [82]. *Pries et al.* [54] revealed that the contact angle measured after laser treatment is 0°. *Singh et al.* [80] on the other hand, observed that laser modification of titanium alloy resulted in decreased surface roughness resulting in coatings with lower wettability than unmodified samples.

Conversely, in paper [76], a Ti: sapphire chirped-pulse regenerative amplification laser system with a central wavelength of 800 nm, was used to modify a Ti4Al6V alloy. Initially, the investigated surfaces were characterized by the hydrophilic nature of the modified surface. In the experiment, the contact angle measurements were performed cyclically for up to 155 days after the modification. The observations showed that the effect of the modified surfaces changed from hydrophilic to hydrophobic. Analogous results were also reported in the work of [69,74,75,81,83] for the contact angle for water and for Hank's balanced salt solution (HBSS) for titanium and titanium alloys.

In the research of [73], an alloy was laser-treated with a femtosecond laser, followed by an additional hydrothermal treatment and oxidation. The application of extra modifications did not change the previously established theory that states that with time, the contact angle of laser-modified surfaces increases. Raimbault et al. [71] pointed out that the increase in the contact angle after a certain time after a laser treatment increases due to the increasing passivation of the surface. Moreover, this paper shows that the environment of repository-modified samples is important due to the changing character of the sample wetting. Keeping the laser-modified TA6V samples in an air environment was associated with a faster change in the surface character from hydrophilic to hydrophobic, while an increase in the wetting angle for samples stored in boiling water was smaller. In the study of [72], it is depicted that the change in surface wettability from The decrease in wetting angle is hydrophilic to hydrophobic after a laser treatment does not adversely affect cell adhesion to the material. A better way to illustrate the bonding of a surface to a water droplet is to use the Cassie Baxter model, which takes into account the idea that the contact between a droplet and a surface is affected by air trapped in the irregularities [84]; therefore, an increase in the wetting angle is not uniquely associated with a weakening of the bond between the material and the cells. In addition, the bond is also affected by mechanical stress, which is related to the amount of cracking on the material surface [72].

The references of [62,74,82,83] indicate that laser-modified surfaces can exhibit anisotropic as well as isotropic properties depending on the laser operating parameters [69]. The anisotropy of a surface is more favorable due to its higher wettability than for isotropic surfaces and the contact angles for anisotropic surfaces will be different for parallel and perpendicular directions [70,78,85]. Surface modification resulting in an anisotropic surface is very important for the ability to control the wettability of materials [69], and in addition, due to the medical use of titanium and its alloys, the development of an anisotropic surface is beneficial because of improved osseointegration conditions [47]. It is indicated that the wettability character of laser-modified surfaces can be controlled and changed by external factors such as heat [77,79]. In the study of [77], the hydrophilic properties of a titanium alloy surface were observed immediately after laser modification. It was shown that it was possible to manipulate the character of the surface wettability by using the heating of the samples and UV radiation, which allowed the reversible transformation of the surface from hydrophilic to hydrophobic and the other way around.

4. Corrosion

Titanium and its alloys are characterized by a very high chemical activity. The Pourbaix diagram shows that titanium is a metal that passivates very quickly [86]. The natural, very good corrosion resistance of titanium and its alloys is indicated [87–91]. Additionally, the corrosion behavior is influenced by the presence of β-stabilizing alloying elements and their role in the stability and thickness of the passive layer formed [92]. An important direction of research is the study of the corrosion resistance of titanium bio-alloys [93,94], and to improve this property, it is necessary to limit as far as possible the release of metal ions into the body, since the possibility of adverse health effects due to the presence of, e.g., V or Al in the body has been identified [42,95,96].

It has been shown that titanium and its alloys are stable materials when exposed to environments that react with their surface, but the problem that arises in the root cause of this material is the development of fatigue corrosion, which occurs as a result of the loads on implants. For example, the stress-shielding effect that occurs in implants is due to the greater hardness of titanium than bone. The literature also states that cyclic loads induced by walking affect the corrosion resistance of titanium materials. Moreover, alloying additives in titanium alloys can have negative effects such as vanadium, aluminum, or chromium, whereas alloys with a density of, for example, niobium or tantalum lead to the formation of oxide layers which improve the root resistance of titanium alloys [97].

Laser modification allows the free manipulation of the surface roughness and wettability, which has a significant impact on the corrosion resistance of a material [98,99]. In 2012 [100], corrosion tests on pure titanium and titanium alloy with aluminum and vanadium were performed and based on the corrosion intensity (lA/cm^2), and it was observed that there was a decrease in the corrosion and corrosion-fatigue behavior of the studied materials after laser modification. It was indicated that the reason for the decrease in the corrosion resistance was the presence of residual stresses and small grain sizes in the modified surface, and the resulting microstructure changes differed in potential. The literature indicates that the grain size is affected by the modification of heat [101].

In contrast, the work of [102] cites results from 1984 (Picraux and Pope) 2000 (Suzuki et al.), and 2002 (Yue et al.), which state that laser modification improves the corrosion resistance in, for example, Hank's solution. Travessa et al. [103] showed that laser modification caused significant metallurgical and chemical changes on the surface of an alloy, including the formation of oxides, which resulted in a significant improvement of the corrosion properties compared to metal not subjected to laser treatment. It was also observed, among other things, in the potentiodynamic polarization curves. Navarro [26] showed that a femtosecond laser modification caused changes in the surface of modified samples with different porosities, resulting in an increased impedance (Figure 5). Tests carried out by [103,104] indicated that a thicker oxide or nitride layer formed after laser modification on a material surface, improved the corrosion resistance of a titanium alloy. It was observed that the surface structure after the laser modification, when there were unevenly distributed phase components in it, affected the weakening of the corrosion resistance of the material. The corrosion resistance test carried out in a Ringer's solution Ti6Al4V titanium alloy [105,106], showed an increase in the corrosion resistance in acidic media. Whereas in the papers of [107,108], an increased corrosion resistance of laser-modified samples in Hank's solution and saline solution was shown.

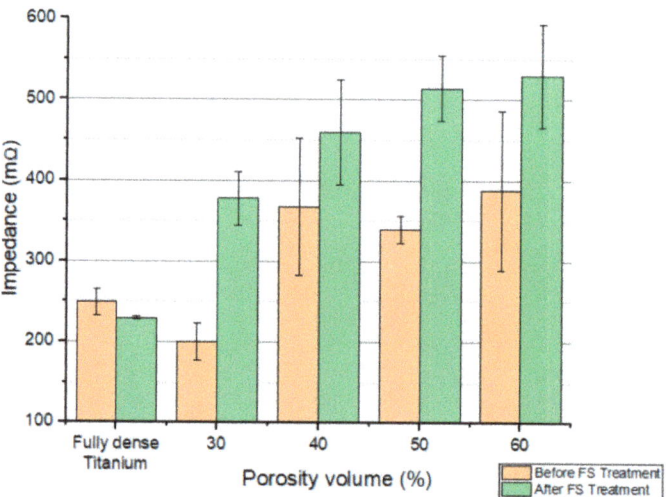

Figure 5. The change of impedance after laser modification for samples with different porosity volume [26].

Researchers in [78,109,110] investigated the effect of laser remelting on pitting and electrochemical corrosion resistance. An analysis of the polarization curves in the work [109] showed an increased resistance to pitting corrosion, which according to the authors was due to a microstructural modification caused by the rapid solidification that occurred during the laser remelting of the surface. Dhara et al. [78] have shown, based on obtained polarization curves, that as a result of the modification, the passive film formed a stable barrier against corrosion. In the paper of [106] it was pointed out that a higher electrochemical resistance was due to the reduction in the volume of the α and β phase, and the thickening of the surface texture. SEM and AFM studies carried out in [108] indicated that the laser modification formed a more reproducible and smoother topography, which increased the corrosion resistance of the material, and a stabilization of the passive layer on the titanium surface was observed, making it less susceptible to further growth [111]. Tests carried out on the chemical composition of the material showed an increase in the presence of nitrides, which acted as a barrier to the ingress of other molecules, and which was considered a condition that could improve the corrosion resistance of a material. An improvement in corrosion resistance by laser ablation was undertaken by [112] in their work, in which they removed the oxide layer, which gave their material a low corrosion resistance, and via the laser ablation, produced a corrosion-resistant oxide layer. This could be observed from an increase in the self-corrosion potential for different energy doses and a decrease in the self-corrosion current, which numerically reflected the corrosion rate of the material; therefore, indicating that the lower the surface corrosion rate the better the corrosion resistance.

5. Hardness

The hardness of titanium and its alloys is reported in the literature to be higher than the hardness of steel on the Vickers scale [113]. For example, titanium alloy Ti6Al4V has a hardness of HV higher (340 HV) than pure titanium (200 HV), which is related to the presence of alloying elements [114,115]. The surface modifications of these materials aim to improve the hardness and mechanical parameters to increase the residence time of an implant in the body [116]. The hardness of the materials used for implants is important because an increase in the material hardness is associated with an increase in the wear resistance [116]. The literature also indicates that the high hardness of a material can adversely affect the behavior of the biomaterial in the body [117]. Laser modification, on the other hand, allows a controlled change in the parameters. Hardness, for example, is widely discussed in the field of orthopedic implants, such as hip and knee implants, because of the need to control the shielding effect of the implant in the bone, and the importance of this parameter, as orthopedic implants are placed under a certain pressure in the body, meaning that the implant must counterbalance this pressure to perform its function properly [118].

A systematic review [119] and research by the authors [10,105,120–122] on the various techniques of the surface modification of titanium have observed an increase in the hardness of materials as a result of laser modification, which has its theoretical basis in [123], where it was indicated that by using a heat source it was possible to control changes in the hardness of materials. Table 4 shows the hardness values obtained for modified titanium alloys and pure titanium using different laser types and parameters. In Figure 6 it is also shown that laser modification improved the nano-hardness of Ti13Nb13Zr alloy samples after a laser modification by an Nd: YAG laser.

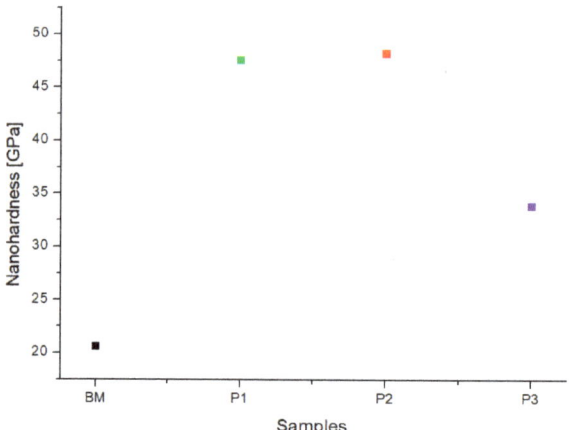

Figure 6. Nano-hardness of laser modified samples. BM: base material, P1: laser modified sample with 800 W and scan rate 60%, P2: laser modified sample with 800 W and scan rate 30%, P3: laser modified sample with 900 W and scan rate 60% [82].

In paper [105], the difference in the hardness value HV between the remelted material and the base material was about 300 HV. The study on the hardness of a titanium alloy was carried out by Khorram et al. [124] which indicated an increase in hardness of 36%. Additionally, Zhang et al. [125] indicated an increase in hardness after laser modification of 60%, while Ushakov et al. [126] determined that the possibility of increasing the Vickers hardness after a laser treatment ranged between 20% and 40%. Moreover, the increase in microhardness was accompanied by an increase in the fracture toughness. The laser modification of porous titanium in the research of [127] resulted in an increase in the hardness at the pore pillars, while an increase in the pore size in the titanium sample resulted in a decrease in the hardness. A significant effect of the pore size on the surface hardening potential by a laser modification was demonstrated.

The change in hardness of samples after laser remelting varies due to the phase transformations occurring during the cooling process [24,109,110]. The increase in material hardness is due to the transformation of the β-phase at high temperatures from about 900 °C to 1050 °C [128], into the α phase and the martensitic α phase, which is hard but very brittle [121,129–131], and the creation of the ω-phase [132,133]. After a laser-induced heat treatment, the formation of a martensitic α-phase was observed as a result of the transformation of the alpha-phase of titanium alloys initially into a beta-phase, and then during the rapid cooling of a material after laser treatment, a martensitic α-phase was formed [134]. Geng et al. [135] indicated that the hardness tests performed after a laser modification of the alloy showed that the measurement of this parameter performed in the β-phase, yielded lower values of hardness than in the α-phase. The results for the β-phase were characterized by large deviations for the elastic modulus due to the small thickness of the β-grains and the presence of α-grains. In studies [136,137] it has been shown that as a result of rapid cooling of a material at the surface after treatment, a martensitic phase with the addition of the β phase is formed, while deep into the material a decrease in the compactness of the martensitic phase is observed, and the α phase tends to become dominant. A comprehensive analysis of the phase transformation in a Ti-64 titanium alloy subjected to an Nd: YAG laser modification was carried out in [138]. The presence of a melted zone and a heat-affected zone after laser treatment were marked and they had different microstructural characteristics. The melted zone was characterized by the presence of martensitic plates throughout, while the heat-affected zone, which was far from the laser source, contained short-rod b particles, martensitic plates and untransformed bulk in its microstructure. It was shown that the change in hardness during laser treatment was influenced by the presence of martensitic plates in the melted zone and a reduction in the grain size. The formation of the martensitic phase limited the diffusion of alloying elements, which directly led to the hardening of the surface structure. The study of [138] presents the preliminary results of values for hardness, using a calculation method and a measurement method. For both methods, the hardness increased with respect to the material without a laser treatment. The hardness for the melted zone was higher than for the heated affected zone and the differences in the hardness values were due to the fact that the calculation method did not take into account the occurrence of coarse α grains in the heated zone, while it was also difficult to take into account the contribution of the individual phases of the material in the zones [138].

The authors of [125] further observed the formation of a large number of dislocations and mesh distortions on the surface of a sample, increasing the hardness of the material. In the melted zone, the formation of nanotwins was observed, which slowed down the dislocation movements, and this had an additional hardening effect on the material [138,139]. It was indicated that reducing the grain size increased the hardness of the material [109]; however, after laser treatment, the grain size of the material increased while the hardness increased, suggesting that the grain size does not significantly affect the hardness as much as the dislocations formed and the phase transformations of the material [140]. In the work of [141], it was indicated that an increase in the amount of oxygen due to a femtosecond laser treatment resulted in an increase in the hardness with a concomitant increase in the brittleness of the material. Applying laser texturing to the surface linearly and performing dimple patterns resulted in an increase in the nano-hardness from 2 GPa for the base material to 4 GPa and 6 GPa, respectively, as tested by nanoindentation. The hardness obtained for each test was dependent on the type of laser selected and the processing parameters chosen. In paper [142], a pulsed laser treatment was performed using two different laser power parameters, and the modified samples were divided into two regions. It was shown that the region of the sample modified with a higher laser power of 3.99 W had a higher hardness than the region modified with a laser power of 1.71 W. In addition, it was shown that for treatment with the lower laser power, no significant difference in the hardness were registered between the remelted layer and the native material. The research of [143] indicated that a laser modification with a low laser power did not increase the hardness of the material as much as in the case of a high laser power, for the reason that a lower power also means less heat and there is not as much formation of the martensitic alpha phase after remelting. A high laser power is directly related to the rapid cooling of the material and as a result the formation of the martensitic alpha phase. In the work of [144], the formation of a TiO_2 rutile and anatase phase was demonstrated, and it was indicated that the amount of oxygen molecules present in the structure depended on the laser modification method, namely, the type of laser and the parameters used. Pan et al. [145], as a result of their conducted research, indicated that the observed increase in the microhardness of the samples subjected to laser treatment was also because with an increase in the laser operating parameters there is an increase in the pressure. It was noted that an overly high value of the laser operating parameters was not able to effectively improve the microhardness of a surface [146], and this was because the yield strength of the titanium alloys had been exceeded [145]. The laser modification primarily increased the microhardness at the surface of the material, and the decrease in the hardness values was a gradient with an increasing test depth [145]. The change of the laser wavelength from 532 nm to 1064 nm showed a huge increase in the surface roughness which had a negative impact on the corrosion properties [147].

Table 4. Microhardness and hardness after laser treatment on titanium and its alloys.

Material	Microhardness and Hardness after Laser Treatment	Type of Laser	Energy/Laser Power	Impact Time	Pulse Duration	Frequency	Scan Speed (Mm/S)	Laser Pulses	Environment	References
Ti35Nb10Ta	3.8 GPa 3.3 GPa untreated material: 3.06 GPa	Nd: YAG	1000 W 1500 W	-	-	-	6.67 mm/s 10 mm/s	-	Helium	[120] [148]
Ti30Nb45n	3.5 GPa 3.3 GPa untreated material 2.44 GPa: 4.9 GPa	Nd: YAG	1000 W 1500 W	-	-	-	the authors 6.67 mm/s 10 mm/s	-	Helium	[120] [149]
Ti6Al4V	3, 4.01 for 150 µJ, 2.59 GPa for 240 µJ 3.84 GPa 294 GPa	Nd: YAG femtosecond laser Nanosecond Laser Shock Peening (LSP)	50 mJ 10–240 µJ 4 J	- - 1 s 3 s	- 290 fs 20 ns	10 Hz 50 kHz 1 H	-	-	- Argon -	[105] [145]
Ti-5Al-2.5Sn	4.02 GPa	Nd: YAG	165 W	-	-	14 Hz	-	500–18,000	-	[124]
Ti6Al4V	2.45–3.43 GPa	-	-	-	-	-	-	-	-	[146]
Cp-Ti	untreated material: 1.08 GPa 2.59 GPa	Nd: YAG	100 W	-	5 ms	20 Hz	8 mm/s	-	-	[122]

6. Wear Resistance and Fatigue Behavior

An appropriate approach to improving wear resistance at present is to design alloys based on their chemical composition. In parallel, several technologies have been developed to prepare a modified layer with a high wear resistance [150]. The literature indicates that the direct effect on wear resistance is related to the hardness of a material [107,151]. One of the few disadvantages of titanium and its alloys is a low wear resistance [115]. For these materials, the need to improve this property is widely discussed [11,152] because the low wear resistance is a limitation of the various applications. Significant differences in the wear resistance of the various titanium alloys are indicated depending on the α, β, or $\alpha\beta$ type [150,153]. The literature points to a low wear resistance, especially in the case of titanium alloys β [92], for example, Ti-5Al-5Mo-5V-1Cr-1Fe [133], Ti–35Nb–7Zr–5Ta [154], and Ti10V2Fe3Al [155].

The literature reports that laser surface modification is a simple method to improve the wear resistance of a material [119,156–159]. In the research by Cheng et al. [133], despite an increased surface hardness after laser modification, no improvement in the wear resistance of the titanium alloy Ti5Al5Mo5V1Cr1Fe was observed.

The papers of [160,161] indicated that laser treatment in a nitrogen environment caused the formation of TiN, which significantly improved the surface hardness and wear resistance of the materials. The increased wear resistance was explained by a decrease in the coefficient of friction [160]. The research of [161] indicated that the relative wear resistance increased by 1.7 times compared to a material without a laser treatment. Moreover, the literature of [158] indicated that a TiN layer formed as a result of remelting in the presence of nitrogen, had a higher hardness and wear resistance than treated titanium alloys.

A study of the effect of laser processing on the wear resistance of pure titanium was carried out in [162], where laser processing was shown to reduce the weight loss of samples during a dry wear test. The study confirmed Archad's theory, that the wear rate is reversely proportional to the hardness [162,163]. The results of [162] for wear resistance indicated that the mass loss for a samples subjected to laser modification was much lower than for material without treatment, and that the wear indices for selected modified samples were practically constant, while for unmodified samples they increased significantly. The samples were subjected to a study of the change in the coefficient of friction during normal loading for increasing loads and the results in the cited work [162] indicated that the pure titanium samples without a modification showed a higher wear and material loss with an increasing normal load, while for modified samples a decrease in the wear was observed with increasing loads. In the work of [164], an increase in the wear resistance was observed in untreated material, while an increase in the load during the experiments lowered the wear resistance. The laser treatment created defects and when the load was applied, the structures were compressed, which increased the wear resistance because the material did not detach from the surface. It was shown that when the maximum temperature reached during the thermal cycle was higher than the melting temperature, a phase change to the martensitic phase occurred, and the fatigue strength and wear resistance of the material were improved; thus, this procedure is also used in the hardening of steels and cast irons [137].

The study results of Zeng et al. [165] also indicated that a laser treatment improved the wear resistance of a material, that the diagram of the dependence of a material loss volume on the wear time for both a non-laser treated and treated material was linear, except for modified samples where a small mass loss was observed, and that it was 37 times lower than the wear of the raw material. The ion release tests carried out showed that the increased wear resistance determined the reduced release of vanadium, which is a toxic element.

In the study of [166], as a result of laser modification, the formation of TiC was observed and it was indicated that the presence of this composite improved the hardness and wear resistance. It was shown that short laser processing times were more advantageous because the melted layer was more homogeneous. Moreover, increasing the laser processing time increased the hardness and surface roughness but the TiC structure was more uneven and hard TiC particles acted as an abrasive tool. The low speed of the laser beam on the material, causing the formation of deeper melts, also had a beneficial effect on increasing the wear resistance of the material.

In paper [167], an XRD study of a Ti834 alloy subjected to modification was carried out, where the fatigue strength (high cycle fatigue) was tested. It was shown that for modified samples there was an increase in the tested strength, which was justified by the formation of compressive stresses after laser shock peening. Jia et al. [168] confirmed that in this type of laser processing, compressive stresses are induced in the sample and increase with an increasing impact time. A 1998 study [169] showed that laser modification combined with previous coating applications allowed for a reduction in the adhesive and abrasive wear.

The researchers of [170,171] compared the treatment of titanium by ion implantation and laser nitriding. It was shown that the use of a CO_2 laser in a nitrogen environment allowed a reduction in the friction between the tested surfaces, with a concomitant decrease in the fatigue strength [171]. Another study [170] evaluated the fretting fatigue behavior of a titanium alloy. It was observed that due to the formation of a heterogeneous, brittle surface after machining during fretting, the surface of the titanium alloy was unable to accept the loads set during fretting. Current literature [172–174] also shows that the use of laser modification is associated with decreasing the fatigue strength in titanium alloys. It is indicated that the fatigue strength is reduced by up to 30% compared to the very good fatigue strength of titanium alloys [172,175]. Additionally, it was indicated that with the increase in surface roughness after laser treatment, the cracks occurring on the surface generate a reduction in the fatigue strength. The resulting surface damage was identified as crack initiation sites [174]. The untreated material had crack initiation sites at the edges of the material, while in the case of the laser-treated material, the initiation site was in the center of the material [173].

7. Conclusions

The literature review focused on the effects of laser modification without material gain on titanium and its alloys. The presented work provides a comprehensive knowledge base on the effects of a fiber laser, Nd: YAG, Yd: YAG laser, and femtosecond laser and the shock peening method on selected properties of the titanium materials used in the medical industry. The paper discusses such properties as the roughness, wettability, resistance to corrosion, wear, and fatigue, as well as the effect of laser modification on material hardness.

1. The first section focused on the surface roughness and wettability, allowing us to assess the impact of laser modification with different types of lasers, which led to the conclusion that the use of this type of modification increases the surface roughness and that it varies depending on the operating parameters of the lasers.
2. The wettability of a surface is a topic that is widely discussed due to the fact that laser modification affects the change in the nature of the surface. Notably, a large impact on the hydrophilicity or hydrophobicity of a surface is the timing of the test in this direction, as well as the environment in which the samples are stored, but depending on the application of the material, there are different requirements, which does not indicate a more advantageous character.
3. Collected publications in the field of corrosion resistance research determine that the action of a laser beam on titanium materials improves the corrosion resistance, which is important because this reduces the release of dangerous elements from the implants.
4. Laser modification alters the micro- and nano-hardness for each type of laser. It is indicated that laser modification allows the process to be carried out in such a way that the hardness obtained after the change is close to that of bone.
5. The effect of laser modification on material wear was presented based on a collection of literature from a wide time range, which allowed the presentation of further opportunities to discuss the selection of optimal laser operating parameters, such as the laser operating power, laser beam density, and pulse duration.

6. In addition, the aspect of wear resistance was discussed, where it was shown that the use of laser modification improves this material property.
7. The presented review of the current literature on the subject provides a theoretical basis for studying the effects of laser processing on titanium and its biostops and for conducting targeted processing in the area where modification is needed to improve implants.
8. The presented review of the current literature related to the effects of laser modification on selected properties of titanium materials and provides a theoretical basis for the researchers' research.
9. The review indicates the need to deepen the research related to the wettability of the surface of titanium materials used in biomedicine, due to the fact that there is no clear indication of which character of the surface is more favorable, and it is necessary to identify the areas of application of a hydrophobic and hydrophilic surface obtained by modification. In addition, it is important to focus on studies related to the durability of materials against wear and fatigue and corrosion because these two properties directly affect the length of stay of an implant in the body.

Author Contributions: Conceptualization, J.S. and M.S.; Methodology, J.S.; Formal analysis, M.S.; Investigation, J.S.; Writing—original draft preparation, J.S.; Writing—review and editing, J.S. and M.S. All authors have read and agreed to the published version of the manuscript.

Funding: This research received no external funding.

Institutional Review Board Statement: Not applicable.

Informed Consent Statement: Not applicable.

Data Availability Statement: Not applicable.

Conflicts of Interest: The authors declare no conflict of interest.

References

1. Bandyopadhyay, A.; Sahasrabudhe, H.; Bose, S. *Laser Surface Modification of Metallic Biomaterials*; Elsevier Ltd.: Amsterdam, The Netherlands, 2016; ISBN 9780081009420.
2. Ardila-Rodríguez, L.A.; Menezes, B.R.C.; Pereira, L.A.; Takahashi, R.J.; Oliveira, A.C.; Travessa, D.N. Surface Modification of Aluminum Alloys with Carbon Nanotubes by Laser Surface Melting. *Surf. Coat. Technol.* **2019**, *377*, 1–11. [CrossRef]
3. Landowski, M. Influence of parameters of laser beam welding on structure of 2205 duplex stainless steel. *Adv. Mater. Sci.* **2019**, *19*, 1–11. [CrossRef]
4. Mao, B.; Siddaiah, A.; Liao, Y.; Menezes, P.L. Laser Surface Texturing and Related Techniques for Enhancing Tribological Performance of Engineering Materials: A Review. *J. Manuf. Process.* **2020**, *53*, 153–173. [CrossRef]
5. Daskalova, A.; Angelova, L.; Carvalho, A.; Trifonov, A.; Nathala, C.; Monteiro, F.; Buchvarov, I. Effect of Surface Modification by Femtosecond Laser on Zirconia Based Ceramics for Screening of Cell-Surface Interaction. *Appl. Surf. Sci.* **2020**, *513*, 145914. [CrossRef]
6. Yang, L.; Ding, Y.; Cheng, B.; He, J.; Wang, G.; Wang, Y. Investigations on Femtosecond Laser Modified Micro-Textured Surface with Anti-Friction Property on Bearing Steel GCr15. *Appl. Surf. Sci.* **2018**, *434*, 831–842. [CrossRef]
7. Temmler, A.; Walochnik, M.A.; Willenborg, E.; Wissenbach, K. Surface Structuring by Remelting of Titanium Alloy Ti6Al4V. *J. Laser Appl.* **2015**, *27*, 1–7. [CrossRef]
8. Laketić, S.; Rakin, M.; Momčilović, M.; Ciganović, J.; Veljović, Đ.; Cvijović-Alagić, I. Surface Modifications of Biometallic Commercially Pure Ti and Ti-13Nb-13Zr Alloy by Picosecond Nd:YAG Laser. *Int. J. Miner. Metall. Mater.* **2021**, *28*, 285–295. [CrossRef]
9. Jażdżewska, M.; Kwidzińska, D.B.; Seyda, W.; Fydrych, D.; Zieleiński, A. Mechanical Properties and Residual Stress Measurements of Grade IV Titanium and Ti-6Al-4V and Ti-13Nb-13Zr Titanium Alloys after Laser Treatment. *Materials* **2021**, *14*, 6316. [CrossRef]
10. Majkowska-Marzec, B.; Sypniewska, J. Microstructure and Mechanical Properties of Laser Surface-Treated Ti13Nb13Zr Alloy with MWCNTs Coatings. *Adv. Mater. Sci.* **2021**, *21*, 5–18. [CrossRef]
11. Zieliński, A.; Jażdżewska, M.; Łubiński, J.; Serbiński, W. Effects of Laser Remelting at Cryogenic Conditions on Microstructure and Wear Resistance of the Ti6Al4V Alloy Applied in Medicine. *Trans. Technol. Publ.* **2012**, *183*, 215–224. [CrossRef]
12. Lisiecki, A.; Klimpel, A. Diode Laser Surface Modification of Ti6Al4V Alloy to Improve Erosion Wear Resistance. *Arch. Mater. Sci. Eng.* **2008**, *32*, 5–12.
13. Amaya-Vazquez, M.R.; Sánchez-Amaya, J.M.; Boukha, Z.; Botana, F.J. Microstructure, Microhardness and Corrosion Resistance of Remelted TiG2 and Ti6Al4V by a High Power Diode Laser. *Corros. Sci.* **2012**, *56*, 36–48. [CrossRef]

14. Lisiecki, A. Hybrid Laser Deposition of Composite WC-Ni Layers with Forced Local Cryogenic Cooling. *Materials* **2021**, *14*, 4312. [CrossRef] [PubMed]
15. Lv, F.; Liang, H.; Xie, D.; Mao, Y.; Wang, C.; Shen, L.; Tian, Z. On the Role of Laser in Situ Re-Melting into Pore Elimination of Ti–6Al–4V Components Fabricated by Selective Laser Melting. *J. Alloys Compd.* **2021**, *854*, 156866. [CrossRef]
16. Temmler, A.; Willenborg, E.; Wissenbach, K. Designing Surfaces by Laser Remelting Designing Surfaces by Laser Remelting. In Proceedings of the ICOMM 2012—International Conference on Micromanufacturing, Evanston, IL, USA, 12–14 March 2014. [CrossRef]
17. Szkodo, M.; Bień, A.; Stanisławska, A. Laser Beam as a Precision Tool to Increase Fatigue Resistance in an Eyelet of Undercarriage Drag Strut. *Int. J. Precis. Eng. Manuf.-Green Technol.* **2022**, *9*, 175–190. [CrossRef]
18. Ukar, E.; Lamikiz, A.; Martínez, S.; Arrizubieta, I. Laser Texturing with Conventional Fiber Laser. *Procedia Eng.* **2015**, *132*, 663–670. [CrossRef]
19. Aragaw, E.M.; Gärtner, E.; Schubert, A.; Stief, P.; Dantan, J.; Etienne, A.; Siadat, A. ScienceDirect ScienceDirect Combined Laser Hardening and Laser Surface Texturing Forming Tool 1. 2379 Combined Laser Hardening and Laser Surface Texturing Existing Products for an Assembly Oriented Product Family Identification Methodology to Analyze The. *Procedia CIRP* **2020**, *94*, 914–918. [CrossRef]
20. Balasubramanian, S.; Muthukumaran, V.; Sathyabalan, P. A Study of the Effect of Process Parameters of Laser Hardening in Carbon Steels. *Int. J. Civ. Eng. Technol.* **2017**, *8*, 201–207.
21. Poulon-Quintin, A.; Watanabe, I.; Watanabe, E.; Bertrand, C. Microstructure and Mechanical Properties of Surface Treated Cast Titanium with Nd:YAG Laser. *Dent. Mater.* **2012**, *28*, 945–951. [CrossRef]
22. Xu, Y.; Liu, W.; Zhang, G.; Li, Z.; Hu, H.; Wang, C.; Zeng, X.; Zhao, S.; Zhang, Y.; Ren, T. Friction Stability and Cellular Behaviors on Laser Textured Ti–6Al–4V Alloy Implants with Bioinspired Micro-Overlapping Structures. *J. Mech. Behav. Biomed. Mater.* **2020**, *109*, 1–14. [CrossRef]
23. Braga, F.J.C.; Marques, R.F.C.; de Filho, E.A.; Guastaldi, A.C. Surface Modification of Ti Dental Implants by Nd:YVO 4 Laser Irradiation. *Appl. Surf. Sci.* **2007**, *253*, 9203–9208. [CrossRef]
24. Menci, G.; Gökhan, A.; Waugh, D.G.; Lawrence, J.; Previtali, B. Applied Surface Science Laser Surface Texturing of β-Ti Alloy for Orthopaedics: Effect of Different Wavelengths and Pulse Durations. *Appl. Surf. Sci.* **2019**, *489*, 175–186. [CrossRef]
25. Yadi, M.; Esfahani, H.; Sheikhi, M.; Mohammadi, M. CaTiO3/α-TCP Coatings on CP-Ti Prepared via Electrospinning and Pulsed Laser Treatment for in-Vitro Bone Tissue Engineering. *Surf. Coat. Technol.* **2020**, *401*, 126256. [CrossRef]
26. Navarro, P.; Olmo, A.; Giner, M.; Rodríguez-Albelo, M.; Rodríguez, Á.; Torres, Y. Electrical Impedance of Surface Modified Porous Titanium Implants with Femtosecond Laser. *Materials* **2022**, *15*, 461. [CrossRef]
27. Jażdżewska, M.; Majkowska-Marzec, B. Hydroxyapatite Deposition on the Laser Modified Ti13Nb13Zr Alloy. *Adv. Mater. Sci.* **2018**, *17*, 5–13. [CrossRef]
28. Dywel, P.; Szczesny, R.; Domanowski, P.; Skowronski, L. Structural and Micromechanical Properties of Nd:YAG Laser Marking Stainless Steel (AISI 304 and AISI 316). *Materials* **2020**, *13*, 2168. [CrossRef] [PubMed]
29. Fujii, H.; Takahashi, K.; Yamashita, Y. Application of Titanium and Its Alloys for Automobile Parts. *Nippon Steel Tech. Rep.* **2003**, *02003*, 70–75. [CrossRef]
30. Assari, A.H.; Eghbali, B. Solid State Diffusion Bonding Characteristics at the Interfaces of Ti and Al Layers. *J. Alloys Compd.* **2019**, *773*, 50–58. [CrossRef]
31. Tabie, V.M.; Li, C.; Saifu, W.; Li, J.; Xu, X. Mechanical Properties of near Alpha Titanium Alloys for High-Temperature Applications - a Review. *Aircr. Eng. Aerosp. Technol.* **2020**, *92*, 521–540. [CrossRef]
32. Gomez-Gallegos, A.; Mandal, P.; Gonzalez, D.; Zuelli, N.; Blackwell, P. Studies on Titanium Alloys for Aerospace Application. *Defect Diffus. Forum* **2018**, *385 DDF*, 419–423. [CrossRef]
33. Elshazli, A.M.; Elshaer, R.N.; Hussein, A.H.A.; Al-Sayed, S.R. Erratum: Elshazli et Al. Laser Surface Modification of TC21 (α/β) Titanium Alloy Using a Direct Energy Deposition (DED) Process. *Micromachines* **2021**, *12*, 739. [CrossRef] [PubMed]
34. Khorasani, A.M.; Goldberg, M.; Doeven, E.H.; Littlefair, G. Titanium in Biomedical Applications—Properties and Fabrication: A Review. *J. Biomater. Tissue Eng.* **2015**, *5*, 593–619. [CrossRef]
35. Laska, A. Parameters of the Electrophoretic Deposition Process and Its Influence on the Morphology of Hydroxyapatite Coatings. Review. *Inżynieria Mater.* **2020**, *1*, 20–25. [CrossRef]
36. Watanabe, I.; McBride, M.; Newton, P.; Kurtz, K.S. Laser Surface Treatment to Improve Mechanical Properties of Cast Titanium. *Dent. Mater.* **2009**, *25*, 629–633. [CrossRef] [PubMed]
37. Shah, F.A.; Johansson, M.L.; Omar, O.; Simonsson, H.; Palmquist, A.; Thomsen, P. Laser-Modified Surface Enhances Osseointegration and Biomechanical Anchorage of Commercially Pure Titanium Implants for Bone-Anchored Hearing Systems. *PLoS ONE* **2016**, *11*, e0157504. [CrossRef] [PubMed]
38. Koizumi, H.; Takeuchi, Y.; Imai, H.; Kawai, T.; Yoneyama, T. Application of Titanium and Titanium Alloys to Fixed Dental Prostheses. *J. Prosthodont. Res.* **2019**, *63*, 266–270. [CrossRef]
39. Wierzchoń, T. Modification of Titanium and Its Alloys Implants by Low Temperature Surface Plasma Treatments for Cardiovascular Applications. *Inżynieria Mater.* **2018**, *1*, 4–13. [CrossRef]
40. Manjaiah, M.; Laubscher, R.F. A Review of the Surface Modifications of Titanium Alloys for Biomedical Applications. *Mater. Tehnol.* **2017**, *51*, 181–193. [CrossRef]

41. Yamaguchi, T.; Hagino, H. Formation of Titanium Carbide Layer by Laser Alloying with a Light-Transmitting Resin. *Opt. Lasers Eng.* **2017**, *88*, 13–19. [CrossRef]
42. Piotrowska, K.; Madej, M.; Ozimina, D. Assessment of tribological properties of ti13nb13zr titanium alloy used in medicine. *Tribologia* **2019**, *285*, 97–106. [CrossRef]
43. Chen, Q.; Thouas, G.A. Metallic Implant Biomaterials. *Mater. Sci. Eng. R Reports* **2015**, *87*, 1–57. [CrossRef]
44. Tian, Y.S.; Chen, C.Z.; Wang, D.Y.; Lei, T.Q. Laser surface modification of titanium alloys—A review. *Surf. Eng. Light Alloy. Alum. Magnes. Titan. Alloy.* **2010**, *12*, 398–443. [CrossRef]
45. Sirdeshmukh, N.; Dongre, G. Laser Micro & Nano Surface Texturing for Enhancing Osseointegration and Antimicrobial Effect of Biomaterials: A Review. *Mater. Today Proc.* **2021**, *44*, 2348–2355. [CrossRef]
46. Kurella, A.; Dahotre, N.B. Review Paper: Surface Modification for Bioimplants: The Role of Laser Surface Engineering. *J. Biomater. Appl.* **2005**, *20*, 1–25. [CrossRef]
47. Xue, X.; Ma, C.; An, H.; Li, Y.; Guan, Y. Corrosion Resistance and Cytocompatibility of Ti-20Zr-10Nb-4Ta Alloy Surface Modified by a Focused Fiber Laser. *Sci. China Mater.* **2018**, *61*, 516–524. [CrossRef]
48. Simões, I.G.; dos Reis, A.C.; da Costa Valente, M.L. Analysis of the Influence of Surface Treatment by High-Power Laser Irradiation on the Surface Properties of Titanium Dental Implants: A Systematic Review. *J. Prosthet. Dent.* **2021**, 1–8. [CrossRef]
49. Abdal-hay, A.; Staples, R.; Alhazaa, A.; Fournier, B.; Al-Gawati, M.; Lee, R.S.; Ivanovski, S. Fabrication of Micropores on Titanium Implants Using Femtosecond Laser Technology: Perpendicular Attachment of Connective Tissues as a Pilot Study. *Opt. Laser Technol.* **2022**, *148*, 107624. [CrossRef]
50. Korovessis, P.G.; Deligianni, D.D. Role of Surface Roughness of Titanium Versus Hydroxyapatite on Human Bone Marrow Cells Response. *J. Spinal Disord. Tech.* **2002**, *15*, 175–183. [CrossRef]
51. Elias, C.N.; Lima, J.H.C.; Valiev, R.; Meyers, M.A. Biomedical Applications of Titanium and Its Alloys Biological Materials Science 46-49. *Biol. Mater. Sci.* **2008**, *30*, 46–49.
52. Achneck, H.E.; Jamiolkowski, R.M.; Jantzen, A.E.; Haseltine, J.M.; Lane, W.O.; Huang, J.K.; Galinat, L.J.; Serpe, M.J.; Lin, F.H.; Li, M.; et al. The Biocompatibility of Titanium Cardiovascular Devices Seeded with Autologous Blood-Derived Endothelial Progenitor Cells. EPC-Seeded Antithrombotic Ti Implants. *Biomaterials* **2011**, *32*, 10–18. [CrossRef]
53. Ramesh, S.; Karunamoorthy, L.; Palanikumar, K. Surface Roughness Analysis in Machining of Titanium Alloy. *Mater. Manuf. Process.* **2008**, *23*, 174–181. [CrossRef]
54. Pires, L.C.; Guastaldi, F.P.S.; Nogueira, A.V.B.; Oliveira, N.T.C.; Guastaldi, A.C.; Cirelli, J.A. Physicochemical, Morphological, and Biological Analyses of Ti-15Mo Alloy Surface Modified by Laser Beam Irradiation. *Lasers Med. Sci.* **2019**, *34*, 537–546. [CrossRef] [PubMed]
55. Telegin, S.V.; Lyasnikova, A.V.; Dudareva, O.A.; Grishina, I.P.; Markelova, O.A.; Lyasnikov, V.N. Laser Modification of the Surface of Titanium: Technology, Properties, and Prospects of Application. *J. Surf. Investig.* **2019**, *13*, 228–231. [CrossRef]
56. Lawrence, J.; Hao, L.; Chew, H.R. On the Correlation between Nd:YAG Laser-Induced Wettability Characteristics Modification and Osteoblast Cell Bioactivity on a Titanium Alloy. *Surf. Coat. Technol.* **2006**, *200*, 5581–5589. [CrossRef]
57. Rafiee, K.; Naffakh-Moosavy, H.; Tamjid, E. The Effect of Laser Frequency on Roughness, Microstructure, Cell Viability and Attachment of Ti6Al4V Alloy. *Mater. Sci. Eng. C* **2020**, *109*, 110637. [CrossRef]
58. Györgyey, Á.; Ungvári, K.; Kecskeméti, G.; Kopniczky, J.; Hopp, B.; Oszkó, A.; Pelsöczi, I.; Rakonczay, Z.; Nagy, K.; Turzó, K. Attachment and Proliferation of Human Osteoblast-like Cells (MG-63) on Laser-Ablated Titanium Implant Material. *Mater. Sci. Eng. C* **2013**, *33*, 4251–4259. [CrossRef]
59. Samanta, A.; Wang, Q.; Singh, G.; Shaw, S.K.; Toor, F.; Ratner, A.; Ding, H. Nanosecond Pulsed Laser Processing Turns Engineering Metal Alloys Antireflective and Superwicking. *J. Manuf. Process.* **2020**, *54*, 28–37. [CrossRef]
60. Laketić, S.; Rakin, M.; Momčilović, M.; Ciganović, J.; Veljović, Đ.; Cvijović-Alagić, I. Influence of Laser Irradiation Parameters on the Ultrafine-Grained Ti[Sbnd]45Nb Alloy Surface Characteristics. *Surf. Coat. Technol.* **2021**, *418*, 127255. [CrossRef]
61. Liu, Q.; Liu, Y.; Li, X.; Dong, G. Pulse Laser-Induced Cell-like Texture on Surface of Titanium Alloy for Tribological Properties Improvement. *Wear* **2021**, *477*, 203784. [CrossRef]
62. Zaifuddin, A.Q.; Zulhilmi, F.; Aiman, M.H.; Quazi, M.M.; Ishak, M. Enhancement of Laser Heating Process by Laser Surface Modification on Titanium Alloy. *J. Mech. Eng. Sci.* **2021**, *15*, 8310–8318. [CrossRef]
63. Schnell, U.G.; Duenow, H.S. Effect of Laser Pulse Overlap and Scanning Line Overlap on Femtosecond Laser-Structured Ti6Al4V Surfaces. *Materials* **2020**, *13*, 969. [CrossRef] [PubMed]
64. Eghbali, N.; Naffakh-Moosavy, H.; Sadeghi Mohammadi, S.; Naderi-Manesh, H. The Influence of Laser Frequency and Groove Distance on Cell Adhesion, Cell Viability, and Antibacterial Characteristics of Ti-6Al-4V Dental Implants Treated by Modern Fiber Engraving Laser. *Dent. Mater.* **2021**, *37*, 547–558. [CrossRef] [PubMed]
65. György, E.; Pérez del Pino, A.; Serra, P.; Morenza, J.L. Influence of the Ambient Gas in Laser Structuring of the Titanium Surface. *Surf. Coat. Technol.* **2004**, *187*, 245–249. [CrossRef]
66. El Mogahzy, Y.E. Finishing Processes for Fibrous Assemblies in Textile Product Design. *Eng. Text.* **2009**, 300–326. [CrossRef]
67. Zheng, Q.; Mao, L.; Shi, Y.; Fu, W.; Hu, Y. Biocompatibility of Ti-6Al-4V Titanium Alloy Implants with Laser Microgrooved Surfaces. *Mater. Technol.* **2020**, 1–10. [CrossRef]
68. Hao, L.; Lawrence, J.; Li, L. Manipulation of the Osteoblast Response to a Ti-6Al-4V Titanium Alloy Using a High Power Diode Laser. *Appl. Surf. Sci.* **2005**, *247*, 602–606. [CrossRef]

69. Cunha, A.; Serro, A.P.; Oliveira, V.; Almeida, A.; Vilar, R.; Durrieu, M.C. Wetting Behaviour of Femtosecond Laser Textured Ti-6Al-4V Surfaces. *Appl. Surf. Sci.* **2013**, *265*, 688–696. [CrossRef]
70. May, A.; Agarwal, N.; Lee, J.; Lambert, M.; Akkan, C.K.; Nothdurft, F.P.; Aktas, O.C. Laser Induced Anisotropic Wetting on Ti-6Al-4V Surfaces. *Mater. Lett.* **2015**, *138*, 21–24. [CrossRef]
71. Raimbault, O.; Benayoun, S.; Anselme, K.; Mauclair, C.; Bourgade, T.; Kietzig, A.M.; Girard-Lauriault, P.L.; Valette, S.; Donnet, C. The Effects of Femtosecond Laser-Textured Ti-6Al-4V on Wettability and Cell Response. *Mater. Sci. Eng. C* **2016**, *69*, 311–320. [CrossRef]
72. Rotella, G.; Orazi, L.; Alfano, M.; Candamano, S.; Gnilitskyi, I. Innovative High-Speed Femtosecond Laser Nano-Patterning for Improved Adhesive Bonding of Ti6Al4V Titanium Alloy. *CIRP J. Manuf. Sci. Technol.* **2017**, *18*, 101–106. [CrossRef]
73. Lu, J.; Huang, T.; Liu, Z.; Zhang, X.; Xiao, R. Long-Term Wettability of Titanium Surfaces by Combined Femtosecond Laser Micro/Nano Structuring and Chemical Treatments. *Appl. Surf. Sci.* **2018**, *459*, 257–262. [CrossRef]
74. Huerta-Murillo, D.; García-Girón, A.; Romano, J.M.; Cardoso, J.T.; Cordovilla, F.; Walker, M.; Dimov, S.S.; Ocaña, J.L. Wettability Modification of Laser-Fabricated Hierarchical Surface Structures in Ti-6Al-4V Titanium Alloy. *Appl. Surf. Sci.* **2019**, *463*, 838–846. [CrossRef]
75. Shaikh, S.; Kedia, S.; Singh, D.; Subramanian, M.; Sinha, S. Surface Texturing of Ti6Al4V Alloy Using Femtosecond Laser for Superior Antibacterial Performance. *J. Laser Appl.* **2019**, *31*, 5081106. [CrossRef]
76. Dou, H.Q.; Liu, H.; Xu, S.; Chen, Y.; Miao, X.; Lü, H.; Jiang, X. Influence of Laser Fluences and Scan Speeds on the Morphologies and Wetting Properties of Titanium Alloy. *Optik (Stuttg)* **2020**, *224*, 165443. [CrossRef]
77. Wang, Q.; Wang, H.; Zhu, Z.; Xiang, N.; Wang, Z.; Sun, G. Switchable Wettability Control of Titanium via Facile Nanosecond Laser-Based Surface Texturing. *Surf. Interfaces* **2021**, *24*, 101122. [CrossRef]
78. Mukherjee, S.; Dhara, S.; Saha, P. Enhanced Corrosion, Tribocorrosion Resistance and Controllable Osteogenic Potential of Stem Cells on Micro-Rippled Ti6Al4V Surfaces Produced by Pulsed Laser Remelting. *J. Manuf. Process.* **2021**, *65*, 119–133. [CrossRef]
79. Wang, Z.; Song, J.; Wang, T.; Wang, H. Laser Texturing for Superwetting Titanium Alloy and Investigation of Its Erosion Resistance. *Coatings* **2021**, *11*, 1547. [CrossRef]
80. Singh, I.; George, S.M.; Tiwari, A.; Ramkumar, J.; Balani, K. Influence of Laser Surface Texturing on the Wettability and Antibacterial Properties of Metallic, Ceramic, and Polymeric Surfaces. *J. Mater. Res.* **2021**, *36*, 3985–3999. [CrossRef]
81. Li, H.; Wang, X.; Zhang, J.; Wang, B.; Breisch, M.; Hartmaier, A.; Rostotskyi, I.; Voznyy, V.; Liu, Y. Experimental Investigation of Laser Surface Texturing and Related Biocompatibility of Pure Titanium. *Int. J. Adv. Manuf. Technol.* **2022**. [CrossRef]
82. Tęczar, P.; Majkowska-Marzec, B. The influence of laser alloying of ti13nb13zr on surface topography and properties. *Adv. Mater. Sci.* **2019**, *19*, 45–55. [CrossRef]
83. Ta, D.V.; Dunn, A.; Wasley, T.J.; Kay, R.W.; Stringer, J.; Smith, P.J.; Connaughton, C.; Shephard, J.D. Nanosecond Laser Textured Superhydrophobic Metallic Surfaces and Their Chemical Sensing Applications. *Appl. Surf. Sci.* **2015**, *357*, 248–254. [CrossRef]
84. Liu, K.; Yao, X.; Jiang, L. Recent Developments in Bio-Inspired Special Wettability. *Chem. Soc. Rev.* **2010**, *39*, 3240–3255. [CrossRef] [PubMed]
85. Mukherjee, S.; Dhara, S.; Saha, P. Laser Surface Remelting of Ti and Its Alloys for Improving Surface Biocompatibility of Orthopaedic Implants. *Mater. Technol.* **2018**, *33*, 106–118. [CrossRef]
86. De Oliveira, V.M.C.A.; Aguiar, C.; Vazquez, A.M.; Robin, A.L.M.; Barboza, M.J.R. Corrosion Behavior Analysis of Plasma-Assited PVD Coated Ti-6Al-4V Alloy in 2 M NaOH Solution. *Mater. Res.* **2017**, *20*, 436–444. [CrossRef]
87. Dinu, M.; Franchi, S.; Pruna, V.; Cotrut, C.M.; Secchi, V.; Santi, M.; Titorencu, I.; Battocchio, C.; Iucci, G.; Vladescu, A. *Ti-Nb-Zr System and Its Surface Biofunctionalization for Biomedical Applications*; Elsevier Inc.: Amsterdam, The Netherlands, 2018; ISBN 9780128124567.
88. Supernak-Marczewska, M.; Ossowska, A.; Strąkowska, P.; Zieliński, A. Nanotubular Oxide Layers and Hydroxyapatite Coatings on Porous Titanium Alloy Ti13Nb13Zr. *Adv. Mater. Sci.* **2018**, *18*, 17–23. [CrossRef]
89. Ferdinandov, N.V.; Gospodinov, D.D.; Ilieva, M.D.; Radev, R.H. Structure and Pitting Corrosion of Ti-6al-4v Alloy and Ti-6al-4v Welds. *ICAMS Proc. Int. Conf. Adv. Mater. Syst.* **2018**, 325–330. [CrossRef]
90. Gu, K.X.; Wang, K.K.; Zheng, J.P.; Chen, L.B.; Wang, J.J. Electrochemical Behavior of Ti–6Al–4V Alloy in Hank's Solution Subjected to Deep Cryogenic Treatment. *Rare Met.* **2018**. [CrossRef]
91. Wang, Y.; Tayyebi, M.; Assari, A. Fracture Toughness, Wear, and Microstructure Properties of Aluminum/Titanium/Steel Multi-Laminated Composites Produced by Cross-Accumulative Roll-Bonding Process. *Arch. Civ. Mech. Eng.* **2022**, *22*, 1–14. [CrossRef]
92. Çaha, I.; Alves, A.C.; Rocha, L.A.; Toptan, F. A Review on Bio-Functionalization of β-Ti Alloys. *J. Bio-Tribo-Corrosion* **2020**, *6*, 1–31. [CrossRef]
93. Dias Corpa Tardelli, J.; Bolfarini, C.; Cândido dos Reis, A. Comparative Analysis of Corrosion Resistance between Beta Titanium and Ti-6Al-4V Alloys: A Systematic Review. *J. Trace Elem. Med. Biol.* **2020**, *62*, 126618. [CrossRef]
94. Pawłowski, Ł.; Bartmański, M.; Mielewczyk-Gryń, A.; Zieliński, A. Effects of Surface Pretreatment of Titanium Substrates on Properties of Electrophoretically Deposited Biopolymer Chitosan/Eudragit e 100 Coatings. *Coatings* **2021**, *11*, 1120. [CrossRef]
95. Surma, M.K.; Adach, M.; Dębowska, P.; Turlej, P.S. Projekt i analiza obliczeniowa implantU. *Aktual. Probl. Biomech.* **2019**, 111–122.
96. Bartmański, M.; Pawłowski, Ł.; Zieliński, A.; Mielewczyk-Gryń, A.; Strugała, G.; Cieślik, B. Electrophoretic Deposition and Characteristics of Chitosan-Nanosilver Composite Coatings on a Nanotubular TiO2 Layer. *Coatings* **2020**, *10*, 245. [CrossRef]

97. Manivasagam, G.; Dhinasekaran, D.; Rajamanickam, A. Biomedical Implants: Corrosion and Its Prevention-A Review. *Recent Pat. Corros. Sci.* **2010**, *2*, 40–54. [CrossRef]
98. Boinovich, L.B.; Gnedenkov, S.V.; Alpysbaeva, D.A.; Egorkin, V.S.; Emelyanenko, A.M.; Sinebryukhov, S.L.; Zaretskaya, A.K. Corrosion Resistance of Composite Coatings on Low-Carbon Steel Containing Hydrophobic and Superhydrophobic Layers in Combination with Oxide Sublayers. *Corros. Sci.* **2012**, *55*, 238–245. [CrossRef]
99. Stepanovska, J.; Matejka, R.; Rosina, J.; Bacakova, L.; Kolarova, H. Treatments for Enhancing the Biocompatibility of Titanium Implants. *Biomed. Pap.* **2020**, *164*, 23–33. [CrossRef]
100. Gil, F.J.; Delgado, L.; Espinar, E.; Llamas, J.M. Corrosion and Corrosion-Fatigue Behavior of Cp-Ti and Ti-6Al-4V Laser-Marked Biomaterials. *J. Mater. Sci. Mater. Med.* **2012**, *23*, 885–890. [CrossRef]
101. Tayyebi, M.; Adhami, M.; Karimi, A.; Rahmatabadi, D.; Alizadeh, M.; Hashemi, R. Effects of Strain Accumulation and Annealing on Interfacial Microstructure and Grain Structure (Mg and Al3Mg2 Layers) of Al/Cu/Mg Multilayered Composite Fabricated by ARB Process. *J. Mater. Res. Technol.* **2021**, *14*, 392–406. [CrossRef]
102. Mohammed, M.T.; Khan, Z.A.; Siddiquee, A.N. Surface Modifications of Titanium Materials for Developing Corrosion Behavior in Human Body Environment: A Review. *Procedia Mater. Sci.* **2014**, *6*, 1610–1618. [CrossRef]
103. Nagle Travessa, D.; Vilas Boas Guedes, G.; Capella de Oliveira, A.; Regina Cardoso, K.; Roche, V.; Moreira Jorge, A. The Effect of Surface Laser Texturing on the Corrosion Performance of the Biocompatible β-Ti12Mo6Zr2Fe Alloy. *Surf. Coat. Technol.* **2021**, *405*. [CrossRef]
104. Zeng, C.; Wen, H.; Hemmasian Ettefagh, A.; Zhang, B.; Gao, J.; Haghshenas, A.; Raush, J.R.; Guo, S.M. Reoxidation Process and Corrosion Behavior of TA15 Alloy by Laser Ablation. *Surf. Coat. Technol.* **2020**, *385*, 125397. [CrossRef]
105. Al-Sayed, S.R.; Abdelfatah, A. Corrosion Behavior of a Laser Surface-Treated Alpha–Beta 6/4 Titanium Alloy. *Metallogr. Microstruct. Anal.* **2020**, *9*, 553–560. [CrossRef]
106. Singh, R.; Tiwari, S.K.; Mishra, S.K.; Dahotre, N.B. Electrochemical and Mechanical Behavior of Laser Processed Ti-6Al-4V Surface in Ringer's Physiological Solution. *J. Mater. Sci. Mater. Med.* **2011**, *22*, 1787–1796. [CrossRef] [PubMed]
107. Kumari, R.; Scharnweber, T.; Pfleging, W.; Besser, H.; Majumdar, J.D. Laser Surface Textured Titanium Alloy (Ti-6Al-4V) - Part II - Studies on Bio-Compatibility. *Appl. Surf. Sci.* **2015**, *357*, 750–758. [CrossRef]
108. Kuczyńska-Zemła, D.; Sotniczuk, A.; Pisarek, M.; Chlanda, A.; Garbacz, H. Corrosion Behavior of Titanium Modified by Direct Laser Interference Lithography. *Surf. Coat. Technol.* **2021**, *418*, 127219. [CrossRef]
109. Sun, Z.; Annergren, I.; Pan, D.; Mai, T.A. Effect of Laser Surface Remelting on the Corrosion Behavior of Commercially Pure Titanium Sheet. *Mater. Sci. Eng. A* **2003**, *345*, 293–300. [CrossRef]
110. Xu, Y.; Li, Z.; Zhang, G.; Wang, G.; Zeng, Z.; Wang, C.; Wang, C.; Zhao, S.; Zhang, Y.; Ren, T. Electrochemical Corrosion and Anisotropic Tribological Properties of Bioinspired Hierarchical Morphologies on Ti-6Al-4V Fabricated by Laser Texturing. *Tribol. Int.* **2019**, *134*, 352–364. [CrossRef]
111. Ali, N.; Mustapa, M.S.; Ghazali, M.I.; Sujitno, T.; Ridha, M. Fatigue Life Prediction of Commercially Pure Titanium after Nitrogen Ion Implantation. *Int. J. Automot. Mech. Eng.* **2013**, *7*, 1005–1013. [CrossRef]
112. Liu, B.W.; Mi, G.Y.; Wang, C.M. Reoxidation Process and Corrosion Behavior of TA15 Alloy by Laser Ablation. *Rare Met.* **2021**, *40*, 865–878. [CrossRef]
113. Baxter, J.W.; Bumby, J.R. Fuzzy Control of a Mobile Robotic Vehicle. *Proc. Inst. Mech. Eng. Part I J. Syst. Control. Eng.* **1995**, *209*, 79–91. [CrossRef]
114. Da Rocha, S.S.; Adabo, G.L.; Henriques, G.E.P.; Nóbilo, M.A.D.A. Vickers Hardness of Cast Commercially Pure Titanium and Ti-6Al-4V Alloy Submitted to Heat Treatments. *Braz. Dent. J.* **2006**, *17*, 126–129. [CrossRef] [PubMed]
115. Sharma, A.; Waddell, J.N.; Li, K.C.; A Sharma, L.; Prior, D.J.; Duncan, W.J. Is Titanium–Zirconium Alloy a Better Alternative to Pure Titanium for Oral Implant? Composition, Mechanical Properties, and Microstructure Analysis. *Saudi Dent. J.* **2021**, *33*, 546–553. [CrossRef] [PubMed]
116. Sharan, J.; Lale, S.V.; Koul, V.; Mishra, M.; Kharbanda, O.P. An Overview of Surface Modifications of Titanium and Its Alloys for Biomedical Applications. *Trends Biomater. Artif. Organs* **2015**, *29*, 176–187.
117. Davis, R.; Singh, A.; Jackson, M.J.; Coelho, R.T.; Prakash, D.; Charalambous, C.P.; Ahmed, W.; da Silva, L.R.R.; Lawrence, A.A. *A Comprehensive Review on Metallic Implant Biomaterials and Their Subtractive Manufacturing*; Springer: London, UK, 2022; Volume 120, ISBN 0123456789.
118. Pan, J.; Prabakaran, S.; Rajan, M. In-Vivo Assessment of Minerals Substituted Hydroxyapatite / Poly Sorbitol Sebacate Glutamate (PSSG) Composite Coating on Titanium Metal Implant for Orthopedic Implantation. *Biomed. Pharmacother.* **2019**, *119*, 109404. [CrossRef]
119. Zhang, L.C.; Chen, L.Y.; Wang, L. Surface Modification of Titanium and Titanium Alloys: Technologies, Developments, and Future Interests. *Adv. Eng. Mater.* **2020**, *22*, 1–37. [CrossRef]
120. Rossi, M.C.; Amado, J.M.; Tobar, M.J.; Vicente, A.; Yañez, A.; Amigó, V. Effect of Alloying Elements on Laser Surface Modification of Powder Metallurgy to Improve Surface Mechanical Properties of Beta Titanium Alloys for Biomedical Application. *J. Mater. Res. Technol.* **2021**, *14*, 1222–1234. [CrossRef]
121. Conradi, M.; Kocijan, A.; Klobčar, D.; Godec, M. Influence of Laser Texturing on Microstructure, Surface and Corrosion Properties of Ti-6al-4v. *Metals* **2020**, *10*, 1504. [CrossRef]

122. Chai, L.; Wu, H.; Zheng, Z.; Guan, H.; Pan, H.; Guo, N.; Song, B. Microstructural Characterization and Hardness Variation of Pure Ti Surface-Treated by Pulsed Laser. *J. Alloys Compd.* **2018**, *741*, 116–122. [CrossRef]
123. Pushp, P.; Dasharath, S.M.; Arati, C. Classification and Applications of Titanium and Its Alloys. *Mater. Today Proc.* **2022**. [CrossRef]
124. Khorram, A.; Davoodi Jamaloei, A.; Jafari, A. Surface Transformation Hardening of Ti-5Al-2.5Sn Alloy by Pulsed Nd:YAG Laser: An Experimental Study. *Int. J. Adv. Manuf. Technol.* **2019**, *100*, 3085–3099. [CrossRef]
125. Zhang, T.; Fan, Q.; Ma, X.; Wang, W.; Wang, K.; Shen, P.; Yang, J.; Wang, L. Effect of Laser Remelting on Microstructural Evolution and Mechanical Properties of Ti-35Nb-2Ta-3Zr Alloy. *Mater. Lett.* **2019**, *253*, 310–313. [CrossRef]
126. Ushakov, I.; Simonov, Y. Alterations in the Microhardness of a Titanium Alloy Affected to a Series of Nanosecond Laser Pulses. *MATEC Web Conf.* **2019**, *298*, 00051. [CrossRef]
127. Trueba, P.; Giner, M.; Rodríguez, Á.; Beltrán, A.M.; Amado, J.M.; Montoya-García, M.J.; Rodríguez-Albelo, L.M.; Torres, Y. Tribo-Mechanical and Cellular Behavior of Superficially Modified Porous Titanium Samples Using Femtosecond Laser. *Surf. Coat. Technol.* **2021**, *422*, 127555. [CrossRef]
128. Omoniyi, P.O.; Akinlabi, E.T.; Mahamood, R.M. Heat Treatments of Ti6Al4V Alloys for Industrial Applications: An Overview. *IOP Conf. Ser. Mater. Sci. Eng.* **2021**, *1107*, 012094. [CrossRef]
129. Xu, Y.F.; Yi, D.Q.; Liu, H.Q.; Wang, B.; Yang, F.L. Age-Hardening Behavior, Microstructural Evolution and Grain Growth Kinetics of Isothermal ω Phase of Ti-Nb-Ta-Zr-Fe Alloy for Biomedical Applications. *Mater. Sci. Eng. A* **2011**, *529*, 326–334. [CrossRef]
130. Zafari, A.; Barati, M.R.; Xia, K. Controlling Martensitic Decomposition during Selective Laser Melting to Achieve Best Ductility in High Strength Ti-6Al-4V. *Mater. Sci. Eng. A* **2019**, *744*, 445–455. [CrossRef]
131. El-Hadad, S.; Nady, M.; Khalifa, W.; Shash, A. Influence of Heat Treatment Conditions on the Mechanical Properties of Ti–6Al–4V Alloy. *Can. Metall. Q.* **2018**, *57*, 186–193. [CrossRef]
132. Yao, Y.; Li, X.; Wang, Y.Y.; Zhao, W.; Li, G.; Liu, R.P. Microstructural Evolution and Mechanical Properties of Ti-Zr Beta Titanium Alloy after Laser Surface Remelting. *J. Alloys Compd.* **2014**, *583*, 43–47. [CrossRef]
133. He, B.; Cheng, X.; Li, J.; Tian, X.J.; Wang, H.M. Effect of Laser Surface Remelting and Low Temperature Aging Treatments on Microstructures and Surface Properties of Ti-55511 Alloy. *Surf. Coat. Technol.* **2017**, *316*, 104–112. [CrossRef]
134. Guo, B.; Jonas, J.J. Dynamic Transformation during the High Temperature Deformation of Titanium Alloys. *J. Alloys Compd.* **2021**, *884*, 161179. [CrossRef]
135. Geng, Y.; McCarthy, É.; Brabazon, D.; Harrison, N. Ti6Al4V Functionally Graded Material via High Power and High Speed Laser Surface Modification. *Surf. Coat. Technol.* **2020**, *398*, 126085. [CrossRef]
136. Moura, C.G.; Carvalho, O.; Gonçalves, L.M.V.; Cerqueira, M.F.; Nascimento, R.; Silva, F. Laser Surface Texturing of Ti-6Al-4V by Nanosecond Laser: Surface Characterization, Ti-Oxide Layer Analysis and Its Electrical Insulation Performance. *Mater. Sci. Eng. C* **2019**, *104*, 109901. [CrossRef] [PubMed]
137. Vilar, R.; Almeida, A. *Laser Surface Treatment of Biomedical Alloys*; Elsevier Ltd.: Amsterdam, The Netherlands, 2016; ISBN 9780081009420.
138. Dai, J.; Wang, T.; Chai, L.; Hu, X.; Zhang, L.; Guo, N. Characterization and Correlation of Microstructure and Hardness of Ti–6Al–4V Sheet Surface-Treated by Pulsed Laser. *J. Alloys Compd.* **2020**, *826*, 154243. [CrossRef]
139. Chai, L.; Chen, K.; Zhi, Y.; Murty, K.L.; Chen, L.Y.; Yang, Z. Nanotwins Induced by Pulsed Laser and Their Hardening Effect in a Zr Alloy. *J. Alloys Compd.* **2018**, *748*, 163–170. [CrossRef]
140. Zhang, T.; Fan, Q.; Ma, X.; Wang, W.; Wang, K.; Shen, P.; Yang, J. Microstructure and Mechanical Properties of Ti-35Nb-2Ta-3Zr Alloy by Laser Quenching. *Front. Mater.* **2019**, *6*, 318. [CrossRef]
141. Pfleging, W.; Kumari, R.; Besser, H.; Scharnweber, T.; Majumdar, J.D. Laser Surface Textured Titanium Alloy (Ti-6Al-4V): Part 1 - Surface Characterization. *Appl. Surf. Sci.* **2015**, *355*, 104–111. [CrossRef]
142. Chen, S.; Usta, A.D.; Eriten, M. Microstructure and Wear Resistance of Ti6Al4V Surfaces Processed by Pulsed Laser. *Surf. Coat. Technol.* **2017**, *315*, 220–231. [CrossRef]
143. Chauhan, A.S.; Jha, J.S.; Telrandhe, S.; Srinivas, V.; Gokhale, A.A.; Mishra, S.K. Laser Surface Treatment of α–β Titanium Alloy to Develop a β-Rich Phase with Very High Hardness. *J. Mater. Process. Technol.* **2021**, *288*, 116873. [CrossRef]
144. Kashyap, V.; Ramkumar, P. Improved Oxygen Diffusion and Overall Surface Characteristics Using Combined Laser Surface Texturing and Heat Treatment Process of Ti6Al4V. *Surf. Coat. Technol.* **2022**, *429*, 127976. [CrossRef]
145. Pan, X.; HE, W.; Cai, Z.; Wang, X.; Liu, P.; Luo, S.; Zhou, L. Investigations on Femtosecond Laser-Induced Surface Modification and Periodic Micropatterning with Anti-Friction Properties on Ti6Al4V Titanium Alloy. *Chin. J. Aeronaut.* **2022**, *35*, 521–537. [CrossRef]
146. Rajesh, P.; Muraleedharan, C.V.; Komath, M.; Varma, H. Laser Surface Modification of Titanium Substrate for Pulsed Laser Deposition of Highly Adherent Hydroxyapatite. *J. Mater. Sci. Mater. Med.* **2011**, *22*, 1671–1679. [CrossRef] [PubMed]
147. Ranjith Kumar, G.; Rajyalakshmi, G. Role of Nano Second Laser Wavelength Embedded Recast Layer and Residual Stress on Electrochemical Corrosion of Titanium Alloy. *Mater. Res. Express* **2019**, *6*, 086583. [CrossRef]
148. Utomo, E.P.; Herbirowo, S.; Puspasari, V.; Thaha, Y.N. Characteristics and Corrosion Behavior of Ti–30nb–5sn Alloys in Histidine Solution with Various Nacl Concentrations. *Int. J. Corros. Scale Inhib.* **2021**, *10*, 592–601. [CrossRef]
149. Arrazola, P.J.; Garay, A.; Iriarte, L.M.; Armendia, M.; Marya, S.; Le Maître, F. Machinability of Titanium Alloys (Ti6Al4V and Ti555.3). *J. Mater. Process. Technol.* **2009**, *209*, 2223–2230. [CrossRef]

150. Shao, L.; Du, Y.; Dai, K.; Wu, H.; Wang, Q.; Liu, J.; Tang, Y. β-Ti Alloys for Orthopedic and Dental Applications: A Review of Progress on Improvement of Properties through Surface Modificatio. *Coatings* **2021**, *11*, 1446. [CrossRef]
151. He, D.; Zheng, S.; Pu, J.; Zhang, G.; Hu, L. Improving Tribological Properties of Titanium Alloys by Combining Laser Surface Texturing and Diamond-like Carbon Film. *Tribol. Int.* **2015**, *82*, 20–27. [CrossRef]
152. Kaur, M.; Singh, K. Review on Titanium and Titanium Based Alloys as Biomaterials for Orthopaedic Applications. *Mater. Sci. Eng. C* **2019**, *102*, 844–862. [CrossRef]
153. Faria, A.C.L.; Rodrigues, R.C.S.; Claro, A.P.R.A.; de Mattos, M.G.C.; Ribeiro, R.F. Wear Resistance of Experimental Titanium Alloys for Dental Applications. *J. Mech. Behav. Biomed. Mater.* **2011**, *4*, 1873–1879. [CrossRef]
154. Vishnu, J.; Sankar, M.; Rack, H.J.; Rao, N.; Singh, A.K.; Manivasagam, G. Effect of Phase Transformations during Aging on Tensile Strength and Ductility of Metastable Beta Titanium Alloy Ti–35Nb–7Zr–5Ta-0.35O for Orthopedic Applications. *Mater. Sci. Eng. A* **2020**, *779*. [CrossRef]
155. Qiu, C.; Liu, Q.; Ding, R. Significant Enhancement in Yield Strength for a Metastable Beta Titanium Alloy by Selective Laser Melting. *Mater. Sci. Eng. A* **2021**, *816*, 141291. [CrossRef]
156. Makuch, N.; Kulka, M.; Dziarski, P.; Przestacki, D. Laser Surface Alloying of Commercially Pure Titanium with Boron and Carbon. *Opt. Lasers Eng.* **2014**, *57*, 64–81. [CrossRef]
157. Hatakeyama, M.; Masahashi, N.; Michiyama, Y.; Inoue, H.; Hanada, S. Wear Resistance of Surface-Modified TiNbSn Alloy. *J. Mater. Sci.* **2021**, *56*, 14333–14347. [CrossRef]
158. Zhang, L.C.; Chen, L.Y. A Review on Biomedical Titanium Alloys: Recent Progress and Prospect. *Adv. Eng. Mater.* **2019**, *21*, 1–29. [CrossRef]
159. Salguero, J.; Del Sol, I.; Vazquez-Martinez, J.M.; Schertzer, M.J.; Iglesias, P. Effect of Laser Parameters on the Tribological Behavior of Ti6Al4V Titanium Microtextures under Lubricated Conditions. *Wear* **2019**, *426–427*, 1272–1279. [CrossRef]
160. Wang, H.; Nett, R.; Gurevich, E.L. The Effect of Laser Nitriding on Surface Characteristics and Wear Resistance of NiTi Alloy with Low Power Fiber Laser. *Appl. Sci.* **2021**, *11*, 515. [CrossRef]
161. Jiang, P.; He, X.L.; Li, X.X.; Yu, L.G.; Wang, H.M. Wear Resistance of a Laser Surface Alloyed Ti-6Al-4V Alloy. *Surf. Coat. Technol.* **2000**, *130*, 24–28. [CrossRef]
162. Bahiraei, M.; Mazaheri, Y.; Sheikhi, M.; Heidarpour, A. Mechanism of TiC Formation in Laser Surface Treatment of the Commercial Pure Titanium Pre-Coated by Carbon Using PVD Process. *J. Alloys Compd.* **2020**, *834*, 155080. [CrossRef]
163. Tabrizi, A.T.; Aghajani, H.; Saghafian, H.; Laleh, F.F. Correction of Archard Equation for Wear Behavior of Modified Pure Titanium. *Tribol. Int.* **2021**, *155*, 106772. [CrossRef]
164. Veiko, V.P.; Odintsova, G.V.; Gazizova, M.Y.; Karlagina, Y.Y.; Manokhin, S.S.; Yatsuk, R.M.; Vasilkov, S.D.; Kolobov, Y.R. The Influence of Laser Micro- and Nanostructuring on the Wear Resistance of Grade-2 Titanium Surface. *Laser Phys.* **2018**, *28*. [CrossRef]
165. Zeng, X.; Wang, W.; Yamaguchi, T.; Nishio, K. Characteristics of Surface Modified Ti-6Al-4V Alloy by a Series of YAG Laser Irradiation. *Opt. Laser Technol.* **2018**, *98*, 106–112. [CrossRef]
166. Mohazzab, B.F.; Jaleh, B.; Fattah-alhosseini, A.; Mahmoudi, F.; Momeni, A. Laser Surface Treatment of Pure Titanium: Microstructural Analysis, Wear Properties, and Corrosion Behavior of Titanium Carbide Coatings in Hank's Physiological Solution. *Surf. Interfaces* **2020**, *20*, 100597. [CrossRef]
167. Jia, W.; Hong, Q.; Zhao, H.; Li, L.; Han, D. Effect of Laser Shock Peening on the Mechanical Properties of a Near-α Titanium Alloy. *Mater. Sci. Eng. A* **2014**, *606*, 354–359. [CrossRef]
168. Jia, W.; Zan, Y.; Mao, C.; Li, S.; Zhou, W.; Li, Q.; Zhang, S.; Ji, V. Microstructure Evolution and Mechanical Properties of a Lamellar Near-α Titanium Alloy Treated by Laser Shock Peening. *Vacuum* **2021**, *184*, 109906. [CrossRef]
169. Langlade, C.; Vannes, A.B.; Krafft, J.M.; Martin, J.R. Surface Modification and Tribological Behaviour of Titanium and Titanium Alloys after YAG-Laser Treatments. *Surf. Coat. Technol.* **1998**, *100–101*, 383–387. [CrossRef]
170. Vadiraj, A.; Kamaraj, M. Fretting Fatigue Behavior of Surface Modified Biomedical Titanium Alloys. *Trans. Indian Inst. Met.* **2010**, *63*, 217–223. [CrossRef]
171. Vadiraj, A.; Kamaraj, M.; Kamachi Mudali, U.; Nath, A.K. Effect of Surface Modified Layers on Fretting Fatigue Damage of Biomedical Titanium Alloys. *Mater. Sci. Technol.* **2006**, *22*, 1119–1125. [CrossRef]
172. Campanelli, L.C. A Review on the Recent Advances Concerning the Fatigue Performance of Titanium Alloys for Orthopedic Applications. *J. Mater. Res.* **2021**, *36*, 151–165. [CrossRef]
173. dos Santos, A.; Campanelli, L.C.; Da Silva, P.S.C.P.; Vilar, R.; de Almeida, M.A.M.; Kuznetsov, A.; Achete, C.A.; Bolfarini, C. Influence of a Femtosecond Laser Surface Modification on the Fatigue Behavior of Ti-6Al4V ELI Alloy. *Mater. Res.* **2019**, *22*. [CrossRef]
174. Potomati, F.; Campanelli, L.C.; Da Silva, P.S.C.P.; Simões, J.G.A.B.; de Lima, M.S.F.; Damião, Á.J.; Bolfarini, C. Assessment of the Fatigue Behavior of Ti-6Al-4V ELI Alloy with Surface Treated by Nd:YAG Laser Irradiation. *Mater. Res.* **2019**, *22*, 1–5. [CrossRef]
175. Liu, W.; Liu, S.; Wang, L. Surface Modification of Biomedical Titanium Alloy: Micromorphology, Microstructure Evolution and Biomedical Applications. *Coatings* **2019**, *9*, 249. [CrossRef]

Article

Development of an Improved YOLOv7-Based Model for Detecting Defects on Strip Steel Surfaces

Rijun Wang [1,2,*], Fulong Liang [1,2], Xiangwei Mou [1,2,*], Lintao Chen [1,2], Xinye Yu [1,2], Zhujing Peng [1] and Hongyang Chen [1]

1. Teachers College for Vocational and Technical Education, Guangxi Normal University, Guilin 541004, China
2. Key Laboratory of AI and Information Processing, Hechi University, Hechi 546300, China
* Correspondence: rijunwang@mailbox.gxnu.edu.cn (R.W.); xwmou@mailbox.gxnu.edu.cn (X.M.)

Abstract: The detection of defects on the surface is of great importance for both the production and the application of strip steel. In order to detect the defects accurately, an improved YOLOv7-based model for detecting strip steel surface defects is developed. To enhances the ability of the model to extract features and identify small features, the ConvNeXt module is introduced to the backbone network structure, and the attention mechanism is embedded in the pooling module. To reduce the size and improves the inference speed of the model, an improved C3 module was used to replace the ELAN module in the head. The experimental results show that, compared with the original models, the mAP of the proposed model reached 82.9% and improved by 6.6%. The proposed model can satisfy the need for accurate detection and identification of strip steel surface defects.

Keywords: defect detection; YOLOv7; deep learning; ConvNeXt; attention pooling module

Citation: Wang, R.; Liang, F.; Mou, X.; Chen, L.; Yu, X.; Peng, Z.; Chen, H. Development of an Improved YOLOv7-Based Model for Detecting Defects on Strip Steel Surfaces. *Coatings* **2023**, *13*, 536. https://doi.org/10.3390/coatings13030536

Academic Editor: Giorgos Skordaris

Received: 28 January 2023
Revised: 18 February 2023
Accepted: 25 February 2023
Published: 1 March 2023

Copyright: © 2023 by the authors. Licensee MDPI, Basel, Switzerland. This article is an open access article distributed under the terms and conditions of the Creative Commons Attribution (CC BY) license (https://creativecommons.org/licenses/by/4.0/).

1. Introduction

As an important raw material of industry, strip steel is widely used in the production of machinery, aerospace, automotive, defense, light industry, etc. [1]. However, limited by the quality of raw materials, production environment, equipment, manual errors, etc., the strip steel can lead to a variety of problems, the most common one being surface defects [2–4]. The surface defects are an important indicator for manufacturers and customers or consumers to judge the quality of strip steel. In general, the surface defects of strip steel, including crazing, inclusion, patches, pitted surface, rolled-in scale, scratches, and these defects have an impact on the aesthetics of the steel, but more importantly, they reduce the strength, toughness, corrosion resistance and wear resistance of the strip steel [5–7]. In addition, the defects will also affect the strip steel sales of enterprises and may even bring personal safety risks to users [8]. Therefore, the detection and identification of strip steel defects have become a hot issue for scholars to study.

The traditional defect detection methods for strip steel surfaces include manual inspection methods, non-destructive testing methods [9], and frequency flash detection methods [10]. Manual inspection requires inspectors to identify a large number of strip steel defects through the naked eye, which needs a lot of labor due to the complex diversity of defects. Secondly, the large amount of repetitive work makes the inspectors prone to visual fatigue, which can lead to missed inspections and false inspections [11]. Due to the traditional methods existing in low efficiency, error, high requirements for the skills of the inspector, and other shortcomings occur. In recent years, based on deep learning, image processing, target detection, and other automated technologies have begun to gradually replace traditional methods. Deep learning-based target detection can obtain higher recognition accuracy and detection speed, which greatly improves the efficiency of defect detection in real factories. Among them, the You Only Look Once (YOLO) algorithm series has become a popular method in the current target detection research field because of its ability to maintain good detection accuracy despite its fast detection speed.

The causes of defects on the strip steel surface are numerous, and the morphology of defects is complicated. According to the characteristics of the defect shapes, the defects can be roughly divided into three categories: point, line, and surface. Typical defects (shown in Figure 1) can be summarized as crazing, inclusion, patches, pitted surface, rolled-in scale, and scratches. The crazing (as shown in Figure 1a) is caused by excessive surface burning, decarburization, loosening, deformation, and a high content of sulfur and phosphorus impurities on the surface during processing. The crazing generally appears as water ripples or fish scale, which is different from the cracks caused by loose oxide skin. The inclusion (as shown in Figure 1b) is usually caused by the presence of inclusions (metallic or non-metallic) during the strip steel rolling. This defect occurs when the inclusions are fractured or exposed. The size of inclusion defects is related to the number of inclusions, and the edges are relatively clear, usually gray-white, yellow, or brown. The formation of patches (as shown in Figure 1c) is related to the incomplete cleaning of iron oxide on the surface of strip steel and also related to the failure to remove the residual liquid in the annealing process in time. The patches are usually black with large and different shapes. The distribution of patches on the surface of strip steel is random. After the strip steel has been rolled, the iron oxide comes off its surface, resulting in a continuous rough surface called a pitted surface (as shown in Figure 1d). The pitted surface usually appears as dents of different sizes and depths, with dotted distribution or periodic distribution. There are two main reasons for the formation of the rolled-in scale (as shown in Figure 1e); one is that foreign materials are on the surface of the roller, which makes the surface of the strip steel bulge during the roll-forming process. The other one is the low hardness of the roller, which causes the strip steel surface material to come off during the roll forming process; the strip steel surface appears depressed. The scratches (as shown in Figure 1f) is due to the action of external forces or scratches by sharp objects during transportation.

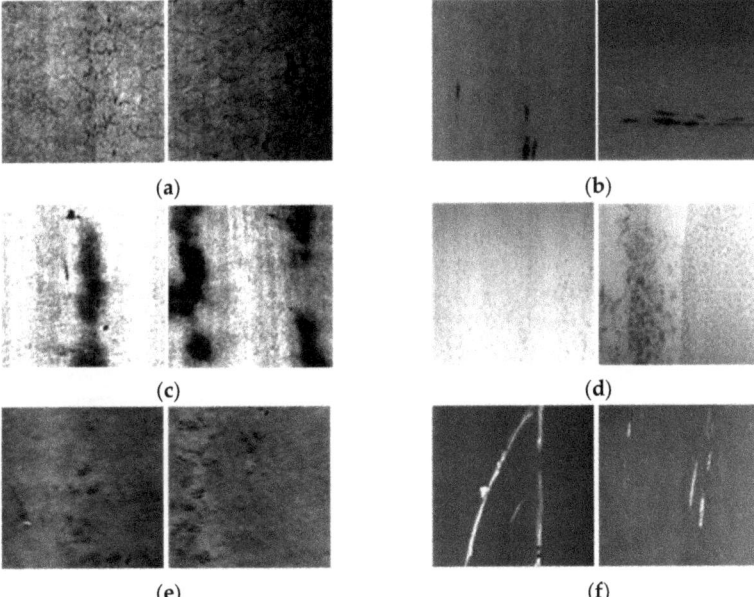

Figure 1. Typical defects of the strip steel. (**a**) crazing; (**b**) inclusion; (**c**) patches; (**d**) pitted surface; (**e**) rolled-in scale; (**f**) scratches.

All of the above defects have a negative impact on the integrity and functionality of the strip steel. As we can see in Figure 1, these defects present a variety of types. Some of the defects are small and vary in size. The same type of defect presents different characteristics due to different causes. Moreover, the distinction between defects is not clear enough. Therefore, it is extremely difficult to identify defects on the surface of the strip steel accurately.

Identifying defects on the strip steel surface is an important criterion for judging the quality of the strip steel and also facilitates the producer in finding the source of the problem for further improvement. It is of great importance for the production and manufacturing of strip steel. However, traditional methods are no longer applicable in today's progressively intelligent era, and the rapid development of computers has brought a great impetus to the field of computer vision. A fast regularity metric for defect detection in non-textured and uniformly textured surfaces is proposed by Tasi et al. [12]. This method is used to detect defects only through a single discriminant feature. It avoids the use of complex classifiers in a high-dimensional feature space. On the other hand, the method does not require learning from a set of defective and non-defective training samples. Liu et al. [13] proposed a new Haar–Weibull variance (HWV) model for unsupervised steel surface defect detection. The anisotropic diffusion model is used to eliminate the influence of patches, and then a new HWV model is developed to characterize the texture distribution of each local patch in the image, thus forming a parametric distribution to extract the background in the image effectively. In order to solve the under-segmentation or over-segmentation problem, a global adaptive percentile threshold method for gradient image is proposed in the literature [14]. Without considering the defect size, this method can adaptively change the percentile used for thresholding and retain the characteristics of defects. A defect detection model using an optimal Gabor filter was proposed by Tong et al. [15]. By using an optimal Gabor filter, the model can significantly reduce the computational cost and operate in real time to solve the problem of fabric detection. Choi et al. [16] applied the Gabor filter to the detection of porous defects in steel plates, and the classification performance of defects was improved with the use of the double threshold method. An entity sparsity tracking (ESP) method for identifying surface defects is proposed by Wang et al. [17] capable of detecting surface defects in an unsupervised manner.

In order to accelerate industrial production, improve product quality and save labor, many researchers have devoted themselves to applying deep learning target recognition methods to industrial production. Current deep learning-based target detection algorithms are basically divided into two categories, namely, the one-stage target detection algorithms and the two-stage detection algorithms. The two-stage target detection algorithm can be roughly divided into two steps. The first step is to locate the target in a candidate frame, and the second step is to make a final prediction of the target. The two-stage target algorithm includes the regional convolutional neural network (R-CNN) [18], Fast R-CNN [19], and Faster R-CNN [20]. Compared to the two-stage target detection algorithm, the one-stage target detection algorithms directly predict the location and category of the target, which is simpler and more direct. The one-stage target algorithm includes single shot multiBox detector (SSD) [21], RetinaNet [22], and YOLO series algorithms [23–30]. In order to meet the requirements of strip steel surface defect detection, an improved model [31] is proposed by combining the improved ResNet50 [32] with Faster R-CNN. The experimental results showed that the accuracy of detection was as high as 98.2%. However, the proposed model has a large number of parameters, which makes the algorithm inference runtime longer. A model of YOLOV4 based on an attention mechanism is proposed by Li et al. [33]. The model has a stronger feature extraction capability; the average accuracy reached 85.41% in detecting four types of strip steel defects. The TRANS module based on Transformer [34] was added to the backbone and detection head of the YOLOv5 model, and an improved Transformer-based YOLOv5 model was proposed by Guo et al. [35]. The test results showed that the average detection accuracy is 75.2%, improving about 18% compared to Faster R-CNN.

In summary, the target detection algorithm based on deep learning can effectively solve the problem of strip defect detection. Based on the above work, an improved YOLOv7-based model for detecting defects on strip steel surfaces is proposed in this paper. By introducing the ConvNeXt module to the backbone network, the ability of the network to extract defect features and accelerate network inference is enhanced. By embedding the CBAM into the MP layer of the model detection head, an attention-pooling structure is formed to enhance the ability to cope with complex and different strip steel surface defects.

The organization of this paper includes the following sections. In Section 1, the significance of the research and the contribution of this paper is given. A detailed summary of the research results related to the field of strip steel defect detection is analyzed, especially based on deep learning. A detailed description of the original You Only Look Once version 7 (YOLOv7) model, the loss function, and label assignment is given in Section 2. In Section 3, an improved model based on YOLOv7 was developed for detecting defects on the strip steel surface is described in detail. In Section 4, the detailed experimental results are described. The conclusion is given in Section 5.

Our contributions are as follows:

1. An improved YOLOv7-based model for detecting defects on strip steel surfaces is proposed.
2. To enhance the network's ability to extract defects features and speed up network inference, the ConvNeXt module is introduced to the backbone network of the YOLOv7 model.
3. To reduce the amount of operations and simplify the network structure, the Efficient Layer Aggregation Network (ELAN) module in the detection head of the YOLOv7 model is replaced by an improved C3 module (C3C2).
4. By embedding the Convolutional Block Attention Module (CBAM) into the maximum pooling (MP) layer of the model detection head, an attention pooling structure is formed to enhance the ability to cope with complex and different strip steel surface defects.

2. Methodology

2.1. YOLOv7 Network Structure

The YOLOv7 [30] (version 0.1) network structure is based on YOLOv5 [27] (version 5.0), which introduces the idea of model re-parameterization. In YOLOv7 network structure, deep supervision technique is added, dynamic label assignment strategy is improved, coarse-to-fine guiding label assignment strategy is proposed, etc. Among them, the role of the model re-parameterization is splitting a whole module into several identical or different module branches during the training process and integrating several branch modules into a fully equivalent module during the inference process. The benefit of model re-parameterization is that better feature representations are obtained, computational and parametric quantities are reduced, and inference speed is improved.

Deep supervision is a common technique used in deep network training. The main idea of deep supervision is to add an auxiliary head in the middle layer of the network. The shallow network weights and auxiliary losses are used as a guide to supervising the backbone network. Thus, the problems of disappearing training gradients and slow convergence of deep neural networks are solved (the YOLOv7 model explored in this paper does not have an auxiliary training head).

The coarse-to-fine guiding label assignment strategy is used to make the label assignment more accurate. The strategy is guided by the prediction results of the lead head to generate coarse-to-fine hierarchical labels. The coarse-to-fine hierarchical labels are used for auxiliary head and lead head learning, respectively.

The overall network structure of YOLOv7 (as shown in Figure 2) is very similar to YOLOv5; the main difference between them is the internal components of the network. Firstly, in the backbone part, the extended efficient layer aggregation network (E-ELAN) and MP structure are used in the backbone part of the network. Secondly, the neck layer and the head layer are merged, still called the head layer. The YOLOv7 network extracts image features mainly through the backbone part of the E-ELAN and MP structure. The authors of the original paper believe that the deeper the network is, the better it is for network learning and convergence. A more efficient network can be built by controlling the shortest and longest gradient paths in the network. Thus, after comparing with VoVNett [36], CSPVoVNet [37], and ELAN [38], the E-ELAN (an extended version based on ELAN) is proposed. The E-ELAN only changes the structure of the computational module, while the structure of the transition layer is completely unchanged. By using the strategy of expand, shuffle, and merge cardinality, the network learning capability is continuously enhanced without destroying the original gradient path. Unlike the previous network structure of YOLO, the MP layer in the YOLOv7 network structure uses both maxpooling and 3×3 convolution with stide = 2 to downsampling. The outputs are concatenated by means of concat, which allows the network to extract features better.

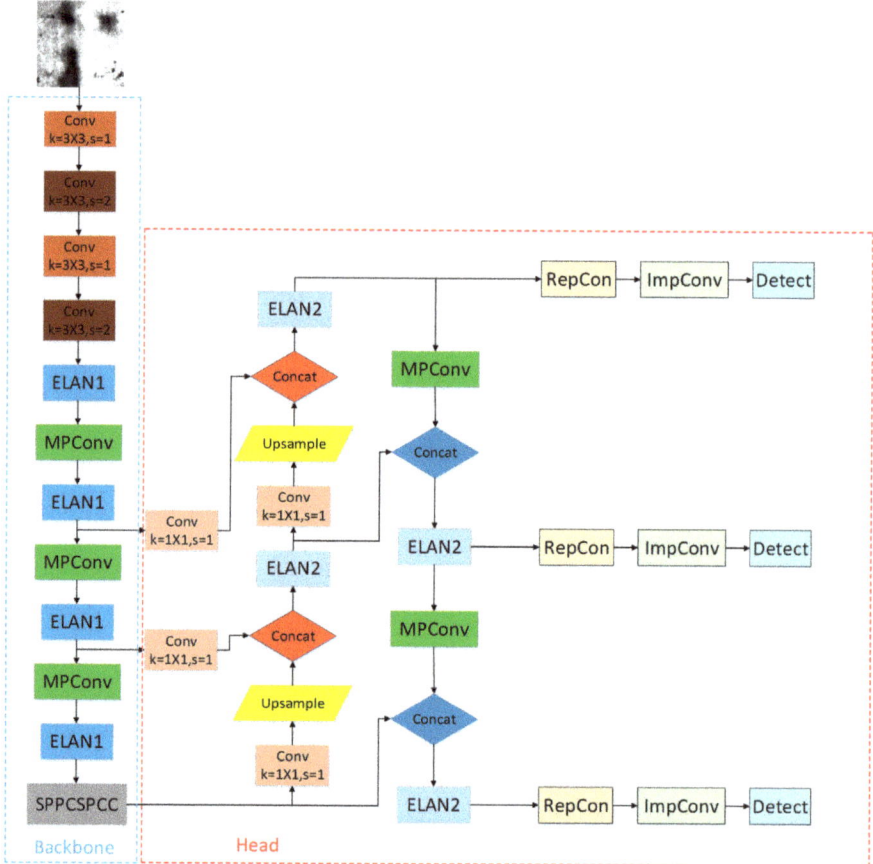

Figure 2. YOLOv7 network structure.

The procedure of strip defect detection with YOLOv7 is as follows:
1. Using a camera with higher resolution to collect pictures of the strip steel with defects on the surface.
2. Using the labelimg tools to process the defects that appear in the strip steel on these images, frame them accurately with a rectangular box and mark the category.
3. Dividing the processed images into the training set, test set, and validation set according to a certain ratio; putting the training set and validation set into the model of YOLOv7 for training and validation; and using the test set to test the model training effect.

2.2. Loss Function and Label Assignment

The overall loss function of YOLOv7 remains the same as YOLOv5. The loss function is divided into three parts: the classification loss L_{cls}, the objective confidence loss L_{obj} and the localization loss L_{loc}.

The binary cross entropy (BCE) loss is used for classification loss L_{cls}, and note that only the classification loss of positive samples is calculated. The objective loss L_{obj} is still BCE loss; note that the *obj* here refers to the complete intersection over union (CIoU) of the target bounding box and GT Box of the network prediction. The objective loss L_{obj} is calculated here for all samples. The localization loss L_{loc} is used as CIoU loss, and note that only the location loss of positive samples is calculated.

Therefore, the loss function of YOLOv7 can be described as follows:

$$Loss = \lambda_1 L_{cls} + \lambda_2 L_{obj} + \lambda_3 L_{loc} \tag{1}$$

where $\lambda_1, \lambda_2, \lambda_3$ are the equilibrium coefficients.

A new method of label assignment is used in the YOLOv7 network structure. Using the prediction of the lead head as a guide, coarse-to-fine hierarchical labels are generated. The labels are used for the learning of the auxiliary head and the lead head, respectively. The lead head has a stronger learning capability, allowing the auxiliary head to learn the information already learned by the lead head directly, and the lead head can focus more on the residual information that has not yet been learned. The details can be seen in Figure 3a,b.

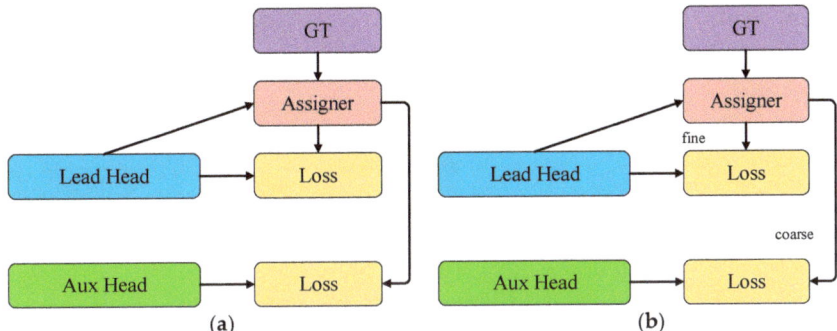

Figure 3. New label assignment method. (**a**) Lead guided assigner; (**b**) Coarse-to-fine guided assigner.

3. Improvement of YOLOv7

The improved YOLOv7 algorithmic network structure architecture proposed in this study is marked in the red box in Figure 4, and detailed information is given in the subsequent two sections.

The specific improvement points of the YOLOv7 algorithm structure in this study are marked with different colored rectangular boxes in Figure 4. The ConvNeXt module in the purple box is added to the head of YOLOv7 to enhance the ability of the model to extract features, and the C3C2 module in the yellow box replaces the ELAN structure in the original structure of YOLOv7 to streamline the model size, and finally, the green CBAM attention mechanism in the green box is embedded in the first MP layer structure to improve the network's ability to identify minor and inconspicuous defects.

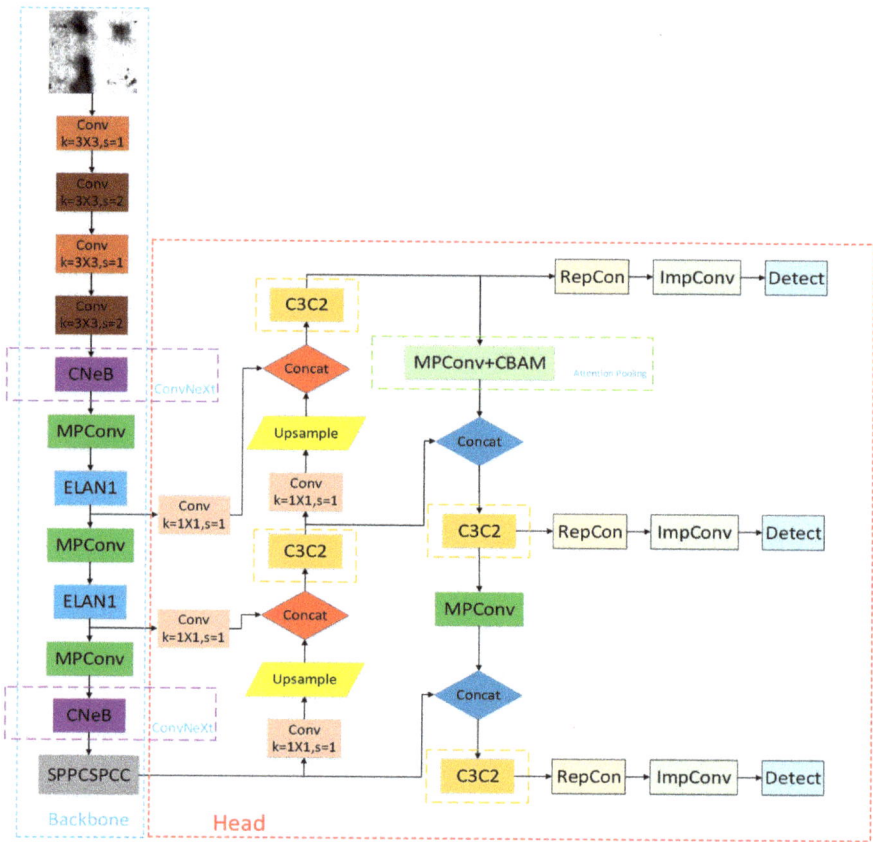

Figure 4. Improved YOLOv7 algorithm network structure.

3.1. ConvNeXt Module

In order to achieve accurate detection of strip steel surface defects, we applied the network model of YOLOv7 to the strip steel dataset. In order to improve the accuracy of the model, we tried to add the newly proposed ConvNeXt [39] convolutional structure module to the backbone of YOLOv7 for better extraction of strip steel surface defect features. The ConvNeXt has four different versions of T/S/B/L, which are configured as follows:

$$\begin{aligned}
&\text{ConvNeXt-T: } C = (96, 192, 384, 768), B = (3, 3, 9, 3) \\
&\text{ConvNeXt-S: } C = (96, 192, 384, 768), B = (3, 3, 27, 3) \\
&\text{ConvNeXt-B: } C = (128, 256, 512, 1024), B = (3, 3, 27, 3) \\
&\text{ConvNeXt-L: } C = (192, 384, 768, 1536), B = (3, 3, 27, 3)
\end{aligned} \quad (2)$$

where C represents the number of input channels in the four stages, and B represents the number of repeated stacking blocks per stage.

The computational complexity, structure size, and the number of input channels of the ConvNeXt increase sequentially from version T to version XL. To avoid breaking the entire continuous downsampling structure in the backbone of YOLOv7, after weighing the module size and the number of output channels, we choose to replace the first and last ELAN modules in its backbone with the ConvNeXt-B module. The structure of the ConvNeXt-B module is shown in Figure 5.

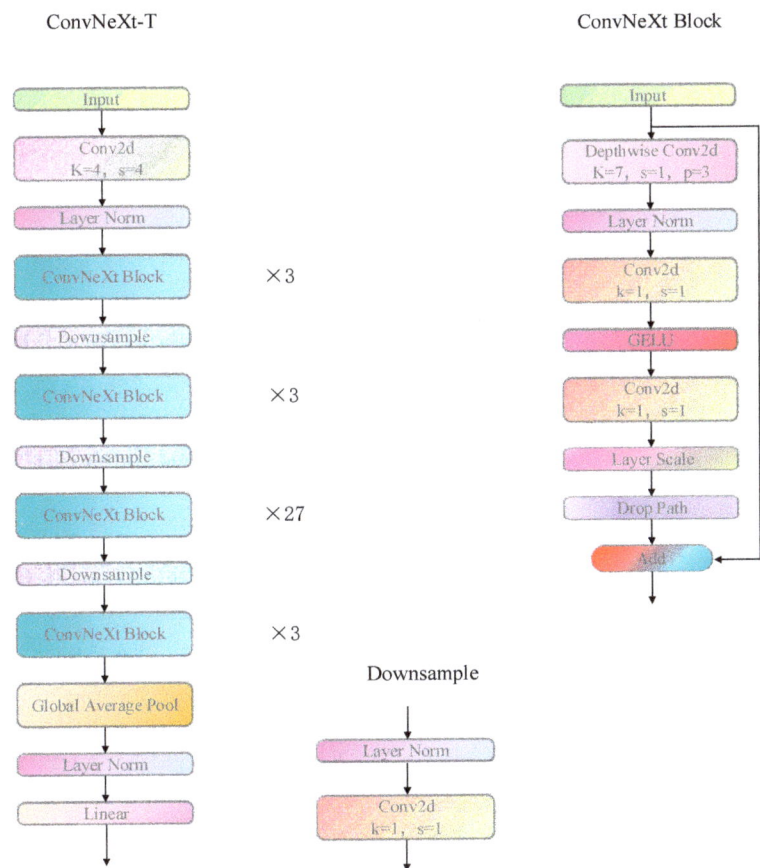

Figure 5. ConvNeXt-B module structure.

The network structure of the ConvNeXt module is a pure convolutional network structure based on the ResNet-50 network structure, which is designed based on the structure of the Swin Transformer [40]. While the ConvNeXt module is benchmarked against the Swin Transformer network structure, it also actively learns from the previous classical network structure. For example, the depthwise convolution structure adopted by the ConvNeXt module is learning from the method of ResNeXt [41]. Through the five comparative experiments of macro design, deep residual learning for image recognition (ResNeXt), inverted bottleneck, large kerner size, and various layer-wise micro designs, the network model is gradually optimized. Finally, ConvNeXt is proposed. With the same floating point of operations (FLOPs), the ConvNeXt has faster inference speed and higher accuracy than Swin Transformer.

3.2. Improvement of C3(C3C2)

In YOLOv7's head part, the ELAN module is used to extract features from the input feature maps. However, for our strip steel surface defect dataset, there are large differences in the size and shape of defects in the same category. Moreover, there are also similarities between different classes of defects. In addition, due to the different material quality of different strip steel samples and the influence of lighting, the gray value of the intra-class defect image will also change. These reasons make it difficult for the network to extract features. The ELAN module in YOLOv7 is not very effective in extracting the features of the image. In addition, the ELAN structure contains more convolutional modules and residual connections, which will bring more computation and reduce the inference speed. Therefore, we try to improve the C3 module in the latest version of YOLOv5 (shown in Figure 6a) into the C3C2 module (shown in Figure 6b) and replace the ELAN module in the head part to further enhance the feature extraction and fusion capability of the YOLOv7 network structure. Meanwhile, the parameter computation is reduced, and the inference speed is improved.

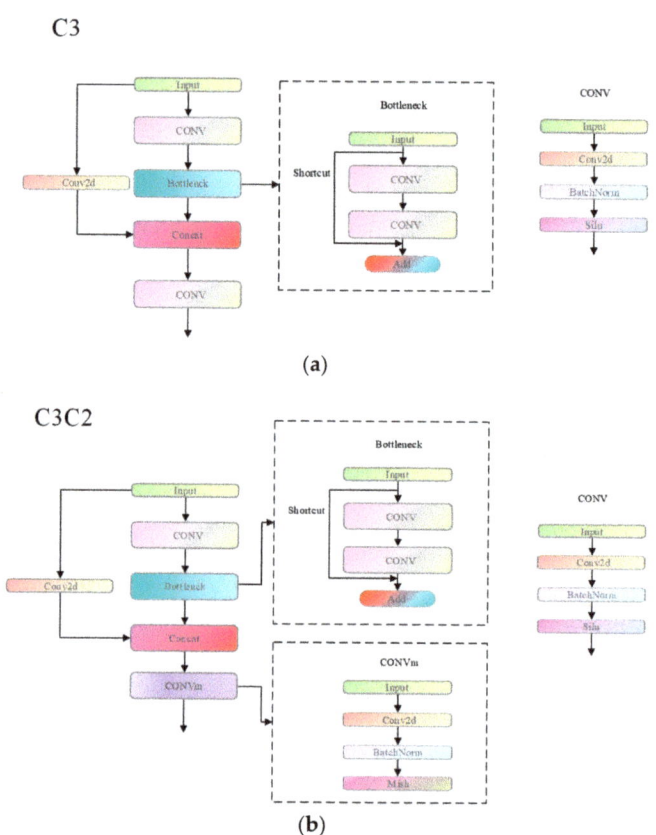

Figure 6. Improvement of C3(C3C2). (**a**) C3 module; (**b**) C3C2 module.

The C3C2 convolution module is inspired by Swin Transformer's network structure. It is based on the original C3 module, the residual branch convolution module is changed to a simple convolution structure, and the batch normalization (BN) layer and activation function layer are removed to reduce the amount of parameter calculation. Given that the Mish activation function [42] has a better ability to suppress overfitting than the Silu activation function, the Mish activation function is more robust to different hyperparameters. In view of the above advantages, we changed the activation function in the final convolution module from the original Silu (swish) to the Mish activation function. As a result, the nonlinear variation of the network is enhanced when the convolution module is input after the final Concat operation.

3.3. Attention Pooling Module

The MP layer structure, as shown in Figure 7a, is used for downsampling in the YOLOv7 head. The feature map will be downsampled in two branches after entering the

MP layer structure, one branch for the Maxpooling layer with a large convolutional kernel of 2 and the other branch for the convolutional kernel of size 3 with a step size of 2 for downsampling. Finally, the output of the two branches will be Concat operation and then output. Since the strip steel defect dataset contains many small and dense defects, which are not easy to identify, we try to add the attention mechanism CBAM [43] to the MP layer structure to build the attention-pooling module, as shown in Figure 7b. As a result, the network is enabled to focus on more important targets by itself and strengthen the ability of the network structure to identify defects.

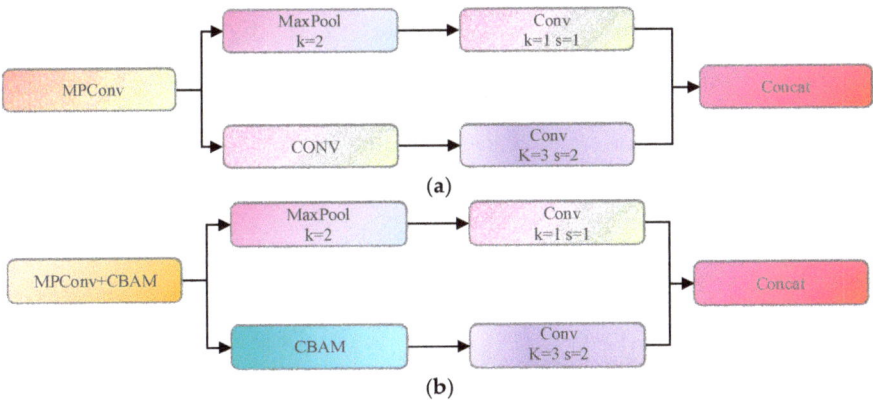

Figure 7. Attention pooling module. (a) MP layer structure; (b) MP layer structure with CBAM.

The CBAM is an attention mechanism module that combines spatial and channel. It can achieve better results than the squeeze-and-excitation networks (SEnets) [44] attention mechanism, which only focuses on channels. The CBAM contains two independent sub-modules, namely, the channel attention module (CAM) and the spatial attention module (SAM). This not only makes the network more capable of focusing on key information but also achieves a plug-and-play effect by weighting attention in both channel and space, respectively.

When the feature map is input to the CBAM, it will first pass through the CAM. In the CAM, the feature map will first pass through two parallel MaxPool and AvgPool layers to compress the feature map into two one-dimensional feature vectors. Next, the two one-dimensional feature vectors are fed into a two-layer shared neural network for activation and channel transformation. The result of the transformation is subjected to an Add operation. Finally, the final channel attention feature is generated by a sigmoid activation operation. The channel attention feature map and the input feature map are multiplied elementwise to generate the input features needed by the SAM. After entering the SAM, the channel-based global max pooling and global average pooling operations are performed to compress the channels of the feature map. Next, the channel-based Concat operation on these two results is carried out. After that, the convolution operation is performed to reduce the dimensionality to one channel. Then, the spatial attention feature map is generated by the sigmoid activation function. Finally, the spatial attention feature is multiplied by the input feature of this module to obtain the final generated feature.

4. Experiment and Result Analysis

In this section, the dataset, evaluation metrics, comparison objects, and methods are described, and the experimental results are analyzed to confirm the usefulness of the improved model.

4.1. Experimental Details and Dataset

The experimental environment used a computer with the following configuration: Windows 11 operating system, Intel (R) Core (TM) i5-9300 CPU with 2.40 GHz processor, and NVIDIA GeForce GTX 1650 graphics card. The experimental dataset of the NEU-DET [45] dataset of Northeastern University consists of six types of defect image data, including crazing, inclusion, patches, pitted surface, rolled-in scale, and scratches. There are 300 samples for each type of defect, for a total of 1800 grayscale images. For the defect detection task, bounding box annotations are also provided in the dataset, indicating the class and location of defects in each image. A total of 1800 images, the image size is 200 × 200, the training set and the test set are divided according to the ratio of 9:1, and then 10% of the training subset is used as the validation set.

4.2. Performance Evaluation

To measure the accuracy of target defect detection, we used two metrics (i.e., average precision (AP) and mean average accuracy (mAP)) as performance evaluation criteria. These are calculated as follows.

$$Precision = \frac{TP}{TP + FP} \tag{3}$$

$$Recall = \frac{TP}{TP + FN} \tag{4}$$

$$AP = \int_0^1 P(R)dR \tag{5}$$

$$mAP = \frac{\sum_{i=1}^{c} AP_i}{c} \tag{6}$$

where the TP is a true-positive defect, the FP is a false-positive defect, and the FN is a false-negative defect. The $P(R)$ is the precision–recall curve, the i is a defect category, and the c is the number of defect categories with a value of six in this experiment.

4.3. Ablation Research

For the ablation study, we designed a detailed algorithm based on YOLOv7 as the baseline model, the CovNeXt network structure is added to the backbone, the ELAN structure is replaced by the C3C2 structure in the detection head, the CBAM is embedded in the MP layer structure to construct the attention pooling structure. In order to reasonably judge whether the proposed improvements are of application value for strip steel defect detection, the joint ablation experiments on the NEU-DET dataset is carried out. The results are listed in Table 1, where the YOLOv7 represents the original YOLOv7 model. The YOLOv7–ConvNeXt-B represents the replacement of the first and last ELAN modules in the backbone of YOLOv7 with the ConvNeXt-B module. The YOLOv7–C3C2 represents the replacement of the original ELAN structure in the head of the YOLOv7 model with our improved C3C2 structure. The YOLOv7–CBAM represents the addition of the CBAM attention mechanism in the first pooling layer of the head, and the Ours represents the proposed algorithm.

As can be seen from Table 1: (1) after adding the ConvNeXt structure to YOLOv7, although the overall mAP value is not improved, it makes the whole network structure have better recognition ability for the crazing, patches, rolled-in scale, and scratches. (2) For the application of the C3C2 module, the size of the network model is reduced from 36.5 MB to 30.2 MB without reducing the overall detection accuracy of the network. The number of parameters of the network is reduced greatly, and the inference speed of the network speeds up. (3) The AP values of inclusion, patches, rolled-in scale, and scratches are also improved. Finally, the YOLOv7–CBAM greatly enhances the sensitivity of the network to line defects and significantly improves the detection capability of the network for both crazing and scratch defects. (4) For our proposed algorithm, the AP values of the four types of defects (crazing, patches, rolled-in scale, and scratches) have been improved more significantly, except for a slight decrease in inclusion defects.

Table 1. Ablation experiments results.

	mAP%	AP%					
		Crazing	Inclusion	Patches	Pitted Surface	Rolled-In Scale	Scratches
YOLOv7	76.3	48.1	76.5	94.8	99.5	67.0	72.0
YOLOv7–ConNeXt-B	76.3	54.1	62.3	96.4	95.6	67.6	82.1
YOLOv7–C3C2	75.5	35.9	76.8	99.1	93.1	71.5	76.5
YOLOv7–CBAM	79.4	63.6	66.7	97.7	99.5	65.5	83.6
Ours	82.9	68.9	68.3	97.8	99.5	73.3	89.3

Note: AP = Average precision; mAP = Mean AP; YOLO = You Only Look Once.

4.4. Contrasting Experiment

The Precision–Recall (P–R) curves of the YOLOv5, original YOLOv7, and improved YOLOv7 algorithm models for the detection of the six defects in the NEU-DET dataset are shown in Figure 8.

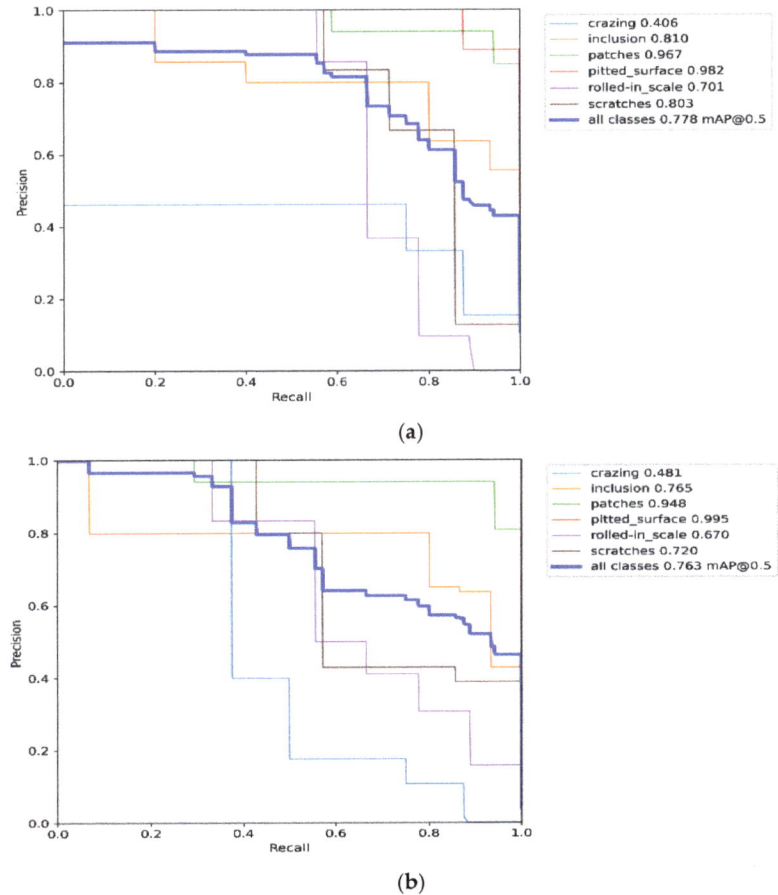

Figure 8. *Cont.*

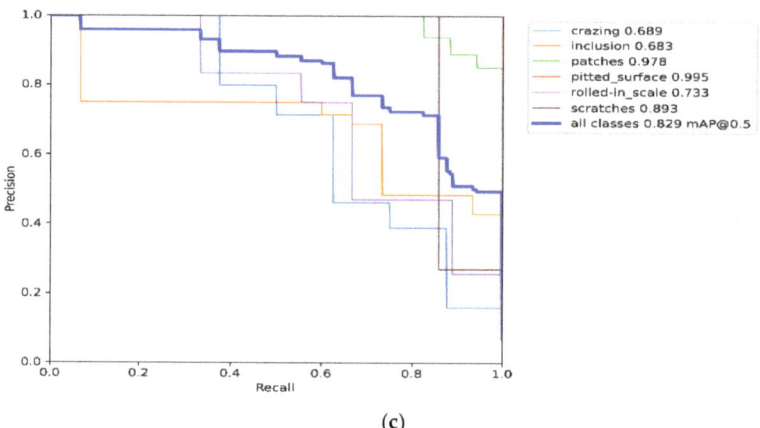

(c)

Figure 8. Precision–recall (P–R) curves. (**a**) YOLOv5; (**b**) YOLOv7; (**c**) Improved YOLOv7.

The P–R curve is an important indicator to measure the performance of the model. In the P–R curves, the larger the area enclosed by the curve, the better the performance. As can be seen from Figure 8, compared with the other two algorithms, our proposed algorithm has better defect detection performance.

Figure 9 represents a comparison of the visualization results of the detection effects of the above models. It is easy to see that the improved YOLOv7 algorithm model is not only able to locate and find all defects more accurately but also make the prediction frame more precise. Meanwhile, compared with the other two models, the improved YOLOv7 algorithm model is able to improve the false and missed detection very well.

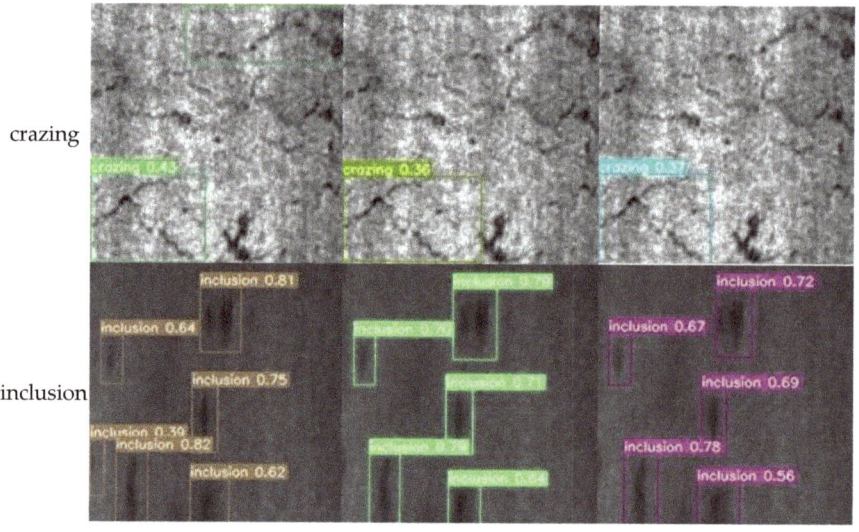

Figure 9. *Cont.*

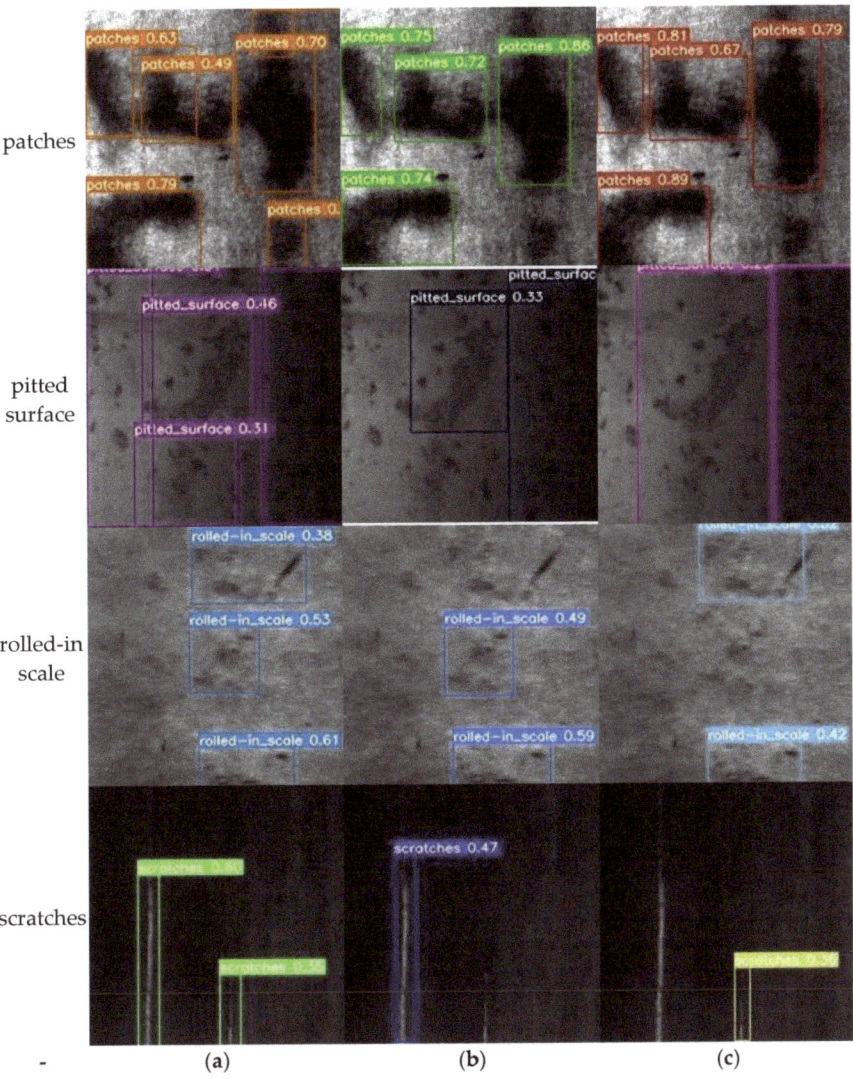

Figure 9. Comparison of detection results: (**a**) our algorithm; (**b**) YOLOv5 algorithm; (**c**) original YOLOv7 algorithm.

In order to more closely verify the superiority of the improved YOLOv7 algorithm compared with other algorithms, a comparison experiment was performed with five other classical algorithms that have been proposed. The experimental results are shown in Table 2. Compared with the other algorithms, the AP values of the proposed algorithm are not as good as the other algorithms for inclusion defects. However, the AP values and mAP values for other typical defects are higher than the other algorithms, and the mAP value is 6.6% higher than the YOLOv7. In summary, the proposed algorithm has a high practicality for the detection of strip defects in industrial production.

Table 2. Comparison results with other algorithms.

	mAP%	AP%					
		Crazing	Inclusion	Patches	Pitted Surface	Rolled-In Scale	Scratches
YOLOv5	77.80	40.60	81.00	96.70	98.20	70.10	80.30
YOLOv7	76.30	48.10	76.50	94.80	99.50	67.00	72.00
YOLOX	73.37	46.06	73.26	86.58	83.55	52.80	97.98
SSD	75.43	62.72	75.63	94.31	71.46	65.89	82.54
RetinaNet	67.56	45.65	68.34	89.99	81.52	58.60	61.27
Ours	82.90	68.90	68.30	97.80	99.50	73.30	89.30

Note: AP = Average precision; mAP = Mean AP; SSD = Single-shot multi-box detection; YOLO = You Only Look Once.

5. Conclusions

Strip steel often inevitably produces some defects due to the level of production technology and the impact of external factors. However, in the image of strip surface defects, the shapes of similar defects are different, and there are similarities between different defects. Undoubtedly, it brings great obstacles to the defect detection algorithm. The conventional target detection algorithm can not well meet the requirements of the actual industrial production process of strip defect detection accuracy rate and inference speed. In this study, an improved YOLOv7-based detection algorithm to meet the requirements of accuracy and inference speed for strip steel defect detection is proposed. The value of this algorithm lies in adding the ConNeXt module to the backbone of YOLOv7, replacing the ELAN structure of the original model with the improved C3C2 structure in the head of YOLOv7, and embedding the CBAM attention module in the original pooling module. Among the six typical defects in the NEU-DET dataset, the proposed algorithm improves by 6.6% in mAP compared to the original model YOLOv7. Compared to the other single-stage target detection algorithms, SSD and RetinaNet, the mAP values increased by 7.47% and 15.34%, respectively. Compared to the YOLOX algorithm and YOLOv5, the detection accuracy of the other five defects is improved, except for the detection accuracy of inclusion defects, which is slightly decreased. Of course, there are still some shortcomings in our proposed algorithm; for example, the defect detection effect for the crazing and inclusion is still weak. We intend to further explore how to improve the algorithm's ability to identify these two types of defects through experiments in the future.

Author Contributions: Methodology, R.W. and X.M.; Software, F.L. and R.W.; Formal analysis, H.C.; Investigation, X.M. and X.Y.; Data curation, L.C. and Z.P.; Writing—original draft, F.L. and R.W.; Writing—review & editing, R.W. and X.M.; Funding acquisition, R.W. and X.M. All authors have read and agreed to the published version of the manuscript.

Funding: This study was co-supported by the industry-university-research innovation fund projects of China University in 2021(No. 2021ITA10018); the fund project of the Key Laboratory of AI and Information Processing (No. 2022GXZDSY101); the Natural Science Foundation Project of Guangxi, China (No. 2018GXNSFAA050026); the Key R&D Program Project of Guangxi, China (No. 2021AB38023); the basic ability improvement project for young and middle-aged teachers of universities in Guangxi, China (No. 2022KY0058).

Institutional Review Board Statement: Not applicable.

Informed Consent Statement: Not applicable.

Data Availability Statement: Data is contained within the article.

Conflicts of Interest: The authors declare no conflict of interest.

References

1. Kou, X.; Liu, S.; Cheng, K.; Qian, Y. Development of a YOLO-V3-based model for detecting defects on steel strip surface. *Measurement* **2021**, *182*, 109454-1–109454-9. [CrossRef]
2. Mordia, R.; Verma, A.K. Visual techniques for defects detection in steel products: A comparative study. *Eng. Fail. Anal.* **2022**, *134*, 106047–106058. [CrossRef]
3. Sun, B.; Cheng, L.; Du, C.-Y.; Zhang, J.-K.; He, Y.-Q.; Cao, G.-M. Effect of Oxide Scale Microstructure on Atmospheric Corrosion Behavior of Hot Rolled Steel Strip. *Coatings* **2021**, *11*, 517. [CrossRef]
4. Shi, H.; Wang, J.; Li, Y. Small sample data enhancement method for strip steel based on improved ACGAN algorithm. *Comput. Integr. Manuf. Syst.* **2023**, 1–12. Available online: https://kns.cnki.net/kcms/detail//11.5946.TP.20230104.1047.004.html (accessed on 12 April 2021).
5. Hao, R.; Lu, B.; Cheng, Y.; Li, X.; Huang, B. A steel surface defect inspection approach towards smart industrial monitoring. *J. Intell. Manuf.* **2021**, *32*, 1833–1843. [CrossRef]
6. Ma, Y.; Zhao, H.; Yan, C.; Feng, H.; Yu, K.; Liu, H. Strip steel surface defect detection method by improved YOLOv5 network. *J. Electron. Meas. Instrum.* **2022**, *36*, 150–157.
7. Liang, X.; Xiao, H. Lightweight strip defect real-time detection algorithm based on SDD-YOLO. *China Meas. Test.* **2023**, 1–8. Available online: https://kns.cnki.net/kcms/detail//51.1714.TB.20230109.1648.002.html (accessed on 12 April 2021).
8. Wu, H.; Lv, Q.; Giovanni, D. Hot-Rolled Steel Strip Surface Inspection Based on Transfer Learning Model. *J. Sens.* **2021**, *2021 Pt 3*, 6637252-1–6637252-8. [CrossRef]
9. Guan, S.; Chang, J.; Shi, H.; Xiao, X.; Li, Z.; Wang, X.; Wang, X. Strip Steel Defect Classification Using the Improved GAN and EfficientNet. *Appl. Artif. Intell.* **2021**, *35*, 1887–1904. [CrossRef]
10. Chu, M.-X.; Liu, X.-P.; Gong, R.-F.; Zhao, J. Multi-class classification method for strip steel surface defects based on support vector machine with adjustable hyper-sphere. *J. Iron Steel Res. Int.* **2018**, *25*, 706–716. [CrossRef]
11. Huang, X.; Sun, S.; Zhang, Y.; Li, B.; Ren, Y.; Zhao, L. Research on the detection method of surface defects of strip steel under uneven illumination. *Mech. Sci. Technol. Aerosp. Eng.* **2023**, 1–8. [CrossRef]
12. Tsai, D.-M.; Chen, M.-C.; Li, W.-C.; Chiu, W.-Y. A fast regularity measure for surface defect detection. *Mach. Vis. Appl.* **2012**, *23*, 869–886. [CrossRef]
13. Liu, K.; Wang, H.; Chen, H.; Qu, E.; Tian, Y.; Sun, H. Steel Surface Defect Detection Using a New Haar-Weibull-Variance Model in Unsupervised Manner. *IEEE Trans. Instrum. Meas.* **2017**, *66*, 2585–2596. [CrossRef]
14. Neogi, N.; Mohanta, D.K.; Dutta, P.K. Defect Detection of Steel Surfaces with Global Adaptive Percentile Thresholding of Gradient Image. *J. Inst. Eng. Ser. B* **2017**, *98*, 557–565. [CrossRef]
15. Tong, L.; Wong, W.K.; Kwong, C.K. Differential evolution-based optimal Gabor filter model for fabric inspection. *Neurocomputing* **2016**, *173*, 1386–1401. [CrossRef]
16. Choi, D.-C.; Jeon, Y.-J.; Kim, S.H.; Moon, S.; Yun, J.P.; Kim, S.W. Detection of Pinholes in Steel Slabs Using Gabor Filter Combination and Morphological Features. *ISIJ Int.* **2017**, *57*, 1045–1053. [CrossRef]
17. Wang, J.; Li, Q.; Gan, J.; Yu, H.; Yang, X. Surface Defect Detection via Entity Sparsity Pursuit with Intrinsic Priors. *IEEE Trans. Ind. Inform.* **2019**, *16*, 141–150. [CrossRef]
18. Girshick, R.; Donahue, J.; Darrell, T.; Malik, J. Rich Feature Hierarchies for Accurate Object Detection and Semantic Segmentation. In Proceedings of the 2014 IEEE Conference on Computer Vision and Pattern Recognition (CVPR 2014), Columbus, OH, USA, 23–28 June 2014; pp. 580–587.
19. Ross, G. Fast R-CNN[A] in: Institute of Electrical and Electronics Engineers. In Proceedings of the 2015 IEEE International Conference on Computer Vision (ICCV 2015), Santiago, Chile, 11–18 December 2015; pp. 1440–1448.
20. Ren, S.; He, K.; Girshick, R.; Sun, J. Faster R-CNN: Towards Real-Time Object Detection with Region Proposal Networks. *IEEE Trans. Pattern Anal. Mach. Intell.* **2017**, *39*, 1137–1149. [CrossRef]
21. Liu, W.; Anguelov, D.; Erhan, D.; Szegedy, C.; Reed, S.; Fu, C.Y.; Berg, A.C. SSD: Single Shot MultiBox Detector. In Proceedings of the Computer Vision—ECCV 2016: 14th European Conference, Amsterdam, The Netherlands, 11–14 October 2016; Springer: Berlin/Heidelberg, Germany, 2016; pp. 21–37.
22. Lin, T.Y.; Goyal, P.; Girshick, R.; He, K.; Dollár, P. Focal Loss for Dense Object Detection. *IEEE Trans. Pattern Anal. Mach. Intell.* **2020**, *42*, 318–327. [CrossRef]
23. Joseph, R.; Santosh, D.; Ross, G.; Ali, F. You Only Look Once: Unified, Real-Time Object Detection. In Proceedings of the 29th IEEE Conference on Computer Vision and Pattern Recognition (CVPR), Las Vegas, NV, USA, 26 June–1 July 2016; Institute of Electrical and Electronics Engineers: Piscataway, NJ, USA, 2016; pp. 779–788.
24. Joseph, R.; Ali, F. YOLO9000: Better, Faster, Stronger. In Proceedings of the 2017 IEEE Conference on Computer Vision and Pattern Recognition (CVPR 2017), Honolulu, HI, USA, 21–26 July 2017; Institute of Electrical and Electronics Engineers: Piscataway, NJ, USA, 2017; pp. 6517–6525.
25. Redmon, J.; Farhadi, A. Yolov3: An incremental improvement. *arXiv* **2018**, arXiv:1804.02767.
26. Bochkovskiy, A.; Wang, C.; Liao, H. YOLOv4: Optimal Speed and Accuracy of Object Detection. *arXiv* **2020**, arXiv:2004.10934.
27. Jocher, G. YOLOv5 Release v5.0. 2021. Available online: https://github.com/ultralytics/yolov5/releases/tag/v5.0 (accessed on 12 April 2021).
28. Ge, Z.; Liu, S.; Wang, F.; Li, Z.; Sun, J. YOLOX: Exceeding YOLO Series in 2021. *arXiv* **2021**, arXiv:2107.08430.

29. Li, C.; Li, L.; Jiang, H.; Weng, K.; Geng, Y.; Li, L.; Wei, X. YOLOv6: A Single-Stage Object Detection Framework for Industrial Applications. *arXiv* **2022**, arXiv:2209.02976.
30. Wang, C.Y.; Bochkovskiy, A.; Liao, H. YOLOv7: Trainable bag-of-freebies sets new state-of-the-art for real-time object detectors. *arXiv* **2021**, arXiv:2207.02696.
31. Wang, S.; Xia, X.; Ye, L.; Yang, B. Automatic Detection and Classification of Steel Surface Defect Using Deep Convolutional Neural Networks. *Met.-Open Access Metall. J.* **2021**, *11*, 388. [CrossRef]
32. He, K.; Zhang, X.; Ren, S.; Sun, J. Deep Residual Learning for Image Recognition. *arXiv* **2015**, arXiv:1512.03385.
33. Li, M.-J.; Wang, H.; Wan, Z.-B. Surface defect detection of steel strips based on improved YOLOv4. *Comput. Electr. Eng.* **2022**, *102*, 45–53. [CrossRef]
34. Dosovitskiy, A.; Beyer, L.; Kolesnikov, A.; Weissenborn, D.; Zhai, X.; Unterthiner, T.; Dehghani, M.; Minderer, M.; Heigold, G.; Gelly, S.; et al. An Image is Worth 16x16 Words: Transformers for Image Recognition at Scale. *arXiv* **2021**, arXiv:2010.11929.
35. Guo, Z.; Wang, C.; Yang, G.; Huang, Z.; Li, G. MSFT-YOLO: Improved YOLOv5 Based on Transformer for Detecting Defects of Steel Surface. *Sensors* **2022**, *22*, 3467. [CrossRef]
36. Lee, Y.; Hwang, J.-W.; Lee, S.; Bae, Y.; Park, J. An Energy and GPU-Computation Efficient Backbone Network for Real-Time Object Detection. In Proceedings of the 2019 IEEE/CVF Conference on Computer Vision and Pattern Recognition Workshops (CVPRW 2019), Long Beach, CA, USA, 16–17 June 2019; Institute of Electrical and Electronics Engineers: Piscataway, NJ, USA, 2016; pp. 752–760.
37. Wang, C.-Y.; Bochkovskiy, A.; Liao, H.-Y.M. Scaled-YOLOv4: Scaling Cross Stage Partial Network. In Proceedings of the 2021 IEEE/CVF Conference on Computer Vision and Pattern Recognition (CVPR 2021), Virtual Conference, 19–25 June 2021; Institute of Electrical and Electronics Engineers: Piscataway, NJ, USA, 2016; pp. 13024–13033.
38. Wang, C.-Y.; Liao, H.-Y.M.; Yeh, I.-H. Designing Network Design Strategies Through Gradient Path Analysis. *arXiv* **2022**, arXiv:2211.04800.
39. Liu, Z.; Mao, H.; Wu, C.Y.; Feichtenhofer, C.; Darrell, T.; Xie, S. A ConvNet for the 2020s. *arXiv* **2022**, arXiv:2201.03545.
40. Liu, Z.; Lin, Y.; Cao, Y.; Hu, H.; Wei, Y.; Zhang, Z.; Lin, S.; Guo, B. Swin Transformer: Hierarchical Vision Transformer using Shifted Windows. *arXiv* **2021**, arXiv:2103.14030.
41. Xie, S.; Girshick, R.; Dollár, P.; Tu, Z.; He, K. Aggregated Residual Transformations for Deep Neural Networks. *arXiv* **2017**, arXiv:1611.05431.
42. Diganta, M. Mish: A Self Regularized Non-Monotonic Activation Function. *arXiv* **2019**, arXiv:1908.08681.
43. Woo, S.; Park, J.; Lee, J.-Y.; Kweon, I.S. CBAM: Convolutional Block Attention Module. *arXiv* **2018**, arXiv:1807.06521.
44. Hu, J.; Shen, L.; Albanie, S.; Sun, G.; Wu, E. Squeeze-and-Excitation Networks. *IEEE Trans. Pattern Anal. Mach. Intell.* **2020**, *42*, 2011–2023. [CrossRef]
45. Song, K.; Yan, Y. A noise robust method based on completed local binary patterns for hot-rolled steel strip surface defects. *Appl. Surf. Sci.* **2013**, *285*, 858–864. [CrossRef]

Disclaimer/Publisher's Note: The statements, opinions and data contained in all publications are solely those of the individual author(s) and contributor(s) and not of MDPI and/or the editor(s). MDPI and/or the editor(s) disclaim responsibility for any injury to people or property resulting from any ideas, methods, instructions or products referred to in the content.

Article

Highly Efficient CuInSe$_2$ Sensitized TiO$_2$ Nanotube Films for Photocathodic Protection of 316 Stainless Steel

Zhanyuan Yang [1], Hong Li [1,2,*], Xingqiang Cui [1], Jinke Zhu [1], Yanhui Li [1,2], Pengfei Zhang [1] and Junru Li [1]

1 College of Mechanical and Electrical Engineering, Qingdao University, Qingdao 266071, China
2 State Key Laboratory of Bio-Fibers and Eco-Textiles, Qingdao University, Qingdao 266071, China
* Correspondence: lhqdio1987@163.com; Tel.: +86-532-8595-3679

Abstract: CuInSe$_2$ nanoparticles were successfully deposited on the surface of TiO$_2$ nanotube arrays (NTAs) by a solvothermal method for the photocathodic protection (PCP) of metals. Compared with TiO$_2$ NTAs, the CuInSe$_2$/TiO$_2$ composites exhibited stronger visible light absorption and higher photoelectric conversion efficiency. After 316 Stainless Steel (SS) was coupled with CuInSe$_2$/TiO$_2$, the potential of 316 SS could drop to -0.90 V. The photocurrent density of CuInSe$_2$/TiO$_2$ connected to 316 SS reached 140 μA cm^{-2}, which was four times that of TiO$_2$ NTAs. The composites exhibited a protective effect in the dark state for more than 8 h after 4 h of visible light illumination. The above could be attributed to increased visible light absorption, the extended lifetime of photogenerated electrons, and generation of oxygen vacancies.

Keywords: TiO$_2$ NTAs; CuInSe$_2$; photocathodic protection; 316 SS; EIS

Citation: Yang, Z.; Li, H.; Cui, X.; Zhu, J.; Li, Y.; Zhang, P.; Li, J. Highly Efficient CuInSe$_2$ Sensitized TiO$_2$ Nanotube Films for Photocathodic Protection of 316 Stainless Steel. *Coatings* **2022**, *12*, 1448. https://doi.org/10.3390/coatings12101448

Academic Editors: Matic Jovičević-Klug, Patricia Jovičević-Klug and László Tóth

Received: 1 September 2022
Accepted: 27 September 2022
Published: 30 September 2022

Publisher's Note: MDPI stays neutral with regard to jurisdictional claims in published maps and institutional affiliations.

Copyright: © 2022 by the authors. Licensee MDPI, Basel, Switzerland. This article is an open access article distributed under the terms and conditions of the Creative Commons Attribution (CC BY) license (https://creativecommons.org/licenses/by/4.0/).

1. Introduction

316 SS is widely used for industrial applications because of its good corrosion resistance and excellent mechanical properties. Nevertheless, stainless steel (SS) is prone to pitting corrosion in Cl$^-$-rich solution [1–3]. Metal corrosion is extremely harmful, causing huge economic losses and even safety accidents every year [4,5]. Many anti-corrosion methods, including corrosion inhibitors [6], anticorrosion coatings [7] and cathodic protection [8], have been developed to inhibit the corrosion of steel. Photocathodic protection (PCP) is a new type of green anti-corrosion technology. Its principle is that when the semiconductor is irradiated by light which possesses higher energy than the band gap (E_g) of the semiconductor, electrons are excited from the valence band (VB) to the conduction band (CB); then, the photogenerated electrons are transferred from the semiconductor to the connected metal to reduce the potential of the metal surface, thereby inhibiting the metal corrosion. Compared with traditional cathodic protection techniques, PCP technology neither consumes energy nor releases metal ions into the environment [9–11]. Therefore, the development of PCP systems to protect metal against substrates is a promising approach for many industrial applications.

TiO$_2$ has attracted great interest as a photoanode material for PCP application due to its stable physical and chemical properties, lack of toxicity, excellent photoelectric properties and low cost [12–14]. Unfortunately, some inherent defects of TiO$_2$ limit its application. First of all, TiO$_2$ is unable to utilize most sunlight (less than 5% of solar energy) due to its wide E_g [15]. Secondly, the photogenerated carriers in TiO$_2$ are easy to recombine, which greatly reduces its photoelectric conversion efficiency [16], making it unable to protect metals in dark environments. Therefore, it is necessary to modify the TiO$_2$, for example, through doping, with metal elements (W [17], Fe [18], Ni [19], etc.) or non-metal elements (N [20], B [21], etc.), and co-sensitizing with narrow gap semiconductors (MoS$_2$ [22], Co(OH)$_2$ [23], FeS$_2$ [24], etc.).

Ternary semiconductors have aroused great attention because of their adjustable E_g and electronic energy level, controllable composition and internal structure [25]. Polymetallic sulfides/selenides have good electrochemical properties due to the synergistic effect of two metal atoms [26,27]. At present, $AgInS_2$ and $AgInSe_2$ have been used to modify TiO_2 and have achieved good PCP effects [28,29]. In addition, selenides have faster heterogeneous electron transfer rates than sulfides because the electronegativity of Se is lower than S [30]. $CuInSe_2$ is regarded as a promising photovoltaic material due to its adjustable band gap, high optical stability and excellent photoelectric conversion efficiency. Therefore, the sensitization of $CuInSe_2$ may improve the composites' efficiency of utilization of sunlight [31]. More importantly, the CB potential (E_{CB}) of $CuInSe_2$ was more negative than that of TiO_2, so it is possible to construct a $CuInSe_2/TiO_2$ heterojunction, which can facilitate the efficient transfer of electrons from $CuInSe_2$ to TiO_2 [32]. Therefore, $CuInSe_2$ may be an ideal semiconductor material for modifying TiO_2. According to previous reports, some researchers have constructed $CuInSe_2/TiO_2$ nanostructures for photocatalytic degradation of organic pollutants [33] and solar cell applications [34]. However, there are no reports on the PCP performance of $CuInSe_2/TiO_2$ for metals.

In this paper, $CuInSe_2/TiO_2$ composites were synthesized by anodic oxidation and solvothermal methods. The morphologies, crystal structures, composition and light absorption properties of the composite materials were studied. The PCP performances of photoanodes were studied by electrochemical test methods. Through density functional theory (DFT) calculations, the electronic structure changes of composites were calculated. The PCP mechanism of $CuInSe_2/TiO_2$ on 316 SS was studied.

2. Materials and Methods

2.1. Chemicals and Reagents

The Ti foils (40 mm × 10 mm × 0.1 mm; 99.9% purity) were purchased from Shanghai Gao Dewei Co., Shanghai, China, and 316 SS was purchased from Shanghai Baosteel Co., Shanghai, China with the following ingredients (wt%): C 0.08%, Si 0.90%, P 0.045%, S 0.029%, Mn 1.80%, Ni 14.00%, Cr 17% and balanced Fe. $CuCl_2·2H_2O$ (99.0%), $InCl_3·4H_2O$ (99.9%), Se powder (99.0%), $Na_2S·9H_2O$ (98%), NaOH (99%) and NaCl (99.5%) were purchased from Sinopharm Chemical Reagent Co., Ltd. (Shanghai, China).

2.2. Fabrication of $CuInSe_2/TiO_2$ Photoelectrodes

Figure 1a illustrates the process for preparation of $CuInSe_2/TiO_2$ nanotube array (NTA) photoanodes. TiO_2 NTAs were synthesized on a Ti foil by electrochemical anodization. The Ti foil was first chemically buffed in a mixed solution of 2.5 mL H_2O, 0.45 g NH_4F, 6.0 mL H_2O_2 and 6.0 mL HNO_3, and then cleaned with first deionized water (DI water) and then absolute ethanol. The cleaned Ti foil and the Pt plate were used as the anode and cathode, respectively. The Ti foil was immersed in the electrolyte solution (0.36 g NH_4F, 4.0 mL H_2O, 60 mL ethylene glycol) for anodic oxidation at 30 V for 30 min, and then rinsed with, respectively, DI water and ethanol. The oxidized Ti foil was placed in a muffle furnace at 450 °C for 2 h, and then TiO_2 NTAs were obtained.

$CuInSe_2$ nanoparticles were synthesized on the TiO_2 by a solvothermal method. $CuCl_2·2H_2O$ (0.1, 0.2, 0.3 mmol), $InCl_3·4H_2O$ (0.1, 0.2, 0.3 mmol) and Se powder (0.2, 0.4, 0.6 mmol) were mixed with 60 mL methanol and magnetically stirred for 20 min. The mixed solution and the as-fabricated TiO_2 were immersed in a Teflon-lined autoclave and sintered at 200 °C for 12 h. Finally, $CuInSe_2/TiO_2$ composites were obtained. The synthesized photoanodes were marked, respectively, as $CuInSe_2/TiO_2$-A, $CuInSe_2/TiO_2$-B and $CuInSe_2/TiO_2$-C.

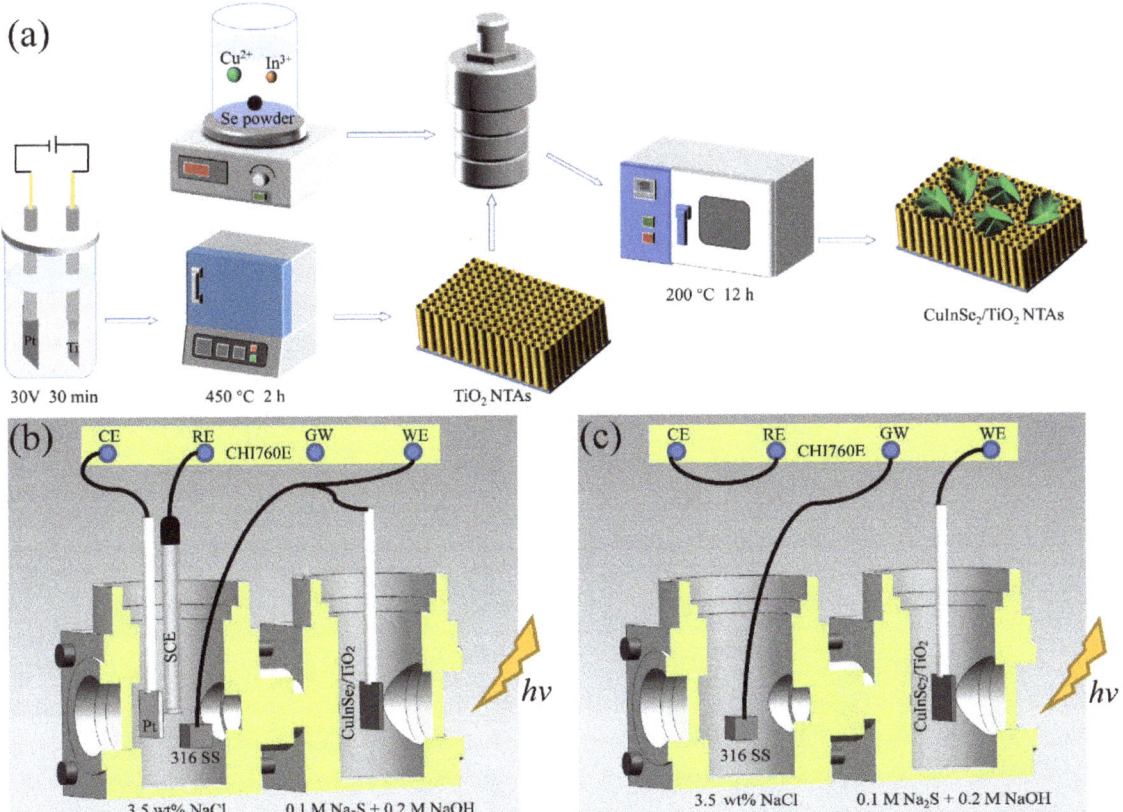

Figure 1. Schematic illustration of (**a**) the synthesis process for the CuInSe$_2$/TiO$_2$ NTAs. Test devices for measuring (**b**) OCP, Tafel and EIS, and (**c**) photocurrent densities.

2.3. Characterization

The morphologies of the photoelectrodes were observed using a scanning electron microscope (SEM, Hitachi SU8220, Tokyo, Japan) with Quantax75 energy-dispersive X-ray spectroscopy (EDS, Hitachi SU8220, Tokyo, Japan). The crystal structures of the photoelectrodes were obtained using X-ray diffraction (XRD, D8-advance, Bruker AXS Co., Karlsruhe, Germany) the Cu Kα radiation. The chemical components and element chemical states were analyzed by X-ray photoelectron spectroscopy (XPS, Thermo Fisher Scientific, Waltham, MA, USA, Al-Kα radiation, 1486.6 eV). The optical properties of the photoelectrode were tested using a UV-Vis diffuse reflectance spectrophotometer (DRS, Hitachi UH4150, Tokyo, Japan). Photoluminescence (PL, Ex: 320 nm) spectra were measured using an FLS980 series fluorescence spectrum scanner. Surface morphologies of 316 SS electrodes were obtained by a metallographic microscope (Axiocam 105 color, Oberkochen, Germany).

2.4. Photoelectrochemical Measurements

All measurements were carried out on an electrochemical workstation (CHI 760E, Chenhua Instrument Co., Ltd., Shanghai, China). The test device was composed of a corrosion cell (3.5 wt% NaCl) and a photoanode cell (0.1 M Na$_2$S + 0.2 M NaOH). The two cells were connected by Nafion film. The 316 SS electrode (1 cm × 1 cm × 1 cm) was encapsulated in epoxy resin with an exposed area of 1 cm^2. A 300 W xenon lamp (PLS-SXE 300C, Beijing Perfectlight Technology Co., Ltd., Beijing, China) with a 420 nm cut-off filter was used as a visible light source to illuminate the photoanode material vertically. The open circuit potentials (OCP), Tafel curves and electrochemical impedance spectroscopy (EIS) were recorded by the two-cell system (Figure 1b). The corrosion potential (E_{corr}) and corrosion current density (J_{corr}) were recorded by Tafel curves which were measured at a scanning rate of 0.5 mV·s^{-1} from −250 to 250 mV vs. OCP. The EIS data were obtained in the frequency range of 10^5–10^{-2} Hz with OCP as the initial potential and the amplitude of AC signal was 5 mV. The Mott–Schottky (M–S) curves were recorded from −1.0 V to 0.5 V with a frequency and an amplitude of 1000 Hz and 10 mV, respectively. The I–V and M–S tests were both carried out in 0.1 M Na_2SO_4 solution. The 316 SS in the corrosion cell and the photoanodes in the photoanode cell were connected by wires to the working electrode (WE). The saturated calomel electrode (SCE) and Pt plate were set as reference electrode (RE) and counter electrode (CE), respectively. The setup used to measure the photocurrent densities is shown in Figure 1c; CE and RE were connected with wires, 316 SS and photoanodes were connected to ground wire and WE, respectively.

3. Results and Discussion

3.1. Morphology and Chemical Compositions

The morphologies of TiO_2 and $CuInSe_2/TiO_2$-B were studied by SEM. Figure 2a,b shows that the average inner diameter and tube length of TiO_2 nanotubes were about 40 and 1600 nm, respectively. The ordered structure of TiO_2 NTAs can promote the separation and transport of e^-/h^+ pairs [35]. As shown in Figure 2c,d, $CuInSe_2$ nanoparticles were successfully loaded on the TiO_2 surface. It is evident that the deposition of the $CuInSe_2$ nanoparticles had no effect on the morphology of the nanotubes. The elemental mapping results of $CuInSe_2/TiO_2$-B show that the composite material is composed of Ti, O, Cu, In and Se elements. It further indicates that $CuInSe_2$ nanoparticles were formed uniformly on the surface of the nanotubes in the composite by the simple solvothermal method.

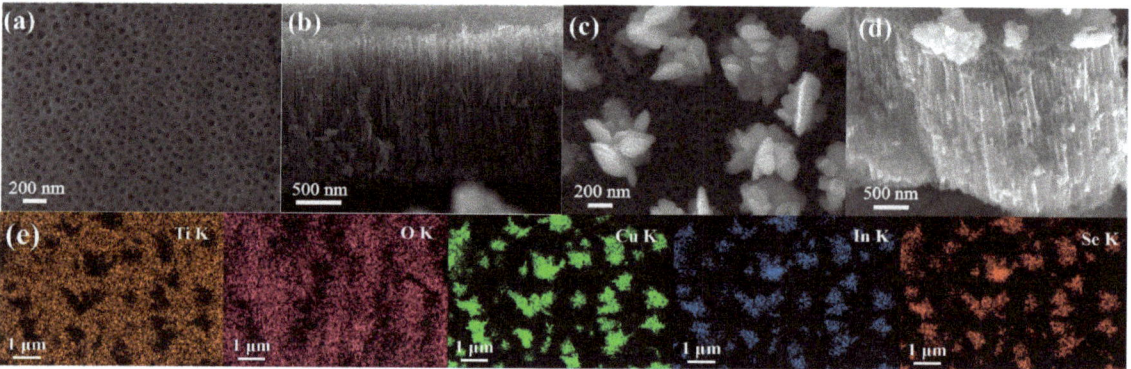

Figure 2. (a) SEM Top-view and (b) cross-section of the TiO_2 NTAs, (c) SEM top-view, (d) cross-section view of the $CuInSe_2/TiO_2$-B NTAs, and (e) EDS elemental mapping of the $CuInSe_2/TiO_2$-B NTAs.

XRD spectra were used to determine the crystal structural information of the synthesized photoanodes. Figure 3 shows the XRD patterns of TiO_2 and $CuInSe_2/TiO_2$-B. For TiO_2, the XRD peaks at 25.3°, 49.5° and 53.9° corresponded to (101), (200) and (105) planes of anatase TiO_2 (JCPDS No. 21-1272), respectively. The other peaks of TiO_2 were derived from the Ti substrate. For $CuInSe_2/TiO_2$-B photoanode, the XRD peaks at 27.3°, 28.3°, 43.9° and 44.2° corresponded to (112), (103), (105) and (301) crystal planes of tetragonal chalcopyrite $CuInSe_2$. No diffraction peak of the impurity phase was detected by XRD. Combined with the results of SEM and XRD, the $CuInSe_2/TiO_2$ photoanode materials with high purity were successfully prepared by anodic oxidation and solvothermal methods.

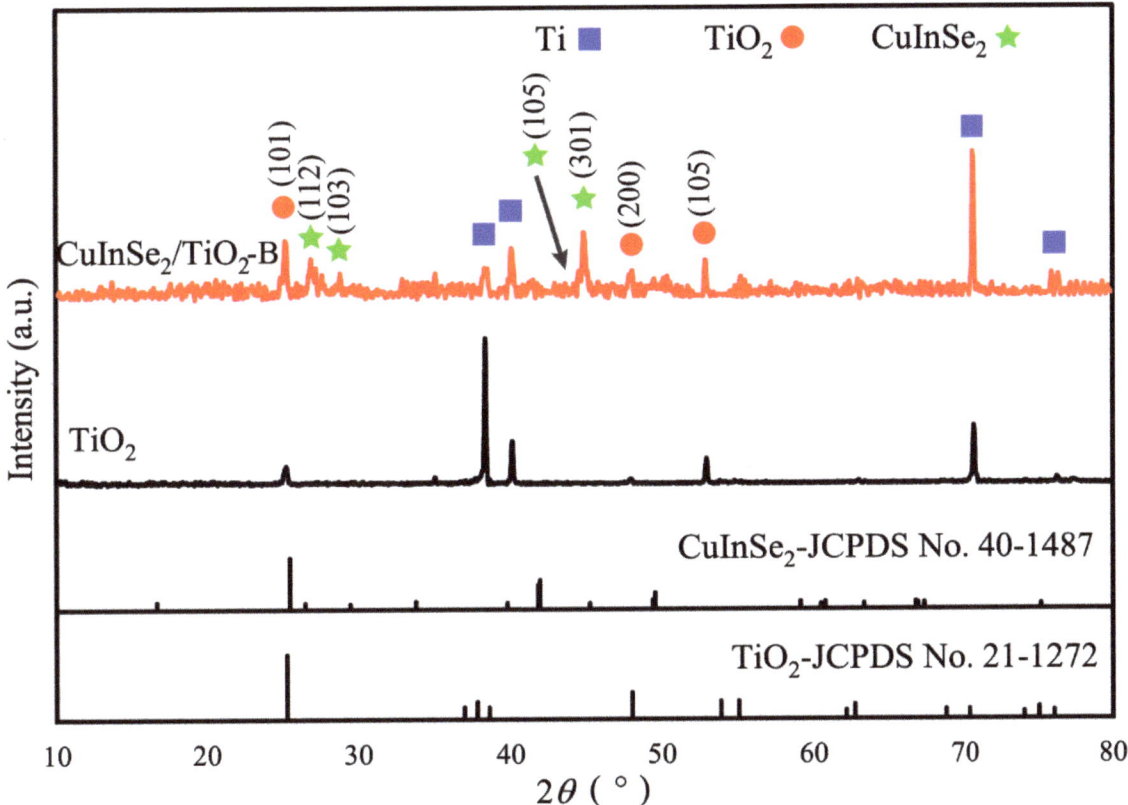

Figure 3. XRD patterns of TiO$_2$ and CuInSe$_2$/TiO$_2$-B.

The elemental chemical state of CuInSe$_2$/TiO$_2$-B was determined by XPS spectra. Figure 4a shows that the dominant elements of the composite material were Se, In, Ti, O, and Cu, in addition to C element. This result is consistent with the EDS mapping results. The peaks at 458.6 and 464.3 eV were in agreement with Ti 2p$_{3/2}$ and Ti 2p$_{1/2}$, respectively, indicating that Ti exists in the form of Ti^{4+}, which is derived from TiO2 [36] (Figure 4b). Figure 4c shows the peaks of 529.8 and 531.3 eV that were attributed to the lattice oxygen (O$_L$) and adsorbed oxygen (O$_A$) [37], respectively. The presence of O$_A$ indicates that oxygen vacancies were generated on the surface of the synthesized sample. Oxygen vacancy can facilitate electron transfer to the material surface, thus reducing the recombination of photogenerated carriers [38]. The binding energies of In 3d$_{5/2}$ and 3d$_{3/2}$ located at 444.5 and 452.4 eV, respectively (Figure 4d), corresponded to the binding energies of In^{3+} from CuInSe$_2$ [39]. As shown in Figure 4e, the peak at 54.2 eV was indexed to Se 3d$_{5/2}$, indicating the presence of Se^{2-} in the composite, and no selenium oxide was formed [40]. Two main peaks with the binding energy peaks at 932.7 and 952.5 eV in Figure 4f were attributed to Cu 2p$_{3/2}$ and Cu 2p$_{1/2}$, respectively. The difference between the two binding energy peaks is 19.8 eV confirming Cu^{2+} was reduced to Cu$^+$ during the process of reaction [41]. The above analysis further demonstrated the successful synthesis of CuInSe$_2$/TiO$_2$ photoanode composites.

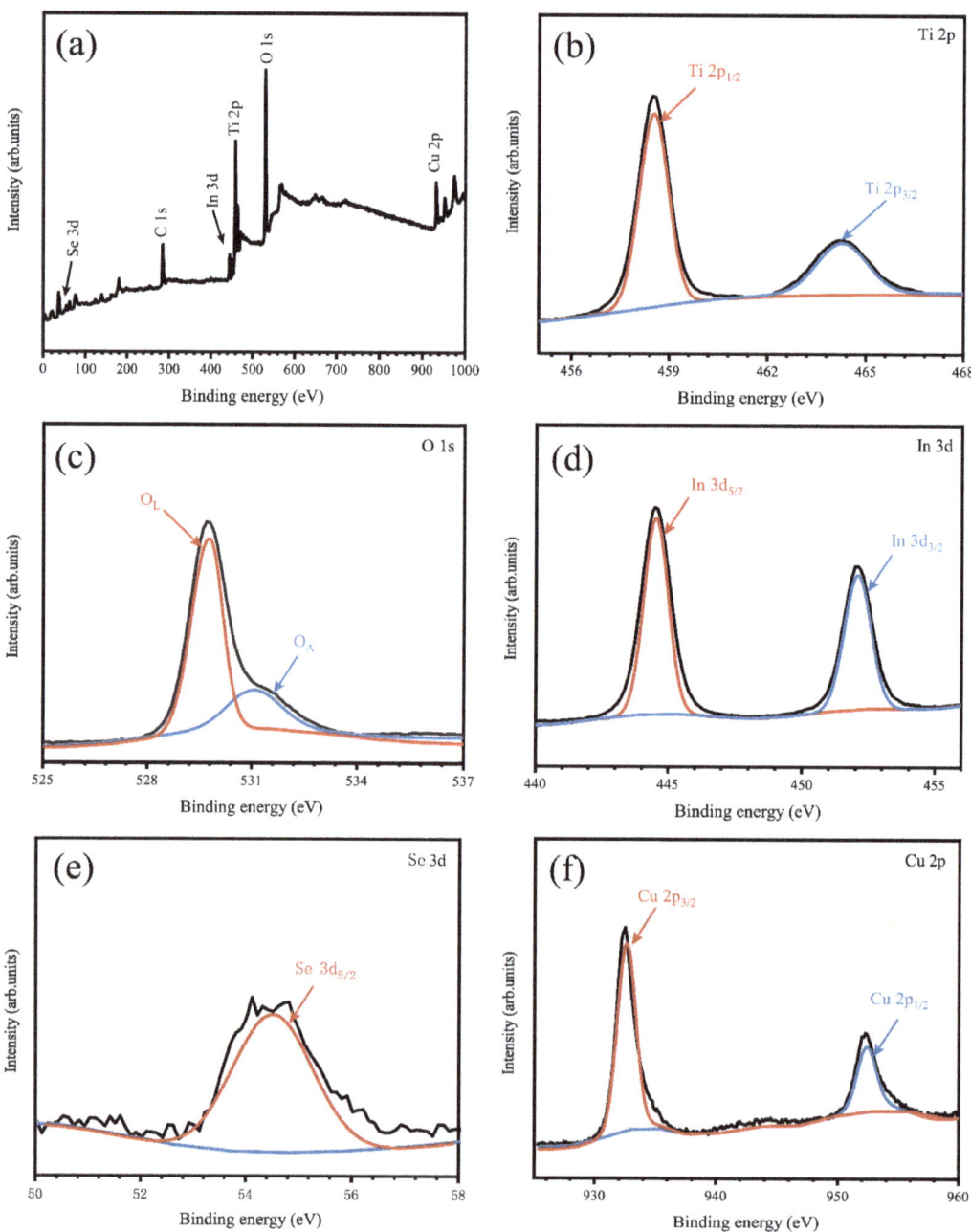

Figure 4. (**a**) The total XPS survey spectrum; high-resolution spectra of (**b**) Ti 2p, (**c**) O 1s, (**d**) In 3d, (**e**) Se 3d and (**f**) Cu 2p of $CuInSe_2/TiO_2$-B NTAs.

3.2. Optical Properties Analysis

The light absorption performances of pure TiO_2 and $CuInSe_2/TiO_2$-B were studied by UV-vis DRS. Figure 5a shows that pure TiO_2 can only absorb UV light with a wavelength less than 370 nm. In addition, the light absorption of TiO$_2$ in the range of 400–800 nm may be due to light scattering caused by cracks in TiO$_2$ [42]. The absorption edge was o red-shifted to 520 nm after depositing CuInSe$_2$ on the TiO$_2$. This indicates that the sensitization of CuInSe$_2$ nanoparticles improved the visible light absorption capacity of TiO$_2$ NTAs.

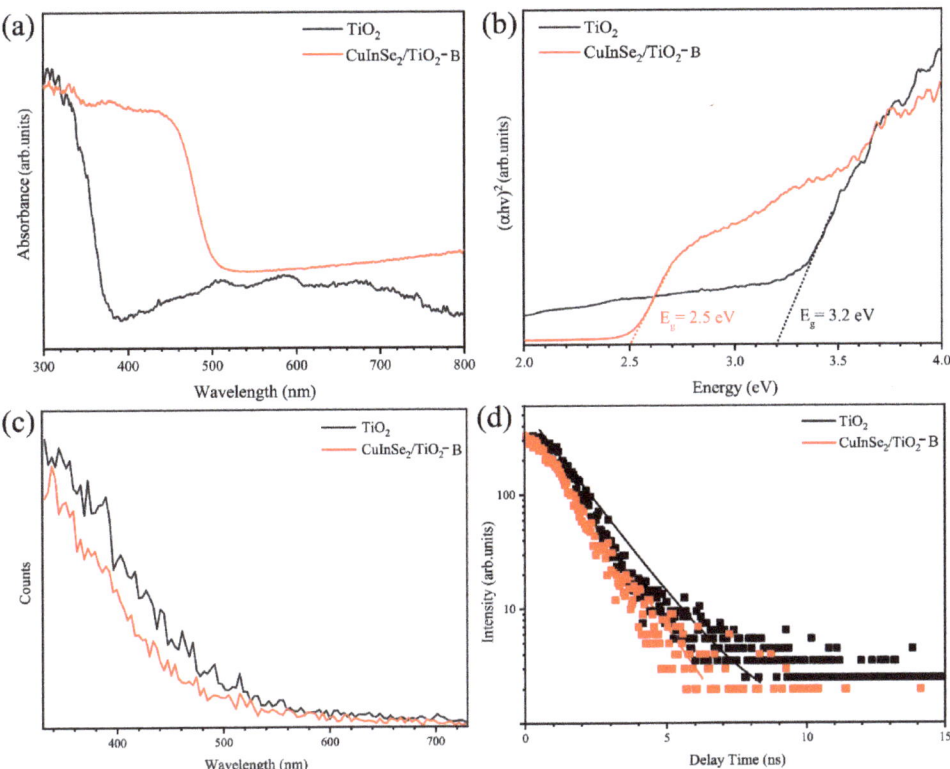

Figure 5. (**a**) UV-vis DRS, (**b**) Tauc plots, (**c**) PL spectra and (**d**) time-resolved PL spectra of the prepared TiO$_2$ and CuInSe$_2$/TiO$_2$-B.

The E_g of the photoanode materials can be calculated by the formula [43]:

$$(\alpha h v)^2 = A(h v - E_g) \qquad (1)$$

where α, h, v, and A stand for absorption coefficient, Planck's constant, optical frequency, and characteristic constant, respectively. Figure 5b shows that the E_g of TiO$_2$ and CuInSe$_2$/TiO$_2$ NTA composites were 3.2 and 2.5 eV, respectively.

The recombination rates of photogenerated carriers of CuInSe$_2$/TiO$_2$ and TiO$_2$ NTAs were analyzed by PL spectra. Figure 5c shows that the peak intensity of TiO$_2$ was larger than that of CuInSe$_2$/TiO$_2$, suggesting the photogenerated carrier recombination rate of CuInSe$_2$/TiO$_2$ decreases. The lifetime of the photogenerated carriers of was evaluated by time-resolved PL spectra (Figure 5d), which were used to characterize the lifetime of photogenerated carriers. The attenuation curve was fitted by the formula:

$$R(t) = B_1 e^{(-t/\tau_1)} + B_2 e^{(-t/\tau_2)} \qquad (2)$$

The emission-decay time-constant values associated with τ_1, τ_2, B_1 and B_2 can be calculated by the formula:

$$R(t) = \left(B_1\tau_1^2 + B_2\tau_2^2\right)/(B_1\tau_1 + B_2\tau_2) \tag{3}$$

Table 1 displays the fitting and calculation results from Figure 5d. The mean electron lifetimes of TiO_2 and $CuInSe_2/TiO_2$ were 1.26 and 1.32 ns, respectively. This indicated that the formation of the $CuInSe_2/TiO_2$ heterojunction inhibited the recombination of e^-/h^+ pairs, thus enhancing the performance of PCP.

Table 1. Fitting results from Figure 5d.

Samples	τ_1	B_1	τ_2	B_2	t
TiO_2	1.26	202.54	1.26	245.59	1.26
$CuInSe_2/TiO_2$	1.32	308.75	1.32	362.61	1.32

3.3. PCP Performance and Stability Evaluation

The PCP performances were evaluated by comparing the values of the photocurrent densities of 316 SS connected to different photoelectrode materials. As shown in Figure 6a, the photocurrent densities of all the photoanodes were almost zero before illumination, indicating that the photoanodes offered no protective effect for 316 SS in the dark state. When the light source was turned on, the photocurrent response of the composite material was considerably improved compared to TiO_2. The composite materials prepared with different precursor concentrations also showed different photocurrent densities, and the $CuInSe_2/TiO_2$-B connected to 316 SS exhibited the largest photocurrent density (140 µA cm^{-2}).

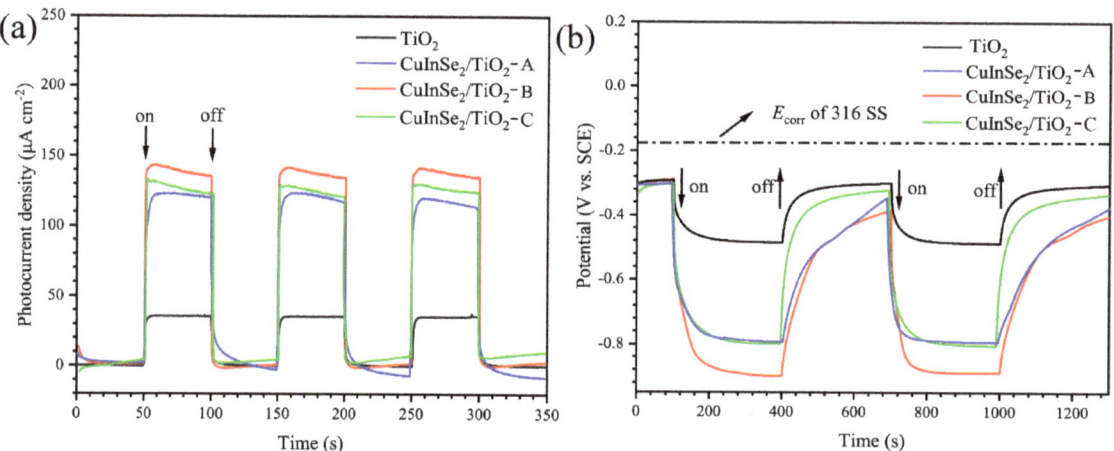

Figure 6. (a) I–t curves and (b) OCP–t curves between the different materials and 316 SS.

The potential change of the metal coupled with the photoelectrode is also a critical parameter for evaluating the properties of PCP. The more the potential drops, the better the performance of PCP [44]. In order to further study the PCP properties of photoelectrode materials, the OCP change curves of 316 SS connected with different photoelectrode materials were tested. As demonstrated in Figure 6b, the potential of 316 SS in 3.5 wt% NaCl was −0.19 V (vs. SCE). When the light source was turned on, the potentials of 316 SS connected with TiO_2 and $CuInSe_2/TiO_2$ NTAs both shifted negatively and then tended to be stable. The potential drops of TiO_2 under visible light irradiation was 0.29 V. For $CuInSe_2/TiO_2$ composites, the $CuInSe_2/TiO_2$-B showed the largest drop (0.71 V), which was consistent with the results of the photocurrent densities. With the increase in precursor concentration, the protective effect of the composite on 316 SS first increased and then decreased, which may be due to the excess of CuInSe$_2$ blocking the pores of the TiO$_2$ NTA and hindering the absorption of visible light. The potentials of the composite were still lower than those of 316 SS after the light source was closed, suggesting that the composite can also protect 316 SS for a period of time in the dark state.

Figure 7a shows the Tafel curves of 316 SS, 316 SS coupled with TiO$_2$ and CuInSe$_2$/TiO$_2$-B NTAs in light and dark states. The fitting data of the Tafel curves are shown in Table 2. In the absence of light, the E_{corr} of 316 SS connected with TiO$_2$ and CuInSe$_2$/TiO$_2$-B negatively shifted to −0.44 and −0.47 V (vs. SCE), respectively, which may be because of the galvanic effect [20]. Under visible light irradiation, the negative shift of the E_{corr} of 316 SS coupled with CuInSe$_2$/TiO$_2$-B was larger than that of TiO$_2$, indicating that CuInSe$_2$/TiO$_2$-B has a better protection effect. In addition, the J_{corr} of 316 SS connected with CuInSe$_2$/TiO$_2$-B was significantly increased compared to TiO$_2$, which may be due to the increased electrochemical reaction rate at the interface caused by the polarization of photogenerated electrons [8].

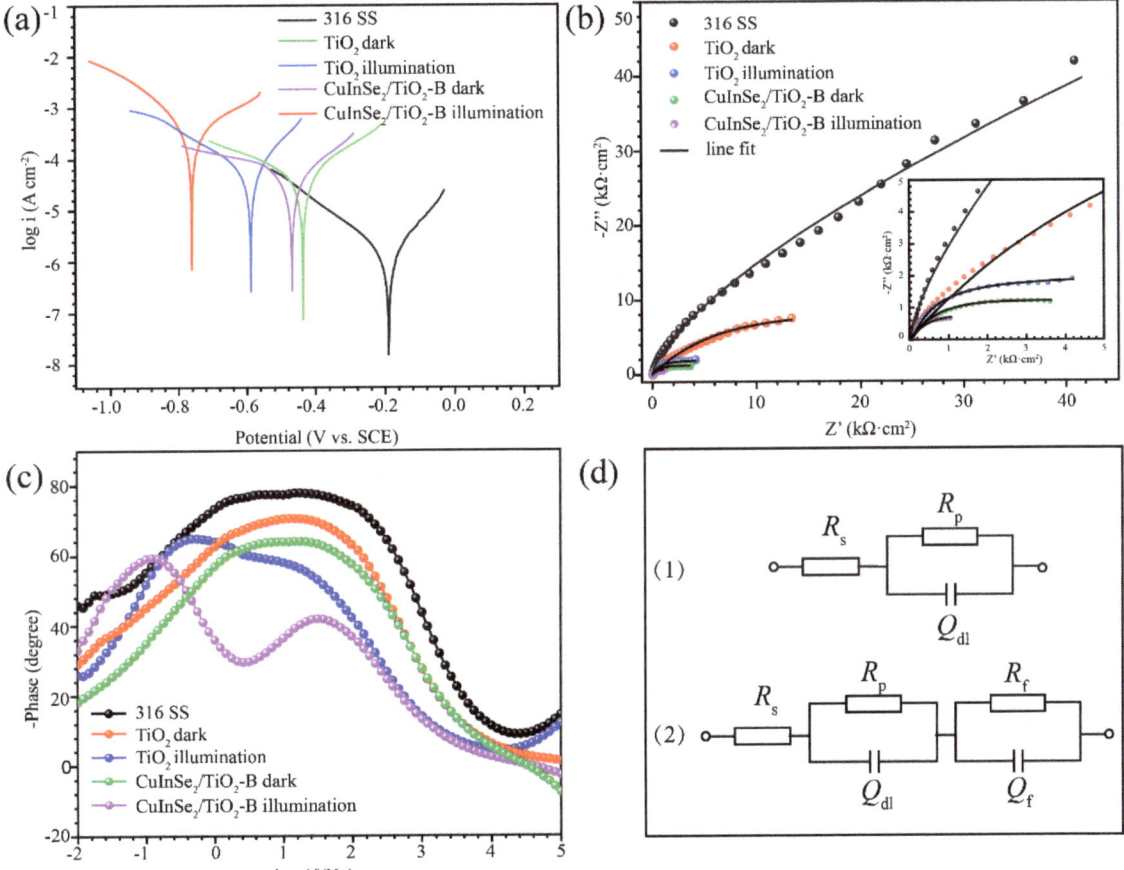

Figure 7. (**a**) Tafel curves, (**b**) Nyquist plots, and (**c**) Bode-phase curves of pure 316 SS and 316 SS coupled with different photoanodes under intermittent visible light; (**d**) the equivalent circuit for fitting the impedance data.

The interfacial properties of CuInSe$_2$/TiO$_2$-B were investigated by EIS. Figure 7b,c displays the Nyquist plots and Bode-phase curves of pure 316 SS and 316 SS connected with different photoanodes under intermittent visible light. The impedance arc radius of 316 SS connected with CuInSe$_2$/TiO$_2$-B was smaller than that of TiO$_2$ under light, which may be because more electrons were transferred from CuInSe$_2$/TiO$_2$-B to 316 SS, thus facilitating the electrochemical reaction rate of the interface [45]. In addition, the resistance arc radius of CuInSe$_2$/TiO$_2$-B NTAs coupled with 316 SS in the dark state was still smaller than that of 316 SS, indicating that CuInSe$_2$/TiO$_2$-B can also protect 316 SS in the dark state. Figure 7d shows the fitted equivalent circuit models from EIS data, where R_s represent solution resistance, R_p and Q_{dl} represent polarization resistance and double-electric-layer capacitance, respectively, and R_f and Q_f represent surface-film resistance and capacitance, respectively. The equivalent circuit model of bare 316 SS can be described as $R_s(R_pQ_{dl})$. The equivalent circuit of the 316 SS coupled with different photoanode materials can be fitted as $R_s(R_pQ_{dl})(R_fQ_f)$. Table 3 shows the electrochemical parameters fitted from the equivalent circuits. The values of R_p can reflect the difficulty of corrosion [46–48]. The smaller value of R_p means more electrons were transferred to 316 SS. The R_p value of 316 SS was significantly reduced after 316 SS was connected to CuInSe$_2$/TiO$_2$-B under visible light, indicating that CuInSe$_2$/TiO$_2$-B had a higher separation efficiency of photoinduced carriers than TiO$_2$.

Table 2. Electrochemical parameters obtained by Figure 7a.

Samples	E_{corr} (V vs. SCE)	J_{corr} (μA cm^{-2})
316 SS	−0.19	1.58
TiO$_2$ [dark]	−0.44	5.22
CuInSe$_2$/TiO$_2$-B [dark]	−0.47	5.31
TiO$_2$ [illumination]	−0.59	32.31
CuInSe$_2$/TiO$_2$-B [illumination]	−0.76	76.26

Table 3. Electrochemical impedance parameters of the as-prepared photoanodes obtained from Figure 7b.

Samples	R_s (Ω·cm^2)	Q_f		R_f (Ω·cm^2)	Q_{dl}		R_p (Ω·cm^2)
		Y_{01} (S^n·Ω^{-1} cm^{-2})	n_1		Y_{02} (S^n·Ω^{-1} cm^{-2})	n_2	
316	5.559	-	-	-	2.186 × 10^{-5}	0.9232	1.001 × 10^5
TiO$_2$ [a]	7.558	4.984 × 10^{-4}	1.00	22.190	3.232 × 10^{-4}	0.7796	4.357 × 10^4
CuInSe$_2$/TiO$_2$-B [a]	4.869	1.912 × 10^{-3}	0.8027	1.948 × 10^3	6.390 × 10^{-4}	0.8369	2.840 × 10^4
TiO$_2$ [b]	3.567	2.879 × 10^{-7}	0.99	3.636	6.784 × 10^{-4}	0.7243	3.783 × 10^3
CuInSe$_2$/TiO$_2$-B [b]	6.677	3.167 × 10^{-3}	0.8916	1.616 × 10^3	9.438 × 10^{-4}	0.7958	20.990

[a] Dark. [b] Visible light illumination.

Figure 8a shows the photoinduced I–V curves of CuInSe$_2$/TiO$_2$-B and TiO$_2$ with visible light turned on and off. The photocurrent densities of the CuInSe$_2$/TiO$_2$-B NTA photoelectrode were higher than those of TiO$_2$. This indicates that the heterojunction structure formed between CuInSe$_2$ and TiO$_2$ can increase the transfer rate of photoelectrons and promote the separation of photogenerated carriers.

Figure 8b–d displays the M–S plots of different photoanodes. The three curves all show a positive slope, suggesting that prepared photoanodes have the characteristics of an n-type semiconductor. The flat band potential (E_{fb}) of a semiconductor can be estimated using C^{-2} = 0 in the M–S curve [49]. Figure 8b–d shows that the E_{fb} of TiO$_2$, CuInSe$_2$ and CuInSe$_2$/TiO$_2$-B were −0.20, −0.68 and −0.41 V (vs. SCE), respectively. Obviously, the E_{fb} of CuInSe$_2$/TiO$_2$-B was more negative than that of TiO$_2$, indicating that the modification of CuInSe$_2$ can promote charge transfer in TiO$_2$ NTAs [50]. The slope of M–S plot is negatively correlated with the charge density [51]. The slope of the CuInSe$_2$/TiO$_2$-B curve was more negative that of TiO$_2$, demonstrating that CuInSe$_2$/TiO$_2$-B had a higher free carrier density and superior photoelectrochemical performance than TiO$_2$.

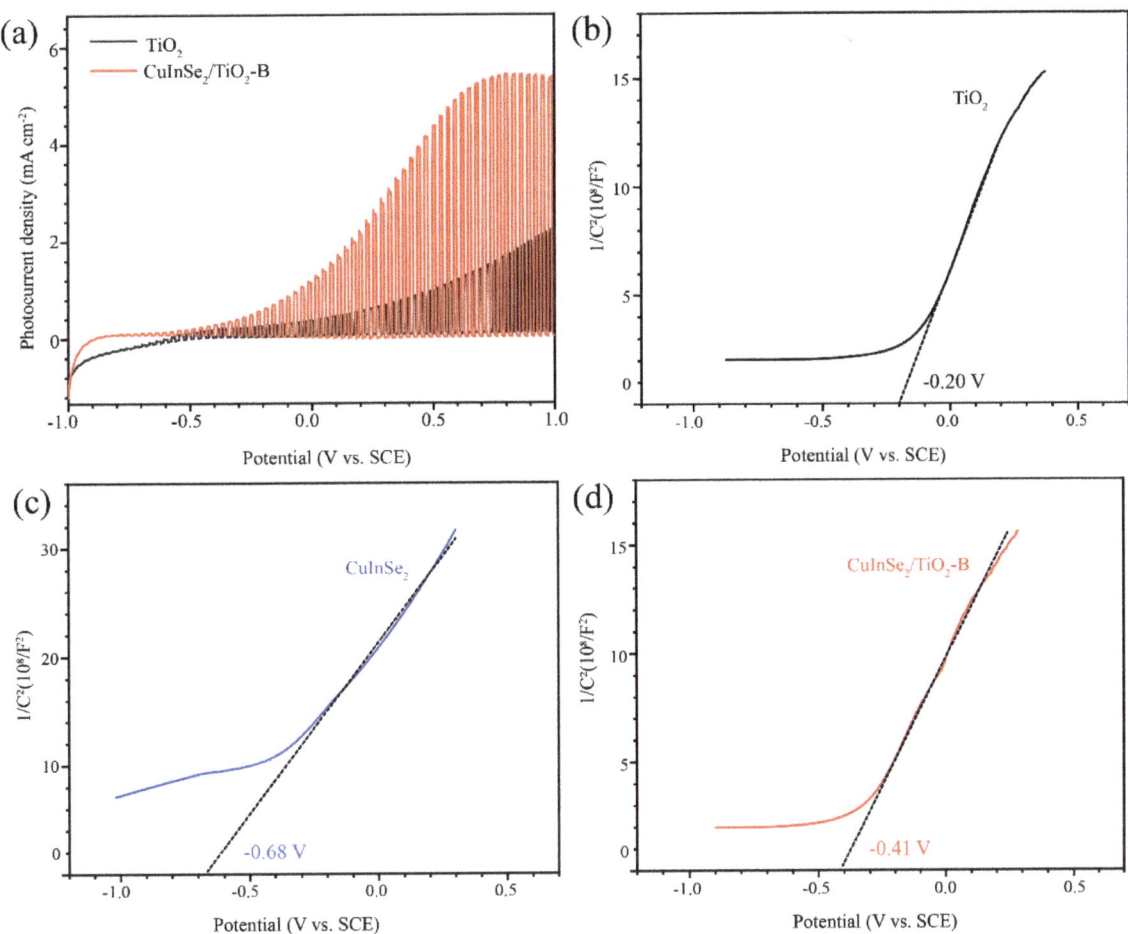

Figure 8. (**a**) I–V curves of the prepared TiO$_2$ and CuInSe$_2$/TiO$_2$-B; (**b**–**d**) M–S plots of TiO$_2$, CuInSe$_2$ and CuInSe$_2$/TiO$_2$-B.

The stability of the photoanodes is important for their PCP applications. Therefore, a long-term OCP test was performed. As shown in Figure 9a, after the CuInSe$_2$/TiO$_2$-B was illuminated for 4 h, the potential of 316 SS coupled with CuInSe$_2$/TiO$_2$-B can be stabilized at −0.82 V. The potentials were still lower than the self-corrosion potential of 316 SS for more than 8 h in the dark state. This may be due to the extended lifetime of photogenerated electrons and the generation of oxygen vacancies. Figure 9b shows the XRD spectra of CuInSe$_2$/TiO$_2$-B before and after long-term OCP measurements. The XRD results showed that the crystal structure of the photoanode did not change after long-term testing. It indicated that CuInSe$_2$/TiO$_2$ composite had superior stability. The surface of the 316 SS before and after the experiment was characterized by metallographic microscopy. According to Figure 9c,e, the surface of 316 SS protected by CuInSe$_2$/TiO$_2$-B was consistent with that before the experiment, while several pitting holes appeared on the surface of the unprotected 316 SS. These demonstrated the protective performance of CuInSe$_2$/TiO$_2$ for 316 SS.

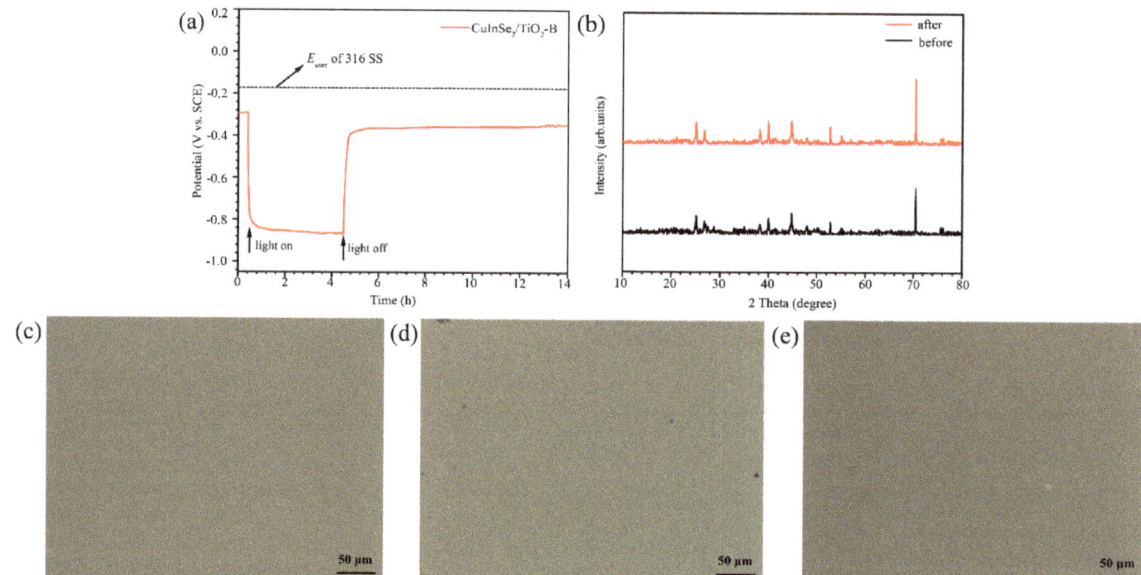

Figure 9. (**a**) Long-term OCP change of 316 SS connected to the CuInSe$_2$/TiO$_2$-B photoanode under on and off visible light illumination; (**b**) XRD images of CuInSe$_2$/TiO$_2$-B before and after long-term OCP measurements; the optical images of the 316 SS (**c**) before the experiment, (**d**) unprotected and (**e**) protected by CuInSe$_2$/TiO$_2$-B for 14 h.

3.4. DFT Analysis and PCP Mechanism

The total electronic density of states (TDOS) and partial electronic density of states (PDOS) of the prepared CuInSe$_2$/TiO$_2$ NTAs and pure TiO$_2$ NTAs were calculated using first-principles density functional theory (calculation methods were provided in the Supplementary Materials) In pure TiO$_2$, the maximum value of valence band (VBM) is provided by the electrons of O 2p state electrons, and the minimum value of conduction band (CBM) is provided by the electrons of Ti 3d state electrons (Figure 10a). The E_g can be estimated by the difference between VBM and CBM. The calculated E_g (2.0 eV) of TiO$_2$ is smaller than the DRS result (3.2 eV), which is due to the generalized gradient approximation (GGA) theory underestimating the Hubbard interaction [52]. In CuInSe$_2$, the primary contribution of VBM is of Cu 3d electrons (Figure 10b), and CBM is mainly provided by Se 4p electrons. When CuInSe$_2$ was deposited on the surface of TiO$_2$ NTAs, the 3d electrons of Cu were hybridized with the 3d electrons of Ti and the 2p electrons of O at VB, which further improved the mobility of the photogenerated carriers. Compared with TiO$_2$, after the formation of the CuInSe$_2$/TiO$_2$ heterostructure, the CBM of TiO$_2$ shifted negatively, and the E_g decreased significantly, indicating the enhanced visible light absorption of the CuInSe$_2$/TiO$_2$ heterostructures. The DFT calculation results are consistent with the DRS results.

The possible mechanism for the improved protection of CuInSe$_2$/TiO$_2$ photoanodes was analyzed (Figure 11). The E_{fb} of TiO$_2$ and CuInSe$_2$ obtained from the M–S curves were −0.20 and −0.68 V (vs. SCE), respectively, which are equal to 0.04 and −0.44 V vs. NHE, pH = 7, respectively. The E_{CB} of an n-type semiconductor is 0.2 eV more negative than E_{fb} [30], so the E_{CB} of TiO$_2$ and CuInSe$_2$ were −0.16 and −0.64 eV, respectively. The E_{VB} of TiO$_2$ and CuInSe$_2$ were 3.04 and 1.13 eV, respectively, obtained from the empirical formula $E_{VB} = E_g + E_{CB}$. Under visible light irradiation, photoelectrons migrated from the CB of CuInSe$_2$ to the CB of TiO$_2$, because the E_{CB} of CuInSe$_2$ is more negative than that of TiO$_2$, and then to the surface of 316 SS. As a result, the potentials of 316 SS were lower than the self-corrosion potential, thereby inhibiting the anodic oxidation reaction of 316 SS. In addition, the photogenerated holes in the VB of TiO$_2$ were transferred to the VB of CuInSe$_2$ and consumed by reaction with the hole scavenger, thereby reducing the recombination of e$^-$/h$^+$ pairs.

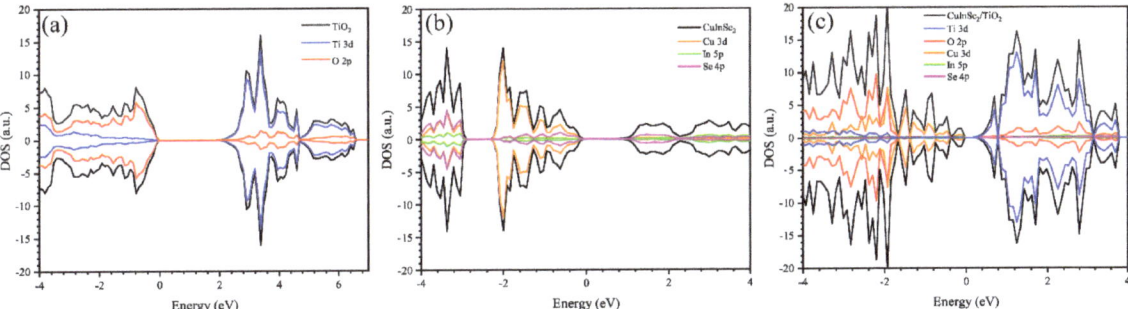

Figure 10. The calculated TDOS and PDOS for (**a**) TiO$_2$, (**b**) CuInSe$_2$ and (**c**) CuInSe$_2$/TiO$_2$.

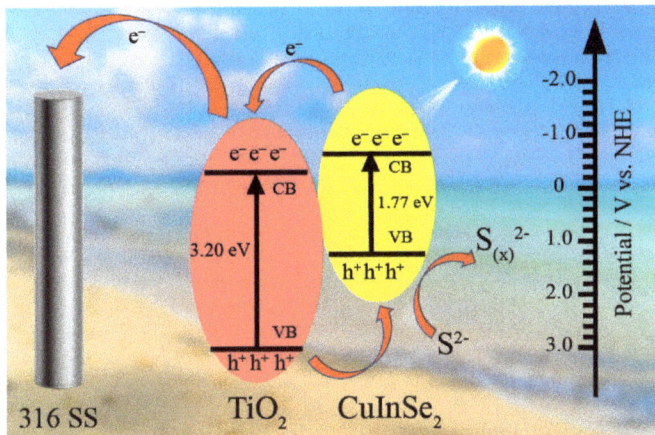

Figure 11. Possible electron-transfer mechanism of the CuInSe$_2$/TiO$_2$.

4. Conclusions

In this study, novel CuInSe$_2$/TiO$_2$ NTA photoanodes were successfully fabricated by electrochemical anodic oxidation and a solvothermal method. A highly efficient heterojunction was formed between tetragonal chalcopyrite CuInSe$_2$ and anatase TiO$_2$. The sensitization of CuInSe$_2$ improved the absorption capacity of the composites to visible light, inhibited the recombination of electron-hole pairs and improved the electron transfer ability. The CuInSe$_2$/TiO$_2$-B NTA photoanode exhibited the best PCP performance. The photocurrent density of the composite connected to 316 SS could reach 140 µA cm^{-2} under visible light, and the potential drops to -0.90 V, which is much lower than the self-corrosion potential of 316 SS (-0.19 V). In addition, the protection effect can still be maintained for more than 8 h after 4 h of visible light irradiation. The optical images of the protected 316 SS fully demonstrated the excellent protection capability of CuInSe$_2$/TiO$_2$ NTAs. Therefore, CuInSe$_2$/TiO$_2$ NTAs show great potential application in the field of PCP.

Supplementary Materials: The following supporting information can be downloaded at: https://www.mdpi.com/article/10.3390/coatings12101448/s1. Supplementary Materials: Computational methods.

Author Contributions: Experiment, writing—original draft, review and editing, Z.Y.; writing—review and editing, supervision, funding acquisition, H.L.; experiment, picture drawing, X.C.; experiment, J.Z.; resources, Y.L.; resources, funding acquisition, P.Z.; resources, J.L. All authors have read and agreed to the published version of the manuscript.

Funding: This work was funded by National Natural Science Foundation of China [grant number: 51801109] and Science and Technology Support Plan for Youth Innovation of Colleges in Shandong Province [grant number: DC2000000891].

Institutional Review Board Statement: Not applicable.

Informed Consent Statement: Not applicable.

Data Availability Statement: The authors confirm that the data supporting the findings of this study are available within the article.

Conflicts of Interest: The authors declare no conflict of interest.

References

1. Akhtar, S.; Matin, A.; Kumar, A.; Brahim, A.; Laoui, I.T. Enhancement of anticorrosion property of 304 stainless steel using silane coatings. *Appl. Surf. Sci.* **2018**, *440*, 1286–1297. [CrossRef]
2. Zhang, X.; Chen, G.; Li, W.; Wu, D. Graphitic carbon nitride homojunction films for photocathodic protection of 316 stainless steel and Q235 carbon steel. *J. Electroanal. Chem.* **2020**, *857*, 113703. [CrossRef]
3. Zhang, B.; Wang, J.; Wu, B.; Guo, X.; Wang, Y.; Chen, D.; Zhang, Y.; Du, K.; Oguzie, E.; Ma, X. Unmasking chloride attack on the passive film of metals. *Nat. Commun.* **2018**, *9*, 2559. [CrossRef] [PubMed]
4. El Ibrahimi, B.; Jmiai, A.; Bazzi, L.; El Issami, S. Amino acids and their derivatives as corrosion inhibitors for metals and alloys. *Arab. J. Chem.* **2020**, *13*, 740–771. [CrossRef]
5. Hou, B.; Li, X.; Ma, X.; Du, C.; Zhang, D.; Zheng, M.; Xu, W.; Lu, D.; Ma, F. The cost of corrosion in China. *NPJ Mater. Degrad.* **2017**, *1*, 4. [CrossRef]
6. Tan, B.; He, J.; Zhang, S.; Xu, C.; Chen, S.; Liu, H.; Li, W. Insight into anti-corrosion nature of Betel leaves water extracts as the novel and eco-friendly inhibitors. *J. Colloid Interface Sci.* **2021**, *585*, 287–301. [CrossRef]
7. Mirhashemihaghighi, S.; Swiatowska, J.; Maurice, V.; Seyeux, A.; Klein, L.; Salmi, E.; Ritala, M.; Marcus, P. Interfacial native oxide effects on the corrosion protection of copper coated with ALD alumina. *Electrochim. Acta* **2016**, *193*, 7–15. [CrossRef]
8. Kear, G.; Barker, B.; Stokes, K.; Walsh, F. Corrosion and impressed current cathodic protection of copper-based materials using a bimetallic rotating cylinder electrode (BRCE). *Corros. Sci.* **2005**, *47*, 1694–1705. [CrossRef]
9. Delfani, F.; Rahbar, N.; Aghanajafi, C.; Heydari, A.; KhalesiDoost, A. Utilization of thermoelectric technology in converting waste heat into electrical power required by an impressed current cathodic protection system. *Appl. Energy* **2021**, *302*, 117561. [CrossRef]
10. Prasad, N.; Pathak, A.; Kundu, S.; Mondal, K. Highly active and efficient hybrid sacrificial anodes based on high p pig iron, Zn and Mg. *J. Electrochem. Soc.* **2021**, *168*, 111504. [CrossRef]
11. Yuan, J.; Tsujikawa, S. Characterization of Sol-Gel-Derived TiO_2 Coatings and Their Photoeffects on Copper Substrates. *J. Electrochem. Soc.* **1995**, *142*, 3444–3450. [CrossRef]
12. Shao, J.; Wang, X.; Xu, H.; Zhao, X.; Niu, J.; Zhang, Z.; Huang, Y.; Duan, J. Photoelectrochemical Performance of SnS_2 Sensitized TiO_2 Nanotube for Protection of 304 Stainless Steel. *J. Electrochem. Soc.* **2021**, *168*, 016511. [CrossRef]
13. Chen, R.; Xu, Y.; Xie, X.; Li, C.; Zhu, W.; Xiang, Q.; Li, G.; Wu, D.; Li, X.; Wang, L. Synthesis of TiO_2 nanotubes/nickel-gallium layered double hydroxide heterostructure for highly-efficient photocathodic anticorrosion of 304 stainless steel. *Surf. Coat. Technol.* **2021**, *424*, 127641. [CrossRef]
14. Jiang, X.; Sun, M.; Chen, Z.; Jing, J.; Lu, G.; Feng, C. Boosted photoinduced cathodic protection performance of $ZnIn_2S_4/TiO_2$ nanoflowerbush with efficient photoelectric conversion in NaCl solution. *J. Alloy Compd.* **2021**, *876*, 160144. [CrossRef]
15. Sood, S.; Mehta, S.; Sinha, A.; Kansal, S. Bi_2O_3/TiO_2 heterostructures: Synthesis, characterization and their application in solar light mediated photocatalyzed degradation of an antibiotic, ofloxacin. *Chem. Eng. J.* **2016**, *290*, 45–52. [CrossRef]
16. Wang, M.; Sun, L.; Lin, Z.; Cai, J.; Xie, K.; Lin, C. p-n Heterojunction photoelectrodes composed of Cu_2O-loaded TiO_2 nanotube arrays with enhanced photoelectrochemical and photoelectrocatalytic activities. *Energy Environ. Sci.* **2013**, *6*, 1211–1220. [CrossRef]
17. Momeni, M.; Khansari-Zadeh, S.; Farrokhpour, H. Fabrication of tungsten-iron-doped TiO_2 nanotubes via anodization: New photoelectrodes for photoelectrochemical cathodic protection under visible light. *SN Appl. Sci.* **2019**, *1*, 1160. [CrossRef]
18. Liu, Y.; Xu, C.; Feng, Z. Characteristics and anticorrosion performance of Fe-doped TiO_2 films by liquid phase deposition method. *Appl. Surf. Sci.* **2014**, *314*, 392–399. [CrossRef]
19. Sun, M.; Chen, Z.; Yu, J. Highly efficient visible light induced photoelectrochemical anticorrosion for 304 SS by Ni-doped TiO_2. *Electrochim. Acta.* **2013**, *109*, 13–19. [CrossRef]
20. Qiu, P.; Sun, X.; Lai, Y.; Gao, P.; Chen, C.; Ge, L. N-doped TiO_2@TiO_2 visible light active film with stable and efficient photocathodic protection performance. *J. Electroanal. Chem.* **2019**, *844*, 91–98. [CrossRef]

21. Momeni, M.; Taghinejad, M.; Ghayeb, Y.; Bagheri, R.; Song, Z. Preparation of various boron-doped TiO_2 nanostructures by in situ anodizing method and investigation of their photoelectrochemical and photocathodic protection properties. *J. Iran. Chem. Soc.* **2019**, *16*, 1839–1851. [CrossRef]
22. Ma, X.; Ma, Z.; Lu, D.; Jiang, Q.; Li, L.; Liao, T.; Hou, B. Enhanced photoelectrochemical cathodic protection performance of MoS_2/TiO_2 nanocomposites for 304 stainless steel under visible light. *J. Mater. Sci. Technol.* **2021**, *64*, 21–28. [CrossRef]
23. Lu, X.; Liu, L.; Ge, J.; Cui, Y.; Wang, F. Morphology controlled synthesis of $Co(OH)_2/TiO_2$ p-n heterojunction photoelectrodes for efficient photocathodic protection of 304 stainless steel. *Appl. Surf. Sci.* **2021**, *537*, 148002. [CrossRef]
24. Wang, N.; Wang, J.; Liu, M.; Ge, C.; Hou, B.; Liu, N.; Ning, Y.; Hu, Y. Preparation of FeS_2/TiO_2 nanocomposite films and study on the performance of photoelectrochemistry cathodic protection. *Sci. Rep.* **2021**, *11*, 7509. [CrossRef] [PubMed]
25. Aldakov, D.; Lefrançois, A.; Reiss, P. Ternary and quaternary metal chalcogenide nanocrystals: Synthesis, properties and applications. *J. Mater. Chem. C* **2013**, *1*, 3756–3776. [CrossRef]
26. Bhosale, R.; Agarkar, S.; Agrawal, I.; Naphade, R.; Ogale, S. Nanophase $CuInS_2$ nanosheets/CuS composite grown by the SILAR method leads to high performance as a counter electrode in dye sensitized solar cells. *Rsc. Adv.* **2014**, *4*, 21989–21996. [CrossRef]
27. Loo, A.; Bonanni, A.; Sofer, Z.; Pumera, M. Exfoliated transition metal dichalcogenides (MoS_2, $MoSe_2$, WS_2, WSe_2): An electrochemical impedance spectroscopic investigation. *Electrochem. Commun.* **2015**, *50*, 39–42. [CrossRef]
28. Li, H.; Song, W.; Cui, X.; Li, Y.; Hou, B.; Zhang, X.; Wang, Y.; Cheng, L.; Zhang, P.; Li, J. $AgInS_2$ and graphene co-sensitized TiO_2 photoanodes for photocathodic protection of Q235 carbon steel under visible light. *Nanotechnology* **2020**, *31*, 305704. [CrossRef]
29. Jiang, X.; Sun, M.; Chen, Z.; Jing, J.; Feng, C. High-efficiency photoelectrochemical cathodic protection performance of the $TiO_2/AgInSe_2/In_2Se_3$ multijunction nanosheet array. *Corros. Sci.* **2020**, *176*, 108901. [CrossRef]
30. Zhao, X.; Yang, Y.; Li, Y.; Cui, Y.; Zhang, X.; Xiao, P. NiCo-selenide as a novel catalyst for water oxidation. *J. Mater. Sci.* **2016**, *51*, 3724–3734. [CrossRef]
31. Jiang, Z.; Feng, L.; Zhu, J.; Liu, B.; Li, X.; Chen, Y.; Khan, S. Construction of a hierarchical $NiFe_2O_4/CuInSe_2$ (p-n) heterojunction: Highly efficient visible-light-driven photocatalyst in the degradation of endocrine disruptors in an aqueous medium. *Ceram. Int.* **2021**, *47*, 8996–9007. [CrossRef]
32. Wu, Z.; Tong, X.; Sheng, P.; Li, W.; Yin, X.; Zou, J.; Cai, Q. Fabrication of high-performance $CuInSe_2$ nanocrystals-modified TiO_2 NTs for photocatalytic degradation applications. *Appl. Surf. Sci.* **2015**, *351*, 309–315. [CrossRef]
33. Liao, Y.; Zhang, H.; Zhong, Z.; Jia, L.; Bai, F.; Li, J.; Zhong, P.; Chen, H.; Zhang, J. Enhanced visible-photocatalytic activity of anodic TiO_2 nanotubes film via decoration with $CuInSe_2$ nanocrystals. *ACS Appl. Mater. Interfaces* **2013**, *5*, 11022–11028. [CrossRef]
34. Yu, Y.; Chien, W.; Ko, Y.; Chen, S. Preparation and characterization of P3HT: $CuInSe_2$: TiO_2 thin film for hybrid solar cell applications. *Thin Solid Film.* **2011**, *520*, 1503–1510. [CrossRef]
35. Zhang, T.; Rahman, Z.; Wei, N.; Liu, Y.; Liang, J.; Wang, D. In situ growth of single-crystal TiO_2 nanorod arrays on Ti substrate: Controllable synthesis and photoelectro-chemical water splitting. *Nano Res.* **2017**, *10*, 1021–1032. [CrossRef]
36. Deng, X.; Zhang, H.; Guo, R.; Cui, Y.; Ma, Q.; Zhang, X.; Cheng, X.; Li, B.; Xie, M.; Cheng, Q. Effect of fabricating parameters on photoelectrocatalytic performance of CeO_2/TiO_2 nanotube arrays photoelectrode. *Sep. Purif. Technol.* **2018**, *193*, 264–273. [CrossRef]
37. Jiang, X.; Sun, M.; Chen, Z.; Jing, J.; Feng, C. An ultrafine hyperbranched CdS/TiO_2 nanolawn photoanode with highly efficient photoelectrochemical performance. *J. Alloy. Compd.* **2020**, *816*, 152533. [CrossRef]
38. Antony, R.; Mathews, T.; Dash, S.; Tyagi, A.; Raj, B. X-ray photoelectron spectroscopic studies of anodically synthesized self aligned TiO_2 nanotube arrays and the effect of electrochemical parameters on tube morphology. *Mater. Chem. Phys.* **2012**, *132*, 957–966. [CrossRef]
39. Anuroop, R.; Pradeep, B. Structural, optical, ac conductivity and dielectric relaxation studies of reactively evaporated In_6Se_7 thin films. *J. Alloy. Compd.* **2017**, *702*, 432–441. [CrossRef]
40. Zhang, H.; Fang, W.; Wang, W.; Qian, N.; Ji, X. Highly Efficient Zn-Cu-In-Se Quantum Dot-Sensitized Solar Cells through Surface Capping with Ascorbic Acid. *ACS Appl. Mater. Interfaces* **2019**, *11*, 6927–6936. [CrossRef]
41. Jia, H.; Cheng, S.; Zhang, H.; Yu, J.; Lai, Y. Band alignment at the Cu_2SnS_3/In_2S_3 interface measured by X-ray photoemission spectroscopy. *Appl. Surf. Sci.* **2015**, *353*, 414–418. [CrossRef]
42. Dai, G.; Yu, J.; Liu, G. Synthesis and Enhanced Visible-Light Photoelectrocatalytic Activity of p-n Junction $BiOI/TiO_2$ Nanotube Arrays. *J. Phys. Chem. C* **2011**, *115*, 7339–7346. [CrossRef]
43. Tauc, J. Optical Properties and Electronic Structure of Amorphous Ge and Si. *Mater. Res. Bull.* **1968**, *3*, 37–46. [CrossRef]
44. Lu, X.; Liu, L.; Xie, X.; Cui, Y.; Oguzie, E.; Wang, F. Synergetic effect of graphene and $Co(OH)_2$ as cocatalysts of TiO_2 nanotubes for enhanced photogenerated cathodic protection. *J. Mater. Sci. Technol.* **2020**, *37*, 55–63. [CrossRef]
45. Sun, Z.; Li, F.; Zhao, M.; Xu, L.; Fang, S. A comparative study on photoelectrochemical performance of TiO_2 photoanodes enhanced by different polyoxometalates. *Electrochem. Commun.* **2013**, *30*, 38–41. [CrossRef]
46. Saha, S.; Dutta, A.; Ghosh, P.; Sukul, D.; Banerjee, P. Adsorption and corrosion inhibition effect of Schiff base molecules on the mild steel surface in 1 M HCl medium: A combined experimental and theoretical approach. *Phys. Chem. Chem. Phys.* **2015**, *17*, 5679–5690. [CrossRef]
47. Solmaz, R.; Kardaş, G.; Çulha, M.; Yazıcı, B.; Erbil, M. Investigation of adsorption and inhibitive effect of 2-mercaptothiazoline on corrosion of mild steel in hydrochloric acid media. *Electrochim. Acta* **2008**, *53*, 5941–5952. [CrossRef]

48. Özcan, M.; Dehri, İ.; Erbil, M. Organic sulphur-containing compounds as corrosion inhibitors for mild steel in acidic media: Correlation between inhibition efficiency and chemical structure. *Appl. Surf. Sci.* **2004**, *236*, 155–164. [CrossRef]
49. Tong, X.; Shen, W.; Chen, X. Enhanced H_2S sensing performance of cobalt doped free-standing TiO_2 nanotube array film and theoretical simulation based on density functional theory. *Appl. Surf. Sci.* **2019**, *469*, 414–422. [CrossRef]
50. Zhou, M.; Hou, Z.; Chen, X. Graphitic-C_3N_4 nanosheets: Synergistic effects of hydrogenation and n/n junctions for enhanced photocatalytic activities. *Dalton Trans.* **2017**, *46*, 10641–10649. [CrossRef]
51. Wang, C.; Long, X.; Wei, S.; Wang, T.; Li, F.; Gao, L.; Hu, Y.; Li, S.; Jin, J. Conformally Coupling CoAl-LDH on Fluorine-Doped Hematite: Surface and Bulk Co-modification for Enhanced Photoelectrochemical Water Oxidation. *ACS Appl. Mater. Interfaces* **2019**, *11*, 29799–29806. [CrossRef] [PubMed]
52. Wang, X.; Lei, J.; Shao, Q.; Li, X.; Ning, X.; Shao, J.; Duan, J.; Hou, B. Preparation of $ZnWO_4/TiO_2$ composite film and its photocathodic protection for 304 stainless steel under visible light. *Nanotechnology* **2019**, *30*, 045710. [CrossRef] [PubMed]

Article

Study on Size Effect of Surface Roughness Based on the 3D Voronoi Model and Establishment of Roughness Prediction Model in Micro-Metal Forming

Juanjuan Han [1,*,†], Wei Zheng [2,†], Qingqiang Chen [1], Jie Sun [1] and Shubo Xu [1]

1. School of Mechanical and Electronic Engineering, Shandong Jianzhu University, Jinan 250101, China
2. School of Mechanical Engineering, Shandong University, Jinan 250061, China
* Correspondence: hjj890912@163.com
† These authors contributed equally to this work.

Abstract: The primary purpose of this paper is to study the size effect of surface roughness and realize the quantitative description of the surface roughness in micro-forming process. This work is a continuation of the previous work by the authors. The effects of the initial surface roughness of the specimen, the grain size, and grain orientations on the surface roughness of micro-upsetting products were investigated. The ratio of the number of grains of the surface layer to the total number of grains was adopted to characterize the size effect. The variation of the size effect on the contact normal pressure during the compression process was also analyzed. And the quantitative description of the evolution law of surface roughness for micro-formed parts was realized. The corresponding micro compression experiment was done in order to testify the prediction model.

Keywords: surface roughness; size effect; grain size; 3D voronoi; micro-forming

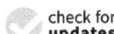

Citation: Han, J.; Zheng, W.; Chen, Q.; Sun, J.; Xu, S. Study on Size Effect of Surface Roughness Based on the 3D Voronoi Model and Establishment of Roughness Prediction Model in Micro-Metal Forming. *Coatings* **2022**, *12*, 1659. https://doi.org/10.3390/coatings12111659

Academic Editors: Matic Jovičević-Klug, Patricia Jovičević-Klug and László Tóth

Received: 6 October 2022
Accepted: 26 October 2022
Published: 1 November 2022

Publisher's Note: MDPI stays neutral with regard to jurisdictional claims in published maps and institutional affiliations.

Copyright: © 2022 by the authors. Licensee MDPI, Basel, Switzerland. This article is an open access article distributed under the terms and conditions of the Creative Commons Attribution (CC BY) license (https://creativecommons.org/licenses/by/4.0/).

1. Introduction

Due to the rapid development of micro-system technology (MST) and micro-electromechanical systems (MEMS), the forming and manufacturing technology of micro-parts is highly regarded. In contrast with the products from a traditional plastic forming process, the products from a micro-forming process have small sizes, high dimensional accuracy, and surface quality requirements due to their high-end applications [1–3]. Due to their miniaturized sizes, size effect is crucial in micro-forming processes [4–6]. During the micro-plastic forming process, the surface roughness does not decrease with the reduction in the geometric size of the product. Meanwhile, mechanical machining is challenging to improve the surface quality due to the difficulty in clamping and operation. Therefore, the surface roughness control for micro-formed products has become one of the critical challenges in micro-forming.

For the macro-formed products, the roughness is affected by various factors such as the initial specimen roughness, the sliding distance between the contact surface and die, the contact normal pressure on the surface, and the properties of the material, etc. [7–9]. Among these factors, the influence of the normal pressure, the initial surface roughness, the sliding distance, and the contact surface area on the surface roughness have been studied in our previous work [10]. However, in the micro-plastic forming process, the material properties change significantly with miniaturization due to the tiny size. One such property is the grain size, and the other is the specimen size. Both of these factors lead to a change in the ratio of thickness to the grain sizes of the specimen [11]. When the size of the workpiece is gradually reduced close to the grain size, the grain number in the workpiece is less, and a "size effect" occurs. The change in grain size and the incongruity of grain orientation become increasingly apparent, resulting in the change in material mechanical properties, impacting the roughness of the workpiece after deformation [12].

Many studies have reported that under uniaxial tensile deformation, the uneven plastic deformation between grain boundaries leads to surface roughness evolution [13,14]. Kubo et al. [14] reported the difference in the crystal orientations because of the stress state on the surface roughness of a low-carbon steel sheet. It is therefore suggested that surface quality after press forming may be further improved by reducing the number of grains with crystal orientation of ND(001) in interstitial-free (IF) steel sheets. The surface quality after press forming could also be improved by reducing the difference in deformation resistances among the grains. The effect of crystal orientations on the inner surface roughness of micro metal tubes in hollow sinking was investigated by Kishimoto et al. [15]. They suggested that the roughness of inner surface can be suppressed by decreasing the wall thickness. One of the crystal orientations was determined to inhibit the increase in internal surface roughness during the hollow sinking. Yoshida [16] found that the surface roughness mainly depends on the grain size of the workpiece through the plane strain tensile simulation process, whereas the ratio of the workpiece thickness to the grain size has little influence on the roughness. Wang et al. [17] found that the grain size and texture composition impact the surface roughness through the bending deformation of copper tubes. At a particular scale, the surface roughness increases with grain sizes. In the biaxial tensile deformation study, Anand et al. [11] suggested that the ratio of sheet thickness to grain size (t/d) significantly affects surface roughness due to the size effect. For thin brass sheets, the ratio of thickness to grain size has a significant role in increasing the roughness during biaxial tension. With the increase in t/d, the surface roughness decreases. Moreover, when $t/d \geq 12$, the surface roughness basically remains unchanged.

To sum up, compared with the macro-formed parts, the roughness control for micro-formed parts is more complicated. The effects of grain size and grain orientation on the roughness should also be considered. However, the roughness prediction for micro formed parts based on the 3D Voronoi model has not yet been reported in the literature. In this work, the effects of specimen initial surface roughness, grain size, and grain orientation on the surface roughness of products after micro-upsetting were investigated. The cause of size effect on surface roughness was also analyzed. Based on our previous work, a surface roughness prediction model taking into account of size effect was established and testified through the corresponding micro-compression experiments.

2. Materials and Methods

2.1. Development of Polycrystalline Finite Element Model with 3D Surface Roughness

Developing a polycrystalline geometric model is the foundation for investigating polycrystalline materials' plastic deformation. The 3D grain cluster generated in this work was based on the Voronoi diagram method [18,19] and the ABAQUS/python script.

In order to describe the information on surface topography more comprehensively by the 3D parameter characterization method, the surface root mean square deviation (S_q) in ISO 25178-2 was used to evaluate the surface roughness [20]. The S_q represents the degree of deviation of the roughness morphology from the reference plane, and its mathematical expression is given in Equation (1) below.

$$S_q = \sqrt{\frac{1}{l_x l_y} \int_0^{l_x} \int_0^{l_y} Z^2(x,y)dxdy} = \sqrt{\frac{1}{mn} \sum_{i=1}^{m} \sum_{j=1}^{n} \left[Z_{ij}(x_i, y_j) - Z_0\right]^2}. \qquad (1)$$

where $Z(x, y)$ is the height of the contour for each point, and l_x and l_y are the measured lengths in X and Y directions, respectively. m and n are the numbers of measurement points in X and Y directions, Z_{ij} is the height for the measuring point, and Z_0 is the centerline height.

According to the self-programmed W–M function [21], the modeling of 3D rough surfaces with different fractal dimensions and scale coefficients was performed by MATLAB R2018a. The corresponding fractal dimension and scale coefficient were determined based on the measured rough surface profile. Figure 1a shows a randomly roughness profile with a surface roughness of 1.6 µm within 0.1 mm × 0.1 mm generated by MATLAB R2018a. Combining the roughness topography with the 3D Voronoi model, a 3D polycrystalline finite element model with initial roughness can be obtained in the finite element software ABAQUS, as shown in Figure 1b.

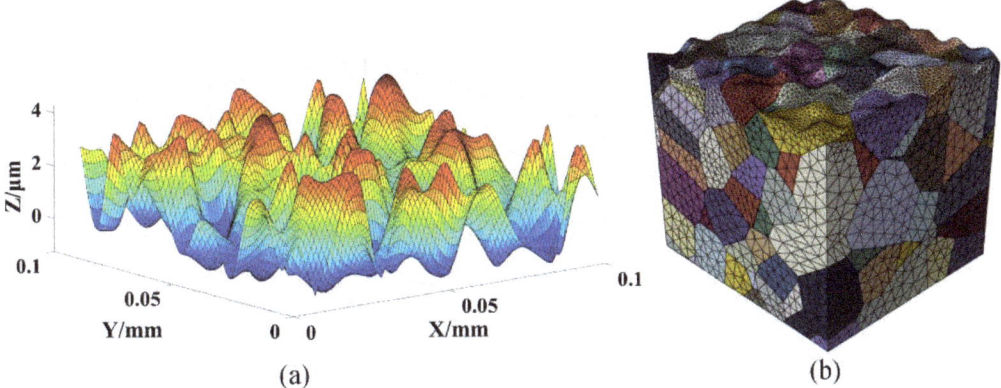

Figure 1. (**a**) The randomly generated roughness profile with a surface roughness of 1.6 µm; and (**b**) The 3D polycrystalline finite element model with initial roughness within 0.1 mm × 0.1 mm × 0.1 mm.

2.2. Development of 3D Polycrystalline Regionalized Constitutive Model at Micro-Scale

Based on the analysis of the microstructure of annealed pure copper, three different grain orientations are existed in the deformation direction: <411>, <100>, and <111> [22]. Table 1 shows the volume fractions and orientation factors for three different grain orientations of pure copper.

Table 1. Volume fractions and orientation factors for three different grain orientations.

	Type <411>	Type <100>	Type <111>
Volume fraction (λ)	0.06	0.22	0.72
orientation factor (M)	2.64	2.45	3.67

For polycrystalline materials, the flow stress is a weighted component of the flow stress of individual grains, given in Equation (2) below.

$$\sigma = M_{in}\tau_i + M_{in}K'd^{\frac{-1}{2}} = \sum_{i=1}^{n} \lambda_i M_i \left(\tau_i + K'd^{\frac{-1}{2}} \right). \quad (2)$$

where λ_i is the volume fraction of the i-th grain, M_i is the orientation factor of the i-th grain, τ_i is the frictional shear stress that restricts the dislocation sliding along the slip plane, $K'd^{-1/2}$ is the resistance shear stress due to dislocation stacking near the grain boundary, d is the grain size of the material, and n is the number of grains. Through the micro-compression experiment of pure copper, the flow stress–strain curves of the specimens with different grain sizes were obtained [18]. Equation (3) shows the representations of $K'(\varepsilon)$ and $\tau_i(\varepsilon)$.

$$\begin{cases} K'(\varepsilon) = 2.688\varepsilon^{0.29} \\ \tau_i(\varepsilon) = 100.1\varepsilon^{0.38} \end{cases}. \quad (3)$$

Moreover, M_{surf} = 2.0 for the surface layer grains [23]. The material constitutive relationship at any grain size can be determined by ignoring the rotation of grains and substituting the values of $M_{<411>}$, $M_{<100>}$, $M_{<111>}$, and M_{surf} in Equation (2).

2.3. Polycrystalline Pure Copper Finite Element Simulation of the Upsetting Process

The micro-upsetting processes of pure copper with a 3D rough surface were simulated and analyzed to investigate the effect of grain size and grain orientations on the specimen surface roughness. In the simulation process, the diameter and the height of the specimen were set to 0.5 mm 1.5 mm, respectively. The grain sizes of the specimens were set to 30 μm, 50 μm, 80 μm, and 120 μm. One-eighth of the specimens were used for simulation, and the numbers of grains in the specimens were 681, 147, 36, and 11. As shown in Table 1, volume fractions for different grain orientations were set as 0.06, 0.22, and 0.72 for <411>, <100>, and <111> orientations, respectively. Four 3D surfaces with different initial roughness values were set in the simulation. The values of the roughness for the four 3D surfaces were 0 μm corresponding to an ideal plane surface, 0.7917 μm, 1.6131 μm, and 3.1152 μm. Three different 3D rough surface contours at the same surface roughness were randomly generated to decrease the experimental error. Figure 2 shows the assignment results with a random grain orientation of the model.

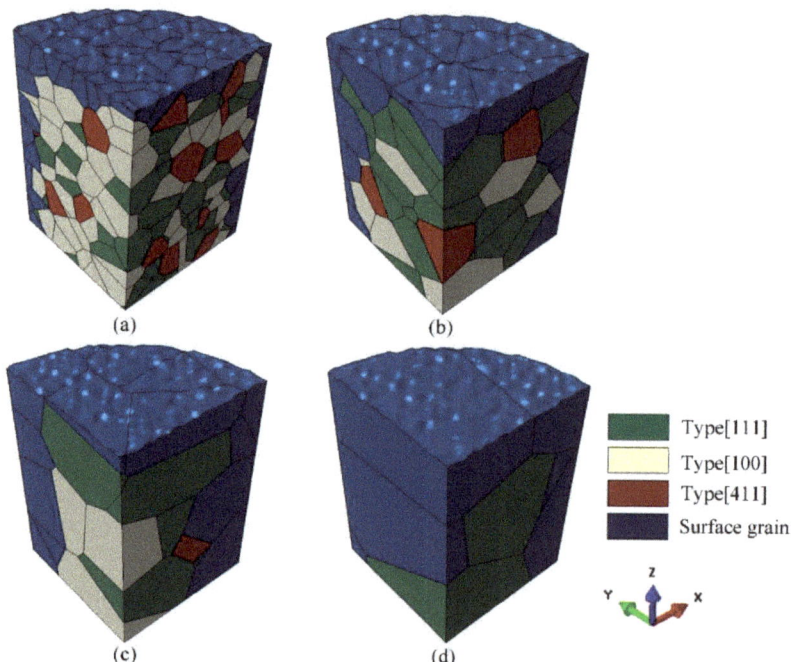

Figure 2. Assignment results for random orientations assignment results of the finite element models with different grain sizes (**a**) d = 30 μm; (**b**) d = 50 μm; (**c**) d = 80 μm; and (**d**) d = 120 μm.

Considering the variation of material properties at the microscale, 10 random grain orientations were generated to investigate the effect of grain orientations on specimen roughness. Figure 3 depicts different grain orientations with the grain size of 50 μm.

The material constitutive model developed in Section 2.2 was adopted during the simulation of the micro-upsetting process. Moreover, the friction model used in this work is the Coulomb friction model based on the Wanheim/Bay friction model transformation [24]. The friction coefficient is a function related to the normal pressure of the contact surface. According to the symmetry condition, the left and lower ends of the specimen were set as the left–right and the upper–lower symmetric boundary condition, respectively. The upper die displacement was set downward along the Z direction with a 20% deformation. After the deformation, the coordinates of the specimen surface contours were extracted by postprocessing. According to Equation (1), the variation of the surface morphology roughness after deformation was calculated. The contact area of the specimen surface was divided by mesh refinement. Figure 4 exhibits the developed simulation model and its mesh generation.

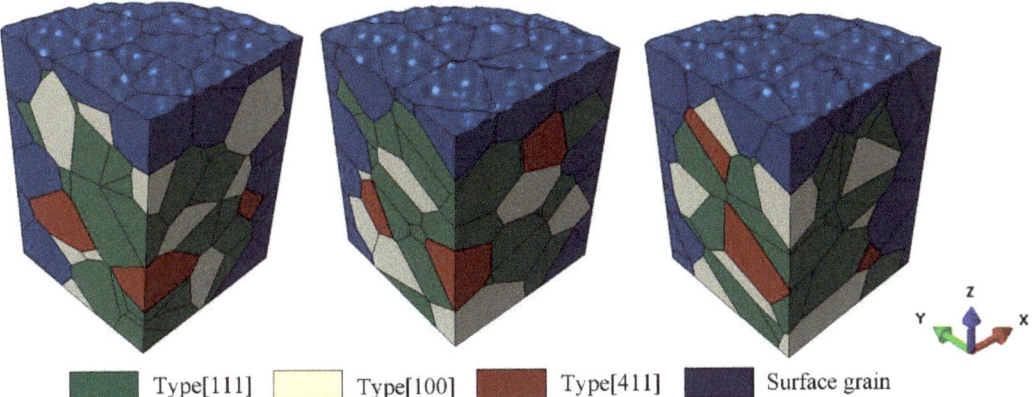

Figure 3. Three different grain orientations with the grain size of 50 μm.

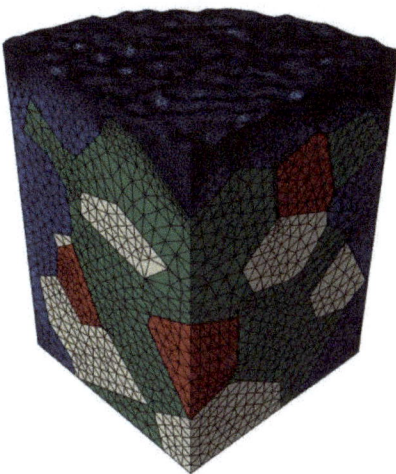

Figure 4. The 3D polycrystal model and its mesh generation.

3. Results and Discussions

3.1. Numerical Analysis Results

Based on the simulation results of the micro-upsetting deformation process, the effects of initial surface roughness, grain sizes, and grain orientations on the evolution of the surface roughness of specimens were analyzed in this work.

3.1.1. Effect of Initial Surface Roughness of Specimen on the Evolution of Contact Surface Roughness

According to the research of Wanheim and Bay [25], when the surface roughness of the mold is fixed, the shear film strength coefficient m_c is mainly determined by the specimen initial roughness. Based on the previous research work of the authors [10,24], the relationship between the friction coefficient and the contact normal pressure under different initial surface roughness is determined. When the reduction in the upsetting is 20%, the equivalent stress and strain distribution of the specimens under different initial surface roughness are shown in Figure 5a,b, respectively. It depicts that the distribution of stress and strain increases with the initial surface roughness. During the deformation process, the actual contact area of the specimen surface with a high degree of initial surface roughness is smaller, resulting in the larger stress and strain of the micro-asperities on the contact surface under the same deformation. The larger the stress and strain of the micro-asperities on the contact surface, the more prone it is to plastic deformation.

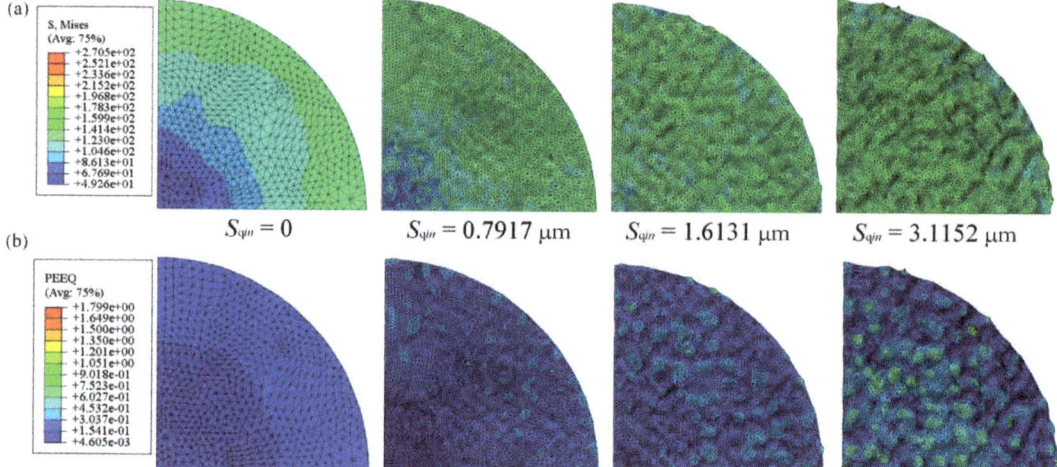

Figure 5. The effect of initial surface roughness on (**a**) the stress distribution and (**b**) the strain distribution under different values of initial surface roughness.

The surface morphologies of deformed specimens under various initial roughness are shown in Figure 6a–d. Section "A" of Figure 6 shows the initial surface morphology, and section "B" reveals the final surface morphology with an upsetting reduction of 20%. When the material undergoes plastic deformation, a single micro asperity can only bear contact pressure proportional to the material's yield stress. The increasing number of micro-asperities that enter the contact state and undergo plastic deformations make the surface flatter. In addition, it was also found that surface roughness is an inevitable phenomenon in the deformation process. Even if the initial surface roughness of the specimen and the die is zero, a specific roughness value will still appear after deformation. This is a result of the orientation difference between adjacent grains leading to the local strain concentration on the specimen surface during the deformation process.

By calculating the coordinates of the surface nodes in the Z-axis direction, the variation of the specimen surface roughness under a different initial roughness was obtained, as described in Figure 7. When upsetting reduction gradually increases from 0% to 20%, that is, the deformation strain ε varied from 0 to 0.2, the surface roughness of the specimen with initial values of 0.7917 μm, 1.6131 μm, and 3.1152 μm gradually decreases to 0.2332 μm, 0.6733 μm, and 1.4709 μm. The roughness change is 0.5585 μm, 0.9398 μm, and 1.6443 μm. It means that with the increase in deformation, if the initial surface roughness of the specimen is high, the variation of the surface roughness after deformation will also be high. From the microscopic view, the actual contact area of the rough surface with fractal characteristics is much smaller than the nominal contact area, leading to a high value of actual contact pressure on the micro-asperities in real contact. With more plastic deformation to the micro-asperities, the workpiece has considerable change in its surface roughness.

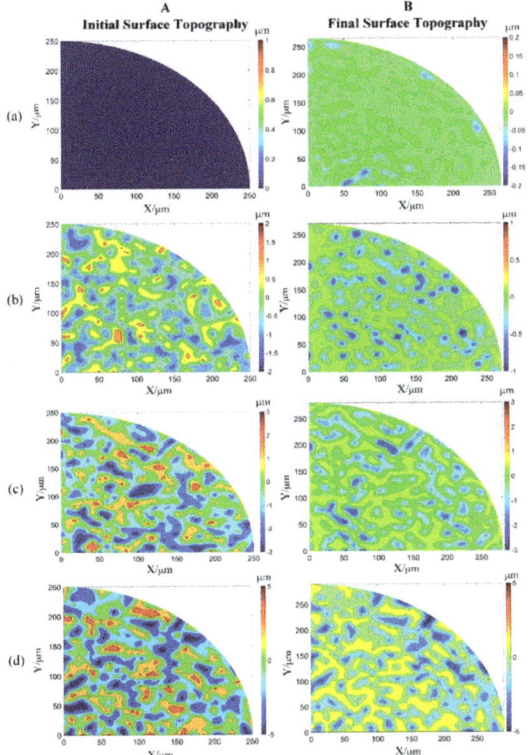

Figure 6. The surface morphology under different initial roughness after micro compression process (**a**) $S_{qin} = 0$ μm; (**b**) $S_{qin} = 0.7917$ μm; (**c**) $S_{qin} = 1.6131$ μm; and (**d**) $S_{qin} = 3.1152$ μm.

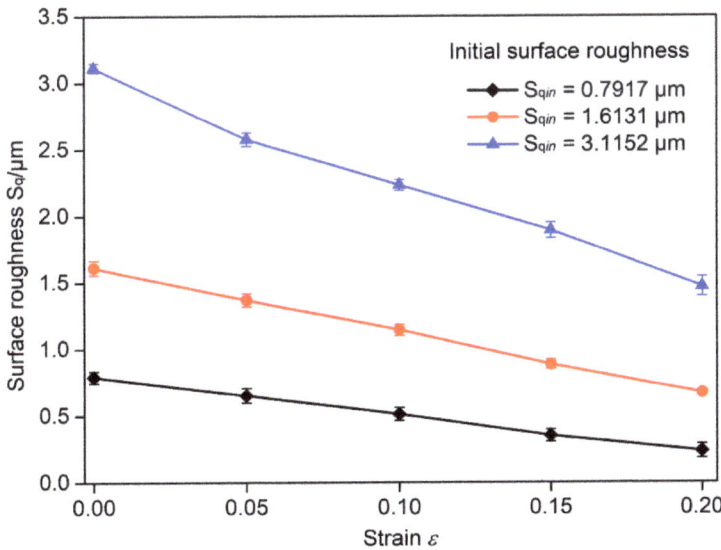

Figure 7. Variation of surface roughness for different values of the initial surface roughness.

3.1.2. Effect of Grain Size on the Evolution of Contact Surface Roughness

Figure 8 exhibits the equivalent stress of deformed specimens for different grain sizes. It shows that with the increase in grain sizes, the flow stress of the material decreases gradually. The coarse grains increase the proportion of the surface layer grains gradually. The constraint of surface grain is smaller than that of internal grain. At the same time, the larger grain sizes weaken the deformation coordination between grains. The specimen deformation is significantly affected by the properties of single grain. Hence, with the coarse grain sizes, the specimen surface roughness become more irregularity after the micro compression process.

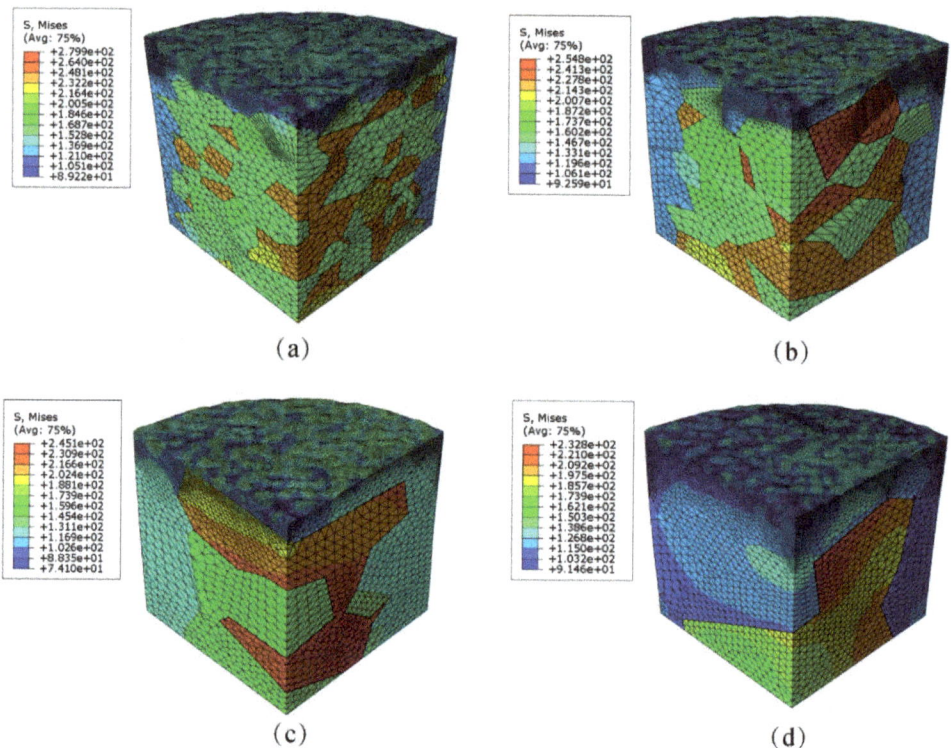

Figure 8. The equivalent stress of the deformed specimen for different grain sizes (**a**) $d = 30$ μm; (**b**) $d = 50$ μm; (**c**) $d = 80$ μm; and (**d**) $d = 120$ μm.

The contact surface morphologies of the deformed specimens with different grain sizes were obtained by extracting the node coordinates of the surface profile in the Z-axis direction. Let us take one of the three random morphologies as an example. Figure 9a–d show the surface morphologies for different grain sizes when the reduction amount is 20%. As the grain size increases, the flattening effect of surface asperities weakened, leading to the rougher surface of the specimen. The surface roughness for different grain sizes was calculated according to Equation (1). The surface roughness of the specimens changed from 1.6131 μm to 0.5327 μm, 0.6733 μm, 0.7512 μm, and 0.8720 μm corresponding to the grain sizes of 30 μm, 50 μm, 80 μm, and 120 μm, respectively, as depicted in Figure 10. Hence, with the increase in grain sizes, the specimen roughness increases gradually. In the deformation process, the grain sizes significantly impact the strain distribution in the material. Under the same deformation conditions, as the grain becomes smaller and smaller, the strain near the grain boundary becomes closer to the inside grain. Compared with the coarse grain, the deformation with fine grain becomes more uniform, and the corresponding surface roughness after deformation is lower.

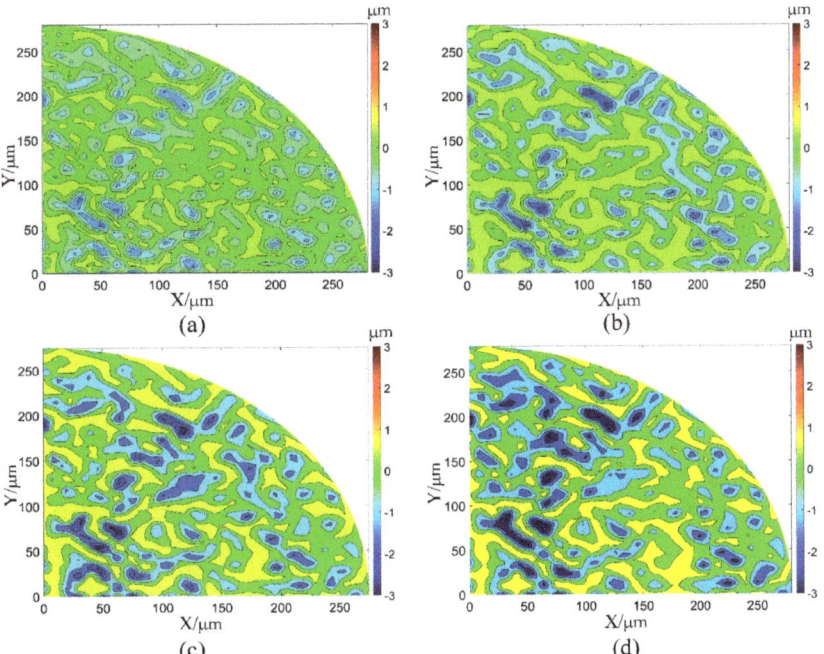

Figure 9. Surface morphology of specimens after deformation for different grain sizes; (**a**) $d = 30$ μm; (**b**) $d = 50$ μm; (**c**) $d = 80$ μm; and (**d**) $d = 120$ μm.

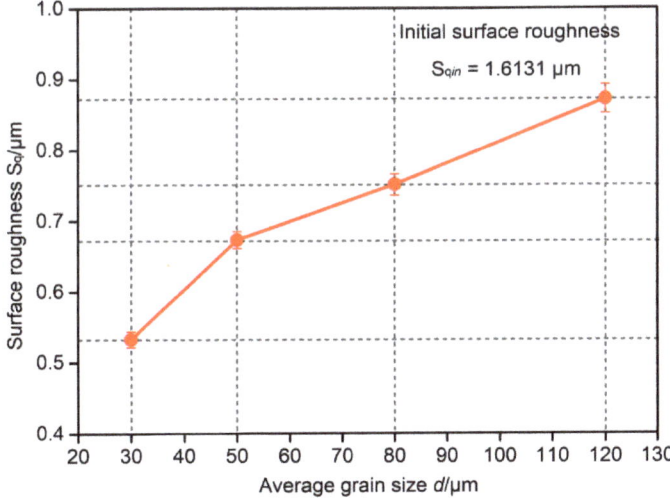

Figure 10. Change in specimen roughness under various grain sizes.

3.1.3. Effect of Grain Orientation on the Evolution of Contact Surface Roughness

Based on the change in material properties at the micro-scale, the 3D Voronoi method was used to develop the finite element models. Ten random grain orientations were generated under the same grain size. The micro-upsetting simulation processes investigated the effect of grain orientations on the specimen roughness. Figure 11 shows the surface morphology of three specimens with different grain orientations (represented by #1, #2, and #3) corresponding to the grain sizes of 50 μm and 120 μm, respectively. The increasing grain sizes resulting in the increase in the uneven degree of stress on the surface micro-asperities. Accordingly, the degree of variation of surface roughness also increases.

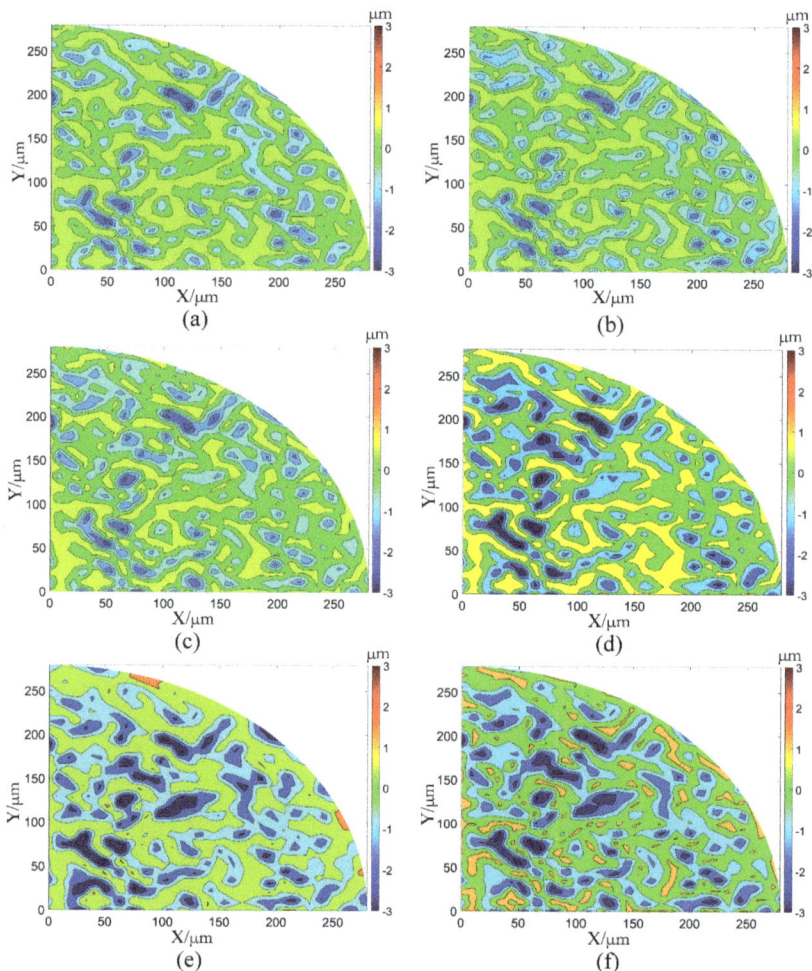

Figure 11. Surface morphology of the deformed specimens with various grain orientations; (**a**) $d = 50$ μm, #1; (**b**) $d = 50$ μm, #2; (**c**) $d = 50$ μm, #3; (**d**) $d = 120$ μm, #1; (**e**) $d = 120$ μm, #2; and (**f**) $d = 120$ μm, #3.

In this paper, Gaussian distribution was used to characterize the effect of grain orientation on surface roughness [26]. Its expression is given below.

$$F(S_q) = \frac{1}{\sqrt{2\pi}\beta} \exp\left(-\frac{(S_q - \gamma)^2}{2\beta^2}\right). \tag{4}$$

where S_q is the value of the roughness, β is the roughness standard deviation of the normal distribution, and γ is the roughness expected value of the normal distribution. With the grain sizes of 50 μm and 120 μm, the probability distributions of specimen surface roughness with random grain orientations are shown in Figure 12a,b. When the average grain size is 50 μm, the average value of surface roughness for different grain orientations is 0.6764 μm. The standard deviation is 0.0166, and the surface roughness distribution is relatively concentrated. Moreover, when the grain size is 120 μm, the average roughness under various grain orientations becomes 0.8602 μm. The standard deviation is 0.1145, and the roughness distribution is relatively discrete. In the micro-forming process, the increase in grain size reduces the number of grains in the surface layer, and the influence of single grain orientation on the deformation process increases. In addition, the coordination between the surface layer grains become worse, which also increases the deformation inhomogeneity. Finally, the difference in heights of the surface profiles and the variation range of surface roughness increases. In short, the effect of grain orientations on the surface roughness of the deformed specimens gradually increases with the increase in the grain size.

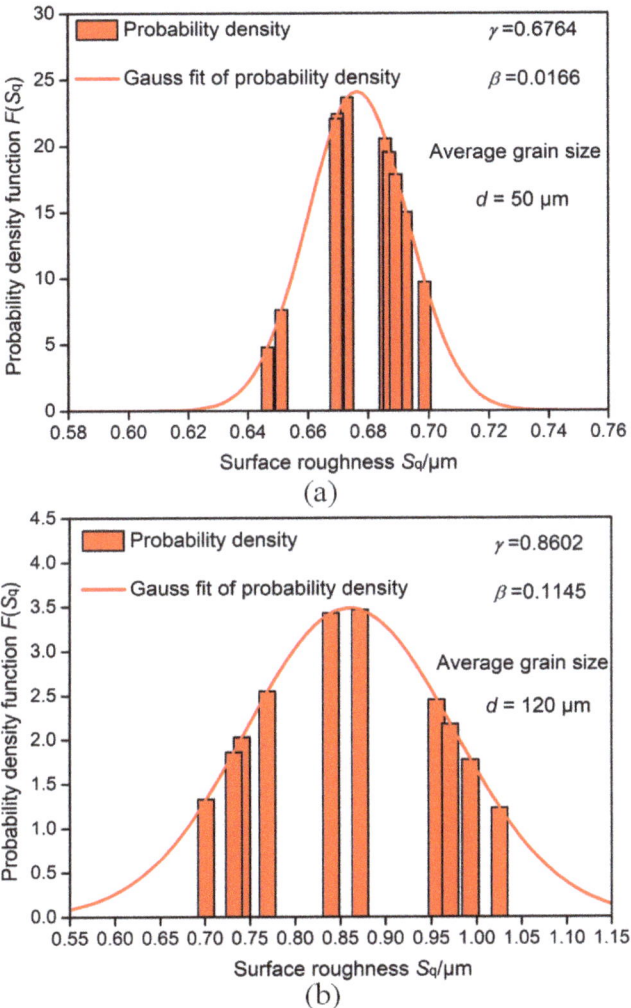

Figure 12. Probability density function of roughness with random grain orientations. (**a**) d = 50 μm; and (**b**) d = 120 μm.

3.2. Quantitative Description of Surface Roughness Variation Caused by Size Effect

According to our previous research on the influencing factors of surface roughness in forming bulk metal [10], the normal pressure on the contact surface directly affects the surface roughness of the resulting deformed product. The discussion in Section 2.2 shows that the increase in grain size in the micro-upsetting process leads to the reduction in the overall flow stress of the material, thus reducing the normal pressure on the contact surface. Therefore, on the basis of the previously developed roughness prediction model for bulk metal forming, in this paper, we investigate the influence of size effect on the contact normal pressure between the specimen and the die by introducing the scale parameters. Then, the quantitative description of the influence of size effect on the surface roughness change was observed.

3.2.1. Influence of Size Effect on Normal Pressure of Contact Surface

The proportion of the surface layer grains numbers to the total grain numbers was introduced to study the influence of size effect on the normal pressure. The relative mathematical is as follows.

$$\delta = 1 - \frac{(D-2d)^2}{D^2}. \tag{5}$$

where D and d are the specimen size and average grain size, respectively, as shown in Figure 13. The range of scale parameter is $0 < \delta < 1$. For micro-formed parts, when the grain size gradually approaches the specimen size ($D > 2d$), the scale parameter gradually approaches 1.

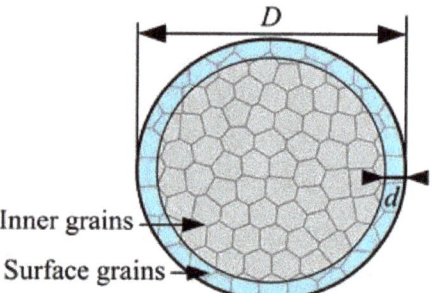

Figure 13. Schematic diagram of specimen diameter and grain size.

Table 2 shows the grain sizes and the corresponding scale parameters for the fixed geometric sizes in simulation process. The model of the materials and the boundary conditions in the simulations were consistent with those in Section 2.2. The deformation reduction for different scale parameters was set as 20%. The average normal pressure of the specimen after deformation was calculated by the simulation results of three random grain orientations for each size parameter. As depicted in Figure 14, a decreasing trend of the average normal pressure exhibits with the increased scale parameters. When the scale parameter δ is below 0.1, the normal pressure has a minor change with the scale parameters. For δ greater than 0.1, the continuous decrease in the flow stress causes the normal pressure on the contact surface to decrease rapidly. As δ larger than 0.8, the variation of the normal pressure with the changing of scale parameters becomes slow. In the plastic deformation process, the increase in contact normal pressure of the specimen leads to a decrease in surface roughness. The size effect of the surface roughness during the micro-compression is mainly determined by the size effect of the normal pressure. For $0.1 < \delta < 0.8$, the specimen roughness increases with the increase in the scale parameters, and there is an apparent size effect. For the $\delta < 0.1$ or $\delta > 0.8$, the size effect of surface roughness becomes insignificant.

Table 2. Grain sizes and the scale parameters in the micro compression simulation process.

Grain size $d/\mu m$	30	50	80	100	120	150	180
Scale parameter δ	0.23	0.36	0.54	0.64	0.73	0.84	0.92

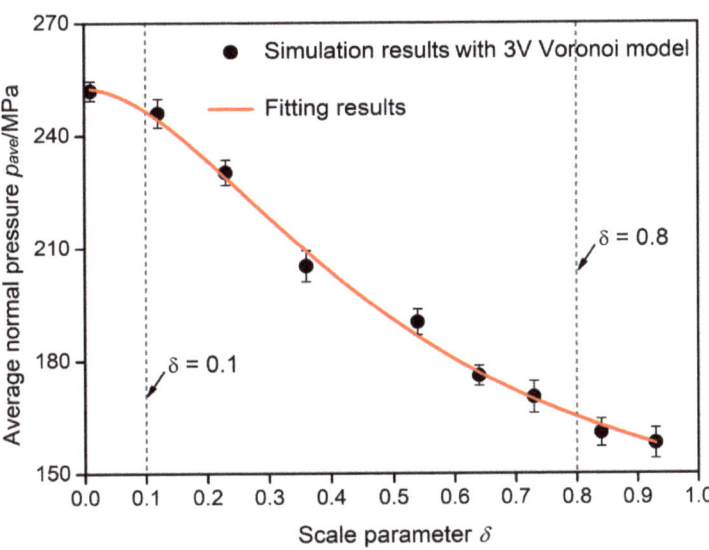

Figure 14. Changing of average normal pressure with scale parameters.

3.2.2. Development of Roughness Prediction Model in Micro-Forming Based on 3D Voronoi Model

A nonlinear fitting was performed to quantify the influence of size effect on surface roughness according to Figure 14. The relationship between the contact normal pressure and the scale parameter was obtained using Equation (6).

$$P_{mic} = 0.496 P_{mac} + \frac{0.154 P_{mac}}{0.304 + \delta^{1.81}}. \tag{6}$$

p_{mic} is the average normal pressure for micro forming with the consideration of size effect, and p_{mac} is the normal pressure for macro forming regardless of size effect. Based on our previous work in the literature [10], a roughness prediction model for micro-formed parts is developed and represented by the equations below.

$$\begin{cases} S_q = S_{qin} - 0.0013 P_{mic} S_{qin} - \frac{0.158}{1+10^{1.12-0.0093 P_{mic}}} \left(1 - 0.0089^{L/P_{mic}}\right) - 0.213 \varphi^{0.611} \\ P_{mic} = 0.496 P_{mac} + \frac{0.154 P_{mac}}{0.304 + \delta^{1.81}} \end{cases} \tag{7}$$

S_q represents the surface roughness of the deformed specimen. L is the sliding distance between the workpiece and the die. φ is the growth rate of the contact surface area. The previous report has detailed all the above parameters [10].

Above all, for the micro-formed parts, the surface roughness of the workpiece is mainly affected by multiple factors. The increase in normal pressure, the relative sliding distance, and surface area growth rate on the contact surface positively improve the workpiece's surface roughness. However, with the increase in grain size, the contact surface's normal pressure decreases gradually, which is detrimental to improve surface quality of micro-formed parts.

3.2.3. Verification of Roughness Prediction Model for Micro-Forming

The micro-compression experiments of pure copper for different grain sizes were carried out to testify the established roughness prediction model d. The cylindrical specimens of pure copper with a diameter of 1.5 mm and a height to diameter ratio of 1.5:1 was selected. The NBD-T1500 vacuum tube furnace was used to conduct two annealing heat treatment processes at 700 °C and 900 °C for one hour under the protection of argon gas. The average grain sizes of the specimens were measured to be 65 μm and 119 μm, as shown in Figure 15. The upper and lower end surfaces of the heat-treated specimens were polished to the same roughness after ultrasonic decontamination. The roughness at various positions on the specimens' surface was measured using the VEECO NT9300 optical contour instrument. Figure 16a–d show the surface roughness profiles in different regions. The average roughness S_{qin} of the pre-treated cylindrical specimen was measured to be 1.57 μm.

Figure 15. Metallography of pure copper specimens with different heat treatments (**a**) 700 °C, d = 65 μm, and (**b**) 900 °C, d = 119 μm.

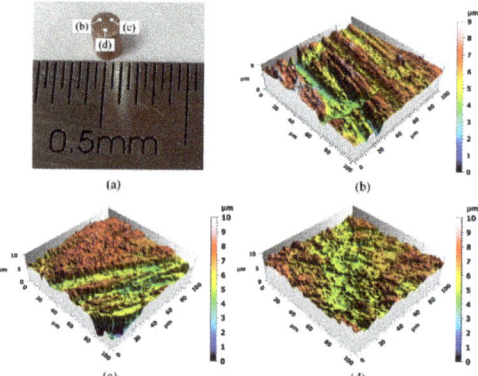

Figure 16. Surface roughness of pretreated cylindrical specimen with (**a**) roughness measurement positions as (**b**) S_{qin} = 1.58 μm, (**c**) S_{qin} = 1.56 μm, and (**d**) S_{qin} = 1.57 μm.

In the micro-upsetting experiment, to reduce the effect of the die surface roughness on the surface quality of the deformed specimen, a set of gaskets with the roughness of 0.2 μm were adopted. Their surface roughness is much lower than the specimen surface roughness. The specimen was placed between two gaskets, and the CMT5105 electronic universal testing machine was used for the upsetting experiment at room temperature under dry friction conditions. Figure 17a,b show the experimental setup and the surface of the upper and lower gaskets used in the experiment, respectively. The deformation was set at 20%, and the low constant strain rate was set at 0.0003/s. Each group of parameters was repeated three times to reduce the experimental error. And four different locations were measured on each group of specimen surface. The surface roughness measurement area and its size diagram are presented in Figure 18.

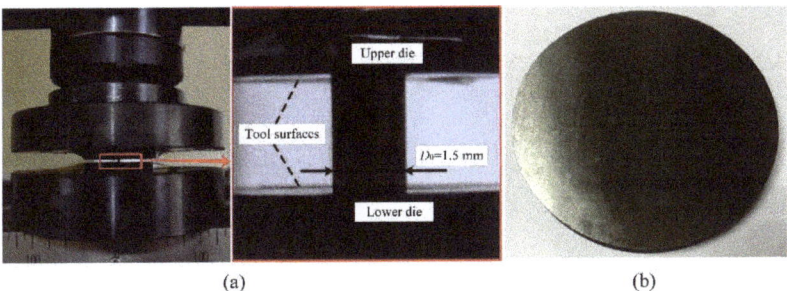

Figure 17. The micro-upsetting experiment with (**a**) experimental setup and (**b**) gasket surface.

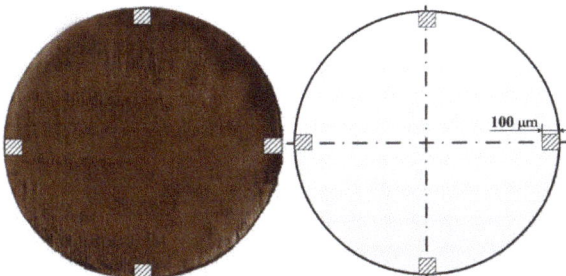

Figure 18. Surface roughness measurement area and its dimensions.

Figure 19a,b present the surface roughness topography of the specimens and their measurement areas after deformation for different grain sizes with a deformation of 20%. The measurement results denoted that the specimen roughness is 0.90 µm with grain sizes of 65 µm, and 1.03 µm with grain sizes of 119 µm. The specimen roughness with larger grain sizes is becoming rougher after deformation under the same deformation conditions.

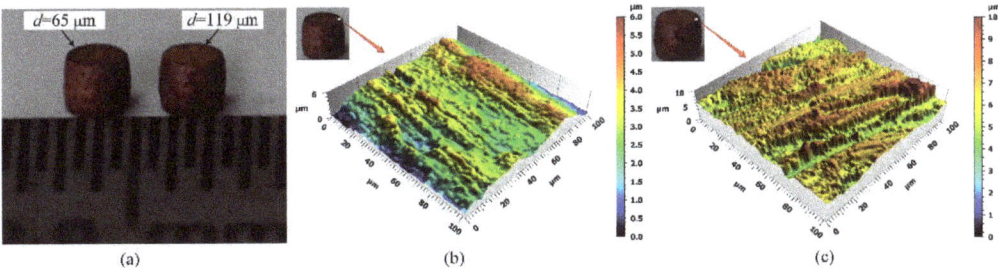

Figure 19. (**a**) Specimens after deformation with different grain sizes; the surface topography of corresponding regions with different grain sizes: (**b**) d = 65 µm, S_q = 0.90 µm; and (**c**) d = 119 µm, S_q = 1.03 µm.

Through the simulation of micro-upsetting process under the same conditions as the experiment, the contact normal pressures were extracted, and the average normal pressure p_{mic} under different grain sizes was calculated. By measuring the surface diameter of the specimen after deformation, the relative sliding distance L and the surface area growth rate φ between the specimen and the die can also be obtained. Combined with Equation (7), the roughness prediction results, the simulation results of the 3D Voronoi model, as well as the experimental results, are given in Figure 20. Because the 3D Voronoi model adopted in the simulation process cannot fully reflect the specimen grain orientations, the prediction results of surface roughness models for two different grain sizes have certain deviations from the experimental values, with deviations of 1.46% and 2.78%, which are within the acceptable error range. Considering the size effect, the established surface roughness prediction model can better predict the surface roughness of formed parts in micro-forming. Meanwhile, the research results in this paper also show that the grain size increase adversely impacts the surface roughness of the micro-formed parts. Consequently, to improve the surface roughness of the micro-formed parts, the grain refinement treatment should be conducted to avoid the increase in the surface roughness due to the large grain size of the bulk.

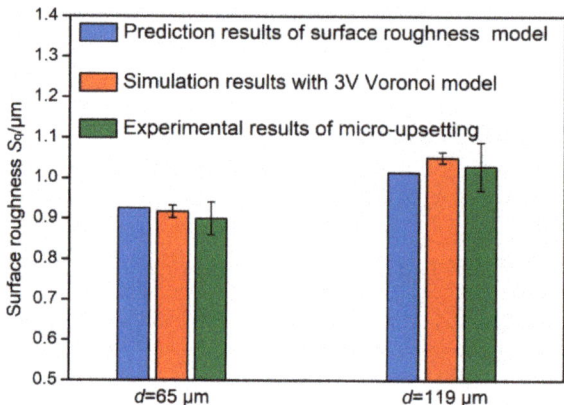

Figure 20. Comparison between model prediction, simulation, and experimental results for different grain sizes.

4. Conclusions

In this paper, in order to realize the quantitative description of surface roughness changes during the deformation process of micro-formed parts, the effects of initial surface roughness, average grain sizes, and grain orientations on the surface roughness of the specimen were analyzed. The influence of the size effect on the specimen surface roughness was quantified and a roughness prediction model with multi-factors for micro-formed parts was developed. The main conclusions of this paper are given below.

(1) The finite element model of polycrystalline pure copper with a 3D rough surface was developed. The simulation results show that the variation in the specimen roughness increases when the increase in initial specimen roughness. Moreover, the increase in grain size leads to the gradual increase in the specimen roughness after deformation. At the same time, the effect of grain orientations on surface roughness enhances with coarse grains.

(2) The scale parameter was adopted to characterize the influence of size effect on the contact normal pressure and the surface roughness. As the scale parameter increases, the average contact normal pressure decreases gradually. For the scale parameter $0.1 < \delta < 0.8$, the surface roughness increases with the increase in scale parameter, which shows an apparent size effect.

(3) A roughness prediction model for micro-formed parts was developed. By comparing with the surface roughness measurement results of micro-upsetting specimens with different grain sizes, the rationality and applicability of the established prediction

model were verified. The established model can be used to predict and control the surface roughness of micro-formed parts, especially for the parts affected by size effect during deformation process.

Author Contributions: Formal analysis, J.H. and W.Z.; Investigation, J.H., W.Z. and Q.C.; Methodology, J.H.; Software, J.H., J.S. and S.X.; Writing—original draft, J.H.; Writing—review & editing, J.H. and W.Z. All authors have read and agreed to the published version of the manuscript.

Funding: This research was funded by Shandong Provincial Natural Science Foundation Grant number. ZR2020QE161 and No. ZR2021ME182, PhD Research Foundation of Shandong Jianzhu University (X20053Z0101).

Institutional Review Board Statement: Not applicable.

Informed Consent Statement: Not applicable.

Data Availability Statement: Not applicable.

Conflicts of Interest: The authors declare no conflict of interest.

References

1. Huang, J.H.; Xu, Z.T.; Li, X.N.; Peng, L.F.; Lai, X.M. An experimental study on a rapid micro imprinting process. *J. Mater. Process. Technol.* **2020**, *283*, 116716. [CrossRef]
2. Guo, N.; Wang, J.; Sun, C.Y.; Zhang, Y.F.; Fu, M.W. Analysis of size dependent earing evolution in micro deep drawing of TWIP steel by using crystal plasticity modeling. *Int. J. Mech. Sci.* **2020**, *165*, 105200. [CrossRef]
3. Luo, L.; Jiang, Z.Y.; Wei, D.B.; Jia, F.H. A study of influence of hydraulic pressure on micro-hydromechanical deep drawing considering size effects and surface roughness. *Wear* **2021**, *477*, 203803. [CrossRef]
4. Behrens, G.; Trier, F.O.; Tetzel, H.; Vollertsen, F. Influence of tool geometry variations on the limiting drawing ratio in micro deep drawing. *Int. J. Mater. Form.* **2016**, *9*, 253–258. [CrossRef]
5. Rathmann, L.; Vollertsen, F. Determination of a contact length dependent friction function in micro metal forming. *J. Mater. Process. Technol.* **2020**, *286*, 116831. [CrossRef]
6. Anand, D.; Kumar, D.R. Effect of sheet thickness and grain size on forming limit diagrams of thin brass sheets. *Adv. Intell. Syst. Comput.* **2019**, *1*, 435–444. [CrossRef]
7. Sail, K.; Aouici, H.; Hassani, S.; Fnides, B.; Belaadi, A.; Naitbouda, A.; Abdi, S. Influence of tribological parameters on S335 steel filing Ti-W-N in dry sliding wear: Prediction model and sliding condition optimization. *Int. J. Adv. Manuf. Technol.* **2017**, *92*, 4057–4071. [CrossRef]
8. Hiegemann, L.; Weddeling, C.; Tekkaya, A.E. Analytical contact pressure model for predicting roughness of ball burnished surfaces. *J. Mater. Process. Technol.* **2016**, *232*, 63–77. [CrossRef]
9. Hiegemann, L.; Weddeling, C.; Khalifa, N.B.; Tekkaya, A.E. Prediction of roughness after ball burnishing of thermally coated surfaces. *J. Mater. Process. Technol.* **2015**, *217*, 193–201. [CrossRef]
10. Han, J.J.; Zhu, J.; Zheng, W.; Wang, G.C. Influence of metal forming parameters on surface roughness and establishment of surface roughness prediction model. *Int. J. Mech. Sci.* **2019**, *163*, 105093. [CrossRef]
11. Anand, D.; Shrivastava, A.; Ravi, K.D. Size Effect on Surface Roughness of Very Thin Brass Sheets in Biaxial Stretching. *Mater. Today Proc.* **2019**, *18*, 2448–2453. [CrossRef]
12. Peng, L.F.; Xu, Z.T.; Gao, Z.Y.; Fu, M.W. A constitutive model for metal plastic deformation at micro/meso scale with consideration of grain orientation and its evolution. *Int. J. Mech. Sci.* **2018**, *138*, 74–85. [CrossRef]
13. Kubo, M.; Nakazawa, Y.; Hama, T.; Takuda, H. Effect of Microstructure on Surface Roughening in Stretch Forming of Steel Sheets. *ISIJ Int.* **2017**, *57*, 2185–2193. [CrossRef]
14. Kubo, M.; Hama, T.; Tsunemi, Y.; Nakazawa, Y.; Takuda, H. Influence of Strain Ratio on Surface Roughening in Biaxial Stretching of IF Steel Sheets. *ISIJ Int.* **2018**, *58*, 704–713. [CrossRef]
15. Kishimoto, T.; Suematsu, S.; Sakaguchi, H.; Tashima, K.; Kajino, S.; Gondo, S.; Suzuki, S. Effect of crystal orientation on inner surface roughness of micro metal tubes in hollow sinking. *Mater. Sci. Eng. A* **2021**, *805*, 140792. [CrossRef]
16. Yoshida, K. Effects of grain-scale heterogeneity on surface roughness and sheet metal necking. *Int. J. Mech. Sci.* **2014**, *83*, 48–56. [CrossRef]
17. Wang, S.W.; Zhang, S.H.; Song, H.W.; Chen, Y. Surface roughness improvement of the bent thin-walled copper tube by controlling the microstructure and texture components. *Procedia Manuf.* **2020**, *50*, 613–617. [CrossRef]
18. Han, J.J.; Zheng, W.; Xu, S.B.; Dang, G.H. The regionalized modelling and simulation of the micro-tensile process based on 3D Voronoi model. *Mater. Today Commun.* **2022**, *31*, 103614. [CrossRef]
19. Sun, F.W.; Meade, E.D.; Dowd, O.N. Strain gradient crystal plasticity modelling of size effects in a hierarchical martensitic steel using the Voronoi tessellation method. *Int. J. Plast.* **2019**, *119*, 215–229. [CrossRef]
20. Stout, K.J. *Development of Methods for the Characterisation of Roughness in Three Dimensions*; Penton Press: London, UK, 2000.

21. Majumdar, A.; Tien, C.L. Fractal characterization and simulation of rough surfaces. *Wear* **1990**, *136*, 313–327. [CrossRef]
22. Huang, X.; Borrego, A.; Pantleon, W. Polycrystal deformation and single crystal deformation dislocation structure and flow stress in copper. *Mater. Sci. Eng. A* **2001**, *319*, 237–241. [CrossRef]
23. Clausen, B.; Lorentzen, T.; Leffers, T. Self-consistent modelling of the plastic deformation of FCC polycrystals and its implications for diffraction measurements of internal stress. *Acta Mater.* **1998**, *46*, 3087–3098. [CrossRef]
24. Han, J.J.; Lin, Y.; Zheng, W.; Wang, G.C. Experimental and numerical investigations on size effect of friction in meso-/microforming without lubricant. *Int. J. Adv. Manuf. Technol.* **2020**, *106*, 4869–4877. [CrossRef]
25. Wanheim, T.; Bay, N.; Petersen, A.S. A theoretically determined model for friction in metal working processes. *Wear* **1974**, *28*, 251–258. [CrossRef]
26. Pan, S.; Han, Y.Y.; Wei, S.; Wei, Y.X.; Xia, L.; Xie, L.; Kong, X.R.; Yu, W. A model based on Gauss Distribution for predicting window behavior in building. *Build. Environ.* **2019**, *149*, 210–219. [CrossRef]

Review

Electrochemical Synthesis of Functional Coatings and Nanomaterials in Molten Salts and Their Application

Yuriy Stulov, Vladimir Dolmatov, Anton Dubrovskiy and Sergey Kuznetsov *

Tananaev Institute of Chemistry of the Federal Research Centre "Kola Science Centre of the Russian Academy of Sciences", 184209 Apatity, Russia
* Correspondence: s.kuznetsov@ksc.ru

Abstract: Nanomaterials are widely used in modern technologies due to their unique properties. Developing methods for their production is one of the most important scientific problems. In this review, the advantages of electrochemical methods for synthesis in molten salts of nanostructured coatings and nanomaterials for different applications were discussed. It was determined that the nanostructured Mo_2C coatings on a molybdenum substrate obtained by galvanostatic electrolysis have a superior catalytic activity for the water-gas shift reaction. The corrosion-resistant and wear-resistant coatings of refractory metal carbides on steels were synthesized by the method of currentless transfer. This method also was used for the production of composite carbon fiber/refractory metal carbide materials, which are efficient electrocatalysts for the decomposition of hydrogen peroxide. The possibility to synthesize GdB_6 nanorods and Si and TaO nanoneedles by potentiostatic electrolysis was shown.

Keywords: electrochemical synthesis; molten salts; nanostructured coatings; refractory carbide coatings; carbon fibers; nanorods; nanoneedles

1. Introduction

In recent years, nanostructured and nanoscale materials have been of great interest because of the unique mechanical, electrical, optical, and chemical properties possessed by such materials due to the small size of their components (e.g., grains, phase inclusions, layers, pores, etc.). The possibility of obtaining nanomaterials with a given structure provides a wide range of opportunities for the creation of new functional materials with unique sets of characteristics [1]. Nanomaterials are of great applied importance for solving problems of energy storage and conversion energy [2–4]; they are also used as catalysts [5–8], electrocatalysts [9], protective coatings and structural materials [10,11].

To the present day, a great number of different methods and approaches have been developed for the synthesis of nanomaterials [12–15]. According to [15] they can be divided into three groups: biological, physical, and chemical. In the current study, we will not describe all methods of nanomaterials synthesis in detail, but we will focus on some of the most common ones, including sol-gel [16–20], micellar [21–23], hydrothermal [24–27], vapor deposition [28–32], ultrasonic [33–37], microwave [38–41], and electrochemical [42–50].

Electrochemical methods of synthesizing nanomaterials and functional materials have a number of advantages over other methods [51]. For example:

- electrolysis of melts using pulsed and reversible currents provides the ability to easily adjust the structure, thickness, porosity, roughness, grain size and texture of electroplated coatings and materials;
- a large number of environments for synthesis (both aqueous and non-aqueous) used for various purposes;
- electrodeposition parameters determined at a laboratory scale can be transferred to the industrial scale;

- high purity of the obtained coatings and materials even if low-quality initial reagents are used, since metals are refined in the process of electrolysis;
- low operating costs and low costs of electrochemical equipment.

On the other hand, the use of molten salts such as chlorides and fluorides of alkali metals as a medium in the electrochemical synthesis of nanomaterials makes it possible to increase the current density due to a high electric conductivity and thereby to intensify the synthesis processes. Electrochemical synthesis in molten salts leads to the appearance of strong electric and magnetic fields in the near-electrode layer (the electric field strength near the cathode is 10^9–10^{10} V m^{-1} [52]). Therefore, the synthesis can proceed under conditions far from the thermodynamical equilibrium, which significantly expands the number of obtained compounds and their phase modifications as compared to other methods of nanomaterial production.

A large number of studies are devoted to obtaining silicon-based nanomaterials in various molten salts using electrochemical methods [53–61]. For example, methods of producing silicon-based nanomaterials for utilization in advanced power engineering are reviewed in [55,57,59–61]. By varying the electrolyte composition, electrolysis parameters and process temperature, it becomes possible to synthesize nanowires and nanotubes.

The synthesis of carbon-based nanomaterials in molten salts has been the subject of many reviews and articles. In review [62], the methods and mechanisms of synthesis of nanomaterials based on graphene and its oxide in various melts are discussed in detail. Reducing the amount of carbon dioxide emissions in the atmosphere has become an urgent problem, and the electroreduction of CO_2 in molten salts can contribute to solving this challenge. In works [63–73], the processes and mechanisms of electroreduction of carbon dioxide in various melts leading to the production of various carbon nanoparticles have been investigated. One study [64] reveals the effect of the electrolysis temperature in molten Li-Na-K carbonates on the shape of carbon nanomaterials. At a process temperature below 850 K, carbon nanosheets are formed on the cathode, and nanotubes are formed at a higher temperature. In the paper [65], the principle possibility of carbon nanotube generation by the electrolysis of molten salts saturated by carbon dioxide was shown. Kushkov et al. [68] found that, depending on the experimental conditions, the electrolysis of molten alkali metal carbonates led to the synthesis of carbon nanotubes or fullerenes C_{60} or C_{70}. Electrochemical methods to synthesize carbon nanotubes and various carbon nanostructures from CO_2 in molten $CaCl_2$-NaCl-CaO were discussed in [72].

For the nanotechnology of carbon materials, it is of great interest to produce nanodiamonds in molten salts. It was shown in [74] that single Li_2CO_3 nanocrystals (<30 nm) can be synthesized by the injection of moist CO_2 into molten LiCl containing Li_2O. The subsequent cathodic polarization of a graphite rod immersed in the molten salt led to the formation of carbon-encapsulated lithium carbonate nanoparticles. Upon heating in ambient-pressure air, these nanostructures acted as high-pressure nanocrucibles, creating diamond crystallites within lithium carbonate.

The electrochemical methods allow to synthesize carbides of different metals, which can be used for surface protection against oxidation or corrosion and as catalysts or electrocatalysts. In studies [75,76] voltammetric investigations of tungsten and carbon-containing melts were carried out to find the conditions for a high-temperature electrochemical synthesis of nanostructured tungsten carbide coatings and nanopowders. Binary tungsten-molybdenum carbide nanopowders were fabricated by high-temperature electrochemical synthesis from tungstate–molybdate–carbonate melts [77].

Analysis of available literature data shows that electrochemical synthesis in molten salts is a promising method for obtaining coatings and nanomaterials.

Of course, this article is not a complete review of all the data available in the literature on the production of coatings and nanomaterials in molten salts. The aim of the present work was to introduce our results on electrochemical synthesis in molten salts of nanostructured coatings and nanomaterials used for different applications.

2. Materials and Methods

In this work, we used chloride-fluoride electrolytes for the synthesis of nanostructured coatings and nanomaterials.

Sodium and potassium chlorides (Prolabo 99.5% min) were recrystallized in distilled water to remove various insoluble impurities. Afterwards they were calcined in a muffle furnace at 773 K. Then, NaCl and KCl were mixed (with a molar ratio of 1:1) and evacuated to a residual pressure of 0.6 Pa at 298, 473, 673, and 823 K; following that, the NaCl and KCl mixture was melted in a purified and dried argon atmosphere.

Sodium fluoride (Prolabo 99.5% min) was refined from impurities using the method of directional crystallization, which is described elsewhere [78].

Li_2CO_3 (Sigma-Aldrich, St. Louis, MI, USA, 99% min) and $Na_2MoO_4 \cdot H_2O$ (Sigma-Aldrich, 99.5% min) were dehydrated for 24 h at 473 K. Chromium trichloride (Sigma-Aldrich, anhydrous, 99.99% min) was used without additional treatment.

Potassium heptafluoroniobate, heptafluorotantalate, and hexafluorosilicate were synthesized at the experimental facility of the Tananaev Institute of Chemistry-Subdivision of the Federal Research Centre, at the Kola Science Centre of the Russian Academy of Sciences. The content of metallic impurities in the synthesized salts did not exceed 10^{-3} wt.% according to mass spectroscopy data. The purification methods of K_2NbF_7, K_2TaF_7, and K_2SiF_6 are described in literature [79–81].

Potassium tetrafluoroborate (Sigma-Aldrich, 99.99% min) was used without additional treatment. GdF_3 was synthesized by heating a mixture of Gd_2O_3 and $[NH_4][HF_2]$ in an argon atmosphere [78]. The procedures for obtaining K_3TaOF_6 and CsCl were described in details elsewhere [82].

A detailed description of the experimental setup, instruments and materials used for synthesis, identification and investigation of the morphology, and properties of the synthesized coatings and nanomaterials are given in [83–85].

Other specific techniques are described in the sections below.

3. Results and Discussions

3.1. Nanostructured Catalytic Coatings Mo_2C

One of the promising areas of hydrogen energy is the direct placement of an integrated device on board a vehicle, which includes a fuel processor in combination with a fuel cell. As a result of the transformation of gasoline or natural gas, hydrogen is obtained, which contains about 12 vol % CO. The carbon monoxide steam reforming reaction (SRR) is used to remove carbon monoxide from the hydrogen-rich gas, since CO is a catalytic poison to the proton exchange membrane of the fuel cell. The high-temperature steam reforming reaction is usually carried out on ferrochrome (Fe_3O_4/Cr_2O_3) as a catalyst at temperatures of 573–723 K, which makes it possible to reduce the CO content to 2 vol % and obtain an additional amount of hydrogen. This product is then subjected to a low-temperature steam reforming reaction using a $Cu/ZnO/Al_2O_3$ catalyst at a temperature range of 433–523 K, which reduces the CO concentration to 0.1% by volume. However, the low-temperature catalyst occupies approximately 70% of the volume of the entire fuel processor catalytic system and is also pyrophoric due to the oxidation of Cu to Cu_2O or CuO, making it potentially hazardous. Strict temperature control is necessary to carry out a low-temperature SRR, which makes the $Cu/ZnO/Al_2O_3$ catalyst impractical.

Thus, the search for a catalyst with high activity at 473–523 K, capable of reducing the CO concentration to 0.1 vol % for use in automobiles and other vehicles, remains relevant. A new class of catalytic systems based on refractory metal carbides can be used for the carbon monoxide steam reforming reaction.

3.1.1. Electrochemical Synthesis

Based on the analysis of the peculiarities of molybdenum and carbon electrodeposition, an NaCl-KCl-Li$_2$CO$_3$-Na$_2$MoO$_4$ chloride-carbonate-molybdate melt was chosen for electrochemical synthesis. As the substrate, molybdenum plates (99.99 wt.% Mo), 40 mm long, 10 mm wide, and 0.1 mm thick were used. Electrochemical synthesis of Mo$_2$C coatings was carried out at the following conditions: temperature 1123 K, cathodic current density 5 mA cm^{-2}, time of electrolysis 7 h, and the glassy carbon (SU-2000) ampoule was used as the anode. X-ray phase and microstructural analyses, scanning electron microscopy, and the BET method were used to characterize the obtained coatings.

It is important that the obtained coatings are β-Mo$_2$C monophase semi-carbide with a hexagonal crystalline lattice (Figure 1). At the same time, bulk Mo$_2$C contains at least several weight percent of cubic molybdenum carbide, which significantly reduces its catalytic activity and stability in the carbon monoxide steam reforming reaction. The formation of hexagonal Mo$_2$C during electrochemical synthesis is the result of specific electrocrystallization conditions (electric field, double layer, and high temperature) [52].

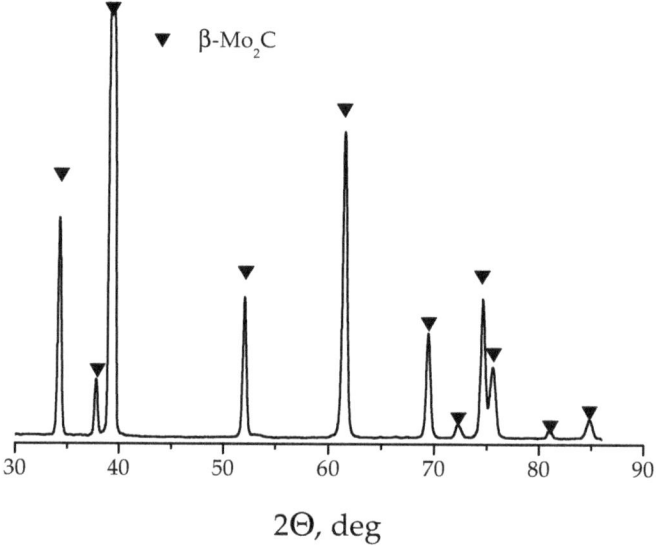

Figure 1. X-ray powder diffraction patterns of the molybdenum carbide coating on molybdenum substrates obtained at 1123 K and a current density of 5 mA cm^{-2} from the NaCl-KCl-Li$_2$CO$_3$ (1.5 wt.%)-Na$_2$MoO$_4$ (8.0 wt.%) melt (solid line), data points indicate the reference sample.

Joint electroreduction of MoO$_4^{2-}$ and CO$_3^{2-}$ ions resulted in the formation of 50 μm thick Mo$_2$C coatings and can be in general described by the reaction:

$$2\text{MoO}_4^{2-} + \text{CO}_3^{2-} + 16e^- \rightarrow \text{Mo}_2\text{C} + 11\text{O}^{2-}. \tag{1}$$

The morphology of the Mo$_2$C coating is presented in Figure 2. The figure shows that the obtained coating is nanostructured, and consists of a large number of pores with sizes ranging from 300 nm to 3 μm and Mo$_2$C nanoneedles.

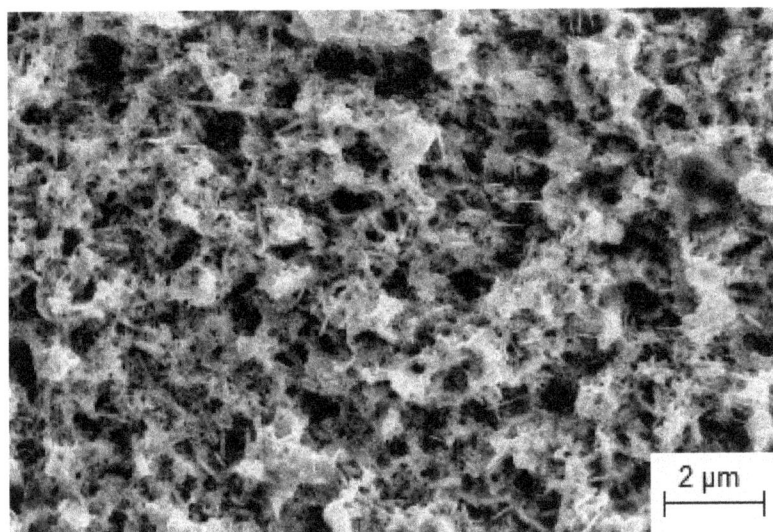

Figure 2. Morphology of molybdenum carbide coating on a molybdenum substrate. T = 1123 K, i = 5 mA cm^{-2}, melt–NaCl-KCl-Li$_2$CO$_3$ (1.5 wt.%)-Na$_2$MoO$_4$ (8.0 wt.%).

3.1.2. Catalytic Activity

The steady-state steam reforming reaction rate for the Mo$_2$C/Mo composition was three orders of magnitude higher than for the Mo$_2$C bulk phase and the commercial Cu/ZnO/Al$_2$O$_3$ catalyst (Figure 3).

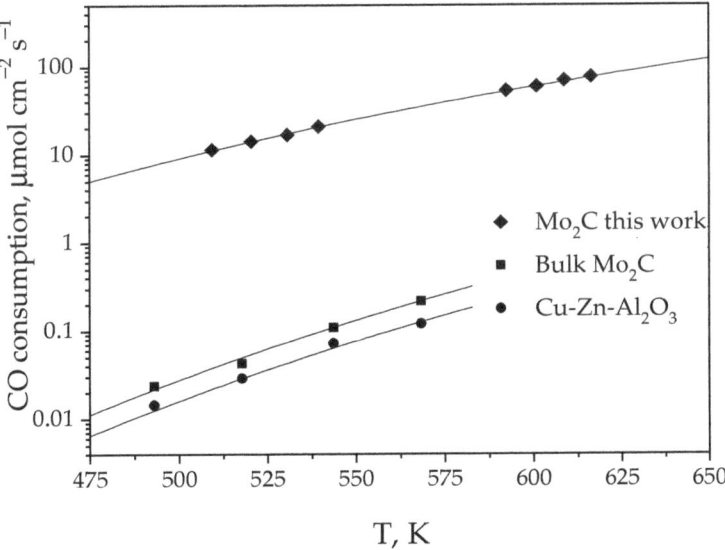

Figure 3. Temperature dependence of the steam reforming reaction rate over different catalysts. Reaction conditions: $p_{CO} = 300$ Pa, $p_{H_2O} = 760$ Pa, $p_{co_2} = 1.2$ kPa, $p_{H_2} = 40$ kPa and balanced helium. The gas flow rate is 50 cm^3 min^{-1} (STP).

The specific surface area (BET surface) of the Mo_2C coating was 38 m^2 g^{-1}. The BET surface of bulk Mo_2C was much higher (61 m^2 g^{-1}) [86]. Despite this, the catalytic activity of the bulk Mo_2C was much lower than that of the catalysts obtained by electrolysis. This can be explained by the presence of cubic modification in the bulk Mo_2C composition, which significantly reduces the SRR rate.

Methane formation was not observed throughout the entire temperature range in which the Mo_2C/Mo coatings were tested. The catalytic activity remained constant over 5000 h of testing. The coatings also remained stable during cyclic temperature tests, while the activity of industrial catalysts decreased.

Thus, by electrochemical synthesis in molten salts, we obtained a new Mo_2C/Mo-based catalytic system for a low-temperature steam reforming reaction. The catalytic activity of the composition, obtained by simultaneous reduction of electroactive MoO_4^{2-} and CO_3^{2-} particles, was three orders of magnitude higher than that of the bulk Mo_2C and commercially available Cu-ZnO-Al_2O_3 catalyst.

3.2. Wear and Corrosion Resistance Coatings on Steels

The main requirements for hard alloys as cutting, milling, and drilling tools are wear resistance, chemical stability at high temperatures, and superior mechanical properties. Current research shows that the local temperature on the surface of these tools increases greatly during tool usage, resulting in oxidation, changes in composition, and surface degradation over time. In addition, these materials sometimes operate in aggressive environments. The corrosion rate of materials is significantly affected by the pH of the environment. Materials based on the refractory metal carbides show a good resistance in acidic electrolytes [87].

The modification of a tool's surface with films and coatings of different compositions can be carried out by many methods. These coatings determine the surface properties [88,89]. Currently NbC and TaC, with the addition of WC, are effectively used as cutting tools [90]. Tantalum carbide is used in the engine and aerospace industries due to its resistance to high-temperature oxidation [91].

Various techniques of obtaining chromium carbide coatings, such as chemical vapor deposition [92], metallothermic reduction of Cr_2O_3/C mixtures [93], and magnetron sputtering [94] were reviewed in [92–96]. It is shown that the deposition of Cr_2C_3 as a coating on steel leads to an increase in the hardness and wear resistance and reduces the friction coefficient [93,94].

In [97], different methods of synthesis of protecting coatings are discussed. It was concluded [97] that coating deposition by molten salt electrolysis is a prospective way for obtaining smooth and uniform coatings on complex-shape samples.

Previously, we have thoroughly studied [98] the processes of alloy formation during electrodeposition of chromium on the surface of steels with the formation of chromium carbide coatings of different compositions.

3.2.1. Currentless Transfer: A Simple Way to Obtain Refractory Carbide Coatings

An alternative method for synthesizing coatings of refractory metal carbides on steels is currentless transfer (CT) [99]. It is important to note that CT is an electrochemical method, and synthesis by CT is subject to the laws of electrochemical thermodynamics, kinetics, and reaction diffusion in solids [98,100–102].

A schematic diagram of CT in molten salts is shown in Figure 4. During the process, metal-salt reactions occur and metal cations with an intermediate oxidation state are formed disproportionately on the surface of steel (or another carbon-containing material) to form a refractory metal carbide and a cation in a higher oxidation state. The driving force of the CT is the Gibbs (ΔG_r) energy of the carbide-forming reaction. The carbide with the smallest (more negative) value of ΔG_r will be formed [99].

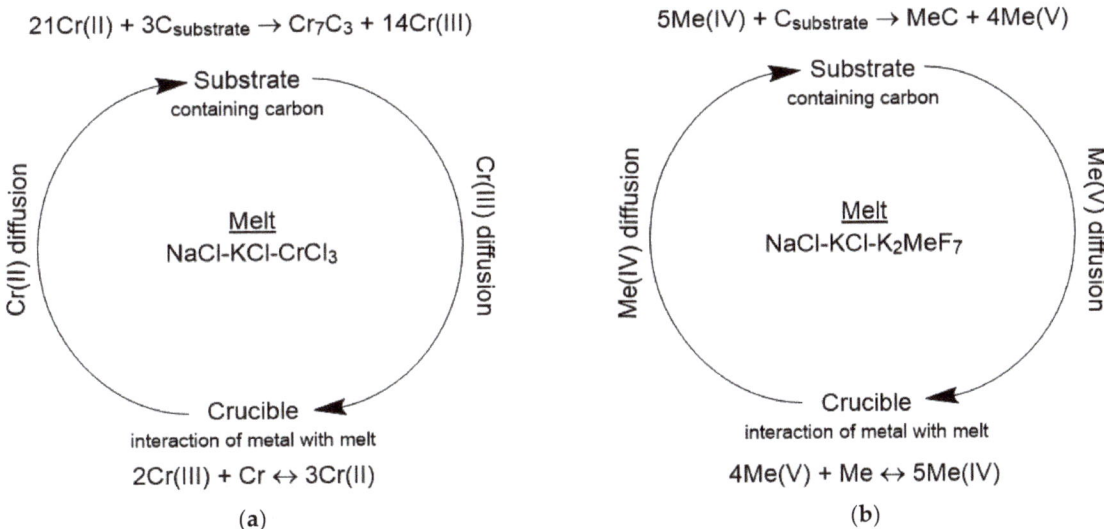

Figure 4. Principle scheme of coating synthesis by the CT in molten salts. (**a**)—Cr$_7$C$_3$ chromium carbide; (**b**)—MeC carbides (Me = Nb, Ta).

Figure 5 shows the temperature dependencies of the Gibbs energy per 1 mole of the product for the formation of chromium, tantalum, and niobium carbides from the elements. It was shown that the formation of Cr$_7$C$_3$, TaC, and NbC carbides is the most preferable. It also follows from the thermodynamic data that the synthesis of chromium carbides on steels is preferable at higher temperatures, as evidenced by the decrease in ΔG_r with increasing temperature. For niobium and tantalum carbides the temperature dependence of ΔG_r is different, and ΔG_r increases (becomes less negative) with increasing temperature. However, with increasing temperature the melt viscosity decreases, and diffusion fluxes of metal complexes in various degrees of oxidation through the electrolyte increase. This leads to the intensification of NbC and TaC formation on steels with increasing temperature despite being less thermodynamically favorable.

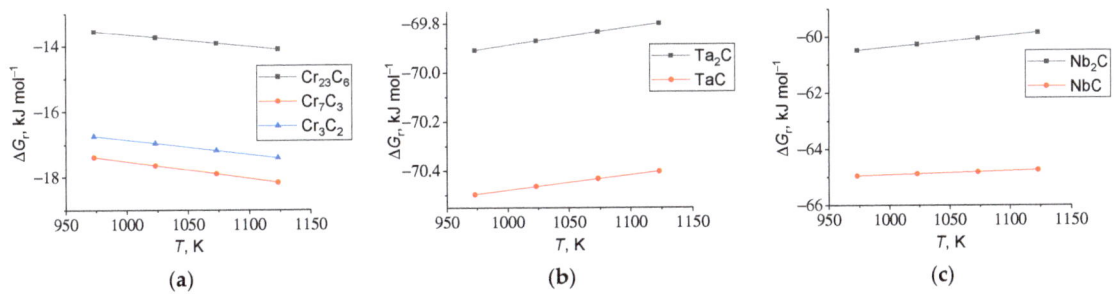

Figure 5. Temperature dependencies of the Gibbs energy for the formation reactions of (**a**)—chromium, (**b**)—tantalum, and (**c**)—niobium carbides from the elements.

Based on these considerations, the optimal conditions for the synthesis of carbide coatings by the CT method were selected. An equimolar NaCl-KCl mixture was used as an electrolyte, the process temperature was 1173 K, the concentrations of refractory metal salts ($CrCl_3$, K_2TaF_7 and K_2NbF_7) were 30 wt.%, and the synthesis time varied from 3 up to 24 h [103].

The obtained coatings were analyzed by XRD, which showed the formation of Cr_7C_3 TaC and NbC coatings (Figure 6), in perfect agreement with the thermodynamic data. The peaks of diffraction patterns corresponding to the substrate are caused by the insignificant thickness of the synthesized coatings.

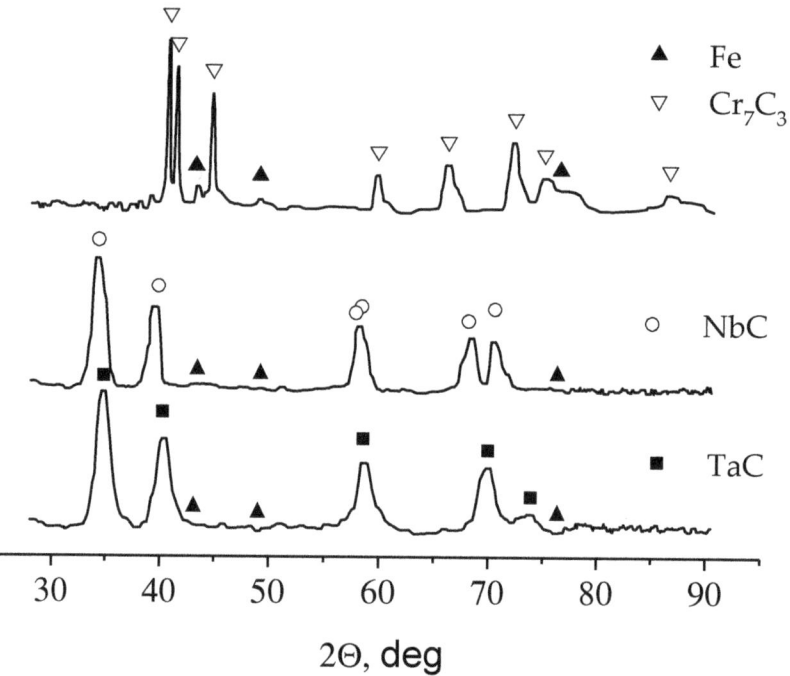

Figure 6. XRD patterns of the refractory carbide coatings on the St3 steel substrate (solid line), data points indicate the reference samples.

3.2.2. Protective Properties of Refractory Carbide Coatings

Cross-section was used to determine the thickness of synthesized refractory metal carbide coatings. In general, the thickness of the coatings obtained by the CT method depends on the duration of the process, the diffusion rate, and the carbon content in the substrate. In our case, the coating thicknesses were 1–2 microns. The morphologies of niobium carbide coatings formed at various synthesis durations on St.3 steel are presented in Figure 7. The coatings were composed of densely packed spherical particles with a diameter of 1–2 μm. If the synthesis time was 3 h (Figure 7a), the coating was not continuous. To obtain continuous coatings, the synthesis time was increased to 8 h (Figure 7b).

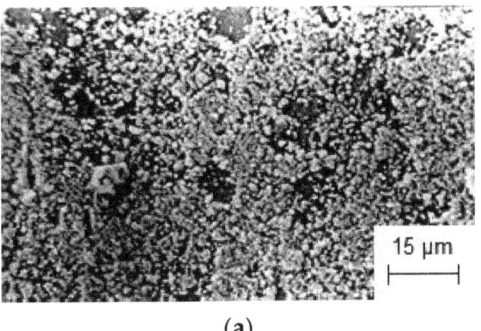 

(a)   (b)

Figure 7. St.3 substrate surface morphology after its (**a**)—3 h exposure and (**b**)—8 h exposure in the NaCl-KCl-K$_2$NbF$_7$ (30 wt.%)—Nb melt at 1173 K.

Samples coated with Cr$_7$C$_3$, NbC and TaC were tested for corrosion resistance in concentrated mineral acids and for wear resistance by the following methods. The corrosion rate was determined by a gravimetric method using cylindrical samples with a length of 30 mm and a diameter of 7 mm. After deposition of the coatings and electrolyte removal, the samples were degreased with refined ethyl alcohol. The corrosion rate was determined at 298 K for 48 h as a mass loss of the sample through a surface area unit per unit of time. At the end of the sample exposure in concentrated mineral acids, corrosion products were removed mechanically, the samples were degreased again, and were weighed on analytical scales with an accuracy of up to 1×10^{-5} g.

The wear rate was measured as the weight loss of the coated sample relative to the original sample, on the SMTs-2 friction machine (LLC "Precision Instruments", Ivanovo, Russia) at a specific load of 5 mPa, in transformer oil, at a slip velocity of 1.2 m s^{-1}, and on a path segment of 2000 m.

The results of the corrosion and wear resistance tests are shown in Figure 8. Refractory metal carbide coatings substantially reduced the corrosion rate of steel samples in concentrated acids, especially for H$_2$SO$_4$ and HCl (Figure 8a). Of course, the quality of obtained coatings (porosity) affects the results to a certain degree. Due to high hardness values of refractory metal carbides and their good adhesion with the substrate, the loss of mass as a result of mechanical wear is reduced by 3–5 times compared with the reference sample (Figure 8b).

Our studies showed that to achieve a good corrosion resistance of samples with Cr$_7$C$_3$, NbC and TaC coatings, the required thickness of protective coatings should be 1–3 µm. At the same time, thin films with a thickness of 100–700 nm obtained at a shorter duration of synthesis are sufficient only to impart resistance to mechanical wear.

The combination of high corrosion and wear resistance makes the investigated coatings a promising material for aggressive media with abrasive wear. These coatings have been tested in industrial facilities. Deposition of niobium carbide coatings on parts of crude oil pumps improved their lifetime by 3–4 times (tested by LLC "New technology", Russia, Nefteyugansk). Tests carried out by LLC "Ecotech" (Russia, Apatity) showed that the coatings of Cr$_7$C$_3$ or TaC on the rubber-cutting knifes made of St3 can increase their wear resistance and increase a tool service life by a factor of 2.0 (Cr$_7$C$_3$) and 2.5 (TaC).

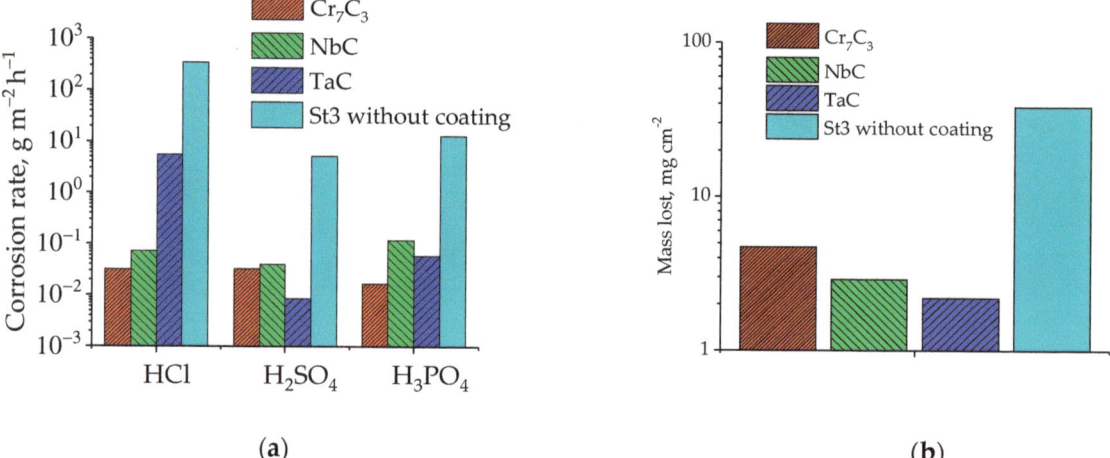

Figure 8. (**a**)—Corrosion rates of steel/refractory metal carbide coatings in concentrated mineral acids and (**b**)–mass loss due to mechanical wear.

3.3. Electrocatalytic Compositions Carbon Fiber/Refractory Metal Carbide

The application of cheap refractory metal carbides as catalysts and electrocatalysts instead of expensive noble metals is preferable for some oxidation or reduction processes [104]. Carbon fibers can serve as a substrate for the electrochemical deposition of coatings and crystals of refractory metal carbides from molten salt media. Obtaining refractory metal carbide-carbon fiber composites using such processes is operationally quite simple and not energy consuming.

Conducting electrocatalytic reactions instead of traditionally catalytic ones has a number of advantages. For example, it is possible to control the reaction kinetics by adjusting the potential, which directly affects the activation energy of the conducted process [105], as well as the environmental friendliness of electrocatalysis since an electron is involved in the reaction as an oxidizing or reducing agent, and there are no hazardous reaction by-products [105,106].

3.3.1. Synthesis of Refractory Metal Carbide Coatings on Carbon Fibers

The synthesis of transition metal carbide/carbon fiber composites was carried out by the CT method described above (Section 3.2.1). X-ray diffraction patterns of all synthesized composites are presented in Figure 9. According to XRD analysis, tantalum carbide had a cubic structure with a border-centered crystal lattice, niobium carbide had a cubic crystal lattice, and crystals of the Mo_2C phase had a hexagonal structure.

Figure 10 shows micrographs of carbon fibers coated with tantalum carbide (a) and a cross-section of a single fiber (b). The micrographs show that the fibers were not spliced together, and the coatings were uniform both in the cross-section and along the fiber.

Carbon fibers with niobium carbide coatings are shown in Figure 11. The diameter of the carbon fibers was approximately 7 ± 1 μm. The thickness of the tantalum and niobium carbide coatings on the carbon fiber was from a few tens of nanometers to a couple hundred nanometers. No other compound phases were detected between the carbon fiber and the coating.

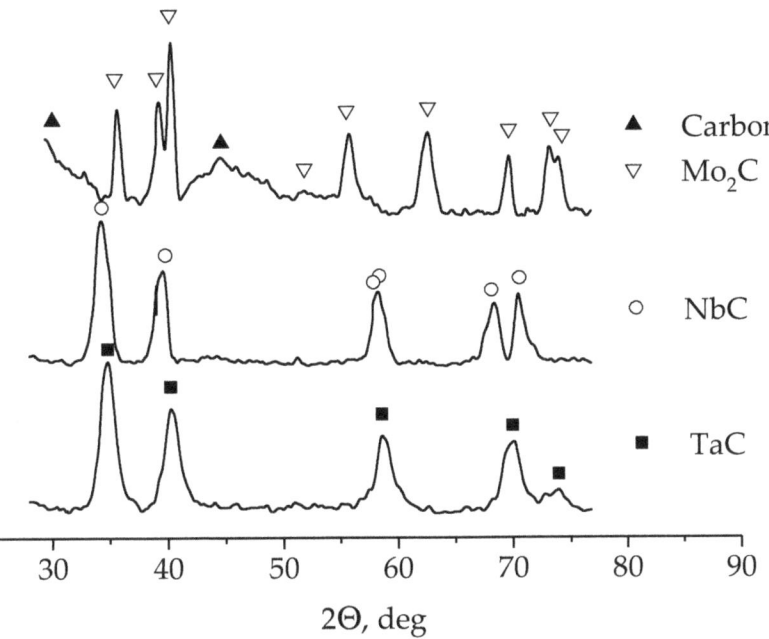

Figure 9. X-ray diffraction patterns of TaC, NbC coatings and Mo_2C crystals obtained by electrochemical methods in molten salts on the surface of carbon fibers (solid line). Data points indicate the reference samples.

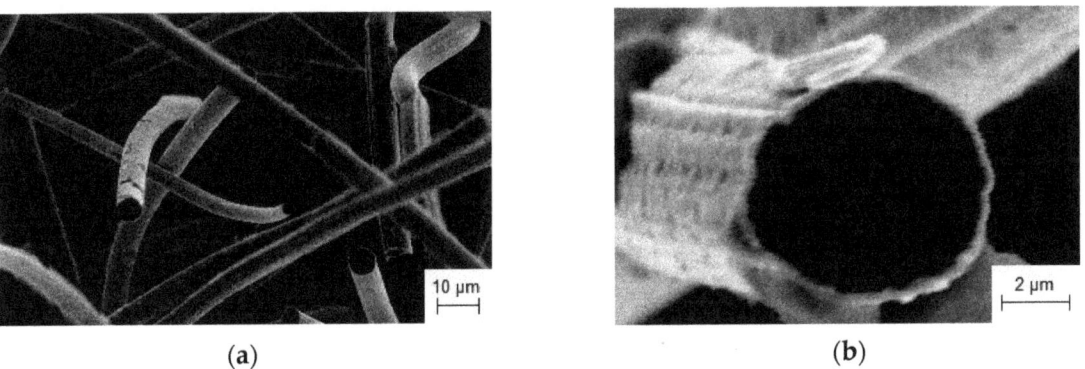

Figure 10. Micrographs of the tantalum carbide-carbon fiber composite at different scale. (**a**)—Overall view and (**b**)—cross-section of a single fiber. Synthesis conditions: CT in NaCl-KCl-K_2TaF_7 (30 wt.%)–Ta melt at 1123 K for 24 h.

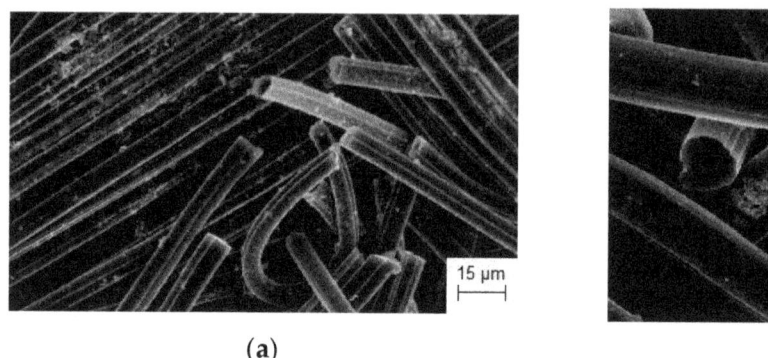

(a)   (b)

Figure 11. Micrographs of the niobium carbide-carbon fiber composite at different scale. (**a**)—Overall view and (**b**)—cross-sections of fibers. Synthesis conditions: CT in NaCl-KCl-K$_2$NbF$_7$ (30 wt.%)–Nb melt at 1123 K for 24 h.

The Mo$_2$C phase was also obtained on carbon fibers by CT in molten salts. The melt in this case contained sodium molybdate, Na$_2$MoO$_4$. It was found that increasing the exposure time of the carbon fiber in the melt leads to an increase in the amount of Mo$_2$C crystals. The Mo$_2$C crystals had a well-defined structure with a crystal size of ~7–21 μm (Figure 12).

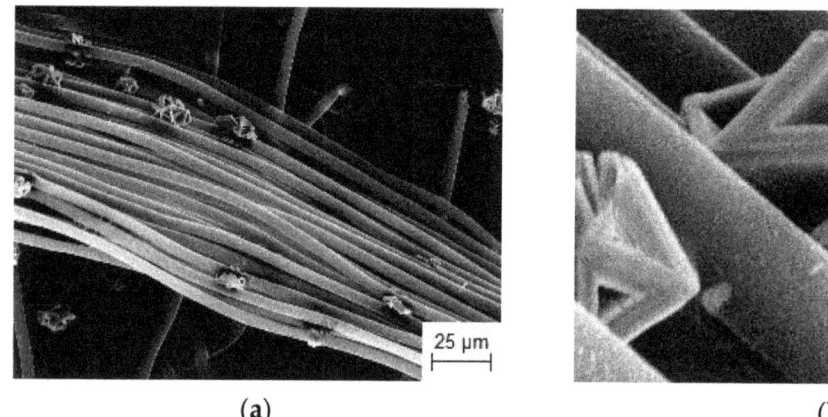
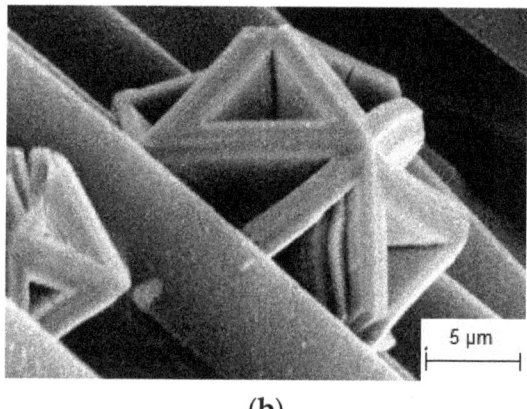

(a)   (b)

Figure 12. Micrographs of the molybdenum carbide-carbon fiber composite at different scale. (**a**)—Overall view and (**b**)—a single Mo$_2$C crystal. Synthesis conditions: CT in NaCl-KCl-Na$_2$MoO$_4$ (15 wt.%)–Mo melt at 1123 K and for 1 h.

3.3.2. Investigation of the Electrocatalytic Activity of Composites Based on Refractory Metal Carbide on Carbon Fiber

The kinetics of electrocatalytic decomposition of hydrogen peroxide on the surface of refractory metal carbides (TaC, NbC, Mo$_2$C) can be investigated by measuring the volume of evolved gases per unit of time. Decomposition of hydrogen peroxide proceeds with the emission of gaseous products: oxygen is emitted at the anode, which was the studied composite material, and hydrogen is emitted at the cathode, which was an uncoated carbon fiber.

To compare the results of studying the kinetics of the hydrogen peroxide decomposition reaction on refractory metal carbide-carbon fiber composites with the kinetics of the same reaction using traditional catalysts as copper and platinum, we used 5 mm^2 metal electrodes. However, it is difficult to estimate the surface area of electrodes made of composites of refractory metal carbides. Their specific surface area is 5–15 m^2 g^{-1}. In all our experiments, the same linear dimensions of the composite electrodes were used, and the same immersion depth in hydrogen peroxide solution was maintained.

The graphical analysis of the kinetic dependences (Figure 13) established the zero order of the electrocatalytic reaction of hydrogen peroxide decomposition [107]. It can be described by the kinetic equation of the form $v = k$, where v is the reaction rate, and k is the reaction rate constant.

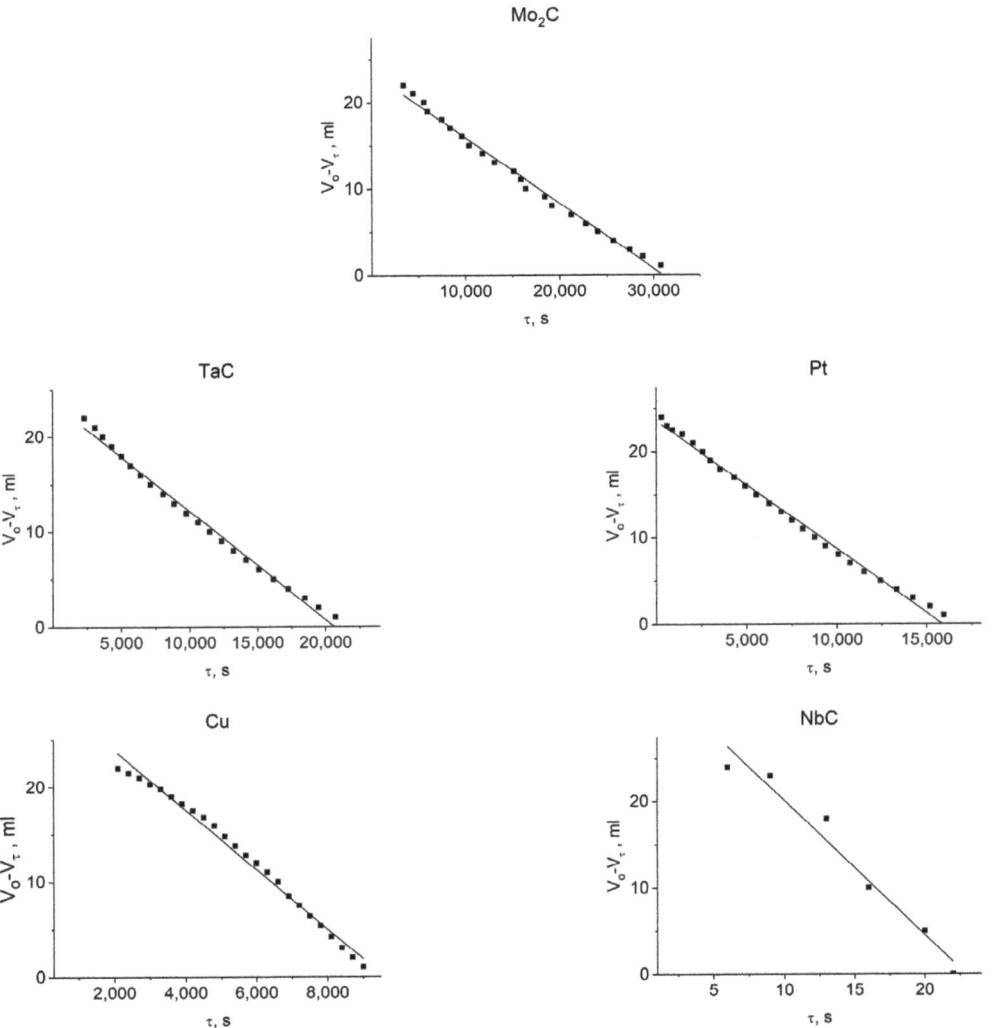

Figure 13. Zero-order kinetic dependencies of the electrocatalytic decomposition of H$_2$O$_2$, taken on the electrodes at the same temperature 303 K.

The activation energies of the studied H_2O_2 decomposition were calculated from experimental data obtained at different temperatures in steps of 10 K. Figure 13 shows the kinetic data at 303 K for all studied composites. Similar dependencies were also observed at higher temperatures, so only one example is presented in this article. It can be seen that the rate of electrocatalytic reaction of hydrogen peroxide decomposition increases in the following series: $Mo_2C/C < TaC/C < Pt < Cu < NbC/C$.

The obtained values of the rate constants for the hydrogen peroxide decomposition at different temperatures were used to calculate activation energies. It was found that the activation energy values for the Mo_2C/C, TaC/C, Pt, Cu and NbC/C electrodes are equal to 107.2, 82.2, 74.8, 48.2, and 37.1 kJ mol^{-1}, respectively.

It should be noted that the activation energy of the hydrogen peroxide decomposition process is given here from the general chemical point of view. Since it is impossible to reliably determine the value of the potential difference between the electrode surface and any point in the solution, we do not have a reliable method for the experimental determination of the activation energy for the specific electrode process. Therefore, the calculated activation energies are given for a formal decomposition process taking place on the anode.

It has been established that the reaction rate does not change over time, in accordance with the zero-order kinetic equation, which has no dependence on the concentration of reacting substances, and the diffusion rate to the surface of electrode is smaller than the rate of chemical transformation.

3.4. Electrochemical Synthesis of Gadolinium Borides Nanorods in Molten Salts

Highly pure rare earth metals and their refractory compounds with oxygen, carbon, nitrogen, and boron are widely used in special alloys in metallurgy, semiconductor electronics and laser techniques; they are used for production of permanent magnets, new types of catalysts, optical glasses, and hydrogen accumulators. Gadolinium boride is especially prospective in nuclear engineering for neutron adsorption. Nanomaterials based on gadolinium hexaboride are promising materials for modern technology due to their unique electric, magnetic, and optical properties. For example, GdB_6 nanorods are characterized by a very low electron work function (~1.5 eV), which makes this nanomaterial very interesting for the fabrication of point electron emitters. GdB_6 nanorods are strong light absorbers in the near infrared region, and they are transparent to visible light.

The traditional manufacturing of rare-earth borides is based on the direct reaction of pure initial components at high temperatures (up to 2700 K), which negatively affects the microstructure of the product. Electrochemical synthesis at moderate temperatures (973–1023 K) is a cost-effective alternative to direct reaction techniques and it has been successfully applied to the production of high-quality refractory borides. Electrochemical techniques provide not only the required stoichiometry and narrow size distribution of the product, but also the designated morphology. Prior to our studies, there was no information on the preparation of GdB_6 nanorods and nanowires in the literature.

For the electrochemical synthesis of GdB_6 it is necessary to know the electrochemical behavior of boron and gadolinium in their joint presence in the melt. Waves on the voltammogram in the $NaCl-KCl-NaF-GdF_3-KBF_4$ melt (Figure 14) reflected the boron reduction from BF_4^- complexes (R_B, the least negative wave at about $-0.8 - -1.0$ V) and gadolinium discharge on a boron deposit (R_{Gd-B}, at potentials -1.5 V and -1.8 V). The ascending section on the voltammogram corresponds to the discharge of gadolinium and alkali metal cations. Significant differences in the electroreduction potentials of boron and gadolinium allows us to make a reasonable assumption about the so-called "kinetic regime" [52] of their compound synthesis in the melt. Thus, the hexaboride can be synthesized only at current densities higher than the limiting current density for a more positive constituent (boron) discharge process. During co-deposition with boron, gadolinium reduced not on a silver surface, but on the layer of already deposited boron. The high depolarization value of this process leads to the formation of boron-gadolinium compound at potentials near −1.5 V and −1.8 V.

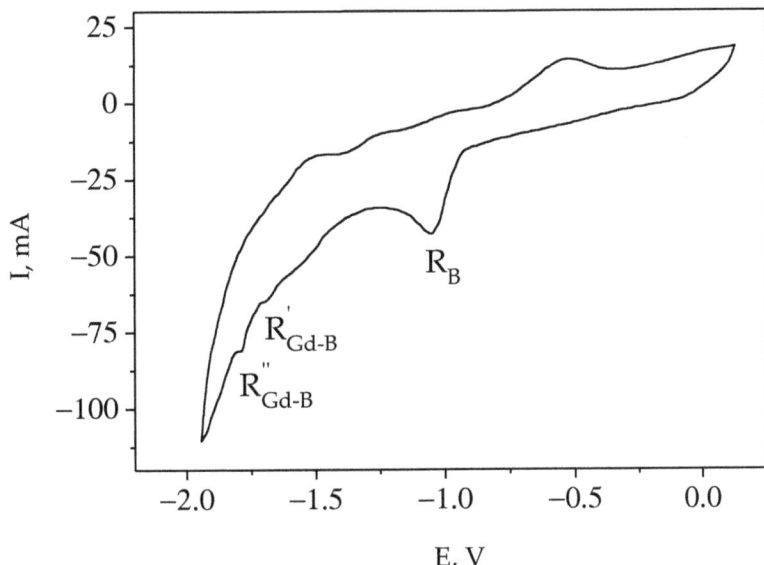

Figure 14. Voltammogram of the KCl-NaCl-NaF (10 wt.%)-GdF$_3$-KBF$_4$ (B/Gd = 2.6) melt, $C_{KBF_4} = 1.8 \times 10^{-5}$ mol cm^{-3}, Ag—working electrode, A = 0.5 cm^2, GC—counter electrode, GC-quasi-reference electrode; ν = 0.5 V s^{-1}, T = 1023 K.

Potentiostatic electrolysis (7 h duration) was carried out, at potentials −1.5 V and −1.8 V. The deposits were scraped off and any electrolyte adhering to the deposits were leached with a warm dilute HCl, then 2% NaOH, and finally washed with distilled water and ethanol. XRD analysis identified the products of electrolysis as gadolinium hexaboride [78]. At the potential of −1.5 V the gadolinium hexaboride was synthesized in the form of coral-like dendrites. Micro- and nanorods were obtained at the potential of −1.8 V (Figure 15). Along with nanorods, a GdB$_6$ nanowire was also formed on the cathode; the length of the nanostructure significantly exceeded its diameter, and, in some cases, the nanowire was bent. The intercalation mechanism was suggested for the formation of nanorods and nanowire [78].

3.5. Synthesis of One-Dimensional Nanostructures: Si and TaO Nanoneedles

One-dimensional nanomaterials, such as nanorods, nanotubes, nanowires, and nanofibers, are promising for catalysis and electrocatalysis [108]. Firstly, this is due to their large surface area with a high concentration of active centers [109]. Secondly, such nanostructures have a unique channel structure, which can be used as a fast pathway for the transfer of electric charges [110,111]. Finally, many adjacent one-dimensional nanostructures create a large number of pores and channels for the supply of reagents and escape of reaction products. Another important advantage of such structures is the possibility of obtaining them directly on the desired substrate without the use of additional binding agents [111].

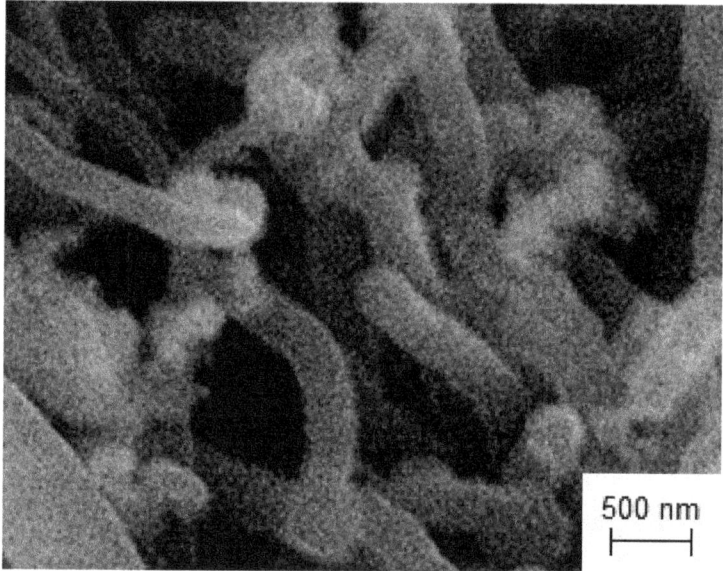

Figure 15. SEM image of the GdB$_6$, synthesized in the KCl-NaCl-NaF (10 wt.%)-GdF$_3$-KBF$_4$ melt at −1.8 V vs. GC quasi-reference electrode.

3.5.1. Electrochemical Synthesis of Si-Nanoneedles in Molten Salts

Silicon technology has created computers, cell phones, and other electronic devices [59,112]. The use of silicon nanoneedles for medical purposes is known [113,114], for example, to grow new blood vessels. Such nanoscale needles are involved in the delivery of genetic material to stimulate blood vessel growth and can deliver drugs directly to living cells.

Electrochemical methods of synthesis make it possible to produce nanoscale silicon, regulating its growth and the size of the final product [56].

A chloride-fluoride melt, NaCl-KCl-NaF (10 wt.%)-K$_2$SiF$_6$, was used for the production of silicon [115]. The cyclic voltammogram of this melt is presented in Figure 16 and in the cathodic semi-cycle, two electroreduction peaks R$_1$ and R$_2$ corresponding to Si(IV) were registered.

Potentiostatic electrolysis at the first wave did not led to the formation of solid product at the cathode, but after washing off the electrolyte remaining on the electrode, needle- and flake-form crystals were observed (Figure 17). The XRD analysis showed that the crystals correspond to elementary silicon [115].

Thus, the recharge process of Si(IV) to Si(II):

$$Si(IV) + 2e \rightarrow Si(II), \qquad (2)$$

is accompanied by the disproportionation reaction [116]:

$$2Si(II) \rightleftarrows Si(IV) + Si. \qquad (3)$$

According to the XRD data, electrolysis at the potentials of the second wave led to the formation of large silicon crystals (Figure 18) during the electroreduction of Si(II) to Si, and fine needle crystals were also observed due the disproportionation reaction (3).

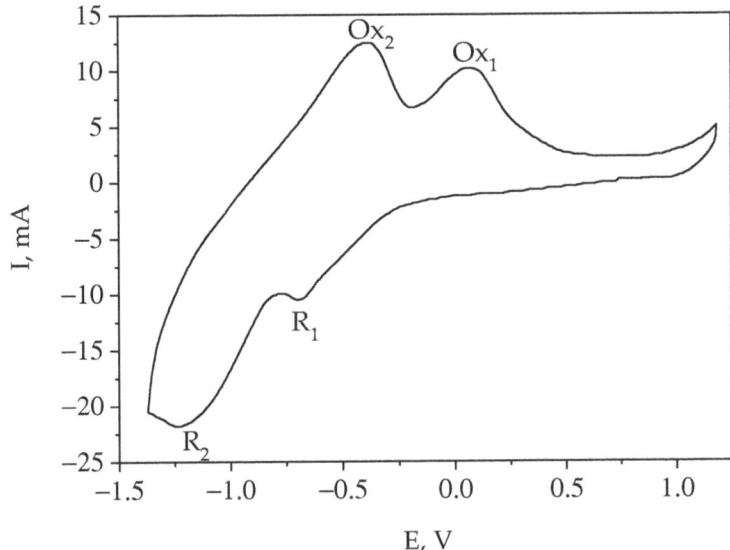

Figure 16. Cyclic voltammogram of the NaCl-KCl-NaF(10 wt.%)–K$_2$SiF$_6$ melt. Concentration of K$_2$SiF$_6$ is 6.87×10^{-5} mol cm^{-3}; v = 0.2 V s^{-1}; T = 1023 K; A = 0.162 cm^2. The quasi-reference electrode was the SU-2000 glassy carbon.

Figure 17. Needle and flake silicon crystals obtained during potentiostatic electrolysis at $E = -0.75$ V vs. the glassy carbon electrode, after rinsing the remaining electrolyte off the electrode after its removal from the melt.

Figure 18. Electron microscope image of silicon crystals on a silver electrode obtained by potentiostatic electrolysis at $E = -1.2$ V vs. the glassy carbon electrode.

3.5.2. Obtaining of TaO Nanoneedles by Electroreduction of K_3TaOF_6 in Molten Salts

Tantalum monoxide can replace the tantalum pentoxide in high-density electric capacitors because of its high dielectric constant and low leakage current [117]. This material has been commercially integrated in capacitors which are used in computer dynamic random-access memory [118].

The formation of TaO was mentioned in several papers [119–123], but in [122,123] the electrocrystallization of tantalum monoxide was not confirmed by XRD analysis.

It was determined that during electroreduction of tantalum monoxofluoride complexes in the $CsCl-K_3TaOF_6$ melt, several phases containing tantalum are formed. The variation of cathodic products obtained at different parameters of electrolysis are presented in Figure 19. As can be seen from the diagrams in Figure 19, tantalum monoxide crystallized at the cathode with other tantalum compounds and did not form at a temperature of 1173 K. The micrograph of tantalum monoxide formed during electrolysis of the $CsCl-K_3TaOF_6$ melt is shown in Figure 20. It demonstrates that tantalum monoxide crystallizes at the cathode as nanoneedles up to 12,000 nm in length with a cross-section of ~100 nm.

Table 1 presents a brief summary of all functional coatings and nanomaterials considered in the present paper, the corresponding synthesis conditions in molten salts, and possible applications.

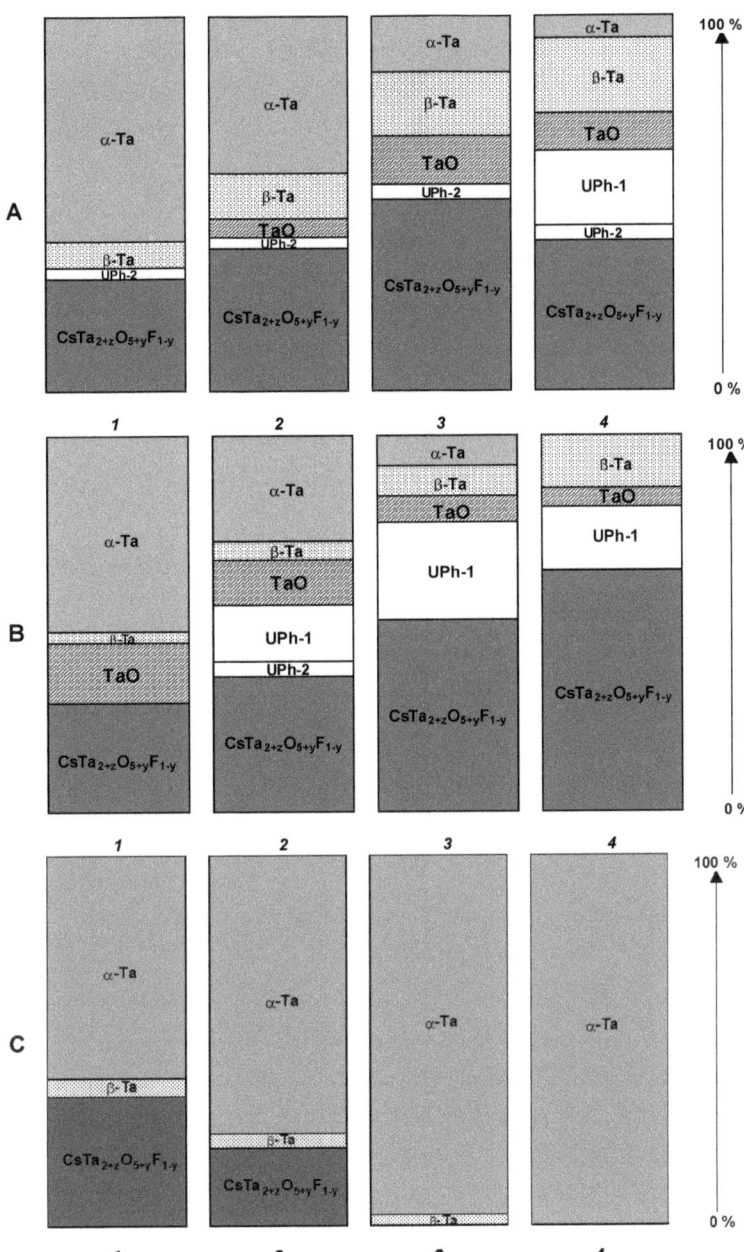

Figure 19. Diagrams of the phase composition of the cathodic products deposited at a current density of 0.15 A cm^{-2} at (**A**) 923 K, (**B**) 1023 K, (**C**) 1173 K. The concentration of K_3TaOF_6 in the CsCl melt was (*1*) 1.25, (*2*) 2.5, (*3*) 5.0, and (*4*) 10.0 wt.%.

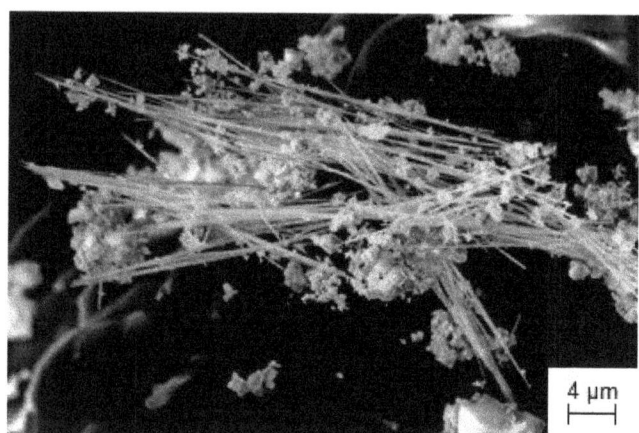

Figure 20. SEM image of TaO obtained by electrolysis of the CsCl-K$_3$TaOF$_6$ (2.5 wt.%) melt at 1023 K and a current density of 0.15 A cm^{-2}.

Table 1. Summary of electrochemical synthesis of functional coatings and nanomaterials in molten salts.

Molten Salt System	Deposition Regime	Experimental Condition	Substrate	Product	Possible Application
NaCl-KCl-Li$_2$CO$_3$-Na$_2$MoO$_4$	Galvanostatic electrolysis	$i = 5$ mA cm^{-2} $\tau = 7$ h $T = 1123$ K	Mo plate	Mo$_2$C nanostructured coating [84]	Catalyst for steam reforming reaction
NaCl-KCl-CrCl$_3$-Cr	Currentless transfer	$\tau = 8$ h $T = 1173$ K	Steel St.3	Cr$_7$C$_3$ coating [98,103]	Protective corrosion- and wear-resistant coating
NaCl-KCl-K$_2$TaF$_7$-Ta	Currentless transfer	$\tau = 8$ h $T = 1173$ K	Steel St.3	TaC coating [103]	Protective corrosion- and wear-resistant coating
NaCl-KCl-K$_2$NbF$_7$-Nb	Currentless transfer	$\tau = 8$ h $T = 1173$ K	Steel St.3	NbC coating [103]	Protective corrosion- and wear-resistant coating
NaCl-KCl-K$_2$TaF$_7$-Ta	Currentless transfer	$\tau = 24$ h $T = 1123$ K	Carbon fibers	TaC coating [107]	Electrocatalyst for H$_2$O$_2$ decomposition
NaCl-KCl-K$_2$NbF$_7$-Nb	Currentless transfer	$\tau = 24$ h $T = 1123$ K	Carbon fibers	NbC coating [107]	Electrocatalyst for H$_2$O$_2$ decomposition
NaCl-KCl-Na$_2$MoO$_4$-Mo	Currentless transfer	$\tau = 1$ h $T = 1123$ K	Carbon fibers	Mo$_2$C crystals [107]	Electrocatalyst for H$_2$O$_2$ decomposition
NaCl-KCl-NaF-GdF$_3$-KBF$_4$	Potentiostatic electrolysis	$E = -1.8$ V vs. GC $\tau = 7$ h $T = 1023$ K	Ag rod	GdB$_6$ nanorods and nanowires [78]	Material for point electron emitters; Material for the neutron adsorption
NaCl-KCl-NaF-K$_2$SiF$_6$	Potentiostatic electrolysis	$E = -0.75$ V vs. GC $\tau = 2$ h $T = 1023$ K	Ag rod	Si nanoneedles [115]	Material for Li-ion batteries; Drug delivery into living cells
CsCl-K$_3$TaOF$_6$	Galvanostatic electrolysis	$i = 0.15$ A cm^{-2} $\tau = 1$ h $T = 1023$ K	Mo rod	TaO nanoneedles [80]	Material for high-density electric capacitors

4. Conclusions

The advantages of high-temperature electrochemical synthesis of nanostructured coatings and various nanomaterials in molten salts were demonstrated:

1. By electrochemical synthesis in molten salts, a new Mo_2C/Mo-based catalytic system for a low-temperature steam reforming reaction was obtained. The catalytic activity of this composition, produced by the simultaneous reduction of electroactive MoO_4^{2-} and CO_3^{2-} species, was three orders of magnitude higher than that of the bulk Mo_2C phase and commercial Cu-ZnO-Al_2O_3 catalyst.
2. The currentless transfer for the synthesis of nanoscale coatings of carbide refractory metals on substrates containing carbon was studied. It was shown that these coatings on steels increase corrosion resistance by several orders of magnitude and increase wear resistance by 3–5 times. Tests carried out by industrial facilities showed that the coatings of Cr_7C_3 or TaC on rubber-cutting knifes made of St3 can improve their wear resistance and increase a tool lifetime by 2.0 (for Cr_7C_3) and 2.5 (for TaC) times.
3. It was found that NbC, TaC, and Mo_2C carbides deposited on carbon fibers by currentless transfer in molten salts can be used as highly active electrocatalysts for hydrogen peroxide decomposition. The kinetics of the electrocatalytic decomposition of H_2O_2 were studied and the following series of electrocatalytic activity was established: $Mo_2C <$ TaC $<$ Pt $<$ Cu $<$ NbC.
4. Using potentiostatic electrolysis, GdB_6 nanorods for different applications were synthesized in the KCl-NaCl-NaF(10 wt.%)-GdF_3-KBF_4 melt.
5. The synthesis of one-dimensional nanomaterials based on Si and TaO for application in modern electronic devices was discussed. Silicon nanoneedles were synthesized by potentiostatic electrolysis in the NaCl–KCl–NaF(10 wt.%)-K_2SiF_6 melt. The possibility to synthesize TaO using the CsCl-K_3TaOF_6 melt was shown. TaO crystallized at the cathode as nanoneedles, together with other tantalum compounds. It was found that TaO can be obtained by the electrolysis of molten salts only at temperatures below 1173 K.

Author Contributions: Investigation, Y.S., V.D. and A.D.; Supervision, S.K.; Visualization, Y.S., V.D. and A.D.; Writing–original draft, Y.S., V.D. and A.D.; Writing–review and editing, S.K. All authors have read and agreed to the published version of the manuscript.

Funding: This research received no external funding.

Institutional Review Board Statement: Not applicable.

Informed Consent Statement: Not applicable.

Data Availability Statement: The data presented in this study are available on request from the corresponding author.

Conflicts of Interest: The authors declare no conflict of interest.

References

1. Gleiter, H. Nanostructured Materials: Basic Concepts and Microstructure. *Acta Mater.* **2000**, *48*, 1–29. [CrossRef]
2. Nazar, L.F.; Goward, G.; Leroux, F.; Duncan, M.; Huang, H.; Kerr, T.; Gaubicher, J. Nanostructured Materials for Energy Storage. *Int. J. Inorg. Mater.* **2001**, *3*, 191–200. [CrossRef]
3. Hone, F.G.; Tegegne, N.A.; Andoshe, D.M. Advanced Materials for Energy Storage Devices. In *Electrode Materials for Energy Storage and Conversion*; CRC Press: Boca Raton, FL, USA, 2021; pp. 71–107. [CrossRef]
4. Kebede, M.A.; Ezema, F.I. Electrode Materials for Energy Storage and Conversion. In *Electrode Materials for Energy Storage and Conversion*; CRC Press: Boca Raton, FL, USA, 2021. [CrossRef]
5. Aoki, Y.; Tominaga, H.; Nagai, M. Hydrogenation of CO on Molybdenum and Cobalt Molybdenum Carbide Catalysts—Mass and Infrared Spectroscopy Studies. *Catal. Today* **2013**, *215*, 169–175. [CrossRef]
6. Ren, H.; Yu, W.; Salciccioli, M.; Chen, Y.; Huang, Y.; Xiong, K.; Vlachos, D.G.; Chen, J.G. Selective Hydrodeoxygenation of Biomass-Derived Oxygenates to Unsaturated Hydrocarbons Using Molybdenum Carbide Catalysts. *ChemSusChem* **2013**, *6*, 798–801. [CrossRef] [PubMed]

7. Schweitzer, N.M.; Schaidle, J.A.; Ezekoye, O.K.; Pan, X.; Linic, S.; Thompson, L.T. High Activity Carbide Supported Catalysts for Water Gas Shift. *J. Am. Chem. Soc.* **2011**, *133*, 2378–2381. [CrossRef]
8. Malyshev, V.; Uskova, N.; Shakhnin, D.; Lukashenko, T.; Antsibor, V.; Ustundag, Z. High-Temperature Electrochemical Synthesis of Nanostructured Coatings of Molybdenum (Tungsten)–Nickel (Cobalt) Alloys and Intermetallic Compounds. In *Selected Proceedings of the 5th International Conference Nanotechnology and Nanomaterials*; Springer: Berlin/Heidelberg, Germany, 2018; pp. 165–176. ISBN 9783319925660.
9. Zhang, W.; Hu, Y.; Ma, L.; Zhu, G.; Wang, Y.; Xue, X.; Chen, R.; Yang, S.; Jin, Z. Progress and Perspective of Electrocatalytic CO_2 Reduction for Renewable Carbonaceous Fuels and Chemicals. *Adv. Sci.* **2018**, *5*, 1700275. [CrossRef]
10. Hanus, M.J.; Harris, A.T. Nanotechnology Innovations for the Construction Industry. *Prog. Mater. Sci.* **2013**, *58*, 1056–1102. [CrossRef]
11. Zhai, W.; Srikanth, N.; Kong, L.B.; Zhou, K. Carbon Nanomaterials in Tribology. *Carbon N. Y.* **2017**, *119*, 150–171. [CrossRef]
12. Pottathara, Y.B.; Thomas, S.; Kalarikkal, N.; Grohens, Y.; Kokol, V. *Nanomaterials Synthesis*; Elsevier: Amsterdam, The Netherlands, 2019; ISBN 9780128157510.
13. Chen, X.; Mao, S.S. Titanium Dioxide Nanomaterials: Synthesis, Properties, Modifications and Applications. *Chem. Rev.* **2007**, *107*, 2891–2959. [CrossRef]
14. Chen, J.S.; Lou, X.W. SnO_2-Based Nanomaterials: Synthesis and Application in Lithium-Ion Batteries. *Small* **2013**, *9*, 1877–1893. [CrossRef]
15. Kolahalam, L.A.; Kasi Viswanath, I.V.; Diwakar, B.S.; Govindh, B.; Reddy, V.; Murthy, Y.L.N. Review on Nanomaterials: Synthesis and Applications. *Mater. Today Proc.* **2019**, *18*, 2182–2190. [CrossRef]
16. Sriondee, M.; Dungsuwan, W.; Thountom, S. Synthesis and Characterization of $Bi_{0.5}(Na_{1-X}K_x)_{0.5}TiO_3$ Powders by Sol–Gel Combustion Method with Glycine Fuel. *Ceram. Int.* **2018**, *44*, S168–S171. [CrossRef]
17. Hao, S.; Lin, T.; Ning, S.; Qi, Y.; Deng, Z.; Wang, Y. Research on Cracking of SiO_2 Nanofilms Prepared by the Sol-Gel Method. *Mater. Sci. Semicond. Process.* **2019**, *91*, 181–187. [CrossRef]
18. Kasi Viswanath, I.V.; Murthy, Y.L.N.; Tata, K.R.; Singh, R. Synthesis and Characterization of Nano Ferrites by Citrate Gel Method. *Int. J. Chem. Sci.* **2013**, *11*, 64–72.
19. Ikesue, A.; Kinoshita, T.; Kamata, K.; Yoshida, K. Fabrication and Optical Properties of High-Performance Polycrystalline Nd:YAG Ceramics for Solid-State Lasers. *J. Am. Ceram. Soc.* **1995**, *78*, 1033–1040. [CrossRef]
20. Yu, S.; Jing, W.; Tang, M.; Xu, T.; Yin, W.; Kang, B. Fabrication of Nd:YAG Transparent Ceramics Using Powders Synthesized by Citrate Sol-Gel Method. *J. Alloys Compd.* **2019**, *772*, 751–759. [CrossRef]
21. Holland, B.T.; Blanford, C.F.; Do, T.; Stein, A. Synthesis of Highly Ordered, Three-Dimensional, Macroporous Structures of Amorphous or Crystalline Inorganic Oxides, Phosphates, and Hybrid Composites. *Chem. Mater.* **1999**, *11*, 795–805. [CrossRef]
22. Diwald, O.; Thompson, T.L.; Zubkov, T.; Goralski, E.G.; Walck, S.D.; Yates, J.T. Photochemical Activity of Nitrogen-Doped Rutile TiO_2(110) in Visible Light. *J. Phys. Chem. B* **2004**, *108*, 6004–6008. [CrossRef]
23. Sasaki, T.; Watanabe, M.; Hashizume, H.; Yamada, H.; Nakazawa, H. Macromolecule-like Aspects for a Colloidal Suspension of an Exfoliated Titanate. Pairwise Association of Nanosheets and Dynamic Reassembling Process Initiated from It. *J. Am. Chem. Soc.* **1996**, *118*, 8329–8335. [CrossRef]
24. Schlüter, M.; Hentzel, T.; Suarez, C.; Koch, M.; Lorenz, W.G.; Böhm, L.; Düring, R.A.; Koinig, K.A.; Bunge, M. Synthesis of Novel Palladium(0) Nanocatalysts by Microorganisms from Heavy-Metal-Influenced High-Alpine Sites for Dehalogenation of Polychlorinated Dioxins. *Chemosphere* **2014**, *117*, 462–470. [CrossRef]
25. Anderson, S.H.; Chung, D.D.L. Exfoliation of Single Crystal Graphite and Graphite Fibers Intercalated with Halogens. *Synth. Met.* **1983**, *8*, 343–349. [CrossRef]
26. Sharma, N.; Sharma, V.; Sharma, S.K.; Sachdev, K. Gas Sensing Behaviour of Green Synthesized Reduced Graphene Oxide (RGO) for H_2 and NO. *Mater. Lett.* **2019**, *236*, 444–447. [CrossRef]
27. Kadam, R.H.; Desai, K.; Shinde, V.S.; Hashim, M.; Shirsath, S.E. Influence of Gd^{3+} Ion Substitution on the $MnCrFeO_4$ for Their Nanoparticle Shape Formation and Magnetic Properties. *J. Alloys Compd.* **2016**, *657*, 487–494. [CrossRef]
28. Shi, Y.; Li, H.; Li, L.J. Recent Advances in Controlled Synthesis of Two-Dimensional Transition Metal Dichalcogenides via Vapour Deposition Techniques. *Chem. Soc. Rev.* **2015**, *44*, 2744–2756. [CrossRef] [PubMed]
29. Ismach, A.; Druzgalski, C.; Penwell, S.; Schwartzberg, A.; Zheng, M.; Javey, A.; Bokor, J.; Zhang, Y. Direct Chemical Vapor Deposition of Graphene on Dielectric Surfaces. *Nano Lett.* **2010**, *10*, 1542–1548. [CrossRef]
30. Kim, K.K.; Hsu, A.; Jia, X.; Kim, S.M.; Shi, Y.; Hofmann, M.; Nezich, D.; Rodriguez-Nieva, J.F.; Dresselhaus, M.; Palacios, T.; et al. Synthesis of Monolayer Hexagonal Boron Nitride on Cu Foil Using Chemical Vapor Deposition. *Nano Lett.* **2012**, *12*, 161–166. [CrossRef]
31. Saito, T.; Chiba, H.; Ito, T.; Ogino, T. Growth of Carbon Hybrid Materials by Grafting on Pre-Grown Carbon Nanotube Surfaces. *Carbon N. Y.* **2010**, *48*, 1305–1311. [CrossRef]
32. Xie, W.; Pang, Z.; Fan, J.; Song, H.; Jiang, F.; Yuan, H.; Li, J.; Ji, Z.; Han, S. Structural Properties of Alq_3 Nanocrystals Prepared by Physical Vapor Deposition and Facile Solution Method. *Int. J. Mod. Phys. B* **2015**, *29*, 1542042. [CrossRef]
33. Maleki, A. Green Oxidation Protocol: Selective Conversions of Alcohols and Alkenes to Aldehydes, Ketones and Epoxides by Using a New Multiwall Carbon Nanotube-Based Hybrid Nanocatalyst via Ultrasound Irradiation. *Ultrason. Sonochem.* **2018**, *40*, 460–464. [CrossRef]

34. Xu, H.; Suslick, K.S. Sonochemical Synthesis of Highly Fluorescent Ag Nanoclusters. *ACS Nano* **2010**, *4*, 3209–3214. [CrossRef] [PubMed]
35. Morel, A.L.; Nikitenko, S.I.; Gionnet, K.; Wattiaux, A.; Lai-Kee-Him, J.; Labrugere, C.; Chevalier, B.; Deleris, G.; Petibois, C.; Brisson, A.; et al. Sonochemical Approach to the Synthesis of Fe_3O_4 @SiO_2 Core—Shell Nanoparticles with Tunable Properties. *ACS Nano* **2008**, *2*, 847–856. [CrossRef]
36. Gedanken, A. Using Sonochemistry for the Fabrication of Nanomaterials. *Ultrason. Sonochem.* **2004**, *11*, 47–55. [CrossRef]
37. Bang, J.H.; Suslick, K.S. Applications of Ultrasound to the Synthesis of Nanostructured Materials. *Adv. Mater.* **2010**, *22*, 1039–1059. [CrossRef]
38. Zuo, P.; Lu, X.; Sun, Z.; Guo, Y.; He, H. A Review on Syntheses, Properties, Characterization and Bioanalytical Applications of Fluorescent Carbon Dots. *Microchim. Acta* **2016**, *183*, 519–542. [CrossRef]
39. Gawande, M.B.; Shelke, S.N.; Zboril, R.; Varma, R.S. Microwave-Assisted Chemistry: Synthetic Applications for Rapid Assembly of Nanomaterials and Organics. *Acc. Chem. Res.* **2014**, *47*, 1338–1348. [CrossRef] [PubMed]
40. Baghbanzadeh, M.; Carbone, L.; Cozzoli, P.D.; Kappe, C.O. Microwave-Assisted Synthesis of Colloidal Inorganic Nanocrystals. *Angew. Chem.-Int. Ed.* **2011**, *50*, 11312–11359. [CrossRef]
41. Bilecka, I.; Niederberger, M. Microwave Chemistry for Inorganic Nanomaterials Synthesis. *Nanoscale* **2010**, *2*, 1358–1374. [CrossRef] [PubMed]
42. Ustarroz, J.; Hammons, J.A.; Altantzis, T.; Hubin, A.; Bals, S.; Terryn, H. A Generalized Electrochemical Aggregative Growth Mechanism. *J. Am. Chem. Soc.* **2013**, *135*, 11550–11561. [CrossRef]
43. Yin, Y.; Jin, Z.; Hou, F. Enhanced Solar Water-Splitting Efficiency Using Core/Sheath Heterostructure CdS/TiO_2 Nanotube Arrays. *Nanotechnology* **2007**, *18*, 495608. [CrossRef]
44. Zhu, T.; Chong, M.N.; Chan, E.S. Nanostructured Tungsten Trioxide Thin Films Synthesized for Photoelectrocatalytic Water Oxidation: A Review. *ChemSusChem* **2014**, *7*, 2974–2997. [CrossRef] [PubMed]
45. Mudring, A.V.; Tang, S. Ionic Liquids for Lanthanide and Actinide Chemistry. *Eur. J. Inorg. Chem.* **2010**, *2010*, 2569–2581. [CrossRef]
46. Chou, S.; Cheng, F.; Chen, J. Electrodeposition Synthesis and Electrochemical Properties of Nanostructured γ-MnO_2 Films. *J. Power Sources* **2006**, *162*, 727–734. [CrossRef]
47. Logeeswaran, V.J.; Kobayashi, N.P.; Islam, M.S.; Wu, W.; Chaturvedi, P.; Fang, N.X.; Wang, S.Y.; Williams, R.S. Ultrasmooth Silver Thin Films Deposited with a Germanium Nucleation Layer. *Nano Lett.* **2009**, *9*, 178–182. [CrossRef] [PubMed]
48. Tepavcevic, S.; Xiong, H.; Stamenkovic, V.R.; Zuo, X.; Balasubramanian, M.; Prakapenka, V.B.; Johnson, C.S.; Rajh, T. Nanostructured Bilayered Vanadium Oxide Electrodes for Rechargeable Sodium-Ion Batteries. *ACS Nano* **2012**, *6*, 530–538. [CrossRef]
49. Abo-Hamad, A.; Hayyan, M.; AlSaadi, M.A.H.; Hashim, M.A. Potential Applications of Deep Eutectic Solvents in Nanotechnology. *Chem. Eng. J.* **2015**, *273*, 551–567. [CrossRef]
50. Shiddiky, M.J.A.; Torriero, A.A.J. Application of Ionic Liquids in Electrochemical Sensing Systems. *Biosens. Bioelectron.* **2011**, *26*, 1775–1787. [CrossRef] [PubMed]
51. Eftekhari, A. *Nanostructured Materials in Electrochemistry*; Wiley: Hoboken, NJ, USA, 2008.
52. Krishtalik, L.I. *Electrode Reactions. Mechanism of Elementary Act*; Nauka: Moscow, Russia, 1982.
53. Islam, M.M.; Abdellaoui, I.; Moslah, C.; Sakurai, T.; Ksibi, M.; Hamzaoui, S.; Akimoto, K. Electrodeposition and Characterization of Silicon Films Obtained through Electrochemical Reduction of SiO_2 Nanoparticles. *Thin Solid Film.* **2018**, *654*, 1–10. [CrossRef]
54. Zou, X.; Ji, L.; Ge, J.; Sadoway, D.R.; Yu, E.T.; Bard, A.J. Electrodeposition of Crystalline Silicon Films from Silicon Dioxide for Low-Cost Photovoltaic Applications. *Nat. Commun.* **2019**, *10*, 5772. [CrossRef]
55. Weng, W.; Xiao, W. Electrodeposited Silicon Nanowires from Silica Dissolved in Molten Salts as a Binder-Free Anode for Lithium-Ion Batteries. *ACS Appl. Energy Mater.* **2019**, *2*, 804–813. [CrossRef]
56. Laptev, M.V.; Isakov, A.V.; Grishenkova, O.V.; Vorob'ev, A.S.; Khudorozhkova, A.O.; Akashev, L.A.; Zaikov, Y.P. Electrodeposition of Thin Silicon Films from the $KF-KCl-KI-K_2SiF_6$ Melt. *J. Electrochem. Soc.* **2020**, *167*, 042506. [CrossRef]
57. Zou, X.; Ji, L.; Pang, Z.; Xu, Q.; Lu, X. Continuous Electrodeposition of Silicon and Germanium Micro/nanowires from Their Oxides Precursors in Molten Salt. *J. Energy Chem.* **2020**, *44*, 147–153. [CrossRef]
58. Yasko, O.; Shakhnin, D.B.; Gab, A.; Malyshev, V.; Gaune-Escard, M. Electrodeposition of Nanostructured Silicon Coatings onto Different Materials from Halide and Halide–Oxide Melts. In *Springer Proceedings in Physics*; Springer Science and Business Media Deutschland GmbH: Berlin, Germany, 2021; Volume 246, pp. 231–235. ISBN 9783030519049.
59. Gevel, T.; Zhuk, S.; Leonova, N.; Leonova, A.; Trofimov, A.; Suzdaltsev, A.; Zaikov, Y. Electrochemical Synthesis of Nano-Sized Silicon from $KCl-K_2SiF_6$ Melts for Powerful Lithium-Ion Batteries. *Appl. Sci.* **2021**, *11*, 10927. [CrossRef]
60. Wang, F.; Ma, Y.; Li, P.; Peng, C.; Yin, H.; Li, W.; Wang, D. Electrochemical Conversion of Silica Nanoparticles to Silicon Nanotubes in Molten Salts: Implications for High-Performance Lithium-Ion Battery Anode. *ACS Appl. Nano Mater.* **2021**, *4*, 7028–7036. [CrossRef]
61. Jing, S.; Xiao, J.; Shen, Y.; Hong, B.; Gu, D.; Xiao, W. Silicate-Mediated Electrolytic Silicon Nanotube from Silica in Molten Salts. *Small* **2022**, *18*, 2203251. [CrossRef]
62. Rezaei, A.; Kamali, A.R. Green Production of Carbon Nanomaterials in Molten Salts, Mechanisms and Applications. *Diam. Relat. Mater.* **2018**, *83*, 146–161. [CrossRef]

63. Hu, L.; Song, Y.; Ge, J.; Zhu, J.; Jiao, S. Capture and Electrochemical Conversion of CO_2 to Ultrathin Graphite Sheets in $CaCl_2$-Based Melts. *J. Mater. Chem. A* **2015**, *3*, 21211–21218. [CrossRef]
64. Wu, H.; Li, Z.; Ji, D.; Liu, Y.; Li, L.; Yuan, D.; Zhang, Z.; Ren, J.; Lefler, M.; Wang, B.; et al. One-Pot Synthesis of Nanostructured Carbon Materials from Carbon Dioxide via Electrolysis in Molten Carbonate Salts. *Carbon N. Y.* **2016**, *106*, 208–217. [CrossRef]
65. Novoselova, I.A.; Oliynyk, N.F.; Volkov, S.V. Electrolytic Production of Carbon Nano-Tubes in Chloride-Oxide Melts under Carbon Dioxide Pressure. In *Hydrogen Materials Science and Chemistry of Carbon Nanomaterials*; Springer: Dordrecht, The Netherlands, 2007; pp. 459–465. ISBN 1402055129.
66. Ijije, H.V.; Lawrence, R.C.; Chen, G.Z. Carbon Electrodeposition in Molten Salts: Electrode Reactions and Applications. *RSC Adv.* **2014**, *4*, 35808–35817. [CrossRef]
67. Mao, X.; Yan, Z.; Sheng, T.; Gao, M.; Zhu, H.; Xiao, W.; Wang, D. Characterization and Adsorption Properties of the Electrolytic Carbon Derived from CO_2 Conversion in Molten Salts. *Carbon N. Y.* **2017**, *111*, 162–172. [CrossRef]
68. Kushkhov, K.B.; Ligidova, M.N.; Ali, J.Z.; Khotov, A.A.; Tlenkopachev, M.R.; Karatsukova, R.K. Electrochemical Processes in Molten Alkaline Metal Carbonates under Carbon Dioxide Overpressure. *Russ. Metall.* **2021**, *2021*, 141–150. [CrossRef]
69. Wang, M.; Kim, Y.; Zhang, L.; Seong, W.K.; Kim, M.; Chatterjee, S.; Wang, M.; Li, Y.; Bakharev, P.V.; Lee, G.; et al. Controllable Electrodeposition of Ordered Carbon Nanowalls on Cu(111) Substrates. *Mater. Today* **2022**, *57*, 75–83. [CrossRef]
70. Kumar, K.; Mondal, S. Fabrication and Characterisation of Carbon Nanotube Reinforced Copper Matrix Nanocomposites. *Can. Metall. Q.* **2022**, *61*, 77–84. [CrossRef]
71. Groult, H.; Le Van, K.; Lantelme, F. Electrodeposition of Carbon-Metal Powders in Alkali Carbonate Melts. *J. Electrochem. Soc.* **2014**, *161*, D3130–D3138. [CrossRef]
72. Hu, L.; Song, Y.; Ge, J.; Zhu, J.; Han, Z.; Jiao, S. Electrochemical Deposition of Carbon Nanotubes from CO_2 in $CaCl_2$–NaCl-Based Melts. *J. Mater. Chem. A* **2017**, *5*, 6219–6225. [CrossRef]
73. Xu, Q.; Schwandt, C.; Chen, G.Z.; Fray, D.J. Electrochemical Investigation of Lithium Intercalation into Graphite from Molten Lithium Chloride. *J. Electroanal. Chem.* **2002**, *530*, 16–22. [CrossRef]
74. Kamali, A.R. Nanocatalytic Conversion of CO_2 into Nanodiamonds. *Carbon N. Y.* **2017**, *123*, 205–215. [CrossRef]
75. Yasko, O.; Malyshev, V.; Gab, A.; Lukashenko, T. Joint Electroreduction of Carbonate and Tungstate Ions as the Base for Tungsten Carbide Nanopowders Synthesis in Ionic Melts. In *Springer Proceedings in Physics*; Springer Science and Business Media, LLC: Berlin, Germany, 2019; Volume 221, pp. 397–402. ISBN 9783030177584.
76. Malyshev, V.; Gab, A.; Shakhnin, D.; Lukashenko, T.; Ishtvanik, O.; Gaune-Escard, M. High-Temperature Electrochemical Synthesis of Nanopowders of Tungsten Carbide in Ionic Melts. In *Springer Proceedings in Physics*; Springer Science and Business Media, LLC: Berlin, Germany, 2018; Volume 214, pp. 311–321. ISBN 9783319925660.
77. Kushkhov, K.B.; Kardanov, A.L.; Adamokova, M.N. Electrochemical Synthesis of Binary Molybdenum-Tungsten Carbides $(Mo,W)_2C$ from Tungstate-Molybdate-Carbonate Melts. *Russ. Metall.* **2013**, *2013*, 79–85. [CrossRef]
78. Bukatova, G.A.; Kuznetsov, S.A. Electrosynthesis of Gadolinium Hexaboride Nanotubes. *Electrochem. Commun.* **2005**, *7*, 637–641. [CrossRef]
79. Kuznetsov, S.A.; Kremenetsky, V.G.; Popova, A.V.; Kremenetskaya, O.V.; Kalinnikov, V.T. Unusual Effect of the Second Coordination Sphere on the Standard Charge Transfer Rate Constants for the Nb(V)/Nb(IV) Redox Couple in Chloride—Fluoride Melts. *Dokl. Phys. Chem.* **2009**, *428*, 209–212. [CrossRef]
80. Grinevitch, V.V.; Kuznetsov, S.A.; Arakcheeva, A.A.; Olyunina, T.V.; Schönleber, A.; Gaune-Escard, M. Electrode and Chemical Reactions during Electrodeposition of Tantalum Products in CsCl Melt. *Electrochim. Acta* **2006**, *51*, 6563–6571. [CrossRef]
81. Kuznetsov, S.A.; Rebrov, E.V.; Mies, M.J.M.; de Croon, M.H.J.M.; Schouten, J.C. Synthesis of Protective Mo–Si–B Coatings in Molten Salts and Their Oxidation Behavior in an Air–Water Mixture. *Surf. Coat. Technol.* **2006**, *201*, 971–978. [CrossRef]
82. Grinevitch, V.V.; Arakcheeva, A.V.; Kuznetsov, S.A. Tantalum Metal of Peculiar Cristal Lattice (b-Ta) as Creation of Electrocrystallization from Molten Salts. In *International Symposium on Ionic Liquids*; DTU Research Database: Carry le Rouet, France, 2003; pp. 277–285.
83. Dubrovskiy, A.R.; Makarova, O.V.; Kuznetsov, S.A. Effect of the Molybdenum Substrate Shape on Mo_2C Coating Electrodeposition. *Coatings* **2018**, *8*, 442. [CrossRef]
84. Dubrovskii, A.R.; Kuznetsov, S.A.; Rebrov, E.V.; Schouten, J.C. Catalytic Coatings of New Generation Based on Mo_2C and a Microstructured Reactor for Steam Conversion of Carbon Monoxide. *Russ. J. Appl. Chem.* **2014**, *87*, 601–607. [CrossRef]
85. Dubrovskiy, A.R.; Okunev, M.A.; Makarova, O.V.; Kuznetsov, S.A. Superconducting Niobium Coatings Deposited on Spherical Substrates in Molten Salts. *Coatings* **2018**, *8*, 213. [CrossRef]
86. Patt, J.; Moon, D.J.; Phillips, C.; Thompson, L. Molybdenum Carbide Catalysts for Water-Gas Shift. *Catal. Lett.* **2000**, *65*, 193–195. [CrossRef]
87. Derakhshandeh, M.R.; Eshraghi, M.J.; Razavi, M. Recent Developments in the New Generation of Hard Coatings Applied on Cemented Carbide Cutting Tools. *Int. J. Refract. Met. Hard Mater.* **2023**, *111*, 106077. [CrossRef]
88. Lee, C.; Danon, Y.; Mulligan, C. Characterization of Niobium, Tantalum and Chromium Sputtered Coatings on Steel Using Eddy Currents. *Surf. Coat. Technol.* **2005**, *200*, 2547–2556. [CrossRef]
89. Lee, Y.J.; Lee, T.H.; Kim, D.Y.; Nersisyan, H.H.; Han, M.H.; Kang, K.S.; Bae, K.K.; Shin, Y.J.; Lee, J.H. Microstructural and Corrosion Characteristics of Tantalum Coatings Prepared by Molten Salt Electrodeposition. *Surf. Coat. Technol.* **2013**, *235*, 819–826. [CrossRef]

90. Emsley, J. *Nature's Building Blocks: An A-Z Guide to the Elements*; Oxford University Press: Oxford, UK, 2003.
91. Baklanova, N.I.; Zima, T.M.; Boronin, A.I.; Kosheev, S.V.; Titov, A.T.; Isaeva, N.V.; Graschenkov, D.V.; Solntsev, S.S. Protective Ceramic Multilayer Coatings for Carbon Fibers. *Surf. Coat. Technol.* **2006**, *201*, 2313–2319. [CrossRef]
92. Wang, S.-C.; Lin, H.-T.; Nayak, P.K.; Chang, S.-Y.; Huang, J.-L. Carbothermal Reduction Process for Synthesis of Nanosized Chromium Carbide via Metal-Organic Vapor Deposition. *Thin Solid Film.* **2010**, *518*, 7360–7365. [CrossRef]
93. Sen, S. Influence of Chromium Carbide Coating on Tribological Performance of Steel. *Mater. Des.* **2006**, *27*, 85–91. [CrossRef]
94. Gómez, M.A.; Romero, J.; Lousa, A.; Esteve, J. Tribological Performance of Chromium/Chromium Carbide Multilayers Deposited by r.f. Magnetron Sputtering. *Surf. Coat. Technol.* **2005**, *200*, 1819–1824. [CrossRef]
95. Cintho, O.M.; Favilla, E.A.P.; Capocchi, J.D.T. Mechanical–Thermal Synthesis of Chromium Carbides. *J. Alloys Compd.* **2007**, *439*, 189–195. [CrossRef]
96. Esteve, J.; Romero, J.; Gómez, M.; Lousa, A. Cathodic Chromium Carbide Coatings for Molding Die Applications. *Surf. Coat. Technol.* **2004**, *188–189*, 506–510. [CrossRef]
97. Shatinsky, V.D.; Nesterenko, F.I. *Protecting Diffusional Coatings*; Naukova Dumka: Kiev, Ukraine, 1988.
98. Stulov, Y.V.; Kuznetsov, S.A. Synthesis of Chromium Carbide Coatings on Carbon Steels in Molten Salts and Their Properties. *Glas. Phys. Chem.* **2014**, *40*, 324–328. [CrossRef]
99. Ilyushchenko, N.G.; Afinogenov, A.I.; Shurov, N.I. *Metal Interactions in Ionic Melts*; Nauka: Moscow, Russia, 1991.
100. Novokreshchenov, Y.V.; Petenev, O.S.; Kositsyn, Y.N. Influence of anodic polarization on spontaneous transfer of metals from salt melts onto articles for protection by thermal chemical treatment. *J. Appl. Chem. USSR* **1986**, *59*, 2492–2494.
101. Malyshev, V.V.; Shakhnin, D.B.; Hab, A.; Kublanovs'kyi, V.S.; Schuster, D. Synthesis of Chromium Silicides in Ionic Melts. *Mater. Sci.* **2020**, *55*, 745–757. [CrossRef]
102. Kovalevskii, A.V.; El'Kin, O.V. The Electrochemical Properties of LiCl-KCl Melt Held in Contact with Samarium. *Russ. J. Phys. Chem. A* **2011**, *85*, 499–502. [CrossRef]
103. Stulov, Y.V.; Dolmatov, V.S.; Dubrovskii, A.R.; Kuznetsov, S.A. Coatings by Refractory Metal Carbides: Deposition from Molten Salts, Properties, Application. *Russ. J. Appl. Chem.* **2017**, *90*, 676–683. [CrossRef]
104. Böhm, H. Adsorption Und Anodische Oxydation von Wasserstoff an Wolframcarbid. *Electrochim. Acta* **1970**, *15*, 1273–1280. [CrossRef]
105. Goti, A.; Cardona, F. Hydrogen Peroxide in Green Oxidation Reactions: Recent Catalytic Processes. In *Green Chemical Reactions*; Springer: Dordrecht, The Netherlands, 2008; pp. 191–212. [CrossRef]
106. Bloom, H. *Electrochemistry: The Past Thirty and the Next Thirty Years*; Springer: Berlin/Heidelberg, Germany, 2012.
107. Dolmatov, V.S.; Kuznetsov, S.A. Synthesis of Refractory Metal Carbides on Carbon Fibers in Molten Salts and Their Electrocatalytic Properties. *J. Electrochem. Soc.* **2021**, *168*, 122501. [CrossRef]
108. Guan, A.; Yang, C.; Quan, Y.; Shen, H.; Cao, N.; Li, T.; Ji, Y.; Zheng, G. One-Dimensional Nanomaterial Electrocatalysts for CO_2 Fixation. *Chem.-An Asian J.* **2019**, *14*, 3969–3980. [CrossRef]
109. Peng, S.; Li, L.; Kong Yoong Lee, J.; Tian, L.; Srinivasan, M.; Adams, S.; Ramakrishna, S. Electrospun Carbon Nanofibers and Their Hybrid Composites as Advanced Materials for Energy Conversion and Storage. *Nano Energy* **2016**, *22*, 361–395. [CrossRef]
110. Wang, H.G.; Yuan, S.; Ma, D.L.; Zhang, X.B.; Yan, J.M. Electrospun Materials for Lithium and Sodium Rechargeable Batteries: From Structure Evolution to Electrochemical Performance. *Energy Environ. Sci.* **2015**, *8*, 1660–1681. [CrossRef]
111. Jin, T.; Han, Q.; Wang, Y.; Jiao, L.; Jin, T.; Han, Q.; Wang, Y.; Jiao, L. 1D Nanomaterials: Design, Synthesis, and Applications in Sodium–Ion Batteries. *Small* **2018**, *14*, 1703086. [CrossRef] [PubMed]
112. Kohmura, Y.; Zhakhovsky, V.; Takei, D.; Suzuki, Y.; Takeuchi, A.; Inoue, I.; Inubushi, Y.; Inogamov, N.; Ishikawa, T.; Yabashi, M. Nano-Structuring of Multi-Layer Material by Single x-Ray Vortex Pulse with Femtosecond Duration. *Appl. Phys. Lett.* **2018**, *112*, 123103. [CrossRef]
113. Salih, O.S.; Al-Akkam, E.J. Microneedles as A Magical Technology to Facilitate Transdermal Drug Delivery: A Review Article. *Int. J. Drug Deliv. Technol.* **2022**, *12*, 896–901. [CrossRef]
114. Pan, J.; Low, K.L.; Ghosh, J.; Jayavelu, S.; Ferdaus, M.M.; Lim, S.Y.; Zamburg, E.; Li, Y.; Tang, B.; Wang, X.; et al. Transfer Learning-Based Artificial Intelligence-Integrated Physical Modeling to Enable Failure Analysis for 3 Nanometer and Smaller Silicon-Based CMOS Transistors. *ACS Appl. Nano Mater.* **2021**, *4*, 6903–6915. [CrossRef]
115. Kuznetsova, S.V.; Dolmatov, V.S.; Kuznetsov, S.A. Voltammetric Study of Electroreduction of Silicon Complexes in a Chloride-Fluoride Melt. *Russ. J. Electrochem.* **2009**, *45*, 742–748. [CrossRef]
116. Nicholson, R.S.; Shain, I. Theory of Stationary Electrode Polarography: Single Scan and Cyclic Methods Applied to Reversible, Irreversible, and Kinetic Systems. *Anal. Chem.* **1964**, *36*, 706–723. [CrossRef]
117. Nakamura, Y.; Asano, I.; Hiratani, M.; Saito, T.; Goto, H. Oxidation-Resistant Amorphous TaN Barrier for MIM-Ta2O5 Capacitors in Giga-Bit DRAMs. In Proceedings of the 2001 Symposium on VLSI Technology. Digest of Technical Papers, Kyoto, Japan, 12–14 June 2001; pp. 39–40. [CrossRef]
118. Matsui, Y.; Hiratani, M.; Kimura, S.; Asano, I. Combining Ta_2O_5 and Nb_2O_5 in Bilayered Structures and Solid Solutions for Use in MIM Capacitors. *J. Electrochem. Soc.* **2005**, *152*, F54. [CrossRef]
119. Fairbrother, F. *The Chemistry of Niobium and Tantalum*; Elsevier Publishing Company: Amsterdam, The Netherlands, 1967; ISBN 978-0444402059.

120. Mozalev, A.; Sakairi, M.; Takahashi, H. Structure, Morphology, and Dielectric Properties of Nanocomposite Oxide Films Formed by Anodizing of Sputter-Deposited Ta-Al Bilayers. *J. Electrochem. Soc.* **2004**, *151*, F257. [CrossRef]
121. Zainulin, Y.G.; Alyamovskii, S.I.; Shveikin, G.P.; Popova, S.V. Possibility of Preparing Zirconium and Tantalum Monoxides at High Pressures and Temperatures. *Zh. Neorg. Khim.* **1978**, *23*, 1155–1157.
122. Lantelme, F.; Barhoun, A.; Li, G.; Besse, J. Electrodeposition of Tantalum in NaCl–KCl–K_2TaF_7 Melts. *J. Electrochem. Soc.* **1992**, *139*, 1249–1255. [CrossRef]
123. Chamelot, P.; Palau, P.; Massot, L.; Savall, A.; Taxil, P. Electrodeposition Processes of Tantalum(V) Species in Molten Fluorides Containing Oxide Ions. *Electrochim. Acta* **2002**, *47*, 3423–3429. [CrossRef]

Disclaimer/Publisher's Note: The statements, opinions and data contained in all publications are solely those of the individual author(s) and contributor(s) and not of MDPI and/or the editor(s). MDPI and/or the editor(s) disclaim responsibility for any injury to people or property resulting from any ideas, methods, instructions or products referred to in the content.

Article

Discrimination of Steel Coatings with Different Degradation Levels by Near-Infrared (NIR) Spectroscopy and Deep Learning

Mingyang Chen [1,2,*], Guangming Lu [1] and Gang Wang [2]

1 School of Computer Science and Technology, Harbin Institute of Technology Shenzhen Graduate School, Shenzhen 518055, China
2 Central Research Institute of Building and Construction (Shenzhen) Co., Ltd., MCC Group, Shenzhen 518066, China
* Correspondence: mc15an@my.fsu.edu

Citation: Chen, M.; Lu, G.; Wang, G. Discrimination of Steel Coatings with Different Degradation Levels by Near-Infrared (NIR) Spectroscopy and Deep Learning. *Coatings* 2022, 12, 1721. https://doi.org/10.3390/coatings12111721

Academic Editor: Matic Jovičević-Klug, Patricia Jovičević-Klug and László Tóth

Received: 14 October 2022
Accepted: 8 November 2022
Published: 11 November 2022

Publisher's Note: MDPI stays neutral with regard to jurisdictional claims in published maps and institutional affiliations.

Copyright: © 2022 by the authors. Licensee MDPI, Basel, Switzerland. This article is an open access article distributed under the terms and conditions of the Creative Commons Attribution (CC BY) license (https://creativecommons.org/licenses/by/4.0/).

Abstract: Assessing the current condition of protective organic coatings on steel structures is an important but challenging task, particularly when it comes to complex structures located in harsh environments. Near-infrared (NIR) spectroscopy is a rapid, low-cost, and nondestructive analytical technique with applications ranging from agriculture, food, and remote sensing to pharmaceuticals. In this study, an objective and reliable NIR-based technique is proposed for the accurate distinction between different coating conditions during their degradation process. In addition, a state-of-the-art deep learning method using a one-dimensional convolutional neural network (1-D CNN) is explored to automatically extract features from the spectrum. The characteristics of the spectrum show a downward trend over the entire wavenumber period, and two major absorption peaks were observed around 5250 and 4400 cm^{-1}. The experimental results indicate that the proposed deep network structure can powerfully extract the complex characteristics inside the spectrum, and the classification accuracy of the training and testing data was 99.84% and 95.23%, respectively, which suggests that NIR spectroscopy coupled with a deep learning algorithm could be used for the rapid and accurate inspection of steel coatings.

Keywords: steel coating assessment; near-infrared spectroscopy; deep learning; degradation

1. Introduction

Assessing the current condition of protective organic coatings on steel structures is an important but challenging task. It is crucial for asset owners to understand when the coating is no longer effective, so the repair can be applied before damage is done to structures due to the failure of protective coatings, and the improvement in maintenance planning can have a significant financial benefit. Common coating assessment practices involve trained inspectors performing close-up visual inspections and using a rating system. However, these practices are not only labor-intensive but also time-consuming and pose significant safety and logistical challenges when structures are located in harsh places [1]. Therefore, an efficient and intelligent evaluation approach is highly required for making informed decisions about protective coating maintenance.

Near-infrared (NIR) spectroscopy is a rapid, nondestructive and relatively inexpensive analytical technique. It characterizes materials based on their absorption intensity in the NIR region of the electromagnetic spectrum, which ranges from 700 to 2500 nm. These optical responses in the NIR region reflect vibrations of molecular functional groups containing atoms like C, N, O, and S or chemical bonds between atoms, such as C=O, C=H and C-O-C, which allows researchers to analyze samples of organic composition [2]. Therefore, the technique has found broad application in agriculture, food, pharmaceuticals, remote sensing, and several other fields [3–5]. For example, Piehl et al. [6] presented the first quantitative and qualitative analysis of plastic contamination in agricultural soil based on the NIR technique. Via Fourier transform infrared (FTIR), they were able to identify and quantify macro and microplastic pieces within the investigated area, which therefore provided important data to determine the extent of contamination concerning agroecosystems. Valand et al. [7] provided an extensive review of the application of NIR spectroscopy for food adulteration and authenticity. The literature has shown that, over the last decades, NIR has proven itself to be a trustworthy technology for examining foods for adulteration and authenticity. It is not only a fast, easy, and generally cost-effective method, but it also can combine with other analytical chemistry techniques to develop validated or standardized methods.

Meanwhile, methods used to extract and process the NIR analytical information to produce quantitative and qualitative models has evolved during the last decade. Sohn et al. [8] proposed a combination model of Savitzky–Golay and a support vector machine (SVM) to classify six different Amaranthus species in Korea, and the result shows that Vis-NIR spectroscopy with an SVM model has the capability to discriminate Amaranthus species with a notable accuracy up to 99.7%. Nawar et al. [9] established a random forest (RF) modeling approach for the quantitative analysis of soil organic carbon (OC). The results suggest that RF regression with spiking provides an accurate prediction of OC under both indoor laboratory and on-site field scanning conditions. Sampaio et al. [10] compared the performance of partial least squares (PLS), interval-PLS, synergy interval-PLS, and moving windows-PLS models, and developed an optimal regression model with high accuracy for rice amylose determination. Recently, the deep learning algorithm has attracted increasing attention for NIR researchers; it has shown great capabilities in creating powerful analytic models based on multilayer abstraction to represent concepts or features [11–13]. Chen et al. [14] proposed a framework for a backpropagation, neural deep learning 1D-CNN to predict the nutrition components in soil samples; Rong et al. [15] applied a simple CNN architecture with a single convolutional layer to distinguish peach varieties; Gholizadehhis et al. [16] examined the capability of vis–NIR spectra coupled with traditional machine learning techniques and a fully connected neural network (FNN) to assess potentially toxic elements in forest soils. The results show that FNN provides better results in the availability of a large sample size. These research works indicate that deep learning can be successfully applied to NIR sensor data analysis.

A lot of research works have been conducted with respect to the application of NIR technology in the field of monitoring or the inspection of steel structure coatings [17–19]. Kishigami et al. [20] recently demonstrated the use of NIR in the estimation of the degree of abrasion of coating thickness. It was found that the observed infrared intensity could be used to estimate the top coating thickness based on the calibration relationship they discovered. Omar et al. [21] developed a novel integrated device based on FTIR and micro-electromechanics to make structural analyses of the epoxy coating of steel pipelines. It shows that their instrument was useful for on-site material analysis, especially in the investigation of the mechanical properties and detection of the distribution of particles inside the material. Raeissi et al. [22] explored the use of a k-means clustering algorithm on the NIR portion of hyperspectral images (NIR-HIS) to provide diagnostic information about the spatial inhomogeneities of the chemical structures of an applied steel coating.

The aim of the study was to present an innovative approach based on NIR spectroscopy as a novel solution to objectively assess the condition of the protective coatings applied to steel structures. Moreover, this method of nondestructive evaluation could provide precise and automatic grading in the assessment of coating degradation due to age or environmental factors. The potential of the approach was shown using a real NIR dataset acquired from prepared coating samples with an accelerated aging process. In addition, a modern convolutional neural network (CNN) was developed to classify the different grades of corrosion based on their NIR data. CNNs have achieved promising performances in such classification tasks due to their large flexibility regarding the dimensionality of the operational layers, their depth and breadth, and their ability to extract strong features about the input data [23]. With the excellent learning ability of CNNs, a spatial distribution of the intensity variations at particular absorption lines in the electromagnetic spectrum of

NIR data can be reconstructed as essential features, which can be used for the identification and for the evaluation of chemical changes that occur as a result of coating degradation.

2. Materials and Data Collection

2.1. Sample Preparation

The protective coating used in this study was based on three layers of a composite coating system that was applied to the world-famous Hongkong-Zhuhai-Macro Bridge (HZMB). The coating system was formulated with zinc-rich epoxy primer, MIO epoxy intermediate paint, and a fluorocarbon topcoat, and each layer has a thickness of 100, 200, and 80 µm, respectively. Moreover, the composite coating was applied to steel plates using a high-pressure airless spraying method, following the specifications of painting and coating for steel structures in China. Chinese steel of grade Q235 was used for the steel plates, of which the dimensions were 100 mm × 50 mm × 10 mm. Figure 1 shows a real image of a steel plate with the composite coating.

Figure 1. Image of a steel plate sample.

2.2. Data Acquisition and Labelling

In order to simulate the behavior of coating degradation, 50 steel plates with a protective composite coating were exposed to a salt spray test device to accelerate corrosion growth. The salt spray test was performed following the instruction of the ISO 9227 standard, and each coated plate was grouped and subjected to a different exposure time: 120, 240, 360, 480, 600, 720, 840, 960, and 1080 h. Each sample was rinsed with water and air-dried before NIR data collection. The data collection was based on FTIR measurements performed using a BRUKER Lumos FTIR microscope. Reflectance spectra were recorded in the range of 8500 to 4000 cm^{-1}. The resolution of the obtained spectra was 3.5 cm^{-1}. For each coated sheet, six points located in the corner and center were measured individually. A total of 300 individual point spectra were recorded per one measurement, and a total of 10 measurements were made based on the different exposure times.

The labelling task was performed shortly after the NIR data collection of each steel sheet, and a human expert was present at the collection site to make an observation of the actual sample surface and assigned a label corresponding to their condition ratings. A four-level rating system is applied by inspectors and, in our study, is described in Table 1. The NIR dataset of a steel coating is then generated and referred to as a matrix of 1991 rows and 1201 columns. Each row represents the spectral information of the individual point from the coated steel sheet; each column (descriptive features) represents the reflectance magnitude of the NIR spectral band with a specific wavelength, which ranges from 8500 to 4000 cm^{-1}; the last column has the condition class labels that are associated with each point. As shown in Table 1, the dataset is programmed to randomly divide into a training set and a testing set with a proportion of 80:20. The training set is used to train the model parameters and the testing set is used to check the classification performance of the trained model.

Table 1. The four-level coating condition rating system.

Coating Condition	Number of Training Set	Number of Testing Set	Description
Level 1	375	99	The coating remains intact
Level 2	566	146	The coating is slightly degraded, with speckled rusting in areas that are less than 1% of the total surface area.
Level 3	356	90	The coating is moderately degraded, with speckled rusting in areas greater than 1% and less than 40% of the total surface area.
Level 4	286	73	The coating is no longer effective, with speckled rusting in areas larger than 40% of the total surface area.
Total	1583	408	/

3. Methodology

3.1. Pre-Processing of NIR Data

Preprocessing of spectral data is the most important step before the subsequent modeling and analyzing. The objective of preprocessing is to remove physical phenomena in the spectra, such as baseline drift, high-frequency noise, and mutual interference between the components [9]. In this study, multiple preprocessing techniques were adopted based on four categories: smoothing, scatter correction, spectral derivatives, and wavelet denoising. Table 2 presented general information about these methods and a comparison of their denoising effect using root mean square error ($RMSE$) and signal-to-noise ratio (SNR). The two metrics are calculated by

$$SNR = 10 \times \log\left(\frac{\sum_{n=1}^{N} f(n)^2}{\sum_{n=1}^{N} [f(n)-\hat{f}(n)]^2}\right) \tag{1}$$

$$RMSE = \sqrt{\frac{1}{n}\sum_{n}[f(n) - \hat{f}(n)]^2} \tag{2}$$

where $f(n)$ is the original spectrum, $\hat{f}(n)$ is the denoised spectrum, and n is the sampling point. It can be seen that SG smoothing, MSC, 1st derivatives, and wavelet denoising with coiflets base have the best denoising effect in their category, respectively. However, the optimal preprocessing method will be determined according to the performance and robustness of the classification model, which is explained in detail in the next section.

Table 2. Comparison of preprocessing method denoising effects.

Categories	Pre-Processing Methods	RMSE	SNR
Smoothing	Mean average (MA) smoothing	0.0545	20.18
	Savitzkygolay (SG) smoothing	0.0112	29.94
Scatter Correction	Multiplicative scatter correction (MSC)	0.0322	28.50
	Standard normal variate (SNV)	0.0322	28.50
Spectral Derivatives	1st Derivatives	0.2208	8.05
	2nd Derivatives	0.2253	7.88
Wavelet Denoising	Haar wavelet	0.2813	5.94
	Daubechies wavelet	0.0178	29.89
	Coiflets wavelet	0.0109	34.16
	Symmlets wavelet	0.0127	32.86

3.2. Architecture of the Proposed CNN Based Model

The CNN network can be variously arranged depending on the designed parameters and depths of the structure as well as the training method of the network. The basic architecture of the 1-D CNN in this paper consists of an input layer, multiple convolutional and pooling layers stacked together, a fully connected layer, and an output layer. A schematic diagram of this process is shown in Figure 2. As the obtained NIR input data represent a one-dimensional vector, the convolutional layer filters the input data with a one-dimensional kernel to obtain subtle feature information. A convolution kernel size of 1×9 is adopted in the first two convolutional layers to quickly obtain rough feature information, and then the convolution kernel with a small size of 1×3 is selected to extract more subtle features. In order to control the shrinkage of the dimensions, full zero padding and upward rounding are adopted in the process of convolution. The output of every convolutional layer is then passed to the pooling layer to reduce the size of the feature map, and the maximum operation is used. The size of the pooling layer is set as 1×2 with a step of 2 in this study. The final feature maps from the pooling layer are then flattened and passed to the fully connected layers at the end of the network. The Rectified Linear Unit (ReLU), the most commonly used activation function, is adapted in this model to implement nonlinear transformations [24]. The model applied Batch Normalization over the output of the convolutional layer to carry out the standardization process and the dropout mechanism is used to alleviate overfitting. The function of cross entropy is selected as the loss function in this study to quantify the difference between two probability distributions [25]. Lastly, the mathematical function of Softmax is implemented for the neural network of multiple classifications. In order to train the proposed CNN model, the learning rate is set to 0.001, the batch size is set to 64, and the epoch of training is set to 100.

3.3. Performance Evaluation

In order to evaluate the overall performance of the proposed model, the following four criteria were used: classification accuracy, precision, recall, and F1 score. These metrics are calculated by

$$\text{Aaccuracy} = \frac{\sum_{i=1}^{n}(TP_i + FN_i)}{\sum_{i=1}^{n}(TP_i + TN_i + FP_i + FN_i)} \quad (3)$$

$$\text{Precision} = \frac{1}{n}\sum_{i=1}^{n}\frac{TP_i}{TP_i + FP_i} \quad (4)$$

$$\text{Recall} = \frac{1}{n}\sum_{i=1}^{n}\frac{TP_i}{TP_i + FN_i} \quad (5)$$

$$F1 = \frac{1}{n}\sum_{i=1}^{n}\frac{2TP_i}{2TP_i + FP_i + FN_i} \tag{6}$$

where TP, TN, FP, and FN are true positive, true negative, false positive, and false negative, respectively. A confusion matrix is provided to demonstrate insight information for the predictions and to comprehend other classification metrics.

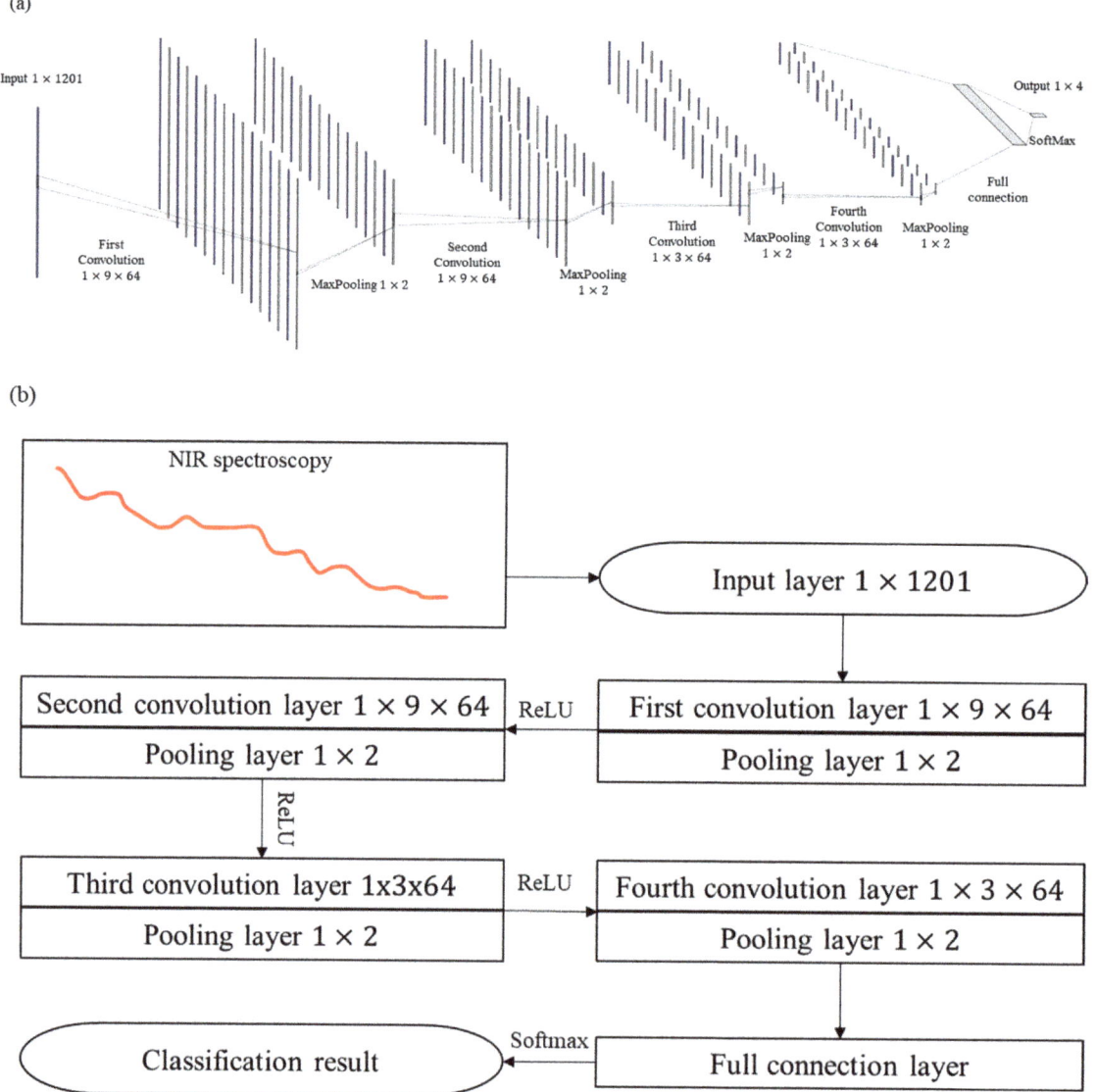

Figure 2. A schematic diagram of the proposed 1D-CNN classification model: (**a**) basic structure; (**b**) architectural details.

3.4. Software Tools

The Python 3.6 program with PyCharm IDE was used to perform spectral extraction, preprocessing, and other analysis models. The 1D-CNN model was programmed using the Pytorch framework, running on the graphics processing unit.

4. Results and Discussion

4.1. Spectral Characteristic

The original and mean NIR spectra of the steel coatings with different grades of corrosion are shown in Figure 3a,b, respectively. Overall, the original spectrum of the four kinds of coating conditions showed a downward trend over the entire wavenumber period. Two major absorption peaks were observed around 5250 and 4400 cm^{-1}. A peak around 5250 cm^{-1} is typically attributed to the hydroxyl ring, which represents a combination of asymmetric stretching and bending of O-H, and the peak around 4400 cm^{-1} represents the combination band of the second overtone of the epoxy ring [22,26,27]. Both characteristics are highly correlated to coating degradation. Another clear trend from Figure 3b was a decrease in the coating degradation in the reflectance of the spectra features. This phenomenon reflected the thickness and total surface coverage of the corrosion compounds within the analysis area [28].

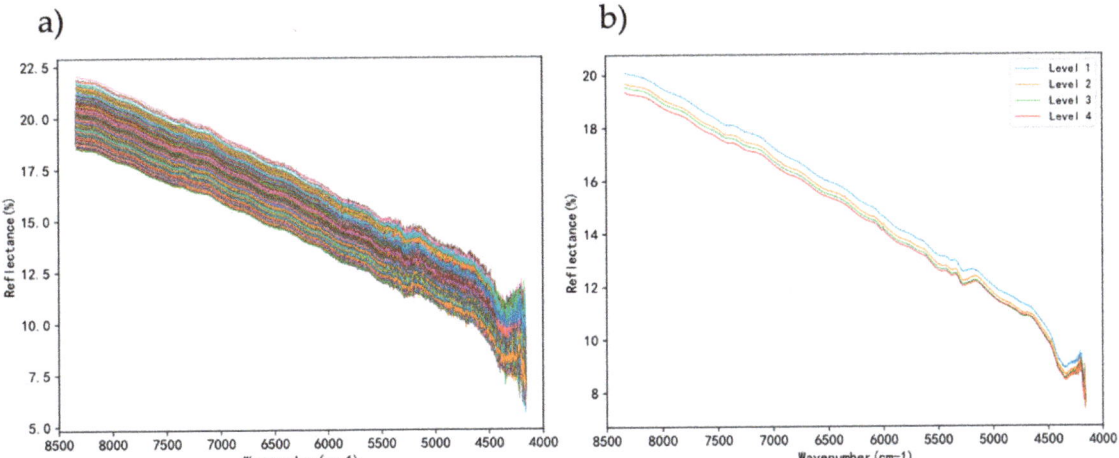

Figure 3. Near-infrared (NIR) spectra of steel coating: (**a**) raw data for all samples; (**b**) mean spectra of different grades of the coating condition.

4.2. Preprocessing Method

NIR spectral data are often interfered with by stray light, noise, baseline drift, and other factors, thus affecting the final qualitative and quantitative analysis results. In this paper, multiple pretreatment methods were used for the comparative analysis, and the following accuracy results of the proposed CNN model are shown in Table 3. Compared with other methods, SG smoothing is more competitive, with a 95.8% accuracy, but the MSC and SNV methods reached close accuracy results of around 95%. Obviously, spectral derivative methods are not suitable for the proposed model. It is worth mentioning that wavelet methods (coiflets and symmlets family) showed an inferior performance of 85–90% accuracy while having better denoising results based on the RMSE and SNR results in Table 2. In summary, SG smoothing was selected as the preprocessing method for the proposed discriminant model.

Table 3. The accuracy results of the 1D-CNN model with different preprocessing methods.

Categories	Pre-Processing Methods	Accuracy (%)
Raw Data	/	91.8
Smoothing	Mean average (MA) smoothing	92.8
	Savitzkygolay (SG) smoothing	95.8
Scatter Correction	Multiplicative scatter correction (MSC)	94.7
	Standard normal variate (SNV)	95.0
Spectral Derivatives	1st Derivatives	80.3
	2nd Derivatives	61.5
Wavelet Denoising	Haar wavelet	55.7
	Daubechies wavelet	85.1
	Coiflets wavelet	91.0
	Symmlets wavelet	84.8

4.3. CNN-Based Steel Coating Condition Assessment Model

The results for loss value and prediction accuracy on the training set and test set are shown in Figures 4 and 5. As can be seen in Figure 4, the loss value shows a trend that drops down rapidly during the first 20 epochs, then decreases slowly until it is steady. In the end, the loss values of the training set and the testing set were 0.759 and 0.829, respectively. In Figure 5, it is observed that the classification performance of the 1D-CNN model shows a trend of rapid rise, gradual and slow increase, and then tends to be stable. After 40 epochs, the accuracy rate of the testing and training sets reach 99.84% and 95.23%, respectively.

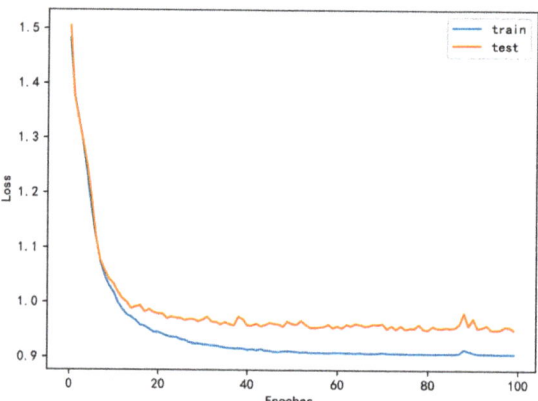

Figure 4. The loss value of the proposed 1D-CNN model.

In order to better explain the prediction results of the 1D-CNN model for each category in the test set, a confusion matrix was introduced and is shown in Figure 6. As can be seen from the confusion matrix, the predictions of a coating with a level 1 condition are all correct, and only one sample with a level 2 coating condition is falsely predicted as being level 3. The coating samples with a level 3 condition show the lowest prediction accuracy, with a total of 14 samples being misclassified; most of them are falsely predicted as a level 4 condition. Similarly, seven coating samples with a level 4 condition are misclassified to level 3. The result of the confusion matrix indicates that most false predictions occur in condition levels 3 and 4.

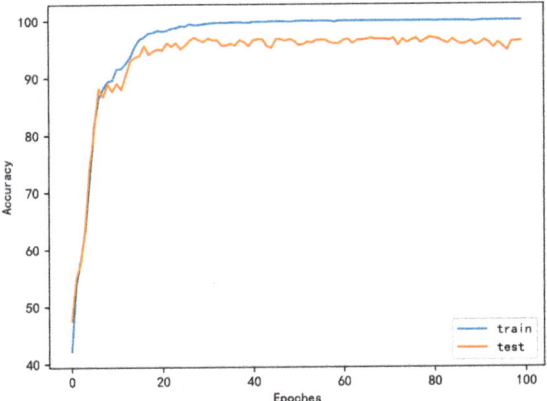

Figure 5. The accuracy of the proposed 1D-CNN model.

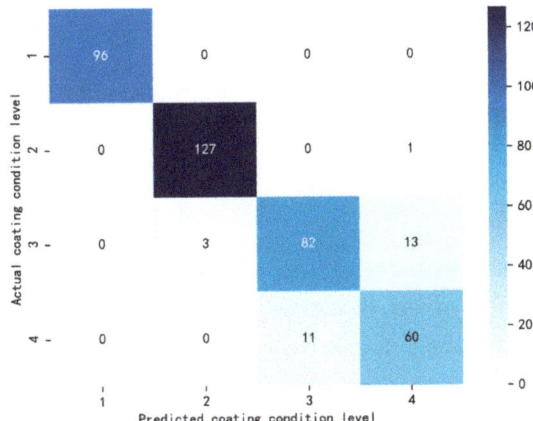

Figure 6. The confusion matrix for the testing set.

The overall performance results are shown in Table 4, which confirms that, for limited high-dimensional data, the proposed 1D-CNN model has excellent classification ability and can discriminate the condition levels of steel coatings. This reveals the superiority of the deep learning model with a high ability for feature extraction and learning over traditional processing.

Table 4. The performance results of the proposed 1D-CNN model.

Performance	Train (%)	Test (%)
Accuracy	99.84	95.23
Precision score	99.83	94.90
Recall score	99.73	94.74
F1 score	99.81	94.48

5. Conclusions

In this paper, a 1D-CNN-based model was developed for the assessment of steel coating conditions using NIR data. Four levels of coating degradation were constructed by artificially-accelerated aging, and the data were collected based on different exposure times. With the current popular research method of deep learning, NIR data can be directly input into the model, extracting feature information from the spectrum and conducting automatic learning. Therefore, an accurate assessment of the coating condition can be achieved. The major findings of this paper can be summarized as follows:

- The characteristics of the spectrum showed a downward trend over the entire wavenumber period, and two major absorption peaks were observed around 5250 and 4400 cm^{-1};
- A decrease in the reflectance of the spectrum features was observed along with the coating degradation process;
- A comparison of the different preprocessing methods indicated that the SG smoothing method was the most suitable method for the proposed model to effectively improve classification performance;
- Based on the above data and pretreatment, the experimental results of the proposed model achieved an overall prediction accuracy of 95.8% and very minimal error measures.

All the findings suggest that the proposed 1D-CNN framework coupled with NIR data has great potential for steel coating assessment and can provide rapid and accurate predictions of coating degradation levels. The next steps in this research should be to collect more coating data under a complex environment to build a comprehensive database; thus, the robustness of the model can be improved.

Author Contributions: Conceptualization, M.C. and G.W.; methodology, M.C. and G.W.; validation, M.C. and G.W.; formal analysis, M.C.; investigation, M.C.; writing—original draft preparation, M.C.; writing—review and editing, G.L. and G.W.; visualization, M.C.; supervision, G.L. and G.W.; funding acquisition, G.W. All authors have read and agreed to the published version of the manuscript.

Funding: This research was funded by the National Key Research and Development Program of China (No. 2019YFB1600702).

Institutional Review Board Statement: Not applicable.

Informed Consent Statement: Not applicable.

Data Availability Statement: Not applicable.

Conflicts of Interest: The authors declare no conflict of interest.

References

1. Huynh, C.P.; Mustapha, S.; Runcie, P.; Porikli, F. Multi-class support vector machines for paint condition assessment on the Sydney Harbour Bridge using hyperspectral imaging. *Struct. Monit. Maint.* **2015**, *2*, 181–197. [CrossRef]
2. Pasquini, C. Near infrared spectroscopy: A mature analytical technique with new perspectives–A review. *Anal Chim. Acta* **2018**, *1026*, 8–36. [CrossRef] [PubMed]
3. Corradini, F.; Bartholomeus, H.; Lwanga, E.H.; Gertsen, H.; Geissen, V. Predicting soil microplastic concentration using vis-NIR spectroscopy. *Sci. Total Environ.* **2019**, *650*, 922–932. [CrossRef] [PubMed]
4. Firmani, P.; De Luca, S.; Bucci, R.; Marini, F.; Biancolillo, A. Near infrared (NIR) spectroscopy-based classification for the authentication of Darjeeling black tea. *Food Control* **2019**, *100*, 292–299. [CrossRef]
5. Hong, Y.; Chen, Y.; Yu, L.; Liu, Y.; Liu, Y.; Zhang, Y.; Liu, Y.; Cheng, H. Combining fractional order derivative and spectral variable selection for organic matter estimation of homogeneous soil samples by VIS–NIR spectroscopy. *Remote Sens.* **2018**, *10*, 479. [CrossRef]
6. Piehl, S.; Leibner, A.; Löder, M.G.J.; Dris, R.; Bogner, C.; Laforsch, C. Identification and quantification of macro- and microplastics on an agricultural farmland. *Sci. Rep.* **2018**, *8*, 17950–17959. [CrossRef] [PubMed]
7. Valand, R.; Tanna, S.; Lawson, G.; Bengtström, L. A review of Fourier Transform Infrared (FTIR) spectroscopy used in food adulteration and authenticity investigations. *Food Addit. Contam. Part A* **2020**, *37*, 19–38. [CrossRef]
8. Sohn, S.; Oh, Y.; Pandian, S.; Lee, Y.; Zaukuu, J.Z.; Kang, H.; Ryu, T.; Cho, W.; Cho, Y.; Shin, E. Identification of Amaranthus species using visible-near-infrared (vis-NIR) spectroscopy and machine learning methods. *Remote Sens.* **2021**, *13*, 4149. [CrossRef]
9. Nawar, S.; Mouazen, A.M. On-line vis-NIR spectroscopy prediction of soil organic carbon using machine learning. *Soil Tillage Res.* **2019**, *190*, 120–127. [CrossRef]
10. Sampaio, P.S.; Soares, A.; Castanho, A.; Almeida, A.S.; Oliveira, J.; Brites, C. Optimization of rice amylose determination by NIR-spectroscopy using PLS chemometrics algorithms. *Food Chem.* **2018**, *242*, 196–204. [CrossRef]

11. Wanderi, K.; Cui, Z. *Organic Fluorescent Nanoprobes with NIR-IIb Characteristics for Deep Learning*; Wiley Online Library: Hoboken, NJ, USA, 2022; p. 20210097.
12. Naqvi, R.A.; Arsalan, M.; Batchuluun, G.; Yoon, H.S.; Park, K.R. Deep learning-based gaze detection system for automobile drivers using a NIR camera sensor. *Sensors* **2018**, *18*, 456. [CrossRef] [PubMed]
13. Nguyen, D.T.; Pham, T.D.; Lee, Y.W.; Park, K.R. Deep learning-based enhanced presentation attack detection for iris recognition by combining features from local and global regions based on NIR camera sensor. *Sensors* **2018**, *18*, 2601. [CrossRef] [PubMed]
14. Chen, H.; Liu, Z.; Gu, J.; Ai, W.; Wen, J.; Cai, K. Quantitative analysis of soil nutrition based on FT-NIR spectroscopy integrated with BP neural deep learning. *Anal. Methods* **2018**, *10*, 5004–5013. [CrossRef]
15. Rong, D.; Wang, H.; Ying, Y.; Zhang, Z.; Zhang, Y. Peach variety detection using VIS-NIR spectroscopy and deep learning. *Comput. Electron. Agric.* **2020**, *175*, 105553. [CrossRef]
16. Gholizadeh, A.; Saberioon, M.; Ben-Dor, E.; Rossel, R.A.V.; Bor Uu Vka, L.V.S. Modelling potentially toxic elements in forest soils with vis–NIR spectra and learning algorithms. *Environ. Pollut.* **2020**, *267*, 115574. [CrossRef]
17. Poliskie, M.; Clevenger, J.O. Fourier transform infrared (FTIR) spectroscopy for coating characterization and failure analysis. *Met. Finish.* **2008**, *106*, 44–47. [CrossRef]
18. Pareja, R.R.; Ibáñez, R.L.; Martín, F.; Ramos-Barrado, J.R.; Leinen, D. Corrosion behaviour of zirconia barrier coatings on galvanized steel. *Surf. Coat. Technol.* **2006**, *200*, 6606–6610. [CrossRef]
19. Caldona, E.B.; Wipf, D.O.; Smith, D.W., Jr. Characterization of a tetrafunctional epoxy-amine coating for corrosion protection of mild steel. *Prog. Org. Coat.* **2021**, *151*, 106045. [CrossRef]
20. Kishigami, S.; Matsumoto, Y.; Ogawa, Y.; Mizokami, Y.; Shiozawa, D.; Sakagami, T.; Hayashi, M.; Arima, N. Quantitative Deterioration Evaluation of Heavy-Duty Anticorrosion Coating by Near-Infrared Spectral Characteristics. *Eng. Proc.* **2021**, *8*, 26.
21. Abdelkarim, O.; Abdellatif, M.H.; Khalil, D.; Bassioni, G. FTIR And Uv In Steel Pipeline Coating Application. *Geomate J.* **2020**, *18*, 130–135. [CrossRef]
22. Raeissi, B.; Bashir, M.A.; Garrett, J.L.; Orlandic, M.; Johansen, T.A.; Skramstad, T.O.R. Detection of different chemical binders in coatings using hyperspectral imaging. *J. Coat. Technol. Res.* **2022**, *19*, 559–574. [CrossRef]
23. Paoletti, M.E.; Haut, J.M.; Plaza, J.; Plaza, A. Deep learning classifiers for hyperspectral imaging: A review. *ISPRS J. Photogramm. Remote Sens.* **2019**, *158*, 279–317. [CrossRef]
24. Dubey, A.K.; Jain, V. Comparative study of convolution neural network's relu and leaky-relu activation functions. In *Applications of Computing, Automation and Wireless Systems in Electrical Engineering*; Springer: Berlin/Heidelberg, Germany, 2019; pp. 873–880.
25. Shore, J.; Johnson, R. Axiomatic derivation of the principle of maximum entropy and the principle of minimum cross-entropy. *IEEE Trans. Inf. Theory* **1980**, *26*, 26–37. [CrossRef]
26. Guerguer, M.; Naamane, S.; Edfouf, Z.; Raccurt, O.; Bouaouine, H. Chemical Degradation and Color Changes of Paint Protective Coatings Used in Solar Glass Mirrors. *Coatings* **2021**, *11*, 476. [CrossRef]
27. González, M.G.; Cabanelas, J.C.; Baselga, J. Applications of FTIR on epoxy resins-identification, monitoring the curing process, phase separation and water uptake. *Infrared Spectrosc. Mater. Sci. Eng. Technol.* **2012**, *2*, 261–284.
28. Deng, F.; Huang, Y.; Azarmi, F. Corrosion behavior evaluation of coated steel using fiber Bragg grating sensors. *Coatings* **2019**, *9*, 55. [CrossRef]

Article

Mechanical, Corrosion, and Wear Properties of TiZrTaNbSn Biomedical High-Entropy Alloys

Xiaohong Wang [1], Tingjun Hu [1], Tengfei Ma [1], Xing Yang [1], Dongdong Zhu [1], Duo Dong [1], Junjian Xiao [1,*] and Xiaohong Yang [2,3,*]

[1] Key Laboratory of Air-Driven Equipment Technology of Zhejiang Province, Quzhou University, Quzhou 324000, China
[2] Academician Expert Workstation, Jinhua Polytechnic, Jinhua 321017, China
[3] Key Laboratory of Crop Harvesting Equipment Technology of Zhejiang Province, Jinhua 321017, China
* Correspondence: xjj919@sohu.com (J.X.); jhyxh656593@163.com (X.Y.); Tel.: +86-139-6703-9758 (J.X.); +86-180-9126-9296 (X.Y.)

Citation: Wang, X.; Hu, T.; Ma, T.; Yang, X.; Zhu, D.; Dong, D.; Xiao, J.; Yang, X. Mechanical, Corrosion, and Wear Properties of TiZrTaNbSn Biomedical High-Entropy Alloys. *Coatings* 2022, *12*, 1795. https://doi.org/10.3390/coatings12121795

Academic Editors: Matic Jovičević-Klug, Patricia Jovičević-Klug and László Tóth

Received: 21 October 2022
Accepted: 21 November 2022
Published: 22 November 2022

Publisher's Note: MDPI stays neutral with regard to jurisdictional claims in published maps and institutional affiliations.

Copyright: © 2022 by the authors. Licensee MDPI, Basel, Switzerland. This article is an open access article distributed under the terms and conditions of the Creative Commons Attribution (CC BY) license (https://creativecommons.org/licenses/by/4.0/).

Abstract: The phase composition, microstructure, mechanical, corrosion, and wear behaviors of the Ti15Zr35Ta10Nb10Sn30 (Sn30) and Ti15Zr30Ta10Nb10Sn35 (Sn35) biomedical high-entropy alloys (BHEAs) were studied. We found that the Ti–Zr–Ta–Nb–Sn BHEAs showed hyper-eutectic and eutectic structures with body-centered cubic (BCC) and face-centered cubic (FCC) solid-solution phases. The Sn30 BHEA exhibited a high Vickers hardness of approximately 501.2 HV, a compressive strength approaching 684.5 MPa, and plastic strain of over 46.6%. Furthermore, the Vickers hardness and compressive strength of Sn35 BHEA are 488.7 HV and 999.2 MPa, respectively, with a large plastic strain of over 49.9%. Moreover, the Sn30 and Sn 35 BHEA friction coefficients are 0.152 and 0.264, respectively. Sn30 BHEA has the smallest and shallowest furrow-groove width, and its wear rate is 0.86 (km/mm^3); at the same time, we observed the delamination phenomenon. Sn35 BHEA has a wear rate value of 0.78 (km/mm^3), and it displays wear debris and the largest–deepest furrow groove. Sn30 BHEA has the highest impedance value, and its corrosion current density I_{corr} is 1.261×10^{-7} (A/cm^2), which is lower than that of Sn35 BHEA (1.265×10^{-6} (A/cm^2)) by 88%, and the passivation current density I_{pass} of Sn30 BHEA and Sn35 BHEA is 4.44×10^{-4} (A/cm^2) and 3.71×10^{-3} (A/cm^2), respectively. Therefore, Sn30 BHEA preferentially produces passive film and has a small corrosion tendency, and its corrosion resistance is considerably better than that of the Sn35 BHEA alloy.

Keywords: biomedical high entropy alloy; TiZrTaNbSn; corrosion resistance; mechanical properties; friction and wear

1. Introduction

Pure Ti has the advantages of non-toxicity, light weight, high strength, good biocompatibility, etc. Therefore, in the 1950s, the United States and the United Kingdom started to apply it for use with living organisms [1]. In the 1960s, Ti alloys (first Ti-6Al-4V [2,3] and later Ti-5Al-2.5Fe and Ti-6Al-7Nb) began to be widely used in clinical practice as a human implant material [4–7]. In the 1970s and 1980s, researchers began to prepare Ti alloys with V-free implants because of its toxic and potentially harmful effects on the human body; furthermore, in the mid-1980s, new types of α+β Ti alloys, namely Ti-5Al-2.5Fe and Ti-6Al-7Nb, were developed in Europe [3]. The mechanical properties of these alloys are similar to Ti-6Al-4V [7], albeit with higher biocompatibility and corrosion resistance properties. However, these alloys still contain Al, which can cause organ damage and harmful symptoms, such as osteomalacia, anemia, and neurotin disorder [4,5]. Based on the above reasons, new types of alloys, which are free of both V and Al but with the addition of Nb, Zr, Ta, and Sn (Ti-13Nb-13Zr, Ti-35Nb-5Ta7Zr, Ti-24Nb-4Zr-7.9Sn), have been developed in recent years [4–6], and their elastic moduli are closer to that of natural human bone, and their strength is also higher than that of pure Ti. Consequently, Ti alloys are being rapidly developed for human implant materials; however, their strength, friction and wear, and corrosion resistance need to be studied further.

Traditional alloy systems usually consist of one or two main elements, and the content of the other elements is much lower. In 2004, Yeh et al. [8–10] first proposed high-entropy alloy, a class of materials containing five or more elements in relatively high concentrations (5–35 at.%) [11,12]. Due to their unique high-entropy, sluggish diffusion, lattice distortion, and cocktail effects [13–15], HEAs show excellent comprehensive properties compared with traditional alloys, such as high hardness [16], high strength [17], corrosion resistance, wear resistance [18,19], etc. At the same time, high-entropy alloys break the traditional alloy design concept of using only one principal element. Despite this, researchers are actively exploring the possibility of applying high-entropy alloys in new products [20], such as biomedical, magnetic, hydrogen storage materials, etc. [21–25]. At present, the comprehensive mechanical properties of medical alloys are still to be improved in clinical practice. Therefore, the design concept of high-entropy alloy can be used to prepare biomedical high-entropy alloy materials with low modulus, high strength, corrosion resistance, and other excellent comprehensive properties to meet demand.

In the past decade, a series of Ti-Zr-Hf-Nb-Ta [21–25], Ti-Mo-Ta-Nb-Zr [26–29], and Ti-Nb-Hf-Ta-Zr-Mo [30,31] HEAs with considerable mechanical and chemical properties suitable for biomedical applications have been designed. Researchers are also improving the performance of HEAs by alloying with O, Si, Al, and Cr [32–35]. However, we noticed that adding Sn to Fe-Co-Cu-Ni(-Mn) HEAs can improve elongation strain and tensile strength by 16.9% and 476.9MPa, respectively [36,37]. In addition, Sn is non-cytotoxic and widely present in β-Ti alloys [38–41]. Previously, we studied the effects of atomic ratios on as-cast microstructural evolution, and the mechanical and electrochemical properties of Ti30Zr20Ta20Nb20Sn10 high-entropy alloy [42]. While its elastic modulus is relatively high, with a value of 110GPa, it does not match the elastic modulus of human bones (30–50 GPa). Moreover, Zr-based $Ti_{0.5}Zr_{1.5}Ta_{0.5}NbSn_{0.2}$ (Ti13.5Zr40.5Ta13.5Nb27Sn5.5) high-entropy alloys display an elastic modulus value of about 40 GPa [43]. Therefore, in our study, we designed a new Zr-based high-entropy Ti-Zr-Ta-Nb-Sn alloy based on metastable β-titanium alloy, which is based on the four elements of Ti-Zr-Ta-Nb. We prepared two kinds of biomedical-grade high-entropy alloys, namely Sn30 BHEA and Sn35 BHEA, by vacuum arc melting, and we systematically studied the feasibility of preparing biomedical-grade high-entropy alloy with five elements (Ti, Zr, Ta, Nb, and Sn), paying specific attention to morphology, compressive strength, electrochemical corrosion, and friction and wear properties. Our work will provide data for the future development of biomedical-grade high-entropy alloy, in addition to guidance for further scientific research on Ti alloys.

2. Experimental

The Ti, Zr, Ta, Nb, and Sn raw materials with a purity of more than 99.9 wt.% were used to prepare BHEAs with the atom ratio of Sn30 BHEA and Sn35 BHEA. The master alloy ingot was arc melted and cooled on a water-cooled copper crucible in a high-purity argon protective atmosphere at least 5 times to ensure the chemical homogeneity. Wire cut electrical discharge machining (WEDM, DHL-500) was used to cut samples from the core region of the master alloy ingot (Buttonhole, maximum diameter Φ32mm, maximum height 16 mm). The surface of the alloy sample was ground with silicon carbide sandpaper up to 2000 grit. The phase composition and microstructure of the alloy sample were characterized via an X-ray diffractometer (XRD, Rigaku D/max-RB) and a Hitachi S-4300 scanning electron microscope and HITACHI SU8010 scanning electron microscope (SEM), and energy dispersive X-ray spectroscopy (EDS) was used to analyze the chemical composition of the HEAs.

Cylinder-shaped (Φ4 mm × 6 mm) samples were cut via WEDM, and used for Vickers hardness and room-temperature compression stress and strain test. The microhardness test was carried out on an HYHVS-1000T Vickers hardness tester using an applied load of

1000 g force (gf) and a dwell time of 15 s. A 370.1 type mechanical testing system (MTS) was employed to record the room-temperature compression stress and strain curves of HEA samples. The compressive strain rate was set as 0.5 mm/min. After the compression test, the lateral surface morphologies of the compressed specimens were examined by SEM.

Corrosion behavior was obtained through the electrochemical workstation (CHI660E) using a three-electrode-cell system. Saturated mercuric chloride was used as reference electrode, platinum electrode as counter electrode, a 1 mm thick sample was embedded in copper wire as working electrode, test temperature was 26 °C, scanning speed was 1 mV/s, and scanning voltage range was from −1.5 to +1.5 V. Before corrosion experiments, the surface of the sample was polished with 2000 grit silicon carbide sandpapers. Then, as-polished specimens were ultrasonically cleaned in deionized water, acetone, and ethanol. Before the potentiodynamic polarization test, the alloy sample was immersed in 3.5% NaCl solution until the open circuit potential (OCP) reached a stable state. After the polarization experiment, TAFEL and IMP were used to measure the polarization curve and AC impedance of Sn30 and Sn35 BHEAs, respectively. Then, Corrview software was used to analyze the Tafel curve and Zview software to analyze the impedance spectrum. The corroded morphologies on the sample surface were examined by SEM, and the composition of the corroded surface was determined by EDS.

When performing the friction and wear test, we ensured each specimen of the two materials was wet-ground and polished using a polishing machine (UNIPOL-1502, Shenyang Kejing Auto-Instrument Co., Ltd., Shenyang, China) with a series of silicon carbide papers of P320, P400, P800, P1200, P1500, P2000, and P3000 grits (Matador Starcke, Germany) under water cooling. Finally, after ultrasonically cleaning for 10 min in deionized water, we fine-polished all specimens with a diamond velour polishing pad under a flowing cerium oxide solution (particle size: 1.5 μm) (Shenyang Kejing Auto-Instrument Co., Ltd., China) before finishing with a mirror-like surface. Then, we tested the wear behaviors of the HEAs by a VHX-2000 tribology tester using a Si_3N_4 ball (4 mm in diameter) as the couple-pair. In our study, the test parameters were as follows: load 10 N, time of 30 min, sliding velocity 600 r/min, and friction reciprocating motion amplitude 2 mm. We recorded the friction coefficient during the sliding process. After the wear test, we determined the wear volume (W_V) of the alloy samples by an MT-500 Probe-type material surface profile measuring instrument. We examined the morphologies and compositions of the HEAs' wear scars by SEM and EDS, respectively.

3. Results

3.1. Phase Composition and Microstructure of Sn30 BHEA and Sn35 BHEA

Figure 1 shows the XRD diffraction patterns of Sn30 BHEA and Sn35 BHEA. We found that there is an obvious peak correspondence relationship by comparing with the standard PDF database. The diffraction peaks of 36.8° and 38.4° correspond to BCC Zr, FCC Nb, and BCC Ti solid solutions, respectively. At the same time, 64.8° corresponds to the FCC Ti solid solution, while 75.9° and 80.2° correspond to the diffraction peaks of the BCC Zr and BCC Ti solid solutions, respectively. We also observed that 41.8° corresponds to the diffraction peak of the BCC Ti solid solution in Sn30 BHEA. Furthermore, the 37.6° (Sn35 BHEA) and 35.9° (Sn30 BHEA) diffraction peaks correspond to the HCP Zr_5Sn_3 phase. Our results show that both alloys contain BCC and FCC phases. Table 1 shows the corresponding thermodynamic parameters. The values of ΔH, ΔS, and δ are in the range where the solid-solution phase is likely to occur. Table 1 shows the calculated values of the entropy–enthalpy quotient parameter (Ω), the valence electron concentration (VEC) criterion, and the mean square deviation of the atomic radius of elements (δ) of these two high-entropy alloys. The thermodynamic parameter VEC further proves that FCC and BCC phases co-exist in the alloy.

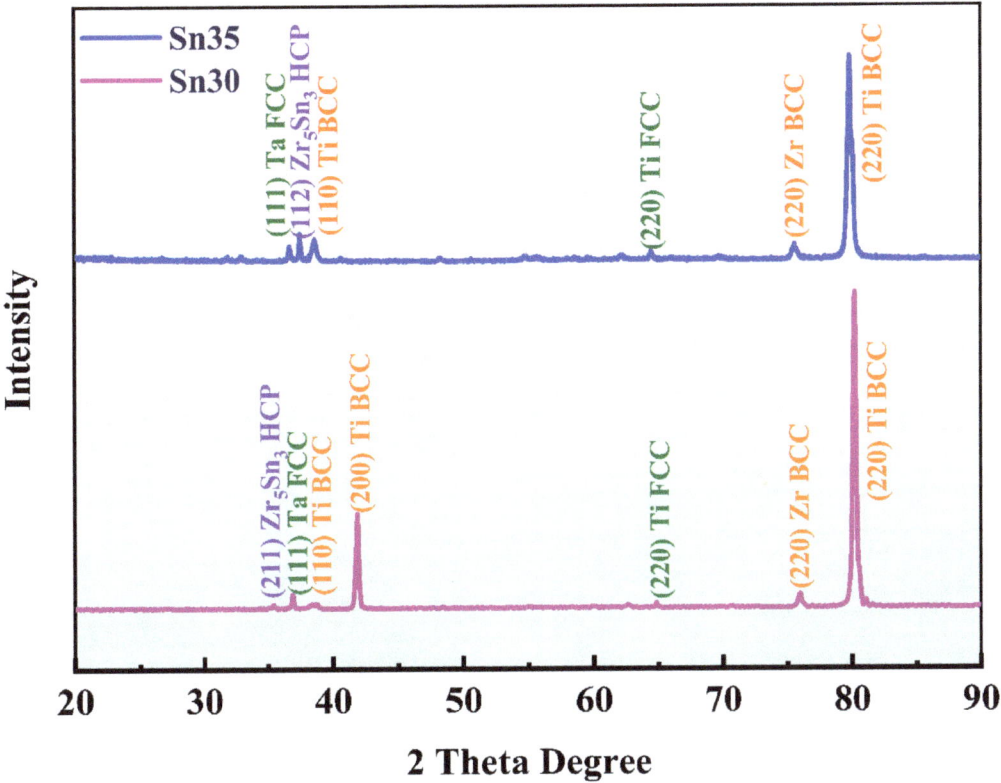

Figure 1. XRD patterns of Sn30 BHEA and Sn35 BHEA.

Table 1. Thermodynamic parameters of Sn30 BHEA and Sn35 BHEA.

Alloy	Ω	δ	ΔH/(KJ/mol)	ΔS/J (K/mol)	VEC
Sn30 BHEA	1.04	3.78	−21.2	12.25	7.2
Sn35 BHEA	0.95	3.64	−22	12.3	7.7

We used SEM and EDS to analyze the chemical compositions and microstructures of HEAs. Figure 2 shows the microstructures of Sn30 BHEA and Sn35 BHEA. Sn30 BHEA is a typical hypoeutectic structure composed of long-strip gray phases, a bright white small-block phase, and an interphase rod eutectic microstructure (Figure 2a,b). Moreover, the eutectic microstructure's volume fraction is 33.8%. Sn35 BHEA is a typical eutectic microstructure with lamellar distribution; however, there are some bright white blocky phases distributed on the dark gray phase. Additionally, the solidification mode looks more equiaxed for Sn35 BHEAs, while it seems more columnar dendritic for Sn30 BHEAs. With changes in atomic ratio, the eutectic structure's content and morphology also considerably changed. In order to analyze the element content of each phase in Figure 2, EDS analysis was carried out on eutectic microstructure (area 1), gray phases (area 2), and bright white blocky phases (area 3). The results are shown in Table 2. We discovered a high content of Zr and Sn in the dark gray phase of both high-entropy alloys. Combined with our XRD analysis results, we deduced that this phase is a Zr solid solution with a BCC crystal structure. The bright-colored small-block area contains a large amount of Ti and Ta, which exceeds the initial composition of the alloy; therefore, it is rich in Ta and Ti FCC phase.

Based on our above analysis, we determined that the bright-colored phase in the remaining eutectic structure is a Ti solid-solution phase with a BCC crystal structure.

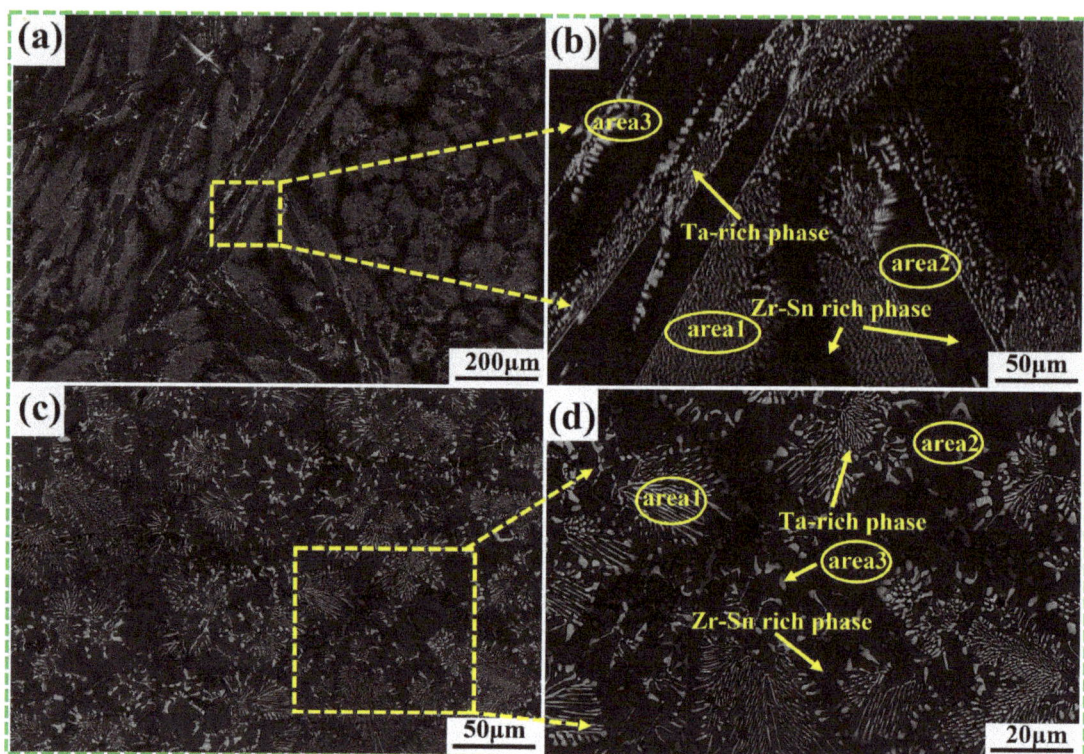

Figure 2. The microstructure of Sn30 BHEA and Sn35 BHEA. (**a,b**) Sn30 BHEA, (**c,d**) Sn35BHEA.

Table 2. EDS analysis of the phase in Figure 2.

Composition	Place	Color	Ti/at%	Zr/at%	Ta/at%	Nb/at	Sn/at%
Sn30 BHEA	area 1	-	19.51	26.56	8.21	16.80	28.93
	area 2	Gray	15.96	39.71	3.75	8.52	32.06
	area 3	Bright white	28.69	19.14	12.07	15.38	24.72
Sn35 BHEA	area 1	-	19.58	27.83	7.82	20.02	24.73
	area 2	Gray	15.69	34.01	0.98	8.83	40.50
	area 3	Bright white	24.55	18.39	14.65	14.55	27.86

To further analyze the element distribution, the map scanning results of its element distribution are shown in Figure 3. Figure 3 shows the map scanning results of the element distribution. The Zr, Nb, and Sn elements are mainly distributed in the dark gray phase, while Ti tends to be uniformly distributed in the dark gray phase, and is eutectic in structure; however, Ti is segregated in the bulk phase close to the eutectic microstructure (Figure 3(a1–a5)). For Sn35 BHEA, the distribution of elements tends to be consistent with that of Sn30 BHEA.

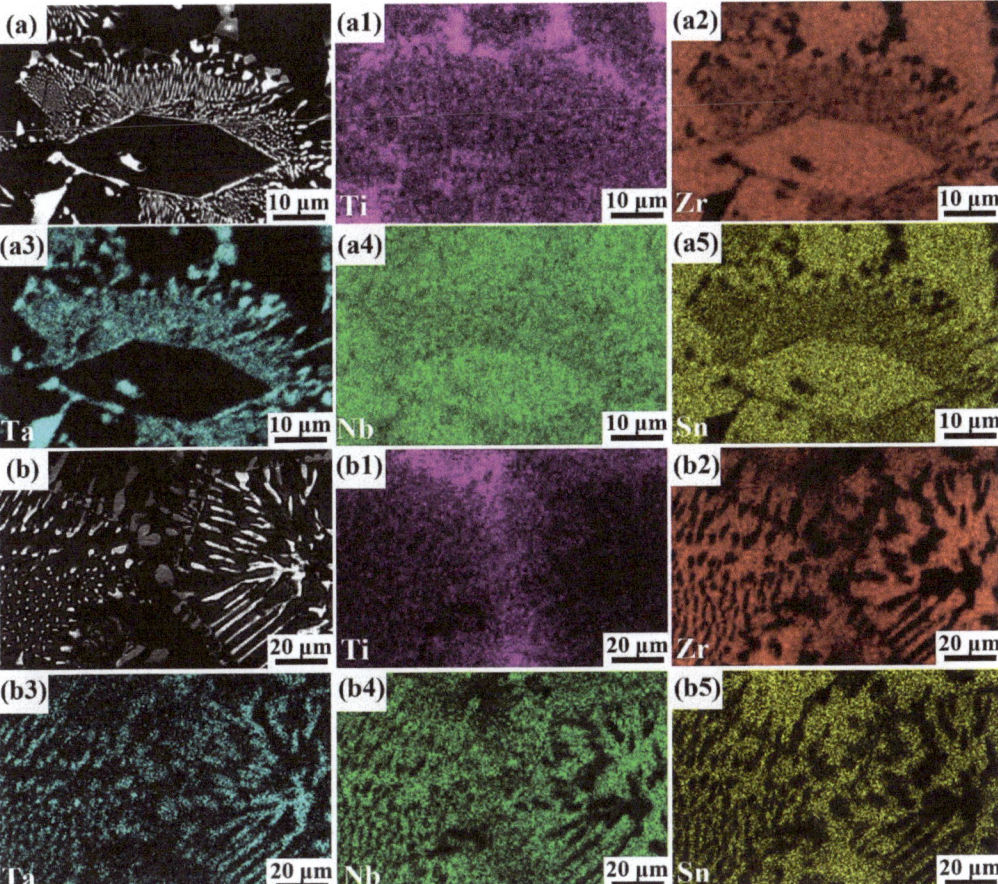

Figure 3. The map scanning of Sn30 BHEA and Sn35 BHEA: (a)–(a5) Sn30 BHEA, (b)–(b5) Sn35 BHEA.

3.2. Mechanical Properties of Sn30 BHEA and Sn35 BHEA

The mechanical properties of Sn30 BHEA and Sn35 BHEA are shown in Figures 4 and 5. Figure 4a shows the compression stress–strain curves of Sn30 BHEA and Sn35 BHEA at room temperature. Compared with Sn35 HEAs, the Sn30 HEAs exhibit double-yielding behavior, which is often observed in shape memory alloys as a stress-induced phase transformation and has relatively lower plasticity [44]. Figure 4b displays the Vickers hardness results for both alloys. Hardness is one of their mechanical properties, and it has a considerable impact on the application of alloys. At the same time, it is also one of the factors that affect alloys' friction and wear properties. Sn30 HEA has the highest hardness level, and the average value of 5 measurements is 501.2HV, while the hardness of alloy Sn35 HEAs is 488.72HV, which is slightly lower. Figure 4c–f show the fracture morphologies of these two high-entropy alloys after compression testing. Figure 4c,d show that Sn30 BHEA's macrofracture morphology is relatively flat, while its micromorphology has river patterns, showing obvious shear failure characteristics. On the gray strip, it also shows typical brittle fracture characteristics. Furthermore, after increasing the magnification, we observed tear edges and dimples on the eutectic structure, showing the mixed characteristics of quasi-cleavage and ductile fractures. Generally, the samples show brittle quasi-cleavage fractures [45]. Figure 4e,f show that the compression fracture surfaces of Sn35 HEAs are uneven and considerably fluctuate, and there are obvious signs of shear tear on the fractures. After increasing the magnification, we observed shallow and slender dimples on the relatively flat shear plane, and we determined that the compression fracture is generally a brittle cleavage fracture. Figure 5 shows the maximum strain and compressive strength values corresponding to the two alloys. The maximum strain value and compressive strength of Sn30 HEAs are 46.6% and 684.5 MPa, respectively, and the maximum strain value and compressive strength of Sn35 HEAs are 49.9% and 999.2 MPa, respectively. Therefore, Sn35 HEAs exhibit better mechanical properties.

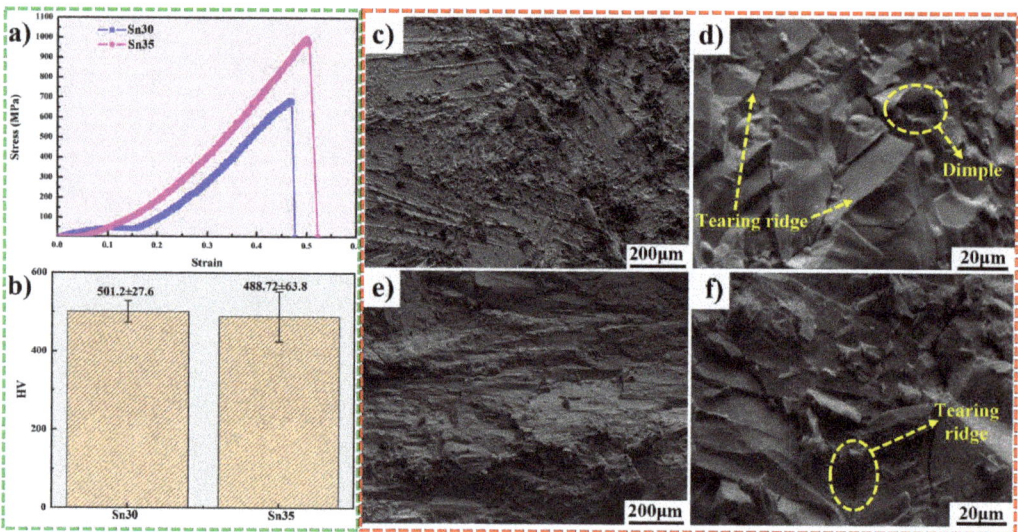

Figure 4. Compressive stress–strain curves and corresponding fracture morphologies of Sn30 BHEA and Sn35 BHEA: (**a**) stress–strain curves, (**b**) HV of Sn30 BHEA and Sn35 BHEA, (**c**,**d**) fracture morphology of Sn30 BHEA, (**e**,**f**) fracture morphology of Sn35 BHEA.

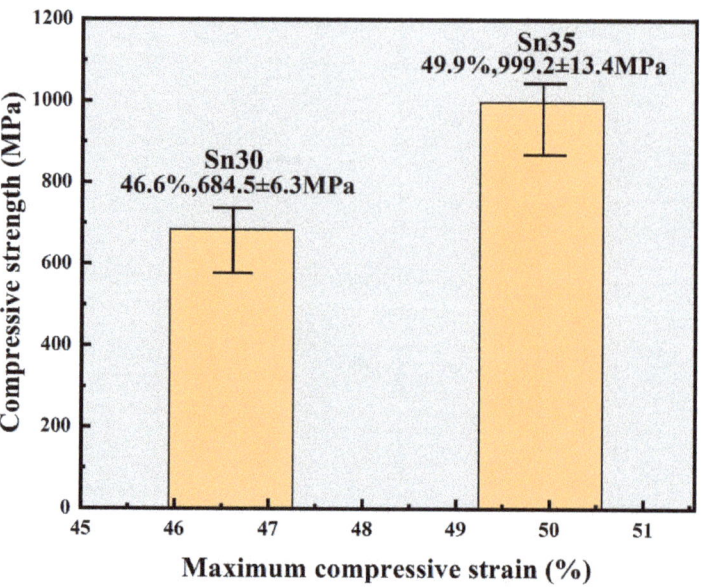

Figure 5. Compression properties of Sn30 BHEA and Sn35 BHEA.

3.3. Friction and Wear Properties of Sn30 BHEA and Sn35 BHEA

As hardness is one of the primary factors that affect the friction and wear properties of the alloys, there are certain requirements regarding the friction and wear properties of materials implanted into the human body. The friction and wear mechanical properties tests of Sn30 BHEA and Sn35 BHEA are shown in Figure 6. Wear marks can be clearly seen in Figure 6a, which shows the sample used in the friction and wear test, including its wear marks and approximate dimensions. Figure 6b displays the friction coefficient curves (COFs). The COFs of Sn30 and Sn35 are 0.152 and 0.264, respectively. Moreover, the friction coefficients first display a sharp increase and then decrease before becoming flat. The main reason is that at the initial stage, the sample and the Si_3N_4 sphere will undergo a point-to-surface contact process. First, the friction coefficient is sensitive to the roughness of the sample surface, causing the friction coefficient to rapidly increase. When the ball makes complete contact with the sample and the contact condition is stable, the friction coefficient tends to be flat. The Sn30 BHEA presented the minimum average COFs. The mechanical properties (hardness, strength, and plasticity) have a considerable effect on tribological properties [46–48]. Normally, the relationship between tribological and mechanical properties can be described as $W = k(P/H)$ (W is wear rate, P is applied load, k is relative to plasticity, and H is hardness) [46]. Figure 4 shows there are no considerable differences in plasticity; however, Sn30 BHEA's hardness is much higher. Hence, the tribological properties of Sn30 BHEA are superior to Sn35 BHEA.

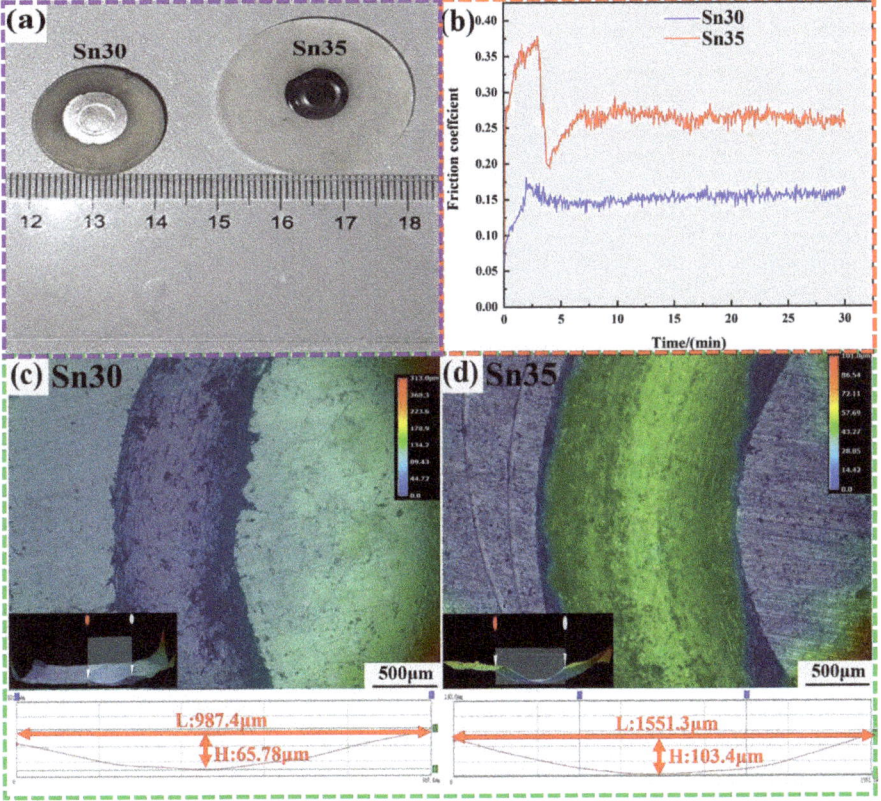

Figure 6. Friction and wear test results of Sn30 BHEA and Sn35 BHEA (**a**) samples, (**b**) friction coefficient curves, and profiles of the worn surfaces for sintered composites and corresponding 2D cross-section profiles of wear tracks: (**c**) Sn30 BHEA, (**d**) Sn35 BHEA.

To further obtain the wear volume (W_v) of the Sn30 BHEA and Sn35 BHEA after the wear test, the section profile morphologies of the worn surfaces were characterized by a Keyence surface profilometer (VHX-2000), and the results are illustrated in Figure 6c,d. Sn30 BHEA showed finer grooves owing to its higher hardness; therefore, it resists abrasive wear (Figure 6c). Abrasive wear was aggravated in Sn35 BHEA, which was confirmed by its widened grooves (Figure 6d). The corresponding wear width and depth in Figure 6a can be obtained directly from the images. Table 3 displays the calculated wear properties, where E is wear resistance, K_v is specific wear rate, and K_2 is linear wear rate. The Kv of Sn30 BHEA is 2.27×10^{-4} (mm^3/nm), the K_2 is 1.163 (mm^3/km), E is 0.86 (km/mm^3), and the W_v is 0.51 mm^3. Furthermore, the Kv of Sn35 BHEA is 2.49×10^{-4} (mm^3/nm), and the values of K_2, E, and W_v are 1.277 (mm^3/km), 0.78 (km/mm^3), and 0.56 mm^3, respectively. A higher material loss rate will also lead to slight work hardening (Figure 7). Therefore, our comprehensive analysis shows that Sn30 BHEA has better friction and wear morphologies.

Table 3. Wear properties of Sn30 BHEA and Sn35 BHEA at room temperature.

Sample	L/m	W_v/(mm^3)	K_v/(mm^3/nm)	K_2/(mm^3/km)	E/(km/mm^3)
Sn30 BHEA	226.2	0.51	2.27×10^{-4}	1.163	0.86
Sn35 BHEA	226.2	0.56	2.49×10^{-4}	1.277	0.78

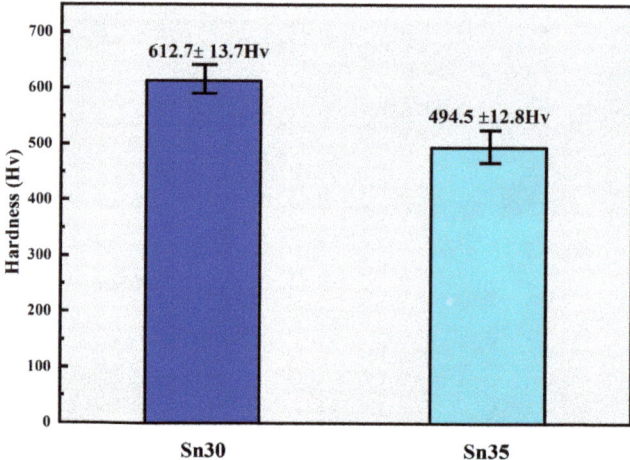

Figure 7. Hardness of the worn surface for the Sn30 BHEA and Sn35 BHEA.

We carried out our SEM characterizations of the worn surfaces to explore the wear mechanisms of Sn30 BHEA and Sn35 BHEA, the results are displayed in Figure 8, and the plastic deformation and delamination traces showed severe wear for both alloys. Furthermore, the wear surfaces of all Sn30 BHEAs and Sn35 BHEAs showed typical furrow characteristics, and there are grooves and wear debris along the sliding direction due to the micro cutting and furrow effects of the Si$_3$N$_4$ ball, which indicates that furrow wear is the main wear mechanism [47,49].

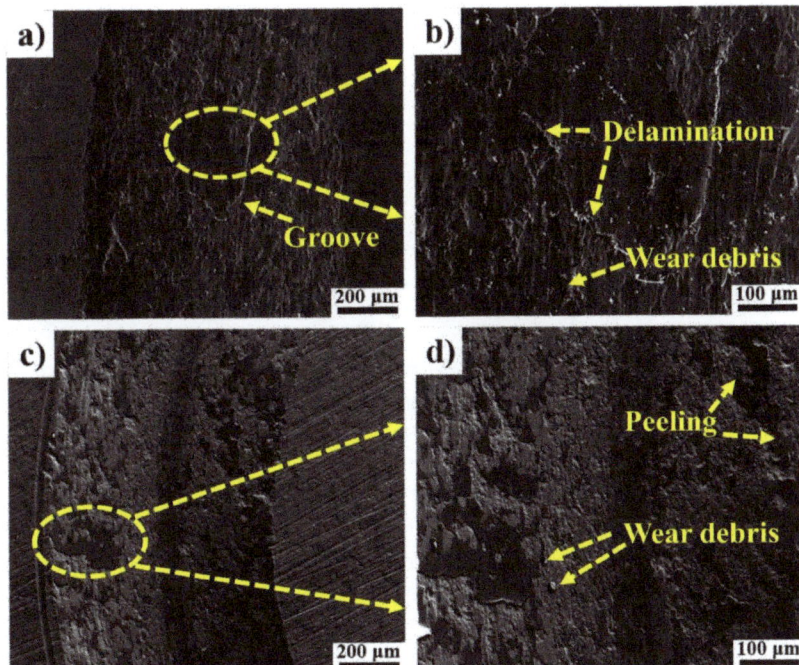

Figure 8. SEM images of Sn30 BHEA and Sn35 BHEA: (**a,b**) Sn30 BHEA, (**c,d**) Sn35 BHEA.

Sn30 BHEA's worn surfaces were smoother with less debris and smaller grooves compared with Sn30 BHEA, as shown in Figure 8a–d. Moreover, Figure 8a shows that the surface morphology of the Sn30 BHEA alloy's friction structure tends to form a regular pit structure, and the wear surface is almost free of wear debris. Figure 8c,d show that there are wear debris generated on Sn35 BHEA's surface morphology, resulting in poor surface quality and cracks. Furthermore, we found severe delamination in Sn35 BHEA. The subsurface generates work hardening owing to its dislocation motion, rearrangement, and grain refinement. However, once deformed, dislocation accumulation occurred to a certain extent, resulting in crack formation and, finally, delamination. The absence of obvious work hardening for Sn30 BHEA (Figure 7) was due to its higher material loss rate. This consecutive deformation can be easily induced through delamination [50].

3.4. Corrosion Property Characterization of Sn30 BHEA and Sn35 BHEA

The corrosion resistance properties of biomedical materials during their implantation into the human body is another aspect that needs to be studied. Figure 9 shows the impedance spectroscopy of Sn30 BHEA and Sn35 BHEA, Figure 9a is the Nyquist diagram, and Figure 9b shows the Bode diagram. The electrochemical AC impedance value can reflect the corrosion resistance of the alloy to a certain extent. The greater the impedance value, the better the corrosion resistance. It can be seen from Figure 9a that the capacitive arc curvature radius of Sn30 BHEA decreases significantly, indicating that among the two high-entropy alloys, Sn30 BHEA has the best corrosion resistance; furthermore, it suggests that the decrease in Sn content and the increase in Zr content will improve the corrosion resistance of Sn30 BHEA and Sn35 BHEA. In the Bode diagram, the phase angle of Sn30 BHEA is close to 75° in a wide frequency range, while the phase angle of Sn35 BHEA is about 60°. Sn30 BHEA has a large resistance value and forms a passive film.

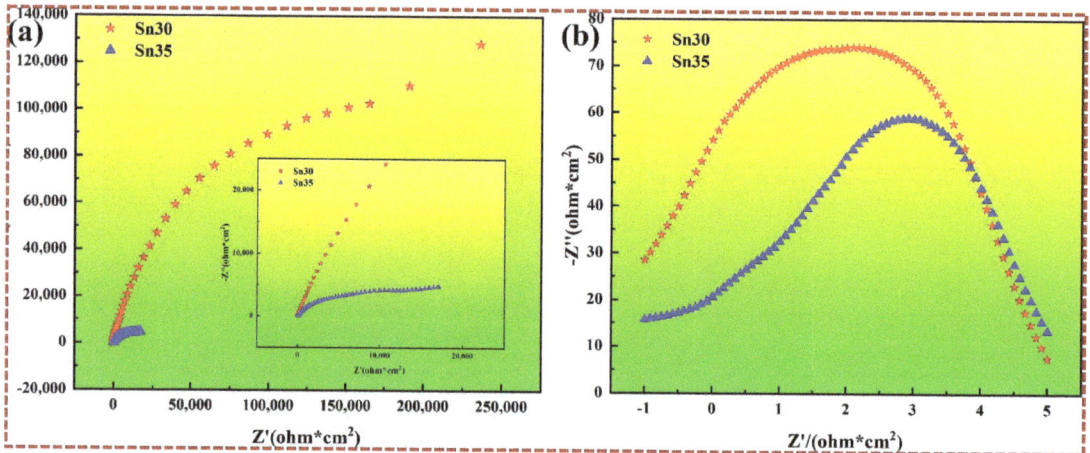

Figure 9. EIS of Sn30 BHEA and Sn35 BHEA: (**a**) Nyquist plots, (**b**) Bode plots.

The potentiodynamic polarization curve and the corrosion morphology of Sn30 BHEA and Sn35 BHEA are shown in Figure 10. The key parameters derived by the Tafel method such as the corrosion current density (I_{corr}) and the corrosion potential (E_{corr}) are listed in Table 4. It was clearly found that Sn30 BHEA exhibited an I_{corr} value smaller than that of Sn35 BHEA. Moreover, for passivation current density I_{pass}, the smaller the value, the easier it is to enter the passivation state. The I_{pass} value of Sn30 BHEA is 4.44×10^{-4} A/cm^{-2}, which is 88% lower than that of Sn35 BHEA—a significant improvement. Another noteworthy fact is that in the anode and cathode area, the value of β_a and β_b of Sn30 BHEA is lower than that of Sn35 BHEA, demonstrating that Sn35 BHEA has stronger corrosion resistance properties. The corrosion potential (E_{corr}) can determine the corrosion trend of the alloy. According to the thermodynamic principle, the smaller the E_{corr}, the greater the corrosion tendency. The E_{corr} of Sn30 HEAs is −0.96, which is about 43.3% smaller than that of Sn35 BHEAs. However, the data that most directly reflect the corrosion resistance of the alloy are the corrosion rate, which can be obtained by the following equation.

$$CR = \frac{0.13 I_{corr} * EW}{d}, \quad (1)$$

where EW is the equivalent weight, and d is the density of the metal (g/cm^3). The calculation results are shown in Table 4. The corrosion rate of Sn30 BHEA is 1.37×10^{-4} mm/a. The corrosion rate of Sn35 BHEA is 1.20×10^{-3} mm/a, which is nearly 88.6% greater than that of ambient pressure. Therefore, Sn30 BHEA has stronger corrosion resistance properties.

Table 4. The key electrochemical parameters of Sn30 BHEA and Sn35 BHEA.

Composition (at.%)	I_{corr} (A/cm^2)	E_{corr} (V)	β_a	β_c	I_{pass} (A/cm^{-2})	Corrosion Rate (mm/a)
Sn30 BHEA	1.261×10^{-7}	−0.96	0.36	0.21	4.44×10^{-4}	1.37×10^{-4}
Sn35 BHEA	1.265×10^{-6}	−0.67	0.76	0.88	3.71×10^{-3}	1.20×10^{-3}

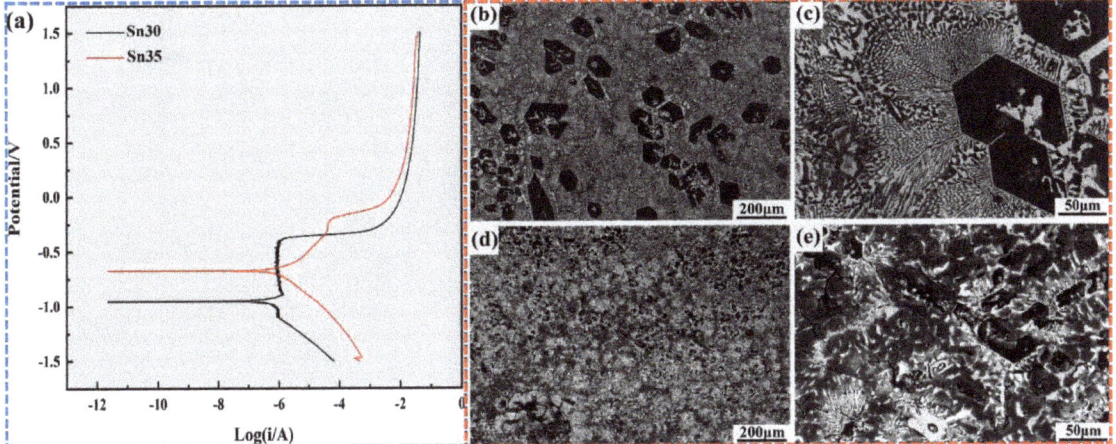

Figure 10. Corrosion properties characterization for Sn30 BHEA and Sn35 BHEA: (**a**) potentiodynamic polarization curves, (**b**,**c**) corrosion morphology of Sn30 BHEA, (**d**,**e**) corrosion morphology of Sn35 BHEA.

The corrosion surface morphology was observed by scanning electron microscopy (SEM) after the potentiodynamic polarization test, as shown in Figure 10b,e. The surfaces of these two high-entropy alloys show certain levels of corrosion after electrochemical corrosion for 2400s and the corrosion degree of (Zr, Sn)-rich phase is high. Meanwhile, the corrosion of Sn35 BHEA also mainly occurs in the gray phase, which is accompanied by many irregular corrosion pits and some cracks. There are also some electrochemical corrosion products on the surface, as shown in Figure 10d,e. The volume fractions of large dendritic Zr-rich zones within Sn30 BHEA and Sn35 BHEA are 77.7% and 90.8% (Figure 11), respectively. Then, from the corrosion morphology and analysis of the above two high-entropy alloys, it can be concluded that the corrosion resistance of Sn30 BHEA is better than that of Sn30 BHEA. However, both alloys exhibit high levels of corrosion resistance [51], indicating that Sn30 BHEA and Sn35 BHEA are a suitable alternative material considering corrosion resistance.

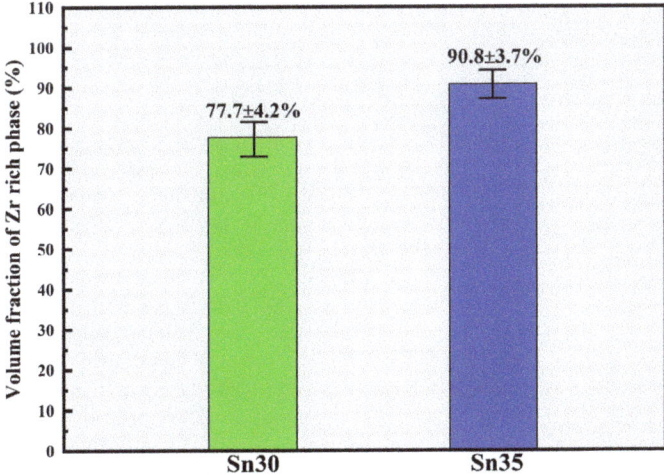

Figure 11. Volume fraction of Zr-rich phase in Sn30 BHEA and Sn35 BHEA.

4. Discussion

According to our SEM results for the surface of the corroded coatings, all of the corroded alloys exhibited similar surface morphologies, i.e., corrosion at the (Zr, Sn)-rich location. The volume fraction of the (Zr, Sn)-rich phase in Sn30 BHEA is smaller than in Sn35 BHEA (Figure 11). Combining the results of our electrochemistry corrosion analysis, we discovered that decreases in the anti-corrosion properties of Sn35 BHEA were due to the increases in the volume fraction of the (Zr, Sn)-rich phase. Moreover, the hardness of Sn30 BHEA is greater than Sn35 BHEA. However, as Sn35 BHEA exhibits a near-eutectic microstructure, it shows excellent plasticity, which, in turn, leads to the best compression strength.

The anti-wear properties are concerned with the alloy's surface quality, hardness, toughness, and loading capability [52–55]. Besides the different anti-wear properties of Sn30 BHEA and Sn35 BHEA, the hardness of each alloy is the key factor related to wear resistance. The hardness of Sn30 BHEA is greater than that of Sn35 BHEA because the volume fraction of the Ta-rich BCC phase is extensively generated (Figure 11). Moreover, during the sliding wear process, the hard Si_3N_4 ball will act similar to a cutter tool and peel off the surface materials, leading to severe plastic deformation and a high wear rate. In addition, the surface's severe wear and damage is accompanied by oxidation under the high-speed wear process. However, for Sn30 BHEA, wear resistance is considerably improved because of the large number of BCC phases generated, which could act as strengthening support sites and carry loads during the wear process. To sum up, Sn30 BHEA shows excellent comprehensive properties. We also conclude that too many Sn elements are not conducive to TiZrTaNbSn HEA design.

5. Conclusions

Taking β-type titanium alloy as the design conception, the comprehensive properties of biomedical high-entropy alloy based on Ti-Zr-Nb-Ta and Sn element were systematically discussed. The main conclusions are as follows:

1. Sn30 BHEA and Sn35 BHEA are typical hypo-eutectic and eutectic structures, respectively, and both alloys are composed of BCC and FCC phases;
2. the two high-entropy alloys have brittle fractures at room temperature. The maximum strain value and compressive strength of Sn30 BHEA are 46.6% and 684.5 MPa, respectively, while the maximum strain value and compressive strength of Sn35 BHEA are 49.9% and 999.2 MPa, respectively. Sn30 HEA's strength and yield strength are better than those of Sn35 HEA;
3. the friction coefficient of Sn30 BHEA is 0.152, the specific and linear wear rates (K_v and K_2, respectively) are 2.27×10^{-4} (mm^3/nm) and 1.163 (mm^3/km), respectively, while the width of the furrow groove is the smallest and shallowest, with almost no wear debris. Furthermore, the friction coefficient of Sn35 BHEA is 0.264, and the values of specific and linear wear (K_v and K_2, respectively), in addition to wear resistance E, are 2.49×10^{-4} (mm^3/nm), 1.277 (mm^3/km), and 0.78 (km/mm^3), respectively. Furthermore, the width of the furrow groove is the largest and deepest, and there are wear debris. In conclusion, the Sn30 HEA has excellent wear resistance and rates compared with Sn35 HEA;
4. Sn30 BHEA has the highest impedance value. The corrosion current density I_{corr} is 1.261×10^{-7} (A/cm^2), which is lower than that of Sn35 BHEA by about 88%. The capacitive arc curvature radius of Sn30 BHEA also considerably decreases. Therefore, Sn30 HEAs preferentially produce passivated film with a small corrosion tendency, indicating that its corrosion resistance is considerably better than that of Sn35 BHEA alloy.

Author Contributions: X.W.: Conceptualization, Investigation, Formal analysis, Writing—original draft. T.H.: Investigation, Writing—original draft. T.M.: Conceptualization. X.Y. (Xing Yang): Investigation. D.Z.: Conceptualization. D.D.: Software. J.X.: Investigation, Methodology, Writing—review and editing. X.Y. (Xiaohong Yang): Conceptualization, Formal analysis, Supervision, Funding acquisition, Writing—review and editing. All authors have read and agreed to the published version of the manuscript.

Funding: This work was supported by the Zhejiang Province Natural Science Foundation of China (No.: LQ20E010003), the National Natural Science Foundation of China (Nos. 52071188, 52171120, 52001262, 52071165), and Research Funding Project of Education Department of Zhejiang Province (No. Y202045618).

Institutional Review Board Statement: Not applicable.

Informed Consent Statement: Not applicable.

Data Availability Statement: The authors confirm that the data supporting the findings of this study are available within the article.

Conflicts of Interest: The authors declare no conflict of interest.

References

1. Chen, Q.; Thouas, G.A. Metallic implant biomaterials. *Mater. Sci. Eng. R Rep.* **2015**, *87*, 1–57. [CrossRef]
2. Niinomi, M. Recent Metallic Materials for Biomedical Applications. *Metall. Mater. Trans. A* **2002**, *33*, 477–486. [CrossRef]
3. Long, M.; Rack, H.J. Titanium alloys in total joint replacement—A materials science perspective. *Biomaterials* **1998**, *19*, 1621–1636. [CrossRef]
4. Okuno, O.; Takayuki, J.; Hamavaka, H. Titanium Alloys in Dental and Medical Field. *J. Lron Steel Inst. Jpn.* **1990**, *76*, 1633–1641. [CrossRef]
5. Bordji, K.; Jouzeau, J.Y.; Mainard, D.; Payan, E.; Netter, P.; Rie, K.T.; Stucky, T.; Hage-Ali, M. Cytocompatibility of Ti-6Al-4V and Ti-5Al-2.5Fe alloys according to three surface treatments, using human fibroblasts and osteoblasts. *Biomaterials* **1996**, *17*, 929–940. [CrossRef]
6. Liu, W.; Liu, S.; Wang, L. Surface Modification of Biomedical Titanium Alloy: Micromorphology, Microstructure Evolution and Biomedical Applications. *Coatings* **2019**, *9*, 249. [CrossRef]
7. Gnedenkov, S.V.; Sinebryukhov, S.L.; Puz', A.V.; Egorkin, V.S.; Kostiv, R.E. In vivo study of osteogenerating properties of calcium-phosphate coating on titanium alloy Ti–6Al–4V. *Bio-Med. Mater. Eng.* **2017**, *27*, 551–560. [CrossRef]
8. Chen, T.K.; Shun, T.T.; Yeh, J.W.; Wong, M.S. Nanostructured nitride films of multi-element high-entropy alloys by reactive DC sputtering. *Surf. Coat. Technol.* **2004**, *188*, 193–200. [CrossRef]
9. Yeh, J.W.; Lin, S.J.; Chin, T.S.; Gan, J.Y.; Chin, S.K.; Shun, T.T.; Tsau, C.H.; Chou, S.Y. Formation of simple crystal structures in Cu-Co-Ni-Cr-Al-Fe-Ti-V alloys with multiprincipal metallic elements. *Metall. Mater. Trans. A* **2004**, *35*, 2533–2536. [CrossRef]
10. Yeh, J.W.; Chen, S.K.; Lin, S.J.; Gan, J.Y.; Chin, S.Y.; Shun, T.T.; Tsau, C.H.; Chang, S.Y. Nanostructured high-entropy al-loys with multiple principal elements: Novel alloy design concepts and outcomes. *Adv. Eng. Mater.* **2004**, *6*, 299. [CrossRef]
11. George, E.P.; Raabe, D.; Ritchie, R.O. High-entropy alloys. *Nat. Rev. Mater.* **2019**, *41578*, 515–534. [CrossRef]
12. Akrami, S.; Edalati, P.; Fuji, M.; Edalati, K. High-entropy ceramics: Review of principles, production and applications. *Mater. Sci. Eng. R Rep.* **2021**, *146*, 100644. [CrossRef]
13. Zhang, Z.Q.; Ketov, S.V.; Fellner, S.; Sheng, H.P.; Mitterer, C.; Song, K.K.; Gammer, C.; Eckert, J. Reactive interdiffusion of an Al film and a CoCrFeNi high-entropy alloy. *Mater. Des.* **2022**, *216*, 110530. [CrossRef]
14. Singh, S.K.; Parashar, A. Atomistic simulations to study crack tip behaviour in multi-elemental alloys. *Eng. Fract. Mech.* **2021**, *243*, 107536. [CrossRef]
15. Chang, C.C.; Hsiao, Y.T.; Chen, Y.L.; Tsai, C.Y.; Lee, Y.J.; Ko, P.H.; Chang, S.Y. Lattice distortion or cocktail effect dominates the performance of Tantalum-based high-entropy nitride coatings. *Appl. Surf. Sci.* **2022**, *577*, 151894. [CrossRef]
16. Chen, C.; Yuan, S.; Chen, J.; Wang, W.; Zhang, W.; Wei, R.; Wang, T.; Zhang, T.; Guan, S.; Li, F. A Co-free Cr-Fe-Ni-Al-Si high-entropy alloy with outstanding corrosion resistance and high hardness fabricated by laser surface melting. *Mater. Lett.* **2022**, *314*, 131882. [CrossRef]
17. Jiao, W.; Li, T.; Chang, X.; Lu, Y.; Yin, G.; Cao, Z.; Li, T. A novel Co-free $Al_{0.75}$CrFeNi eutectic high-entropy alloy with superior mechanical properties. *J. Alloys Compd.* **2022**, *902*, 163814. [CrossRef]
18. Cui, P.; Bao, Z.; Liu, Y.; Zhou, F.; Lai, Z.; Zhou, Y.; Zhu, J. Corrosion behavior and mechanism of dual phase $Fe_{1.125}Ni_{1.06}$CrAl high-entropy alloy. *Corros. Sci.* **2022**, *201*, 110276. [CrossRef]
19. Kong, D.; Wang, W.; Zhang, T.; Guo, J. Effect of superheating on microstructure and wear resistance of $Al_{1.8}$CrCuFeNi$_2$ high-entropy alloy. *Mater. Lett.* **2022**, *311*, 131613. [CrossRef]
20. Miracle, D.B.; Senkov, O.N. A critical review of high entropy alloys and related concepts. *Acta Mater.* **2017**, *122*, 448. [CrossRef]

21. Yuan, Y.; Wu, Y.; Yang, Z.; Liang, X.; Lei, Z.; Huang, H.; Wang, H.; Liu, X.; An, K.; Wu, W.; et al. Formation, structure and properties of biocompatible TiZrHfNbTa high-entropy alloys. *Mater. Res. Lett.* **2019**, *7*, 225–231. [CrossRef]
22. Senkov, O.N.; Scott, J.M.; Senkova, S.V.; Miracle, D.B.; Woodward, C.F. Microstructure and room temperature properties of a high-entropy TaNbHfZrTi alloy. *J. Alloys Compd.* **2011**, *509*, 6043–6048. [CrossRef]
23. Dirras, G.; Lilensten, L.; Djemia, P.; Laurent-Brocq, M.; Tingaud, D.; Couzinie, J.P.; Perrie're, L.; Chauveau, T.; Guillot, I. Elastic and plastic properties of as-cast equimolar TiHfZrTaNb high-entropy alloy. *Mater. Sci. Eng. A* **2016**, *654*, 30–38. [CrossRef]
24. Motallebzadeh, A.; Peighambardoust, N.S.; Sheikh, S.; Murakami, H.; Guo, S. Demircan Canadinc Microstructural, mechanical and electrochemical characterization of TiZrTaHfNb and $Ti_{1.5}ZrTa_{0.5}Hf_{0.5}Nb_{0.5}$ refractory high-entropy alloys for biomedical applications. *Intermetallics* **2019**, *113*, 106572. [CrossRef]
25. Yang, W.; Liu, Y.; Pang, S.; Liaw, P.K.; Zhang, T. Bio-corrosion behavior and in vitro biocompatibility of equimolar TiZrHfNbTa high-entropy alloy. *Intermetallics* **2020**, *124*, 106845. [CrossRef]
26. Wang, S.; Xu, J. TiZrNbTaMo high-entropy alloy designed for orthopedic implants: As-cast microstructure and mechanical properties. *Mater. Sci. Eng. C* **2017**, *73*, 80–89. [CrossRef] [PubMed]
27. Mitsuharu, T.; Takeshi, N.; Takao, H.; Aira, M.; Aiko, S.; Takayoshi, N. Novel TiNbTaZrMo high-entropy alloys for metallic biomaterials. *Scr. Mater.* **2017**, *129*, 65–68.
28. Hua, N.; Wang, W.; Wang, Q.; Ye, Y.; Lin, S.; Zhang, L.; Guo, Q.; Brechtl, J.; Liaw, P.K. Mechanical, corrosion, and wear properties of biomedical Ti-Zr-Nb-Ta-Mo high entropy alloys. *J. Alloys Compd.* **2021**, *861*, 157997. [CrossRef]
29. Takao, H.; Takeshi, N.; Mitsuharu, T.; Aira, M.; Takayoshi, N. Development of non-equiatomic Ti-Nb-Ta-Zr-Mo high-entropy alloys for metallic biomaterials. *Scr. Mater.* **2019**, *172*, 83–87.
30. Juan, C.C.; Tsai, M.H.; Tsai, C.W.; Lin, C.M.; Wang, W.R.; Yang, C.C.; Chen, S.K.; Lin, S.J.; Yeh, J.W. Enhanced mechanical properties of HfMoTaTiZr and HfMoNbTaTiZr refractory high-entropy alloys. *Intermetallics* **2015**, *62*, 76. [CrossRef]
31. Yuuka, L.; Takeshi, N.; Aira, M.; Pan, W.; Kei, A.; Takanoshi, N. Design and development of Ti-Zr-Hf-Nb-Ta-Mo high-entropy alloys for metallic biomaterials. *Mater. Des.* **2021**, *202*, 109548.
32. Zhang, Y.; Zuo, T.; Tang, Z.; Gao, M.C.; Dahmen, K.A.; Liaw, P.K.; Lu, Z.P. Microstructures and properties of high-entropy alloys. *Prog. Mater. Sci. Forum* **2014**, *61*, 1–93. [CrossRef]
33. Lei, Z.; Liu, X.; Wu, Y.; Wang, H.; Jiang, S.; Wang, S.; Hui, X.; Wu, Y.; Gault, B.; Kontis, P.; et al. Enhanced strength and ductility in a high-entropy alloy via ordered oxygen complexes. *Nature* **2018**, *563*, 546–550. [CrossRef] [PubMed]
34. Mariana, C.; Jithin, V.; Pramote, T.; Mihaela, P.M.; Maria, K.; Geetha, M.; Annett, G. Tailoring biocompatible Ti-Zr-Nb-Hf-Si metallic glasses based on high-entropy alloys design approach. *Mater. Sci. Eng. C* **2021**, *121*, 111733.
35. Huang, L.; Long, M.; Liu, W.; Li, S. Effects of Cr on microstructure, mechanical properties and hydrogen desorption behaviors of ZrTiNbMoCr high entropy alloys. *Mater. Lett.* **2021**, *293*, 129718. [CrossRef]
36. Liu, L.; Zhu, J.B.; Zhang, C.; Li, J.C.; Jiang, Q. Microstructure and the properties of FeCoCuNiSnx high entropy alloys. *Mater. Sci. Eng. A* **2012**, *548*, 64–68. [CrossRef]
37. Liu, L.; Zhu, J.B.; Li, L.; Li, J.C.; Jiang, Q. Microstructure and tensile properties of FeMnNiCuCoSnx high entropy alloys. *Mater. Des.* **2013**, *44*, 223. [CrossRef]
38. Gutierrez-Moreno, J.J.; Guo, Y.; Georgarakis, K.; Yavari, A.R.; Evangelakis, G.A.; Lekka, C.E. The role of Sn doping in the ß-type Ti-25at%Nb alloys: Experiment and ab initio calculations. *J. Alloys Compd.* **2014**, *615*, 676–679. [CrossRef]
39. Miura, K.; Yamada, N.; Hanada, S.; Jung, T.; Itoi, E. The bone tissue compatibility of a new Ti-Nb-Sn alloy with a low Young's modulus. *Acta Biomater.* **2011**, *7*, 2320–2326. [CrossRef]
40. Moraes, P.E.L.; Contieri, R.J.; Lopes, E.S.N.; Robin, A.; Caram, R. Effects of Sn addition on the microstructure, mechanical properties and corrosion behavior of Ti-Nb-Sn alloys. *Mater. Charact.* **2014**, *96*, 273–281. [CrossRef]
41. Zheng, Y.F.; Wang, B.L.; Wang, J.G.; Li, C.; Zhao, L.C. Corrosion behaviour of Ti-Nb-Sn shape memory alloys in different simulated body solutions. *Mater. Sci. Eng. A* **2006**, *438*, 891–895. [CrossRef]
42. Ma, T.; Li, Q.; Jin, Y.; Zhao, X.; Wang, X.; Dong, D.; Zhu, D.D. As-cast microstructure and properties of none-qual atomic ratio TiZrTaNbSn high entropy alloys. *China Foundry* **2022**, 1–8.
43. Han, L.; Mu, J.; Zhou, Y.; Zhu, Z.; Zhang, H. Effect of Heat Treatment Temperature on Microstructure and Mechanical Properties of $Ti_{0.5}Zr_{1.5}NbTa_{0.5}Sn_{0.2}$ High-Entropy Alloy. *Acta Met. Sin* **2022**, *582*, 1159–1168.
44. Jung, Y.; Lee, K.; Hong, S.J.; Lee, J.K.; Han, J.; Kim, K.B.; Liaw, P.K.; Lee, C.O.L.; Song, G. Investigation of phase-transformation path in TiZrHf(VNbTa)x refractory high-entropy alloys and its effect on mechanical property. *J. Alloys Compd.* **2021**, *886*, 161187. [CrossRef]
45. Gurel, S.; Yagci, M.B.; Canadinc, D.; Gerstein, G.; Bal, B.; Maier, H.J. Fracture behavior of novel biomedical Ti-based high entropy alloys under impact loading. *Mater. Sci. Eng. A* **2021**, *1*, 803. [CrossRef]
46. Cheng, Y.; Yang, J.; Zhang, X.; Zhong, H.; Ma, J.; Li, F.; Fu, L.; Bi, Q.; Li, J.; Liu, W. High temperature tribological behavior of a Ti-46Al-2Cr-2Nb intermetallics. *Intermetallics* **2012**, *31*, 120–126. [CrossRef]
47. García-Martínez, E.; Miguel, V.; Martínez, A.; Naranjo, J.A.; Coello, J. Tribological characterization of tribosystem Ti48Al2Cr2Nb-coated/uncoated carbide tools at different temperatures. *Wear* **2021**, *484–485*, 203992. [CrossRef]
48. Mengis, L.; Grimme, C.; Galetz, M.C. Tribological properties of the uncoated and aluminized Ti–48Al–2Cr–2Nb TiAl alloy at high temperatures. *Wear* **2021**, *477*, 203818. [CrossRef]

49. Shi, X.; Yao, J.; Xu, Z.; Zhai, W.; Song, S.; Wang, M.; Zhang, Q. Tribological performance of TiAl matrix self-lubricating composites containing Ag, Ti_3SiC_2 and BaF_2/CaF_2 tested from room temperature to 600 °C. *Mater. Des.* **2014**, *53*, 620–633. [CrossRef]
50. Wang, Z.; Yan, Y.; Wu, Y.; Su, Y.; Qiao, L. Repassivation and dry sliding wear behavior of equiatomic medium entropy TiZr (Hf, Ta, Nb) alloys. *Mater. Lett.* **2022**, *312*, 131643. [CrossRef]
51. Tang, Z.; Huang, L.; He, W.; Liaw, P. Alloying and processing effects on the Aqueous corrosion behavior of high entropy alloys. *Entropy* **2014**, *16*, 895–911. [CrossRef]
52. Allahyarzadeh, M.H.; Aliofkhazraei, M.; Rouhaghdam, A.R.S.; Torabinejad, V. Structure and wettability of pulsed electrodeposited Ni-W-Cu-(α-alumina) nanocomposite. *Surf. Coat. Technol.* **2016**, *307*, 525–533. [CrossRef]
53. Allahyarzadeh, M.H.; Aliofkhazraeia, M.; Rouhaghdam, A.S.; Alimadadi, H.; Torabinejad, V. Mechanical properties and load bearing capability of nanocrystalline nickel-tungsten multilayered coatings. *Surf. Coat. Technol.* **2020**, *386*, 125472. [CrossRef]
54. Rupert, T.J.; Schuh, C.A. Sliding wear of nanocrystalline Ni–W: Structural evolution and the apparent breakdown of archard scaling. *Acta Mater.* **2010**, *58*, 4137–4148. [CrossRef]
55. Maharana, H.S.; Mondal, K. Manifestation of Hall–Petch breakdown in nanocrystalline electrodeposited $Ni-MoS_2$ coating and its structure dependent wear resistance behavior. *Surf. Coat. Technol.* **2021**, *410*, 126950. [CrossRef]

Article

High Temperature Oxidation of Enamel Coated Low-Alloyed Steel 16Mo3 in Water Vapor

Germain Boissonnet [1], Ewa Rzad [2], Romain Troncy [1], Tomasz Dudziak [2] and Fernando Pedraza [1,*]

[1] Laboratoire des Sciences de l'Ingénieur pour l'Environnement, Université de La Rochelle LaSIE, UMR-CNRS 7356, Avenue Michel Crépeau, CEDEX 1, 17042 La Rochelle, France
[2] Centre for Corrosion Studies, Łukasiewicz Research Network—Krakow Institute of Technology, 73 Zakopiańska St., 30-418 Krakow, Poland
* Correspondence: fpedraza@univ-lr.fr

Abstract: New types of ceramic coatings based on SiO_2-Na_2O-B_2O_3-TiO_2 oxide phases were investigated as protection for boiler steel in power generation systems. Low-alloyed Cr-Mo 16Mo3 steel was coated with different compositions of enamel coatings to assess the protective potential of these coatings under water vapor at high temperatures. Oxidation at 650 °C for 50 h in Ar + water vapor was performed in a TGA apparatus to investigate the oxidation kinetics. The results indicate that the ceramic coatings provided a high degree of protection for the steel exposed to such conditions compared to the uncoated 16Mo3 steel. Furthermore, despite the formation of cracks in the coatings, no spallation from the steel surface was observed. Interconnected porosity in the coatings is suspected to provoke interfacial degradation.

Keywords: ceramic coatings; 16Mo3 steel; high temperature oxidation; TGA; water vapor

Citation: Boissonnet, G.; Rzad, E.; Troncy, R.; Dudziak, T.; Pedraza, F. High Temperature Oxidation of Enamel Coated Low-Alloyed Steel 16Mo3 in Water Vapor. *Coatings* 2023, 13, 342. https://doi.org/10.3390/coatings13020342

Academic Editors: Matic Jovičević-Klug, Patricia Jovičević-Klug and László Tóth

Received: 28 December 2022
Revised: 29 January 2023
Accepted: 30 January 2023
Published: 2 February 2023

Copyright: © 2023 by the authors. Licensee MDPI, Basel, Switzerland. This article is an open access article distributed under the terms and conditions of the Creative Commons Attribution (CC BY) license (https://creativecommons.org/licenses/by/4.0/).

1. Introduction

Energy consumption increases yearly due to the high demand for electricity worldwide; therefore, new systems for the high-temperature protection of structural steels employed in the energy sector must be developed. However, the operating conditions in different plants are particularly aggressive, and different types of corrosion at high temperatures occur. For instance, in most thermal plants, molten sulfate and chloride derivative salts appear that markedly attack the low Cr-containing steel grades, as demonstrated by Abu-warda et al. [1]. Similarly, an extensive sulfidation attack has been recently reported in the typical 16Mo3 boiler steel [2]. Since the use of noble materials is not economically interesting, many various coatings have been proposed in the open literature whether to fight against steam or fireside corrosion. Amongst the former, Al and Al/Si slurry diffusion coatings were demonstrated to withstand long exposures of 100% steam [3] even at high pressures and long exposures [4]. However, such diffusion coatings are relatively brittle and tensile cracks may appear under high pressures allowing steam to penetrate into the substrate material. In addition, the potential interdiffusion of the substrate and coating elements may lower their use for extended periods of time.

The fight against fireside corrosion (and erosion) is mostly conducted through various derivate techniques of thermal spray to produce overlay coatings on the different metal alloys (steels and nickel-based alloys) as reported in the comprehensive reviews of Dhand et al. [5] and of Kumara and collab. [6]. In low-alloy steels, such as 16Mo3, T21, T22 and the alike, various studies have focused on different coating alternatives. For instance, Galetz et al. focused on the use of cladding, self-fluxing, flame spray, and high velocity oxy-fuel (HVOF) to improve the resistance against molten salts of various low-alloy steels [7]. It was concluded that the introduction of Mo and Si in the coatings was beneficial, yet the most protective coatings were those with the lowest porosity and the thickest ones, i.e., the overlay welded alloys and the self-fluxing spray coatings were the most promising despite their greater cost against the flame-sprayed or HVOF coatings. Jafari and collab. [8] employed high-velocity air–fuel (HVAF) on 16Mo3 boiler steel against KCl-induced hot corrosion in air. When compared with more noble materials (AISI 304 and Sanicro25), the authors found a better performance of the Ni-Al coatings than the Ni-Cr-based ones because the alumina scale grown in the former was denser than the chromia one formed in the latter. This impeded the diffusion of the aggressive chlorine species. Indeed, the chlorides have been reported to induce a great attack on the chromia, also forming NiCr coatings when produced by HVOF [1].

In essence, such thermal spray coating systems show some degree of porosity, depending on the technology used, where ashes can accumulate and initiate accelerated degradation processes of boiler tubes [9,10]. To overcome the problem, new systems based on ceramic compounds are being developed for the high-temperature protection of boiler steels. One of the very first works related to ceramic coatings for steels to be used at a high-temperature regime was presented by Harrison et al. [11]. The coatings were designed for use at temperatures as high as 670 °C while displaying outstanding properties such as high resistance to chipping under repeated thermal shock and protection of the metal against oxidation during prolonged exposures. These 70-µm thick coatings were prepared using a mixture of a special grade of calcined aluminum oxide with a conventional type of ground-coat frit in water that is applied to the steel surface before drying and firing according to well-known methods in the ceramic industry. The development of ceramic coatings for high-temperature applications accelerated since the '70s due to the development of deposition techniques [12]. In 1980, ceramic coatings were applied to adiabatic engines [13]. First, ceramic coatings were employed in gas turbine blades and then in pistons, cylinder linings, valves, and piston crown surfaces [14]. For such applications, ceramic coatings are mainly used for the protection of the base alloys against hot corrosion, oxidation, and wear degradation. One of the most advanced ceramic coatings produced recently are the coatings reducing the based metal temperature (Ni-based superalloys), known as the Thermal Barrier Coatings (TBCs) [15]. Those types of coatings are used in aero-jet engines, as well as in gas turbine technologies [16,17]. The state-of-the-art TBCs are generally based on yttria-stabilized zirconia (YSZ) deposited by Electron Beam Physical Vapor Deposition (EB-PVD) or APS process [18,19]. As a decent alternative, mullite can replace zirconia as a TBC for high-temperature gradient fields [20]. In the energy sector, where temperatures are much lower and the surface area is much larger than in aero jet engines, the gas atmosphere is incomparably worse, and hence, different and cheaper solutions are required. Generally, carbides like silicon carbide (SiC) and tungsten carbide (WC) are used as dispersoids when the hardness and wear resistance of the coatings on boiler tubes against fireside corrosion are the major requirements. Nevertheless, oxides such as TiO_2, SiO_2, Al_2O_3, etc., are used when resistance to hot corrosion and oxidation at high temperatures is required [21]. Therefore, to get a better understanding of advanced ceramic materials for further development in particular engineering applications, e.g., the energy sector and boiler protection, extensive research is essential for evaluating the microstructural and corrosion resistance properties of such coatings. Therefore, this research investigates the oxidation resistance at high temperatures and in the presence of water vapor of new ceramic enamel coatings based on SiO_2-Na_2O-B_2O_3-TiO_2 oxide phases that are applied on low-alloyed 16Mo3 steel.

2. Experimental Method
2.1. Materials and Coatings

The 16Mo3 steel (0.12 C, 0.4 Mn, <0.35 Si, 0.025 S, 0.025 P, 0.3 Cr, 0.3 Mo, wt%, bal. Fe) samples of 7–9 × 12 × 4–5 mm^3 were ground using SiC P600 grid paper prior to coating application. Then, the enamel ceramic coatings were applied in a two-step process that is described elsewhere [2]. The different oxide compositions of the enamel coatings of the study are described in Table 1. Their purity was higher than 99.9%, according to the supplier (Sigma-Aldrich). However, for clarity purposes, the samples will be referenced as samples A to G in this work.

Table 1. Enamel-producing recipes (wt.%) for ceramic coatings development used in this work.

Oxide	Coating Symbol						
	A	B	C	D	E	F	G
SiO_2	54.13	51.72	56.80	55.76	63.17	45.11	39.79
Al_2O_3	-	-	2.89	-	2.22	-	-
B_2O_3	18.10	15.79	7.22	2.02	8.32	15.08	12.14
CaO	-	-	4.34	3.01	2.00	-	-
Na_2O	13.30	13.80	11.46	10.02	12.33	11.08	10.62
K_2O	-	1.50	9.18	3.25	1.91	-	1.16
ZnO	-	-	-	-	3.05	-	-
Li_2O	0.62	1.74	-	5.04	-	0.52	1.34
TiO_2	6.14	7.53	5.69	1.99	3.05	5.11	5.79
BaO	1.99	2.51	-	-	-	1.66	1.93
ZrO_2	-	-	-	14.94	-	-	-
F	3.68	2.18	-	1.99	1.85	3.07	1.67
CoO	0.41	0.54	1.21	0.99	1.05	0.34	0.41
MnO	1.02	1.61	0.52	0.43	0.45	0.85	1.24
NiO	0.61	1.08	0.69	0.57	0.60	0.51	0.83
Cr_2O_3	-	-	-	-	-	16.67	23.08

One shall note that after the application of the selected enamel on the steel substrate (16Mo3) a firing step at 880 °C for about 20 min was conducted. Visual and microscopic assessment of the surface showed no presence of capillary cracks for the coatings of investigation.

2.2. Oxidation in Water Vapor

The oxidation of the different samples was carried out using a Setsys Evo 1750 thermobalance (0.1 µg accuracy, SETARAM, Caluire-et-Cuire, France) under a wet Ar atmosphere containing water vapor at 10 vol.%. The water vapor was created using the Wetsys module (SETARAM, Caluire-et-Cuire, France) and transferred to the bottom of the thermal enclosure of the thermobalance via a heated transfer pipe. To assess the different kinetic behaviors of the samples, the specific mass gain ($\frac{\Delta m}{S}$); where Δm is the mass gain of the sample and S its surface) was plotted on log-log diagram as a function of the time to retrieve the oxidation parameters described by the law of kinetics of Equation (1):

$$\frac{\Delta m}{S} = k \times t^n \quad (1)$$

where k and n correspond to the kinetic parameters that are usually employed to describe and compare the different oxidation regimes in the transient stable period that arrives ahead of any slope change or breakaway. On the one hand, if "n" is comprised between 0.9 and 1.1, the oxidation kinetic is described by the simple linear Equation (2):

$$\frac{\Delta m}{S} = k_l \times t \quad (2)$$

where k_l corresponds to the linear rate constant in g·cm^{-2}·s^{-1}. On the other hand, if "n" is comprised between 0.35 to 0.65 a parabolic law allows the determination of the parabolic rate constant k_p in is g·cm^{-4}·s^{-2} following the Equation (3):

$$\frac{\Delta m}{S} = \sqrt{k_p \times t + A} \quad (3)$$

where A is a constant that depends on the transient period [22].

2.3. Characterization

The characterization of the materials before and after testing at high temperatures was conducted by scanning electron microscopy (SEM Quanta 200F, FEI, Hillsboro, OR, USA) in a SCIOS FEI dual-beam apparatus coupled to energy dispersive spectrometry (EDS, FEI, Hillsboro, OR, USA) from EDAX [2]. Secondary and backscattered electron images were taken at different magnifications yet only the most representative are included in this paper. The preparation of the cross-section of the samples included gentle polishing with increasing SiC papers (Struers) and final 1-µm diamond polishing with a Struers suspension following conventional metallographic protocols. The polished cross-sections were finally rinsed with water and ethanol and dried. The crystal phase identification was realized by X-ray diffraction (XRD, BRUKER, Karlsruhe, Germany) in a Bruker AXS D8 Advance using the λ_{Cu} radiation in symmetric θ–2θ mode. The local phase analyses were made possible with Raman micro-spectrometry (Jobin Yvon LabRam HR800, HORIBA, Tulln, Austria) using a laser of λ = 632 nm. The porosity in the coatings before and after the water vapor tests was investigated by image analyses using Image J software (version 1.54b) (see highlighted contours in yellow on cross-section micrographs).

3. Results

3.1. Coatings

Figure 1 gathers the images of the cross-sections of the enamel coatings as observed in the backscattered electrons mode of the SEM, while Table 2 summarizes the main features. It can be observed that all the coatings homogeneously covered the steel substrate irrespective of their composition and variable thickness. Furthermore, except for the negligible porosity of A and B, all the remaining coatings displayed very tiny bubbles (C, D, F, and G) to small (B and D) and coarse pores (E).

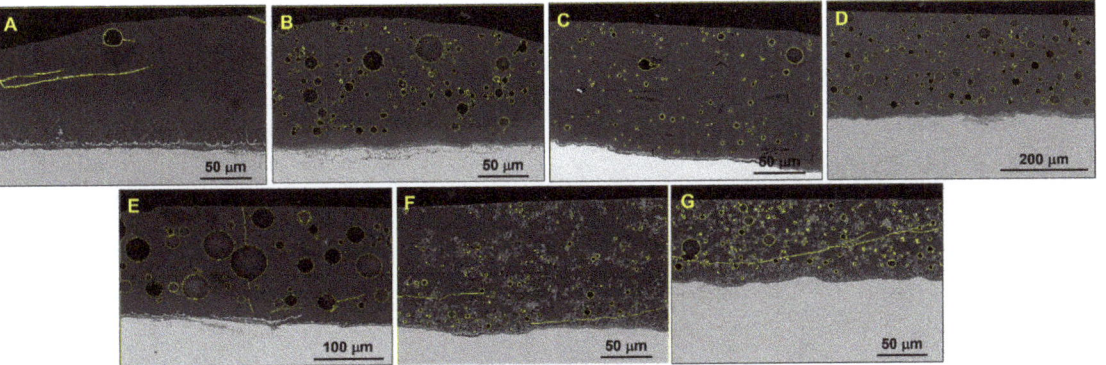

Figure 1. SEM cross-section images in the backscattered mode of the glass enamel coatings (coatings (**A–G**)) in the as-fabricated condition (pores and cracks are highlighted in yellow).

3.2. Oxidation in Wet Air

Figure 2 shows the evolution of the specific mass gains of the uncoated and coated steels with time at 650 °C in Ar-10 vol%H_2O. It can be noted that the coatings dramatically decreased (8 to 16 times) the mass gain of the uncoated steel (Figure 2a). The coatings themselves evolved very differently upon oxidation (Figure 2b). For instance, a small yet continuous mass uptake occurred with A, but the mass gain of B increased significantly and then slowed down. Other coatings (C, D, and F) tended to exhibit some kind of breakaway oxidation after about 5 h. At the end of the test, the highest mass gain after 48 h of annealing in Ar-10%H_2O at 650 °C was reported for C (0.85 mg·cm^{-2}) and the lowest mass gain was recorded for ceramic sample A (0.42 mg·cm^{-2}). However, the final specific mass gain is not sufficient to describe the oxidation behavior of each specimen. For a better understanding of the kinetic behavior of each sample, the oxidation parameters (k, n) retrieved from the kinetic law of Equation (1) were calculated at different time intervals for each specimen that depends on the different transient periods observed and are gathered in Table 3.

Table 2. Summary of the main microstructural features of the enamel coatings on low-alloyed 16Mo3 boiler steel (the vol.% of porosity is based on image analyses of the cross-sections and should be considered as comparative values).

	Thickness (μm)	Porosity (vol.%)	Cracks (Orientation)	Coating-Substrate Interface
A	100 ± 25	1%	Parallel	Oxide and corrosion layer
B	80 ± 15	3%	None	Thin oxide and corrosion layer
C	110 ± 25	14%	None	Continuous thin oxide layer
D	230 ± 15	25%	None	Thin oxide and corrosion layer
E	170 ± 20	34%	Normal and parallel	Thin oxide and corrosion layer
F	150 ± 10	15%	Parallel	Continuous thin oxide layer
G	90 ± 5	18%	Parallel	Continuous thin oxide layer

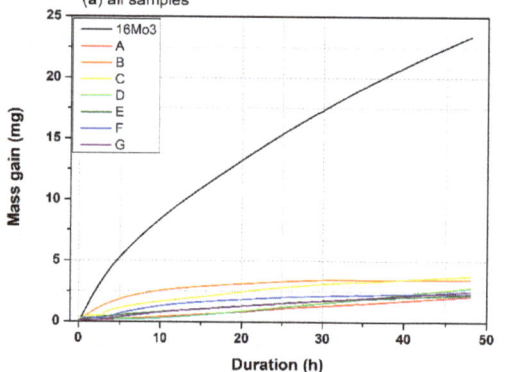 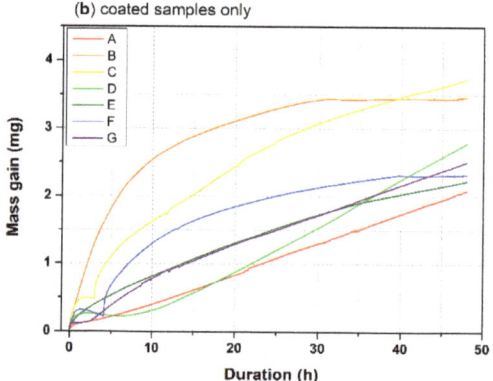

Figure 2. Specific mass gain curves obtained by TGA of (**a**) the raw material (reference 16Mo3) and coated samples exposed to Ar-10%H$_2$O at 650 °C for 48 h; (**b**) presents the data of the coated samples only.

As expected from Figure 2, after a linear growth from the onset of oxidation to 4 h, the uncoated 16Mo3 shows fast oxidation kinetics described by a parabolic constant "k_p" of $3.17 \cdot 10^{-10}$ g$^2 \cdot$cm$^{-4} \cdot$s^{-1} that is approximately two orders of magnitude higher than that the coated samples. The discrepancy in the oxidation behavior of the different coated specimens is evident from the values of Table 3 after an initial short transition period. Indeed, A, D, and G exhibit continuous linear oxidation after 7 h to 13 h, while B, C, E, and F show parabolic oxidation behavior that stabilizes after 4 to 11 h of oxidation. The peculiar breakaway behavior observed in C and F can be related to the appearance of cracks in the coating or in the oxide but without any spallation. The cracks would allow direct access of the substrate to the water vapor provoking an accelerated attack. The B and F samples did not gain further mass after some oxidation period, which could be related to a very protective oxide scale and/or a densification of the coating that prevented further oxidation.

Figure 3 shows the XRD analyses performed on unexposed and exposed samples while Figure 4 and Table 4 show the results of the Raman analyses. The 16Mo3 substrate is covered with a lepidocrocite γ-FeOOH and goethite Fe(III)OOH oxide layer in the as-received conditions that further transforms into hematite α-Fe$_2$O$_3$ after exposure to 650 °C for 48 h in the Ar-10%H$_2$O atmosphere. In contrast, the Raman spectra do not display any significant difference between the as-deposited and the oxidized enamel coatings, while the X-ray patterns of Figure 3b indicate the formation of some new phases attributed to the crystallization of silicate compounds.

Table 3. Oxidation parameters "n" and "k" of the uncoated and coated samples as a function of different time intervals.

Material	Time Interval; Oxidation Parameters "n" and "k" *,**			
16Mo3	0–1 h transition	1–4 h n = 1: linear * $k_l = 8.12 \cdot 10^{-8}$	4–48 h n = 0.65: parabolic ** $k_p = 3.17 \cdot 10^{-10}$	–
A	0–1 h n = 0.49: parabolic ** $k_p = 1.20 \cdot 10^{-13}$	1–7 h transition	7–48 h n = 1.05: linear * $k_l = 2.51 \cdot 10^{-9}$	–
B	0–20 min n = 0.51: parabolic ** $k_p = 1.20 \cdot 10^{-13}$	20–30 min transition	30 min–7 h n = 0.75: sub-linear * $k_l = 2.32 \cdot 10^{-8}$	7–48 h n = 0.3: ~parabolic ** $k_p = 1.64 \cdot 10^{-12}$
C	0–1 h n = 0.45: parabolic ** $k_p = 2.28 \cdot 10^{-12}$	1–4 h breakaway without spallation	4–48 h n = 0.55: parabolic ** $k_p = 4.80 \cdot 10^{-12}$	–
D	0–10 h transition	10–48 h n = 1.4: super-linear * $k_l = 4.31 \cdot 10^{-9}$	–	–
E	0–30 min transition	30 min–11 h n = 0.55: parabolic ** $k_p = 1.39 \cdot 10^{-12}$	11–48 h n = 0.65: parabolic ** $k_p = 2.66 \cdot 10^{-12}$	–
F	0–30 min n = 0.56: parabolic ** $k_p = 1.69 \cdot 10^{-12}$	30 min–4 h breakaway without spallation	4–40 h n = 0.53: parabolic ** $k_p = 1.66 \cdot 10^{-12}$	40–48 h evaporation
G	0–3 h transition	3–13 h n = 1.2: super-linear * $k_l = 4.37 \cdot 10^{-9}$	13–48 h n = 0.74: linear * $k_l = 2.49 \cdot 10^{-9}$	–

* k_l, linear constant of oxidation in g·cm^{-2}·s^{-1}; ** k_p, parabolic constant of oxidation in g^2·cm^{-4}·s^{-1}.

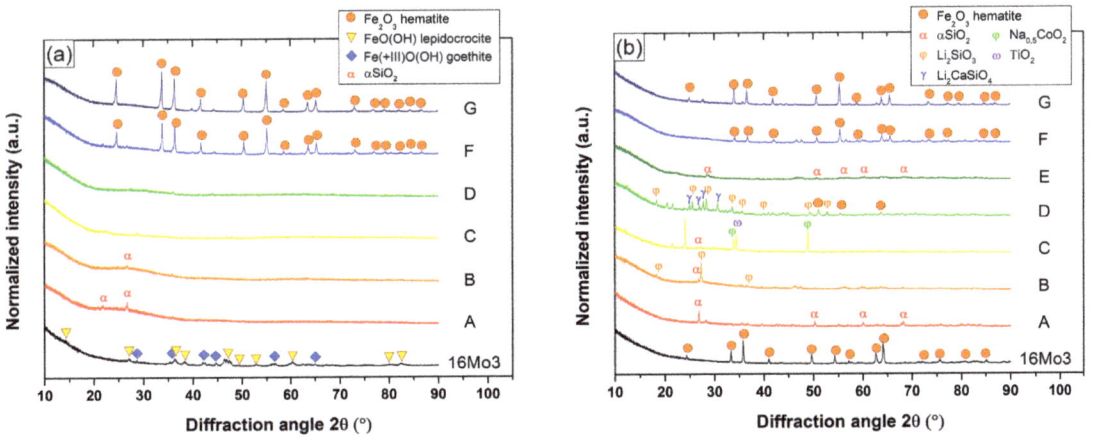

Figure 3. XRD diffraction plots for (**a**) the initial samples and (**b**) the oxidized samples at 650 °C for 48 h in Ar-10%H$_2$O atmosphere.

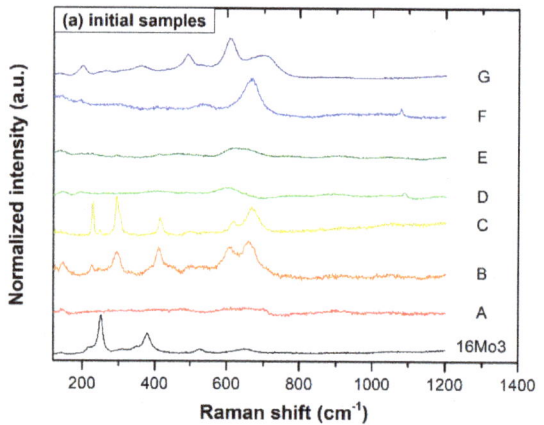
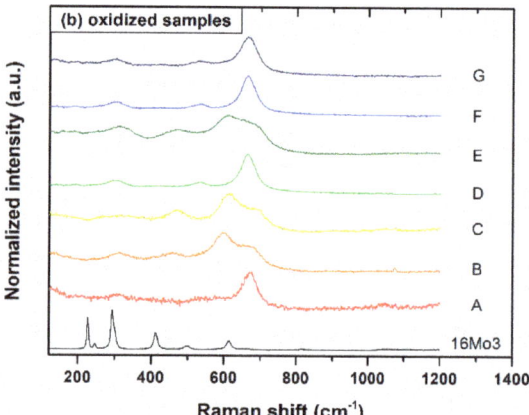

Figure 4. Raman spectra for (**a**) initial and (**b**) oxidized samples at 650 °C for 48 h in Ar-10%H_2O atmosphere.

Figure 5 shows SEM images of the exposed surfaces of the uncoated 16Mo3 steel and the coated samples in an Ar-10%H_2O atmosphere for 48 h at 650 °C. For uncoated 16Mo3 (Figure 5A), some cracks as well as two types of microstructures are observed with oxides exhibiting a platelet-like morphology on the external surface and round-shaped oxides underneath that are revealed as spallation occurred. For the coated samples, various microstructures were observed on the surfaces exposed to the Ar-10%H_2O atmosphere for 48 h at 650 °C. For A, B, and E, small pores were observed (Figure 5B,C,F). Only D samples exhibited partial spallation of the coating (Figure 5E). This spallation probably occurred upon cooling, as no evidence of spallation was observed during the TGA analysis. In the case of F (Figure 5G), some tiny bright precipitates were observed on top of the coatings, whereas the precipitates observed for the other coatings appeared to be embedded in the silica matrix.

Figure 6 shows the cross-section SEM images of the uncoated and coated 16Mo3 steel exposed to Ar-10%H_2O atmosphere at 650 °C for 48 h, while Table 5 summarizes the major features after oxidation compared to the as-fabricated coatings. Irrespective of the coating thickness that may differ from one batch to the other, it is interesting to observe that barely any significant interfacial oxide scale grew between the coating and the substrate. In contrast, the uncoated 16Mo3 developed a very thick (~120 µm) dual oxide layer separated by a porous interlayer. The upper sublayer contains cracks.

Table 4. Raman peaks position and corresponding phases.

	Peaks Position (cm^{-1})							
Initial Samples:	16Mo3	A	B	C	D	E	F	G
	~220 [1]	-	~142 [3]	~228 [2]	-	-	~301 [3]	~198 [2]
	~250 [1]		~223 [2]	~245 [2]			~533 [3]	~264 [2]
	~307 [1]		~243 [2]	~292 [2]			~662 [3]	~359 [2]
	~347 [1]		~295 [2]	~412 [2]				~431 [2]
	~379 [1]		~411 [2]	~496 [2]				~536 [3]
	~529 [1]		~500 [2]	~616 [2]				~607 [2]
	~644 [1]		~535 [3]	~667 [3]				~695 [3]
			~605 [2]					
			~660 [3]					

Table 4. Cont.

Oxidized samples:	16Mo3	Peaks Position (cm^{-1})						
		A	B	C	D	E	F	G
	~227 [2] ~248 [2] ~295 [2] ~413 [2] ~500 [2] ~615 [2] ~658 [2]	~308 [3] ~538 [3] ~667 [3]	~311 [3] ~460 [3] ~600 [2] ~674 [3]	~465 [3] ~612 [3] ~682 [3]	~304 [3] ~533 [3] ~662 [3]	~312 [3] ~467 [3] ~610 [2] ~675 [3]	~304 [3] ~535 [3] ~664 [3]	~304 [3] ~534 [3] ~664 [3]

Phases: [1] Lepidocrocite FeO(OH); [2] Hematite Fe_2O_3; [3] Magnetite Fe_3O_4.

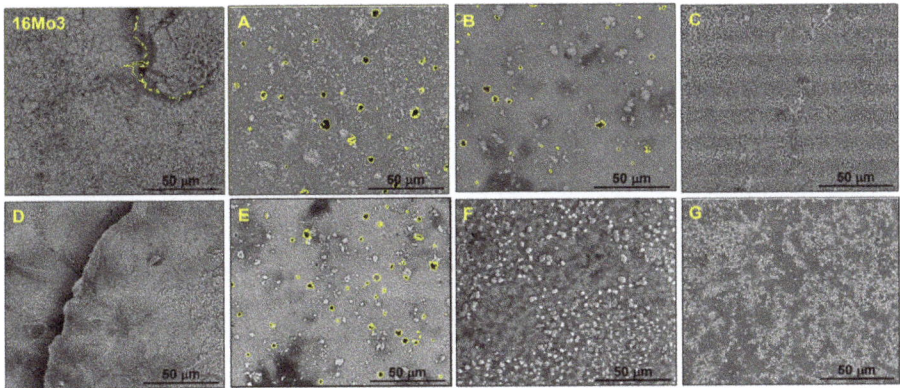

Figure 5. SEM images of the surfaces of uncoated 16Mo3 and coated samples ((A–G) coatings) exposed at 650 °C for 48 h in Ar-10%H$_2$O atmosphere (pores and cracks are highlighted in yellow). Note that there were no oxide scales formed at the top of the enamel coatings.

In the case of the coatings, it can be noted that the porosity of most coatings did not change with oxidation time and that no cracks were found either. However, the interfacial reaction zone between the coating and the substrate extended while new bright contrasted phases formed at the coating/gas interface. For instance, the porosity of B also vanished, but a significant interfacial reaction similar to that of A before oxidation occurred. The porosity still remained in all other coatings, but interestingly, the cracks mostly disappeared, except in D, which developed significant cracking.

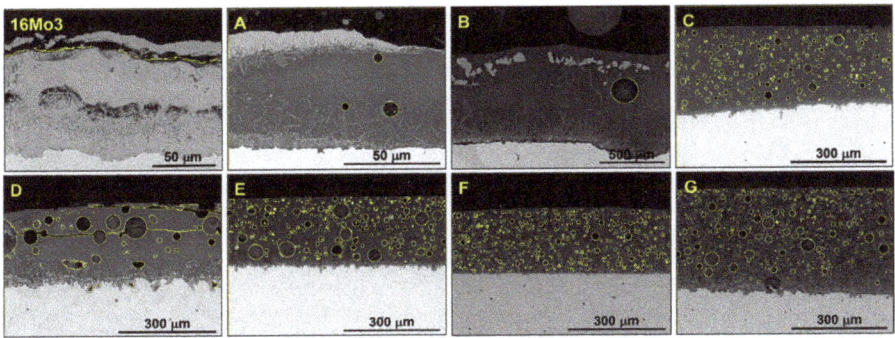

Figure 6. SEM images of cross sections of uncoated 16Mo3 and coated samples (A–G) exposed at 650 °C for 48 h in Ar-10%H$_2$O atmosphere (pores and cracks are highlighted in yellow).

Table 5. Summary of the coating microstructural features of the enamel coatings on low-alloyed 16Mo3 boiler steel after the oxidation at 650 °C for 48 h in wet Ar (the vol.% of porosity is based on image analyses of the cross-sections and should be considered as comparative values). Significant changes compared with the coatings in as-deposited conditions (Table 2) are highlighted using italic font.

	Thickness (μm)	Porosity (vol.%)	Cracks (Orientation)	Coating-Substrate Interface
A	50–60	2%	Parallel	Thick oxide and corrosion layer
B	30–40	3%	None	Thick oxide and corrosion layer
C	280	14%	None	Thick oxide and corrosion layer
D	110–120	25%	Parallel	Thin oxide and corrosion layer
E	225	34%	None	Thin oxide and corrosion layer
F	200	15%	None	Continuous thin oxide layer
G	300	18%	None	Continuous thin oxide layer

4. Discussion

4.1. Coatings

In glass enamel coatings like the ones studied here, SiO_2 and B_2O_3 ensure the formation of the network while Na_2O, K_2O, Li_2O, and CaO are the network modifiers; in particular, Li_2O and Na_2O are strong modifiers. The other oxides provide an intermediate effect, e.g., Al_2O_3 and Cr_2O_3 can take part as network formers or modifiers depending on the surrounding environment and coordination with bonding and non-bonding oxygen anions [23]. In this work, the intention was simply to study different chemistries to obtain a variety of coatings that adhere to the substrate. Such adherence results from the corrosion/oxidation of the melt with the steel substrate in all our coatings [24]. Yet, the reactivity of the melt clearly depends on the initial composition of the ceramic frit because either very thin and continuous oxide layers formed (C, D, F, G), a significant interfacial reaction occurred (A), or a mixed situation with corrosion and broken oxide layers (B, E) was observed in Figure 1. The presence of bubbles and porosity in the enamel coatings has been reported in various other works [25] and is ascribed to the gas evolution upon the firing process of the ceramic powder mixture [26]. In contrast, the main differences between the coatings relate to the presence or absence of cracks in the as-fabricated conditions. The cracks mostly depend on the thickness and the thermal expansion coefficients, in particular the latter [27]. This is demonstrated by comparing, e.g., G, which has 90 μm thickness and cracks running parallel to the thickness, with D, whose thickness is more than twice the previous one (230 μm) but shows no cracking.

4.2. Oxidation

The 16Mo3 steel was selected in this work to show the potential efficiency of the ceramic coatings exposed to harsh conditions. The high-temperature oxidation of the steel was already reported in [28]. However, to the best of our knowledge, no study focused on the effects of water vapor that regularly appears in combustion atmospheres. It was anticipated that this poorly alloyed steel will require additional protection with inexpensive coatings like the enamel ones proposed here. This is clearly reflected in Figure 2a, where the uncoated steel underwent a 6-fold increase of mass gain after 48 h of exposure to wet air at 650 °C compared to any of the coatings. Indeed, before oxidation, the uncoated steel is covered with lepidocrocite γ-FeOOH and goethite α-FeOOH despite the initial grinding of the surface of the steel with SiC P600 sandpaper to leave a rust-free surface with a homogenous roughness. Yet, the kinetics of the formation of lepidocrocite are very fast [29], and it has great thermodynamic stability ($\Delta H_f°$ (lepidocrocite) = -549.4 ± 1.4 kJ.mol^{-1}) before it transforms into goethite, which is even more stable ($\Delta H_f° = -560.7 \pm 1.2$ kJ.mol^{-1}) [30]. Lepidocrocite is similar to goethite as they are polymorphs of the same composition with different structures; the lepidocrocite possesses an orthorhombic structure with space group A_{mam}. The thermal dehydration of both goethite and lepidocrocite results in the hematite and maghemite formation by topotactic transformation (oxidation of magnetite to maghemite), where the magnetite phase oxidizes into the maghemite phase by natural weathering or other processes with the conversion of all Fe^{2+} ions into Fe^{3+} ions [31]. Goethite α-FeOOH, lepidocrocite γ-FeOOH, hematite α-Fe_2O_3, and maghemite γ-Fe_2O_3 are the constituents of magnetite Fe_3O_4 for ferric oxides and oxyhydroxides. Therefore, the oxidation process results in iron vacancies in the crystal lattice due to the partial removal of iron to compensate for the positive charges. The resulting multilayered scale is very similar to the ones observed in other very low-Cr boiler steels such as 13CrMo4 [32] and is believed to arise from the presence of iron vacancies on the FeO scale that foster diffusion, hence provoking the growth of thick oxide layers. As postulated by T. Dudziak, the growth of Fe_2O_3 and of Fe_3O_4 for which the defects are respectively found in the oxygen sublattice and interstitially (support iron ion diffusion) will sustain further diffusion upon oxidation [33].

In the case of the coatings, different oxidation behaviors have been observed in Figure 2b, but no real surface oxide formation occurs (Figure 5). For instance, only coatings B and F seem to follow a parabolic growth with a sharp oxygen uptake during the first stages and a subsequent transition till the interfusion of the cations and oxygen anion through a supposedly single and even oxide scale. Therefore, the oxygen uptake is slowed down with time. In contrast, all the other coatings seem to undergo linear oxidation yet at different rates according to their different slopes. Among the latter, coatings C and D undergo the fastest growth while coatings A and G exhibit the lowest mass gains. Since the TGA curves did not indicate any spallation, the oxidation phenomena can be attributed to either chemical reasons or to the cracking of the coatings. As for the former, it has been noted in Table 1 that the C and D coatings have the lowest amount of Na_2O whereas the best-behaving coatings (B and F) contain the highest B_2O_3 concentrations. B_2O_3 is known to be a strong glass network former whereas Na_2O has a strong ionic bonding that is easy to break and therefore strongly modifies the Si-O-Si bonds [34]. As a result, the glassy enamel coating loses the relative toughness and develops cracks, as shown in Figure 6 for D. The reasons why C does not crack can be thus related to the greater content of the network stabilizer B_2O_3 and of K_2O and TiO_2, whose role is to depolymerize the glass and open the network [34]. Yet, the cracks observed in the D coatings run parallel to the gas/coating interface and should not allow the attack of the substrate underneath. Therefore, the increasing mass gain with time shall derive from the very coarse porosity that allows penetration of the water vapor to the coating/substrate interface. This hypothesis is confirmed by paying particular attention to such areas with the presence of pores surrounded by oxides (Figure 6). Such pores can be ascribed to the evaporation of Cr as $CrO_2(OH)_{2(g)}$ at 650 °C, but there is very little Cr in the steel. Therefore, the presence of oxides in those pores is more likely due to the dissociation of water vapor resulting in $H_{2(g)}$ that reduces the iron oxide and fosters the diffusion of metal or cationic iron. Simultaneously, the water vapor can be transported and form oxide around the pores [35]. Therefore, the coatings that underwent densification with the oxidation temperature like coating B underwent the lowest mass gains. It thus derives that glass enamel coatings with sufficiently dense microstructures should be designed to provide a barrier effect against oxidation.

5. Conclusions

Different enamel coatings based on SiO_2-Na_2O-B_2O_3-TiO_2 oxide phases were investigated as protection for low-alloyed 16Mo3 steel designed for boiler steel in power generation systems. The oxidation in water vapor (Ar + 10 vol.%) at 650 °C for 48 h of these coated samples indicates that the ceramic coatings provided a high degree of protection to the steel exposed to such conditions compared to the uncoated 16Mo3 steel. Furthermore, despite the formation of cracks in some coatings, no spallation from the steel surface was observed. As a matter of fact, the enamel displayed a high adherence to the substrate due to a partial dissolution of the substrate by the glass during the elaboration process that led to the bonding of the ceramic to the metal. The extensive attack for coating D was related to the interconnected porosity in the coating that allowed access of water vapor to the 16Mo3 steel. An accurate assessment of the porosity of the tested coatings should be thus conducted in the future to evaluate such a hypothesis.

Author Contributions: Conceptualization, F.P. and T.D.; methodology, F.P. and G.B.; software, G.B.; validation, F.P., T.D., R.T. and E.R.; formal analysis, G.B.; investigation, G.B. and R.T.; resources, E.R. and R.T.; data curation, F.P., G.B. and T.D.; writing—original draft preparation, F.P., G.B. and T.D.; writing—review and editing, F.P. and G.B.; visualization, F.P., G.B. and T.D.; supervision, F.P. and T.D.; project administration, F.P. and T.D.; funding acquisition, F.P., T.D. All authors have read and agreed to the published version of the manuscript.

Funding: Part of this work was supported by the Campus France- Nawa Poland exchange program "Polonium" with reference number 46878ZJ. In addition, the work was partly financially supported by Ministry of Higher Education in Poland, in the internal project in Łukasiewicz—Krakow Institute of Technology, under the number: 9011/00.

Institutional Review Board Statement: Not applicable.

Informed Consent Statement: Not applicable.

Data Availability Statement: All data that support the findings of this study are included within the article.

Acknowledgments: Part of this work was supported by the Campus France- Nawa Poland exchange program "Polonium" with reference number 46878ZJ. In addition, the work was partly financially supported by Ministry of Higher Education in Poland, in the internal project in Łukasiewicz—Krakow Institute of Technology, under the number: 9011/00.

Conflicts of Interest: The authors declare no conflict of interest.

References

1. Abu-Warda, N.; López, A.J.; Pedraza, F.; Utrilla, M.V. Corrosion Behavior of T24, T92, VM12, and AISI 304 Steels Exposed to KCl–NaCl–K_2SO_4–Na_2SO_4 Salt Mixtures. *Mater. Corros.* **2021**, *72*, 936–950. [CrossRef]
2. Rząd, E.; Dudziak, T.; Polczyk, T.; Boroń, Ł.; Figiel, P.; Oziębło, A.; Chmielewska, D.; Synowiec, B.; Pichniarczyk, P. Sulfidation of Ceramic-Based Coatings Deposited on Low-Alloyed Steel 16Mo3 Exposed at High Temperature. *J. Mater. Eng. Perform.* **2021**, *30*, 8538–8550. [CrossRef]
3. Boulesteix, C.; Pedraza, F.; Proy, M.; Lasanta, I.; de Miguel, T.; Illana, A.; Pérez, F.J. Steam Oxidation Resistance of Slurry Aluminum and Aluminum/Silicon Coatings on Steel for Ultrasupercritical Steam Turbines. *Oxid. Met.* **2017**, *87*, 469–479. [CrossRef]
4. Boulesteix, C.; Kolarik, V.; Pedraza, F. Steam Oxidation of Aluminide Coatings under High Pressure and for Long Exposures. *Corros. Sci.* **2018**, *144*, 328–338. [CrossRef]
5. Dhand, D.; Kumar, P.; Grewal, J.S. A Review of Thermal Spray Coatings for Protection of Steels from Degradation in Coal Fired Power Plants. *Corros. Rev.* **2021**, *39*, 243–268. [CrossRef]
6. Kumar, S.; Kumar, M.; Handa, A. Combating Hot Corrosion of Boiler Tubes—A Study. *Eng. Fail. Anal.* **2018**, *94*, 379–395. [CrossRef]
7. Galetz, M.C.; Bauer, J.T.; Schütze, M.; Noguchi, M.; Cho, H. Resistance of Coatings for Boiler Components of Waste-to-Energy Plants to Salt Melts Containing Copper Compounds. *J. Therm. Spray Technol.* **2013**, *22*, 828–837. [CrossRef]
8. Jafari, R.; Sadeghimeresht, E.; Farahani, T.S.; Huhtakangas, M.; Markocsan, N.; Joshi, S. KCl-Induced High-Temperature Corrosion Behavior of HVAF-Sprayed Ni-Based Coatings in Ambient Air. *J. Therm. Spray Technol.* **2018**, *27*, 500–511. [CrossRef]
9. Dudziak, T.; Hussain, T.; Simms, N.J.; Syed, A.U.; Oakey, J.E. Fireside Corrosion Degradation of Ferritic Alloys at 600 °C in Oxy-Fired Conditions. *Corros. Sci.* **2014**, *79*, 184–191. [CrossRef]
10. Tejero-Martin, D.; Rezvani Rad, M.; McDonald, A.; Hussain, T. Beyond Traditional Coatings: A Review on Thermal-Sprayed Functional and Smart Coatings. *J. Therm. Spray Technol.* **2019**, *28*, 598–644. [CrossRef]
11. Harrison, W.N.; Moore, D.G.; Richmond, J.C. Ceramic Coatings for High-Temperature Protection of Steel. *J. Res. Natl. Bur. Stand.* **1947**, *38*, 293. [CrossRef]
12. Pavan, C.M.; Narendra, B.B.R. Review of Ceramic Coating on Mild Steel Methods, Applications and Opportunities. *Int. J. Adv. Sci. Res. Eng.* **2018**, *4*, 44–49. [CrossRef]
13. Kasala, S.; Vidyavathy, M. Advanced Ceramic Coatings on Stainless Steel: A Review of Research, Methods, Materials, Applications and Opportunities. *Int. J. Adv. Eng. Technol.* **2016**, *7*, 126–141.
14. Adraider, Y.; Pang, Y.X.; Nabhani, F.; Hodgson, S.N.; Sharp, M.C.; Al-Waidh, A. Laser-Induced Deposition of Alumina Ceramic Coating on Stainless Steel from Dry Thin Films for Surface Modification. *Ceram. Int.* **2014**, *40*, 6151–6156. [CrossRef]

15. Vural, M.; Zeytin, S.; Ucisik, A.H. Plasma-Sprayed Oxide Ceramics on Steel Substrates. *Surf. Coat. Technol.* **1997**, *97*, 347–354. [CrossRef]
16. Seraffon, M.; Simms, N.J.; Nicholls, J.R.; Sumner, J.; Nunn, J. Performance of Thermal Barrier Coatings in Industrial Gas Turbine Conditions. *Mater. High Temp.* **2011**, *28*, 309–314. [CrossRef]
17. Sumner, J.; Encinas-Oropesa, A.; Simms, N.J.; Oakey, J.E. High Temperature Oxidation and Corrosion of Gas Turbine Component Materials in Burner Rig Exposures. *Mater. High Temp.* **2011**, *28*, 369–376. [CrossRef]
18. Peters, M.; Leyens, C.; Schulz, U.; Kaysser, W.A. EB-PVD Thermal Barrier Coatings for Aeroengines and Gas Turbines. *Adv. Eng. Mater.* **2001**, *3*, 193–204. [CrossRef]
19. Wells, J.; Chapman, N.; Sumner, J.; Walker, P. The Use of APS Thermal Barrier Coatings in Corrosive Environments. *Oxid. Met.* **2017**, *88*, 97–108. [CrossRef]
20. Ramaswamy, P.; Seetharamu, S.; Rao, K.J.; Varma, K.B.R. Thermal Shock Characteristics of Plasma Sprayed Mullite Coatings. *J. Spray Technol.* **1998**, *7*, 497–504. [CrossRef]
21. Shen, D.; Li, M.; Gu, W.; Wang, Y.; Xing, G.; Yu, B.; Cao, G.; Nash, P. A Novel Method of Preparation of Metal Ceramic Coatings. *J. Mater. Process. Technol.* **2009**, *209*, 2676–2680. [CrossRef]
22. Pieraggi, B. Calculations of Parabolic Reaction Rate Constants. *Oxid. Met.* **1987**, *27*, 177–185. [CrossRef]
23. Schäfer, G. Degradation of Glass Linings and Coatings. In *Shreir's Corrosion*; Elsevier: Amsterdam, The Netherlands, 2010; Volume 3, pp. 2319–2329. ISBN 978-0-444-52787-5.
24. Berdzenishvili, I.G. Functional Corrosion-Resistant Enamel Coatings and Their Adherence Strength. *Acta Phys. Pol. A* **2012**, *121*, 178–180. [CrossRef]
25. Chen, M.; Li, W.; Shen, M.; Zhu, S.; Wang, F. Glass Coatings on Stainless Steels for High-Temperature Oxidation Protection: Mechanisms. *Corros. Sci.* **2014**, *82*, 316–327. [CrossRef]
26. Rossi, S.; Russo, F.; Calovi, M. Durability of Vitreous Enamel Coatings and Their Resistance to Abrasion, Chemicals, and Corrosion: A Review. *J. Coat. Technol. Res.* **2021**, *18*, 39–52. [CrossRef]
27. Son, Y.-K.; Lee, K.H.; Yang, K.-S.; Ko, D.-C.; Kim, B.-M. Prediction of Residual Stress and Deformation of Enameled Steel. *Int. J. Precis. Eng. Manuf.* **2015**, *16*, 1647–1653. [CrossRef]
28. Dudziak, T.; Jura, K. High Temperature Corrosion of Low Alloyed Steel in Air and Salt Mist Atmospheres. *Trans. Foundry Res. Inst.* **2016**, *56*, 77–85. [CrossRef]
29. Xiao, K.; Dong, C.; Li, X.; Wang, F. Corrosion Products and Formation Mechanism during Initial Stage of Atmospheric Corrosion of Carbon Steel. *J. Iron Steel Resist. Int.* **2008**, *15*, 42–48. [CrossRef]
30. Majzlan, J.; Grevel, K.-D.; Navrotsky, A. Thermodynamics of Fe Oxides: Part II. Enthalpies of Formation and Relative Stability of Goethite (α-FeOOH), Lepidocrocite (γ-FeOOH), and Maghemite (γ-Fe2O3). *Am. Mineral.* **2003**, *88*, 855–859. [CrossRef]
31. Kinebuchi, I.; Kyono, A. Study on Magnetite Oxidation Using Synchrotron X-Ray Diffraction and X-Ray Absorption Spectroscopy: Vacancy Ordering Transition in Maghemite (γ-Fe$_2$O$_3$). *J. Mineral. Petrol. Sci.* **2021**, *116*, 211–219. [CrossRef]
32. Pérez, F.J.; Otero, E.; Hierro, M.P.; Gómez, C.; Pedraza, F.; de Segovia, J.L.; Román, E. Corrosion Protection of 13CrMo 44 Heat-Resistant Ferritic Steel by Silicon and Cerium Ion Implantation for High-Temperature Applications. *Surf. Coat. Technol.* **1998**, *108–109*, 121–126. [CrossRef]
33. Dudziak, T. *Steam Oxidation of Fe-Based Materials High Temperature Corrosion*; Ahmad, Z., Ed.; IntechOpen: London, UK, 2016; pp. 1–25, ISBN 978-953-51-4727-5.
34. Mysen, B.O.; Richet, P. *Silicate Glasses and Melts: Properties and Structure*; Elsevier: Paris, France, 2005; Volume 10.
35. Saunders, S.R.J.; Monteiro, M.; Rizzo, F. The Oxidation Behaviour of Metals and Alloys at High Temperatures in Atmospheres Containing Water Vapour: A Review. *Prog. Mater. Sci.* **2008**, *53*, 775–837. [CrossRef]

Disclaimer/Publisher's Note: The statements, opinions and data contained in all publications are solely those of the individual author(s) and contributor(s) and not of MDPI and/or the editor(s). MDPI and/or the editor(s) disclaim responsibility for any injury to people or property resulting from any ideas, methods, instructions or products referred to in the content.

Article

Experimental Study on Fatigue Performance of Welded Hollow Spherical Joints Reinforced by CFRP

Yutong Duan, Honggang Lei * and Shihong Jin

College of Civil Engineering, Taiyuan University of Technology, Taiyuan 030024, China
* Correspondence: lhgang168@126.com

Abstract: The risk of fatigue failure of welded hollow spherical joints (WHSJs) under alternating loads increases due to the inherent defects, the disrepair, and the demand for tonnage upgrades, of suspension cranes. The finite element analysis results revealed that the ranking of the stress concentration factor at the WHSJ was as follows: weld toe in steel tube of tube–ball connection weld > weld toe in steel tube of tube–endplate connection weld > weld toe in sphere of tube–ball connection weld > weld toe in plate of tube–endplate connection weld. Moreover, the peak stress at the weld of the tube–sphere connection was reduced by 32.93% after CFRP bonding reinforcement, which was beneficial for improving the fatigue performance. In this study, 16 full-scale specimens of Q235B WHSJs were tested by an MTS fatigue testing machine to study the strengthening effect of CFRP on the fatigue performance. It was found that the fatigue fracture of WHSJs was transferred from the tube–sphere connection weld to the tube–endplate connection after CFRP reinforcement. According to the fitted S-N curves, the fatigue strength could be increased by 13.26%–18.19% when the cycle number increased from 10,000 to 5,000,000.

Keywords: welded hollow spherical joints; CFRP strengthening method; fatigue performance; S-N curve

Citation: Duan, Y.; Lei, H.; Jin, S. Experimental Study on Fatigue Performance of Welded Hollow Spherical Joints Reinforced by CFRP. Coatings 2022, 12, 1585. https://doi.org/10.3390/coatings12101585

Academic Editors: Matic Jovičević-Klug, Patricia Jovičević-Klug and László Tóth

Received: 26 September 2022
Accepted: 18 October 2022
Published: 20 October 2022

Publisher's Note: MDPI stays neutral with regard to jurisdictional claims in published maps and institutional affiliations.

Copyright: © 2022 by the authors. Licensee MDPI, Basel, Switzerland. This article is an open access article distributed under the terms and conditions of the Creative Commons Attribution (CC BY) license (https:// creativecommons.org/licenses/by/ 4.0/).

1. Introduction

The welded hollow sphere joint grid structure has a simple structure, excellent mechanical properties, and unique advantages, especially in terms of construction cost [1]. As an important part of the grid structure, the welded hollow sphere joint (WHSJ) has been widely used in public buildings such as gymnasiums, exhibition centers, waiting halls, industrial buildings, industrial plants, hangars, and dry coal sheds, etc. [2].

With the development of industrialization, industrial buildings are faced with increasing volume demand and lifting equipment upgrades. The increases in suspension crane level, load, and use frequency, render welded hollow spherical joint grid structures more prone to fatigue damage. The damage positions are shown in Figure 1: the weld toe on the spherical surface of the steel pipe or cross-plate connection (①, ③), the weld toe on the pipe surface of the steel pipe connection (②), and the high-strength bolts for cross-plate–beam connections (④). In related research on the fatigue performance at the spherical weld toe of welded hollow ball joints, Lei [3] established the calculation formula of stress concentration factor (SCF) by means of the thin shell theory, with a numerical range of 2.38–3.09. Fatigue tests of 15 specimens were completed, and the allowable nominal stress amplitude with $n = 2 \times 10^6$ was 31.4 MPa. Yan [4] determined the SCF using ANSYS finite element analysis and a static tensile test with a numerical range of 2.06–4.86. The fatigue test of 29 welded hollow ball full-scale specimens was carried out, and the allowable nominal stress amplitude with $n = 2 \times 10^6$ was 22.3 MPa. Jiao et al. [5] took the cross-plate–welded hollow sphere joint connection as the research object, performed 25 welding ball fatigue tests, and obtained that the allowable nominal stress amplitude with $n = 2 \times 10^6$ was 19.7 MPa. In a study of the weld toe in steel tubes, Zhang [6] performed a fatigue test on 38 specimens and used infrared thermography to predict the position of fatigue fracture. The allowable nominal stress amplitude with $n = 2 \times 10^6$ was 56.96 MPa.

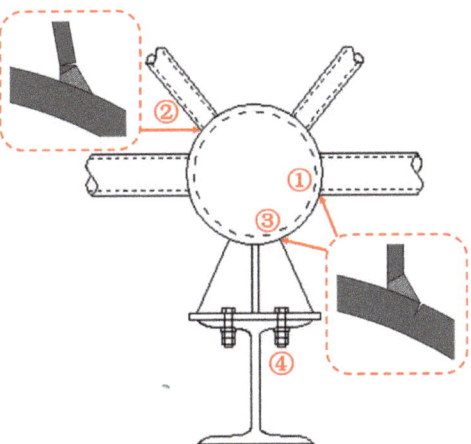

Figure 1. Schematic diagram of fatigue failure location. (①) Spherical surface of the steel pipe connections; (②) Pipe surface of the steel pipe connection; (③) Spherical surface of the cross-plate connection; (④) High-strength bolts.

The experimental results above indicate that the allowable stress amplitude at the spherical weld toe of the WHSJ is only 19.7–31.4 MPa, which is lower than the lowest class Z14 (36 MPa) in the standard for the design of steel structures (GB50017-2017) [7]. The allowable stress amplitude at the weld toe of the steel pipe is 56.96 MPa, which is close to class Z10 (56 MPa) in GB50017-2017 [7]. This indicates that there are significant fatigue hazards in grid structures with such a low fatigue strength. However, the research on the reinforcement of WHSJ is more focused on the static performance [8–11], and the improvement of fatigue performance is rarely involved, which proved that research into improving its fatigue performance fundamentally is extremely urgent.

In studies on the fatigue performance of welding structures reinforced by CFRP, the test objects included butt joints in steel plates [12–15], cruciform welded joints [16–19], square pipe butt joints [20], K-shaped [21,22] and T-shaped [23] tube joints, and welded H-beams [24–26], which were characterized by a thick parent metal, relatively few weld defects, and an easy reinforcement operation. As shown in Table 1, the use of CFRP paste reinforcement could double the fatigue life or fatigue strength of components, and the reinforcement effect is remarkable. Compared with steel plates, the parent metal of WHSJ is a spherical surface, which results in more severe geometric discontinuities and stress concentration, whereas the effect of reinforcement with CFRP pasting is unknown. In this study, CFRP was applied to strengthen the welded hollow spherical joints, and the constant amplitude fatigue test of reinforced and unreinforced specimens was conducted. Based on the obtained S-N curves, the fatigue designing method was proposed and the effects of reinforcement were discussed, laying a foundation for the improvement of the fatigue performance of in-service welded hollow sphere grid structure joints.

Table 1. Fatigue performance improvement percentage of welded joint reinforced by CFRP.

Component Type	Life Improvement (%)
Butt joint	418 [12]; 97.19 [13]
Cruciform welded joint	57–66.5 [16]; 218 [19]
Tubular welded joint	408.82 [20]; 138 [21]
Welded steel beam	213.28 [24]; 93.7 [26]

2. Stress Concentration Analysis

2.1. SCF of WHSJ

In order to obtain the stress distribution state of the nodes in the welded hollow spherical joint grid structure, an FE model with the size of the specimen from Zhang [6] (specification 1, where the hollow ball is stronger than the tube) was built to analyze the stress concentration. A tensile surface load of 1 N/mm^2 was applied to the top of the joint. The calculation results revealed four parts with significant stress concentration in this type of WHSJ (Figure 2). Under the same load, severe stress concentration appeared on the weld toe in the steel tube, due to the tube–sphere connection weld (Location 1) and tube–endplate connection weld (Location 3), with a maximum stress of 5.69 and 5.45 MPa, respectively. On the other hand, the SCF of the weld toe in the sphere (Location 2) and endplate (Location 4) was smaller, with a maximum stress of 4.25 and 0.72 MPa, respectively. Under the action of cyclic load, fatigue failure first occurred at Location 1, followed by Location 3, Location 2, and Location 4, which is consistent with the test failure locations. Therefore, the primary reinforcement area for the fatigue performance of WHSJs would be the weld toe in the steel pipe of the tube–ball connection weld, while the secondary area of concern would be the weld toe in the steel pipe of the tube–endplate connection weld.

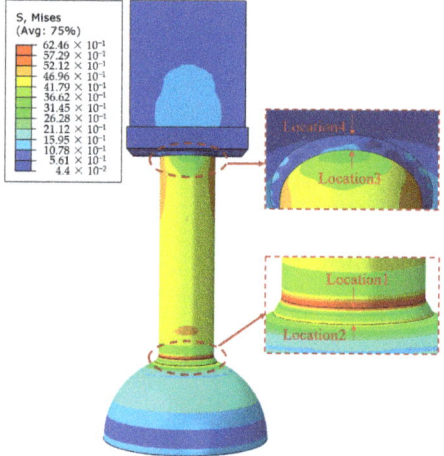

Figure 2. Results of stress concentration analysis.

2.2. SCF of CFRP-Strengthened WHSJ

The basic principle of the fatigue performance of the welded structure reinforced with CFRP is to change the force transmission path of the original structure, such that part of the load is borne by CFRP, thereby reducing the stress state of the reinforced part. Research results have shown that CFRP can effectively reduce the SCF at the weld toe, which decreases significantly with the increase in the number of carbon cloth layers [22]. As shown in Figure 3a, taking account of the symmetry of WHSJs, only one-eighth of the tube–spherical joints were analyzed, where the vertical boundaries of the sphere and the tube surface were fixed constraints. An axial tensile surface load of 150 MPa was applied to the end of the steel tube, and a single layer of CFRP was arranged in the range of 100 mm above and below the weld of the tube–sphere connection. Welds, pipes, and hollow balls used the same material parameters, with an elastic modulus of 206 GPa and a Poisson's ratio of 0.3. The mesh types of CFRP and steel products were M3D4R and C3D8I, respectively, and the mesh at the weld toe of the tube–ball connection was encrypted (Figure 3b).

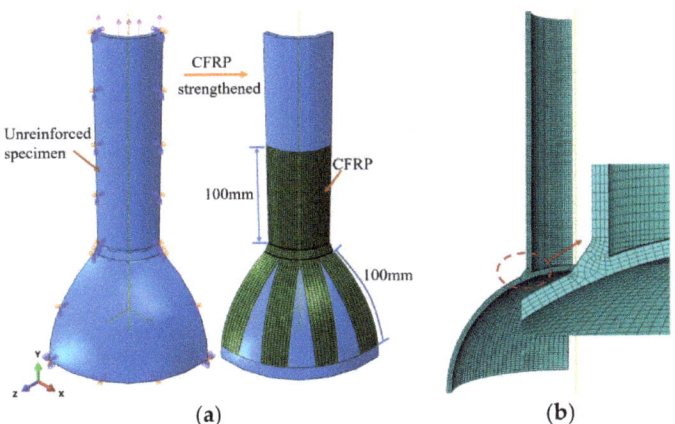

Figure 3. FE modeling process. (**a**) Finite element model; (**b**) Mesh division.

2.3. FEA Results

Under the action of the axial tensile surface load (Figure 4a), the stress concentration at the weld toe of the original tube–sphere joint was the most severe. The calculation results of Von Mises stress at the weld toe extracted along the circumferential direction are shown in Figure 4b. The maximum stress at the weld toe of the unreinforced WHSJ was 329.8 MPa, which changed to 221.2 MPa when reinforced by CFRP, representing a 32.93% reduction in maximum stress. The finite element analysis results show that CFRP bonding could improve the stress concentration at the weld toe, thereby improving its fatigue performance.

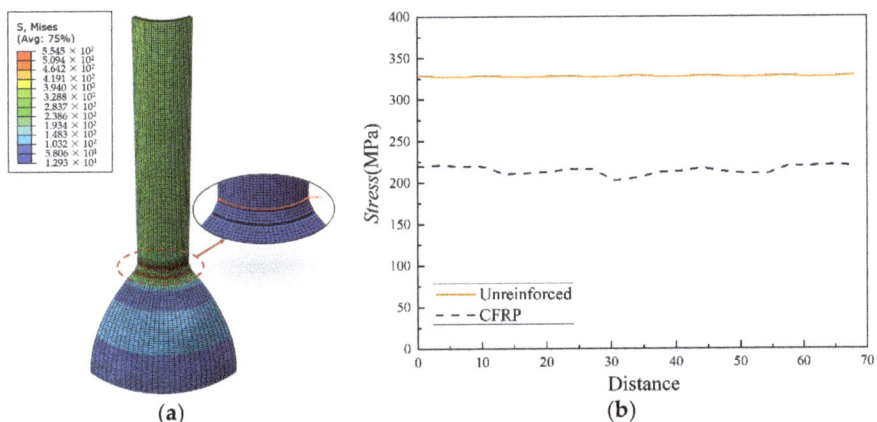

Figure 4. FEA results. (**a**) Stress concentration; (**b**) Stress concentration.

3. Experimental Program

3.1. Specimen Design

A total of 16 WHSJ fatigue tests were carried out, including four unreinforced specimens and 12 CFRP-reinforced specimens. Technical specifications for space frame structures (JGJ 7-2010) [27] stipulate that the ratio of the outer diameter D of the welded hollow sphere (WSH) to the outer diameter d of the steel pipe in the grid structure should be 2.4–3.0, the ratio of the wall thickness of the hollow sphere t to steel tube t_c should be 1.5–2.0, and the ratio of D to t should be 25–45. According to the requirements of the specification, the material of the test piece was selected as Q235B, while the pipe was high-frequency welded with a diameter of 75.5 mm and a thickness of 3.75 mm. The diameter of the hollow sphere was 200 mm, and its thickness was 8 mm. The welding grades were no lower than second grade; the parameters of the specimen are shown in Figure 5. Grade I carbon fiber cloth and matching impregnating adhesive produced by Carbon Composites Co., Ltd. (Tianjin, China) were selected as the reinforcement material; its mechanical properties are shown in Table 2.

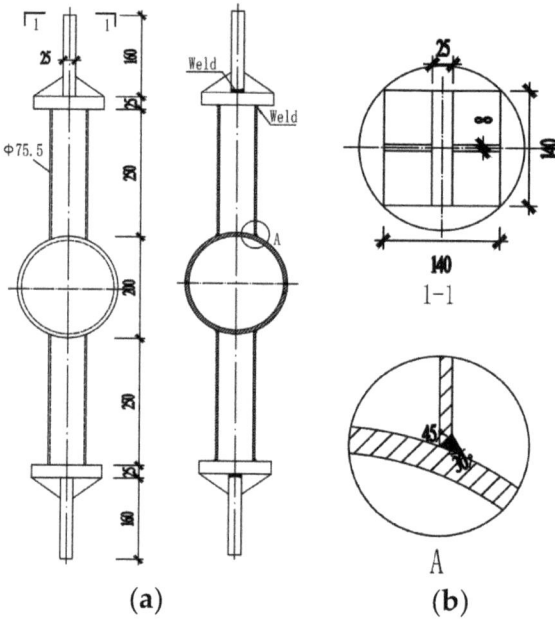

Figure 5. Size of specimens. (**a**) Front view; (**b**) Vertical view and welding details.

Table 2. Mechanical properties of test materials.

Material	Elastic Modulus (GPa)	Yield Strength (MPa)	Tensile Strength (MPa)	Elongation (%)
CFRP	240	—	3512.7	1.7
Adhesive	2.9	—	60.1	3.40
Tube	206	407	518	22.5

3.2. Visual Examination

3.2.1. Size Recheck

The sphere thickness t, tube outer diameter d, and thickness t_c of all specimens were measured using an ultrasonic thickness gauge and vernier caliper. The provisions in the standard for the acceptance of construction quality of steel structures (GB 50205-2020) [28] and welded steel tubes for general construction (SY/T 5786-2016) [29] indicate that the allowable deviation of the thinning amount of the ball wall thickness is a minimum of $0.18\,t$, 1.5 mm if $t \leq 10$, while the allowable deviation of d is $\pm 1\%\,d$ when $60.3 \leq d < 355.6$ and that of t_c is $\pm 10\%\,t_c$ when $3 \leq t_c < 12$. The test results are shown in Table 3. The wall thickness reductions of the welding ball and pipe size in this batch of specimens were within the allowable range of the specification; thus, they were considered acceptable.

Table 3. Measurement results of specimen size (mm).

Test Pieces	t			d			t_c		
	Test Value	Error	Upper	Lower	Error		Upper	Lower	Error
QP-76	7.28	−9.04%	76.10	76.05	0.86%		3.72	3.72	−1.01%

Note: The size of the connecting steel tubes at both ends of the welding ball was measured and divided into an upper branch pipe and lower branch pipe according to the relative position.

3.2.2. Weld Appearance

JGJ 7-2010 [27] stipulates that tube–sphere butt welds should be second grade, and that all welds should be subject to visual inspection. GB 50205-2020 [28] specifies that no cracks, incomplete welding, root shrinkage, arc scratches, poor joints, pores, or slag inclusions are allowed in the appearance of secondary welds that require fatigue checking. On the other hand, undercut is allowed if its length is less than a minimum of 0.05 t, 0.3 mm or its continuous undercut length is 100 mm and both sides of the weld undercut total length are ≤10% of the weld.

This batch of test pieces was assembled manually by welding in the factory, truly simulating the on-site construction process of welding hollow spheres. The appearance checking results of the welds in all the test pieces in accordance with the above provisions are shown in Figure 6, highlighting that the welds of individual specimens had more or less apparent defects such as undercut and poor joints. The results of the weld survey showed that the construction quality of the welded hollow ball grid structure was uneven, especially for the in-service structures which were built at a poor construction level; the welding quality was possibly not up to standard but still functional.

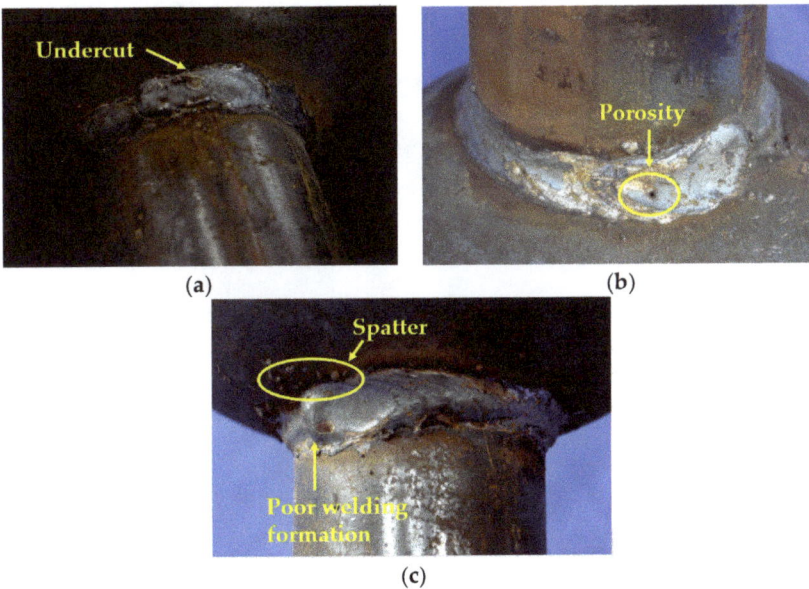

Figure 6. Defects in weld appearance. (**a**) Undercut; (**b**) Porosity; (**c**) Spatter and poor welding formation.

3.2.3. Coaxiality

A large number of manual operations relating to the welded hollow spherical grid structure can easily lead to angle deviation between the branch pipe and the spherical joint during the production process, such that the node domain is not completely in the state of absolute axial force. The additional bending moment generated by the eccentricity of the branch pipe has a great influence on the fatigue strength of the structure. As shown in Figure 7, the analysis of the tube–sphere coaxiality of the specimen revealed the following problems in this batch of specimens:

(1) In-plane and out-of-plane tilt of branch pipe,
(2) Branch pipe not perpendicular to endplate,
(3) Projection line of the chuck on endplate surface not perpendicular to endplate edge,
(4) Branch pipe centroid not coinciding with endplate centroid.

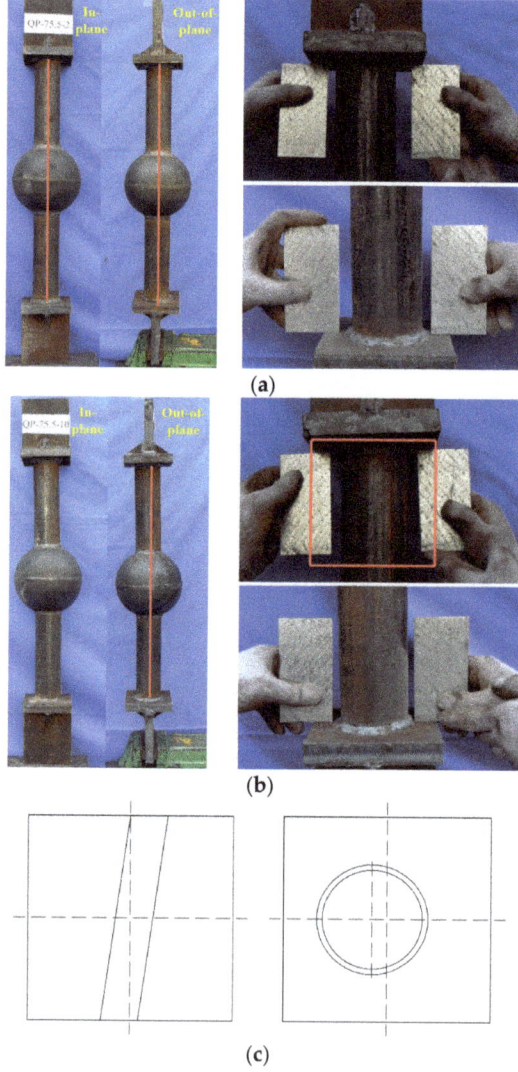

Figure 7. Coaxiality of specimen. (**a**) Tube ball basically centered, with endplate basically vertical. (**b**) Tube not centered and endplate not vertical. (**c**) Deviation of centroid of endplate and branch pipe.

3.3. Specimen Construction

This batch of specimens was divided into two categories: a nonreinforced welding ball control group (hereinafter referred to as the 'control group'), and a CFRP-reinforced welding ball experimental group. According to the census results of the appearance of the specimens, they were divided into qualified specimens ('experimental group A') and unqualified specimens ('experimental group B'). The two types of specimens whose initial states were close to each other were tested for comparison. The specimens in the control group were only polished and smoothed on the surface and sprayed with black matte paint, and the specimens in the experimental group were reinforced on the basis of the control group.

The reinforced part was ground until the metallic luster was exposed, and then steel glue was used to screed the weld surface to make it as smooth as possible. Then, special CFRP impregnating glue was configured according to the proportions and painted evenly on the surface of the test piece. Finally, the carbon cloth and test piece were tightly pasted together, pressing for 10–15 min. The production process and finished product are shown in Figures 8 and 9.

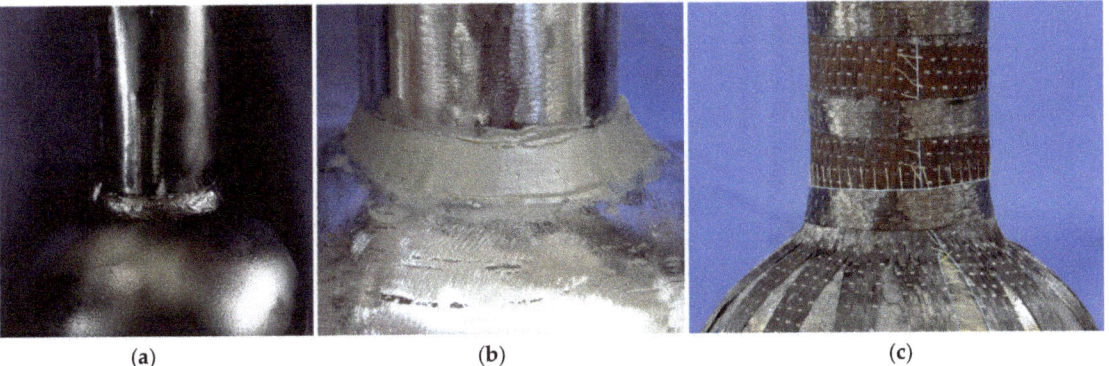

Figure 8. Strengthening process of specimens reinforced with CFRP. (**a**) Polishing (**b**) Screeding (**c**) Pasting.

Figure 9. Specimen bonded with CFRP.

3.4. Loading Scheme

The instruments used in the test were an MTS Landmark370.50 fatigue testing machine and the supporting test system MTS MPE. The maximum displacement loading measurement accuracy of the test machine was 0.00028 mm, the load range was ±15–500 N, and the maximum loading frequency was 100 Hz. This machine can perform sinusoidal, triangular, square wave, custom waveform, and load spectrum fatigue loading. The clamping diameter of the test fixture was in the range of 25–55 mm, meeting the design requirements of this study (Figure 10).

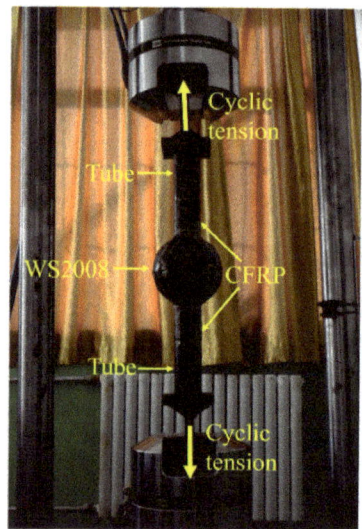

Figure 10. Test setup.

The stress amplitude $\Delta\sigma$ of the lower chord member near the suspension point in the welded hollow spherical grid structure with a suspension crane was the largest; the stress ratio ρ of the fatigue test was set as 0.1. According to the design strength of Q235 steel [7], the maximum stress σ_{max} in this test was 215 MPa. The lower chord was controlled by the dead load, and the stress was relatively large under the static state. Therefore, the minimum stress σ_{min} in this test was determined to be 130 MPa; the loading scheme is shown in Table 4.

Table 4. Constant-amplitude fatigue test loading system of QP-76.

Loading Grade	N_{max}	N_{min}	σ_{max}	σ_{min}	$\Delta\sigma$	ρ
1	181.74	18.17	215	21.5	193.5	0.1
2	152.15	15.22	180	18	162	0.1
3	143.70	14.37	170	17	153	0.1
4	126.79	12.68	150	15	135	0.1
5	118.34	11.83	140	14	126	0.1
6	109.89	10.99	130	13	117	0.1

Note: N_{max} and N_{min} are the maximum and minimum loading forces, respectively. σ_{max} and σ_{min} are the nominal stress of the steel pipe where $\sigma_{max} = N_{max}/A_s$ and $\sigma_{min} = N_{min}/A_s$. Stress range $\Delta\sigma = \sigma_{max} - \sigma_{min}$ and $\rho = \sigma_{max}/\sigma_{min}$.

4. Results and Discussion

4.1. Failure Mode

A total of 16 experimental data were obtained from QP-76, including four data in the control group and 12 data in the experimental group. As shown in Figure 11, the test results of the control group were consistent with the finite element analysis, and the fatigue fractures were all generated at Location 1, being recorded as failure mode 1. The fatigue failure modes of the specimens in the experimental group were divided into two categories. Experimental group A was consistent with the finite element analysis results, with fracture occurring at Location 2. The fatigue life of the specimens was higher than that of the control group, being recorded as failure mode 2, and the test components were qualified specimens. The fracture location of experimental group B was the same as that of the control group, and the fatigue life of the specimen had no increase, being recorded as failure mode 3; the test components were unqualified specimens.

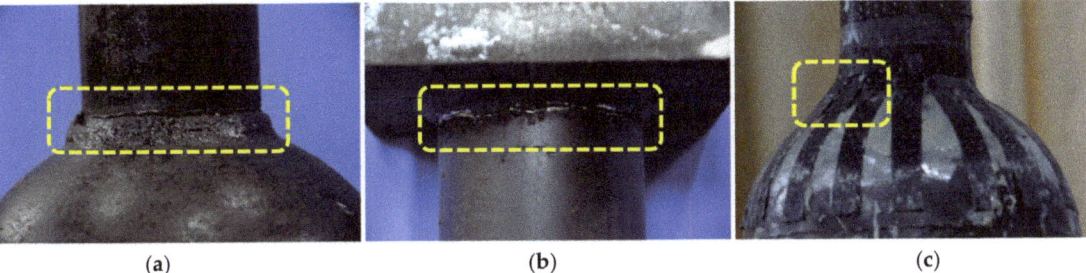

Figure 11. Typical failure mode of QP-76. (**a**) Failure mode 1 (**b**) Failure mode 2 (**c**) Failure mode 3.

Failure mode 1 (Figure 12) was the same as the test results of Zhang [6], occurring at the weld toe of the pipe surface. When the test was stopped, the fracture length basically accounted for half of the tube's entire circumference. The crack originated from the undercut of the tube surface and completely penetrated along the wall thickness direction of the steel tube, before cracking circumferentially along the weld toe of the tube surface. The angle between the section and the horizontal plane was about 45° after the specimen was broken, and obvious fatigue sources could be observed on the fracture surface.

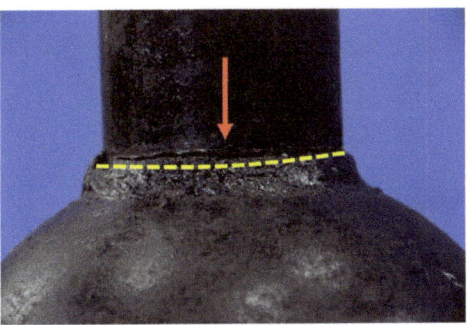 

Figure 12. Failure mode 1.

The fatigue failure position of the test piece in experimental group A was transferred from the tube–sphere connection weld to the pipe–endplate connection weld, and fracture occurred at the pipe surface weld toe, as seen in Figure 13. As shown in the FE analysis (Section 2.1), the most severe stress of WHSJ was concentrated at the weld of the tube–ball connection. After CFRP was used to strengthen the tube–sphere weld, the force transmission mode of the joints changed from 'steel tube→weld' to 'steel tube→weld + CFRP', which changed the stress field around the crack tip and reduced its stress intensity factor. Compared with the control group, the most severe stress of the component was then changed from the tube–ball connection weld to the tube–endplate connection weld, and the fatigue life was improved.

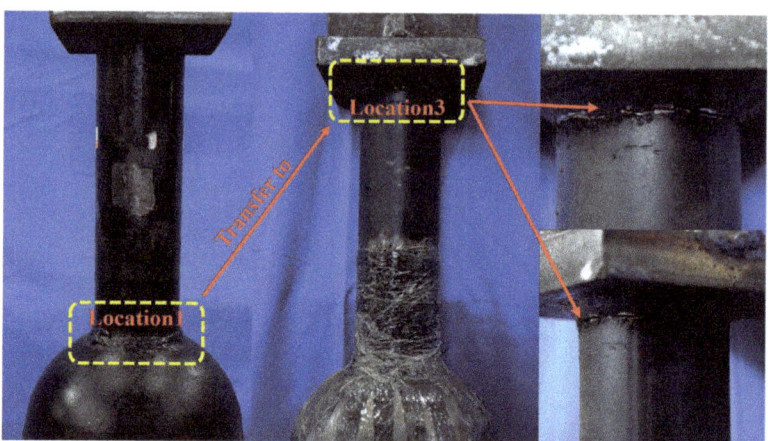

Figure 13. Failure mode 2.

For experimental group B, the failure position of the specimen still occurred at the weld of the tube–sphere connection after reinforcement, similar to the control group. When the crack propagated, the CFRP and the weld were completely peeled off, thus negating the effects of the CFRP. Hence, the fatigue life was not improved compared with the control group.

4.2. Data Statistics

The relationship between the load and the fatigue life of a member can be described by the S-N curve, which is commonly represented by the following power function expression:

$$\sigma^m N = C \tag{1}$$

where σ is the stress range and N is the fatigue life. C and m are fatigue constants that can be obtained through testing.

Taking the logarithm of both sides of Equation (1), the following expression can be obtained:

$$lgN = -mlg\sigma + lgC \tag{2}$$

According to the Equation (2), taking the stress range $\Delta\sigma$ as the variable, the test life under different failure modes in the test results was fitted in the form of a power S-N curve and a double logarithmic $lg\sigma$–lgN curve. The results are shown in Figures 14–16. The fitting results of failure mode 1 and the calculation formula of the corresponding diagonal part at a 97.7% guarantee rate are as follows:

$$\Delta\sigma_1 = 1190 \times N^{-0.17} \tag{3}$$

$$lgN_1 = 14.48 - 4.19lg\Delta\sigma - 0.57 \tag{4}$$

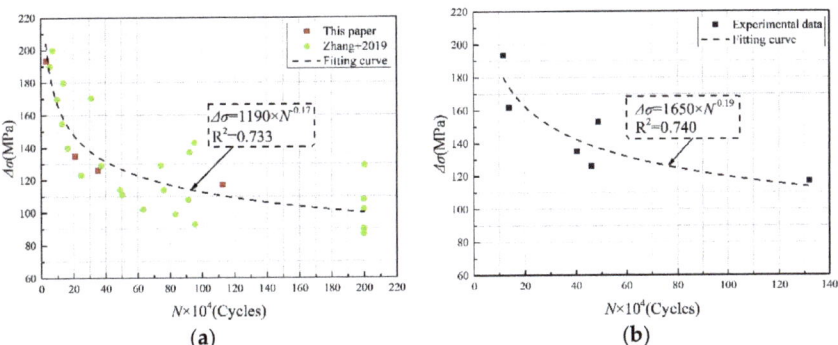

Figure 14. Power S-N curve fitting with $\Delta\sigma$ as variable. (**a**) Unreinforced specimen (**b**) Strengthened with CFRP.

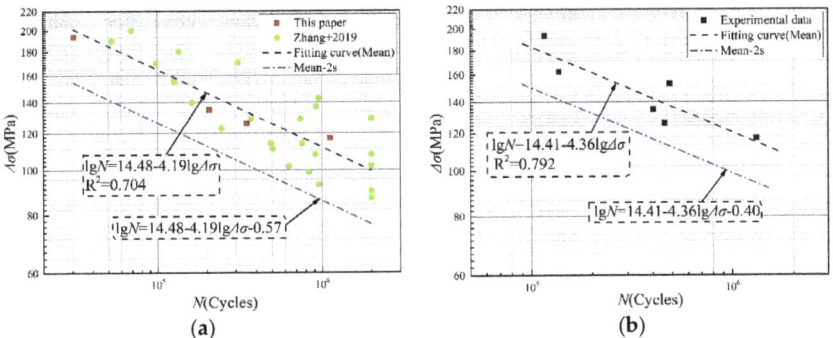

Figure 15. Double-logarithmic S-N curve fitting with $\Delta\sigma$ as variable. (**a**) Unreinforced specimen (**b**) Strengthened with CFRP.

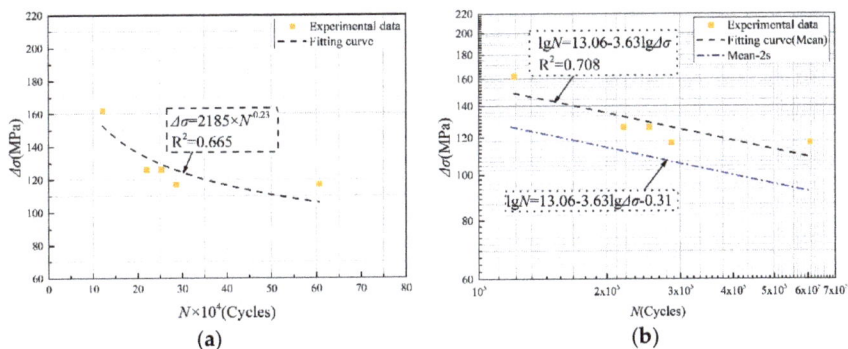

Figure 16. S-N curve of failure mode 3. (**a**) Power (**b**) Double logarithmic.

The corresponding expressions for failure mode 2 are as follows:

$$\Delta\sigma_2 = 1650 \times N^{-0.19} \tag{5}$$

$$lgN_2 = 14.41 - 4.36 lg\Delta\sigma - 0.40 \tag{6}$$

The corresponding expressions for failure mode 3 are as follows:

$$\Delta\sigma_3 = 2185 \times N^{-0.23} \quad (7)$$

$$lgN_3 = 13.06 - 3.63 lg\Delta\sigma - 0.31 \quad (8)$$

4.3. Contrastive Analysis

The test results showed that, for the qualified specimens, the failure position after CFRP reinforcement was transferred to the steel tube–endplate connection. Under the same stress amplitude, the fatigue life of the specimens after CFRP reinforcement could be increased by 17.36%–279.56%. Secondly, no fatigue failure occurred at the reinforcement position when the test stopped. The increase in fatigue life based on the test results was a conservative calculation result. Thirdly, the increase in fatigue life became greater for a larger loading stress amplitude, indicating a better reinforcement effect according to the test results. Lastly, the initial defects of unqualified specimens were obvious, and there were visible neutral deviations in the upper and lower branches in experimental group B. The appearance quality of the weld at the tube–ball joint of the specimen could not meet the requirements of the secondary weld, and the fatigue performance was not improved after CFRP reinforcement.

The test fitting curves are summarized in Figure 17. It can be obtained that after the fatigue failure position was transferred from Location 1 to Location 3, the fatigue strength of the qualified specimen was increased by 13.26%–16.94% between 10,000 and 1,000,000 cycles. Secondly, when the number of cycles reached 2,000,000, the fatigue strength increased by 17.48%. Lastly, when the number of cycles reached 5,000,000, the fatigue strength increased by 18.19%.

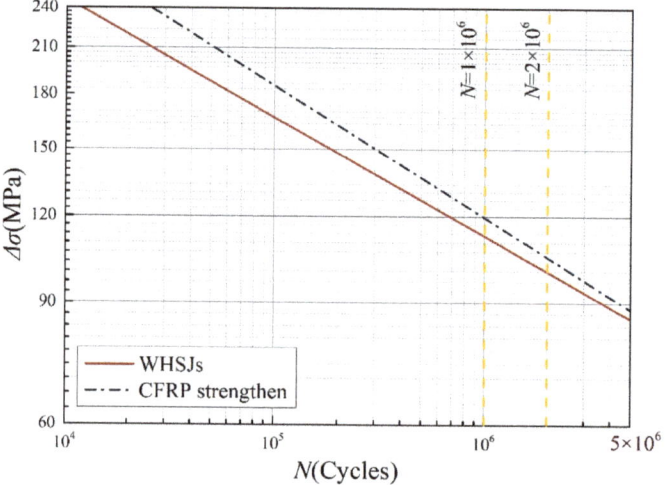

Figure 17. Summary of fitting curves.

The verification test research carried out in this paper demonstrates that the use of CFRP to strengthen the fatigue performance was effective. Nevertheless, the reinforcement effect of CFRP on WSHJ is slightly lower than other welding forms.

5. Fatigue Design Method

Currently, the common fatigue design method is still based on the nominal stress amplitude method. Therefore, two types of fatigue design methods, including the weld toe in the steel pipe of WHSJ (Location 1), and tube–endplate connection (Location 3), were established according to the following formulas:

$$\Delta\sigma_{nom} \leq [\Delta\sigma] \qquad (9)$$

$$[\Delta\sigma] = \left(\frac{C}{n}\right)^{1/\beta} \qquad (10)$$

where $\Delta\sigma_{nom}$ represents the nominal stress amplitude of the steel pipe, and $[\Delta\sigma]$ is the allowable stress amplitude. C and β are the fatigue design parameters that need to be calculated, and n is the fatigue stress cycle number.

For failure mode 1, the fatigue design parameters at the weld toe in the steel tube of the tube–sphere connection were $C = 7.998 \times 10^{13}$ and $\beta = 4.188$, whereas they were $C = 1.02 \times 10^{14}$ and $\beta = 4.360$ in failure mode 2. For the unqualified specimens in failure mode 3, these parameters were $C = 5.60 \times 10^{12}$ and $\beta = 3.626$. For the tube–ball joint weld, the lower limit of allowable stress amplitude of the qualified specimen with $n = 2 \times 10^6$ as the base period and taking 97.7% survival probability was 65.35 MPa, whereas it decreased to 59.95 MPa for the unqualified specimen. For the tube–endplate connection weld, the lower limit of the allowable stress amplitude with the same base period and survival probability was 58.73 MPa, which is lower than the Z8 (71 MPa) of the flange–butt weld connection in the specification [7].

6. Conclusions

In this paper, a finite element simulation and experimental study were conducted to evaluate the fatigue performance of CFRP reinforcement in a welded hollow spherical joint grid structure, and the following conclusions were obtained:

(1) The ranking of the SCF at the weld toe of the WHSJs was as follows: weld toe in steel tube of tube–ball connection weld (Location 1) > weld toe in steel tube of tube–endplate connection weld (Location 3) > weld toe in sphere of tube–ball connection weld (Location 2) > weld toe in plate of tube–endplate connection weld (Location 4). The maximum stress at Location 1 was reduced by 32.93% after CFRP pasting.

(2) The fatigue fracture of the unreinforced WHSJ occurred at Location 1, whereas the failure position was transferred to Location 3 after reinforcement, which was consistent with the finite element simulation results.

(3) The S-N curves at the weld toe of the WHSJs before and after reinforcement were as follows:

$$lgN_1 = 14.48 - 4.19 lg\Delta\sigma - 0.57, \; R^2 = 0.704.$$

$$lgN_2 = 14.41 - 4.36 lg\Delta\sigma - 0.40, \; R^2 = 0.792.$$

Taking $n = 2 \times 10^6$ as the reference period, the lower limit of allowable stress amplitudes under 97.7% survival probability were 65.35 MPa and 58.73 MPa, respectively.

(4) No fatigue failure appeared in the reinforced area of WHSJs. Thus, the increase in fatigue life of WHSJs was conservative on the basis of the test results.

Author Contributions: Conceptualization, Y.D. and H.L.; methodology, Y.D.; software, Y.D.; validation, Y.D. and S.J.; resources, S.J.; data curation, Y.D.; writing—original draft preparation, Y.D.; writing—review and editing, H.L.; visualization, Y.D.; supervision, H.L. and S.J.; project administration, Y.D. and S.J.; funding acquisition, H.L. All authors have read and agreed to the published version of the manuscript.

Funding: This research was funded by National Nature Science Foundation of China grant number No. 51578357 and No. 52278198.

Institutional Review Board Statement: Not applicable.

Informed Consent Statement: Not applicable.

Data Availability Statement: Not applicable.

Conflicts of Interest: The authors declare no conflict of interest.

References

1. Liu, X. Present situation in development of planar space grid structures in China. *Steel Constr.* **1994**, *9*, 13–20. (In Chinese)
2. Lei, H. Research status and trend of welded hollow spherical joints grid structures. In Proceedings of the Sixth Annual Conference of Spatial Structure of Chinese Society of Civil Engineering, Beijing, China, 2–6 December 1992; p. 5. (In Chinese).
3. Lei, H. Research on static and fatigue behaviour of welded hollow spherical joints in space trusses. *J. Build. Struct.* **1993**, *14*, 2–7. (In Chinese)
4. Yan, Y. *Theory Analysis and Testing Study on Fatigue Properties of the Steel Pipe-Welded Hollow Spherical Joints in Space Latticed Structure*; Taiyuan University of Technology: Taiyuan, China, 2013. (In Chinese)
5. Jiao, J.F.; Lei, H.G.; Chen, Y.-F. Experimental Study on Variable-Amplitude Fatigue of Welded Cross Plate-Hollow Sphere Joints in Grid Structures. *Adv. Mater. Sci. Eng.* **2018**, *2018*, 8431584. [CrossRef]
6. Zhang, J. *Experimental and Theoretical Research on Constant Amplitude Fatigue Properties of Weld Toe in Steel Tube of Welded Hollow Spherical Joints in Grid Structures*; Taiyuan University of Technology: Taiyuan, China, 2019. (In Chinese)
7. GB 50017-2017; Standard for Design of Steel Structures. China Architecture and Building Press: Beijing, China, 2017. (In Chinese)
8. Xu, X.; Shu, T.; Zheng, J.; Luo, Y. Experimental and numerical study on compressive behavior of welded hollow spherical joints with external stiffeners. *J. Constr. Steel Res.* **2022**, *188*, 107034. [CrossRef]
9. Zang, Q.; Liu, H.; Li, Y.; Chen, Z.; Li, X. Mechanical properties of welded hollow spherical joints reinforced with external ribs under axial tension. *Spat. Struct.* **2022**, *28*, 71–78. (In Chinese)
10. Li, X. *Study on Welded Hollow Spherical Joints Strengthened while Under Load*; Tianjin University: Tianjin, China, 2021. (In Chinese)
11. Li, C. *Experimental Research and Finite Element Analysis on the Welded Hollow Spherical Joints Strengthened by Steel Thimble out Side the Ball*; Tianjin University: Tianjin, China, 2021. (In Chinese)
12. Chen, T.; Huang, C.; Hu, L.; Song, X. Experimental study on mixed-mode fatigue behavior of center cracked steel plates repaired with CFRP materials. *Thin-Walled Struct.* **2019**, *135*, 486–493. [CrossRef]
13. Zhao, E. *Testing Analysis of Fatigue Durability of Welded Steel Structures Reinforced by CFRP*; Hefei University of Technology: Hefei, China, 2011. (In Chinese)
14. Zhang, N.; Yue, Q.R.; Yang, Y.X.; Hu, L.Q.; Peng, F.M.; Cai, P.; Zhao, Y.; Wei, G.Z.; Zhang, Y.S. Research on the fatigue tests of steel structure member reinforced with CFRP. *Ind. Constr.* **2004**, *4*, 19–21. (In Chinese)
15. Tong, L.; Yu, Q.; Zhao, X.-L. Experimental study on fatigue behavior of butt-welded thin-walled steel plates strengthened using CFRP sheets. *Thin-Walled Struct.* **2020**, *147*, 106471. [CrossRef]
16. Jie, Z.; Wang, K.; Liang, S. Residual stress influence on fatigue crack propagation of CFRP strengthened welded joints. *J. Constr. Steel Res.* **2022**, *196*, 107443. [CrossRef]
17. Chen, T.; Zhao, X.L.; Gu, X.L.; Xiao, Z.G. Numerical analysis on fatigue crack growth life of non-load-carrying cruciform welded joints repaired with FRP materials. *Compos. Part B Eng.* **2014**, *56*, 171–177. [CrossRef]
18. Jie, Z.; Wang, W.; Fang, R.; Zhuge, P.; Ding, Y. Stress intensity factor and fatigue analysis of cracked cruciform welded joints strengthened by CFRP sheets considering the welding residual stress. *Thin-Walled Struct.* **2020**, *154*, 106818. [CrossRef]
19. Chen, T.; Yu, Q.Q.; Gu, X.L.; Zhao, X.L. Study on fatigue behavior of strengthened non-load-carrying cruciform welded joints using carbon fiber sheets. *Int. J. Struct. Stab. Dyn.* **2012**, *12*, 179–194. [CrossRef]
20. Amraei, M.; Jiao, H.; Zhao, X.L.; Tong, L.W. Fatigue testing of butt-welded high strength square hollow sections strengthened with CFRP. *Thin-Walled Struct.* **2017**, *120*, 260–268. [CrossRef]
21. Tong, L.; Xu, G.; Zhao, X.L.; Yan, Y. Fatigue tests and design of CFRP-strengthened CHS gap K-joints. *Thin-Walled Struct.* **2021**, *163*, 107694. [CrossRef]
22. Mohamed, H.S.; Zhang, L.; Shao, Y.B.; Yang, X.S.; Shaheen, M.A.; Suleiman, M.F. Stress concentration factors of CFRP-reinforced tubular K-joints via Zero Point Structural Stress Approach. *Mar. Struct.* **2022**, *84*, 103239. [CrossRef]
23. Xiao, Z.; Zhao, X.L.; Mashiri, F.R.; Xu, B. Fatigue Experiments on CFRP Repaired Welded Thin-Walled Rhsto-Rhs Cross-Beam Connection. In Proceedings of the 10th International Symposium on Structural Engineering for Young Experts (ISSEYE10), Changsha, China, 19–21 October 2008; pp. 971–978.
24. Wang, J. *Research on Fatigue of Steel Crane Beam Strengthened with CFRP*; Shijiazhuang Tiedao University: Shijiazhuang, China, 2021. (In Chinese)
25. Zhao, F. *FE Analysis of Welded Steel Beam Strengthened by CFRP*; Hefei University of Technology: Hefei, China, 2009. (In Chinese)
26. Xu, S. *Research of the Fatigue Test of Welded Steel Structure Reinforced by CFRP*; Hefei University of Technology: Hefei, China, 2008. (In Chinese)
27. JGJ-2010; Technical Specification for Space Frame Structures. China Architecture and Building Press: Beijing, China, 2010. (In Chinese)
28. GB 50205-2020; Standard for Acceptance of Construction Quality of Steel Structures. China Planning Press: Beijing, China, 2020. (In Chinese)
29. SY/T 5768-2016; Welded Steel Tubes for General Construction. National Energy Administration: Beijing, China, 2016. (In Chinese)

Article

Effects of Sodium Carbonate and Calcium Oxide on Roasting Denitrification of Recycled Aluminum Dross with High Nitrogen Content

Hongjun Ni [1,2], Chunyu Lu [1], Yu Zhang [1], Xingxing Wang [1], Yu Zhu [1], Shuaishuai Lv [1,*] and Jiaqiao Zhang [1]

1. School of Mechanical Engineering, Nantong University, Nantong 226019, China; ni.hj@ntu.edu.cn (H.N.); lcy@stmail.ntu.edu.cn (C.L.); 1910310020@stmail.ntu.edu.cn (Y.Z.); wangxx@ntu.edu.cn (X.W.); zhu.y@ntu.edu.cn (Y.Z.); 1810310038@stmail.ntu.edu.cn (J.Z.)
2. Jiangsu Aluminum Dross Solid Waste Harmless Treatment and Resource Utilization Engineering Research Center, Nantong 226019, China
* Correspondence: lvshuaishuai@ntu.edu.cn

Abstract: Aluminum dross is solid waste produced by the aluminum industry. It has certain toxicity and needs to be treated innocuously. The effect of sodium carbonate and calcium oxide on the denitrification efficiency of high nitrogen aluminum dross roasting was studied in this paper. By means of XRD, SEM and other characterization methods, the optimum technological parameters for calcination denitrification of the two additives were explored. The test results show that both additives can effectively improve the efficiency of aluminum dross roasting denitrification, and the effect of sodium carbonate is better. When the mass ratio of sodium carbonate to aluminum dross is 0.6, the roasting temperature is 1000 °C and the roasting time is 4 h, the denitrification rate can reach 91.32%.

Keywords: roasting denitrification; additive; process parameter

1. Introduction

Aluminum dross is solid waste produced in the process of aluminum production. It is mainly composed of non-molten impurities, oxides and various additives floating on the surface of aluminum melt during smelting [1]. According to the source of aluminum dross, aluminum dross can be divided into primary aluminum dross and secondary aluminum dross. Primary aluminum dross usually refers to the scum insoluble in aluminum liquid generated in the process of electrolytic aluminum and cast aluminum. The content of metallic aluminum in primary aluminum dross is high up to 30–70%. In addition, it also contains fluorinated salt, aluminum oxide, aluminum nitride and other substances. The color is generally gray-white, also known as white aluminum dross. Secondary aluminum dross usually refers to the waste residue generated in secondary aluminum industry processes such as remelting primary aluminum dross or recovering aluminum alloy from waste aluminum. A variety of salt solvents need to be added in the process of recovering metal aluminum in the secondary aluminum industry. Therefore, the composition of secondary aluminum dross is more complex than that of primary aluminum dross, mainly including metal aluminum, aluminum oxide, aluminum nitride, aluminum carbide, chloride, etc. The content of metal aluminum is less, which can be reduced to less than 10%. The secondary aluminum dross is generally black, also known as black dross.

In 2020, the global output of recycled aluminum will reach 34.71 million tons, accounting for 33.10% of the total amount of primary aluminum and recycled aluminum [2]. The output of recycled aluminum in developed countries has generally exceeded the output of primary aluminum, such as the United States, which has accounted for more than 80% of the total. While the recycled aluminum industry provides aluminum materials for economic construction, a large amount of aluminum dross, namely recycled aluminum dross, is also produced in the regeneration process [3]. Recycled aluminum dross generally belongs to secondary aluminum dross. There are few heavy metals in aluminum dross slag, but there are still heavy metals such as Se, Cr and Pb [4]. A large amount of accumulation will cause heavy metal pollution of surrounding soil and groundwater [5]. Soluble salts such as NaCl and KCl contained in aluminum dross will enter the soil with rainwater and cause soil salinization [6]. Fluoride in aluminum dross easily causes fluoride pollution, resulting in muddy soil when it is wet. At the same time, aluminum dross contains many chemical reactive substances, which makes it dangerous and toxic. For example, AlAs in aluminum dross will react with water to produce AsH_3 gas, which will lead to hydrogen arsenide poisoning of close contacts and endanger the health of nearby personnel [7]. Metal aluminum and aluminum carbide will react with water to produce H_2 and CH_4 combustible gases and release a large amount of heat, which is easy to cause a fire [8]. In particular, aluminum dross is rich in AlN, which can react with water in the air and slowly release the unpleasant irritant gas NH_3, seriously polluting the atmosphere [9]. Human beings have a very sensitive sense of smell for ammonia. The concentration of ammonia in the air reaches 15.2 mg/m^3, and human beings have an obvious sense of stimulation [10]. When the ammonia concentration is 1290.2 mg/m^3, humans can cause severe cough, seriously damage the respiratory system and may die within 30 min [11]. Therefore, aluminum dross is recognized as hazardous waste in the National Hazardous Wastes Catalogue (2021 Edition) [12] and the European Hazardous Waste List [13].

At present, the treatment methods for recycled aluminum dross are mainly stockpiling and landfill [14], which will cause irreparable damage to the environment. Therefore, it is necessary to find a harmless treatment method with low cost and high efficiency to realize the harmless treatment of recycled aluminum dross. The content of fluoride in recycled aluminum dross is low, and its properties are relatively stable, which will not affect the performance of resource utilization products, and even fluoride can be removed in the process of resource utilization [15]. Recycled aluminum dross contains a large amount of aluminum nitride, and its properties are very unstable [16]. However, the existing aluminum dross recycling technology cannot directly remove AlN, so the recycled aluminum dross needs to be pretreated for nitrogen removal during the recycling process [17].

This paper aims to explore the effect of sodium carbonate and calcium oxide on the denitrification efficiency of aluminum dross roasting. The effects of additive ratio, calcination temperature and calcination time on the denitrification effect of high nitrogen content aluminum dross were studied, and the best process parameters were found to provide a reference for the study of the harmless treatment process of aluminum dross.

2. Experiment
2.1. Raw Material, Reagent and Instrument

Raw material: Recycled aluminum dross with high nitrogen concentration comes from Jiangsu Haiguang Metal Co., Ltd. (Suqian, China), which is the aluminum dross with high nitrogen concentration obtained after recycling part of metal aluminum with aluminum dross. The production flow chart is shown in Figure 1.

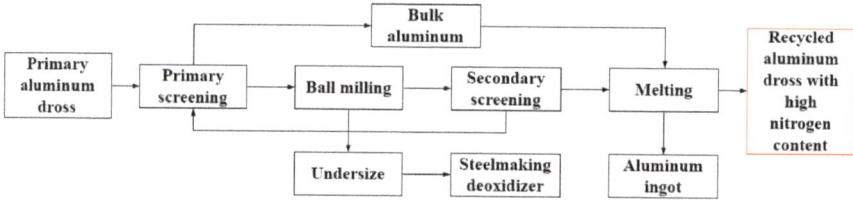

Figure 1. Production flow chart of aluminum dross.

Reagent: sodium carbonate (analytical grade, Jiangsu Qiangsheng Functional Chemical Co., Ltd., Suzhou, China), calcium oxide (analytical grade, Tianjin Damao Chemical Reagent Factory, Tianjin, China), sodium hydroxide (analytical grade, Xilong Chemical Co., Ltd. Shantou, China), hydrochloric acid (analytical grade, Shanghai Lingfeng Chemical Reagent Co., Ltd., Shanghai, China), methyl red (indicator, Shanghai Yuanye Biotechnology Co., Ltd., Shanghai, China), methylene blue (indicator, COOLABER SCIENCE&TECHNOLOGY) and boric acid (chemical pure, Shanghai Merrill Chemical Technology Co., Ltd., Shanghai, China).

Instrument: Program-controlled batch-type furnace (SXL-1230, Hangzhou Zhuochi Instrument Co., Ltd., Hangzhou, China), universal electric furnace (DK-98-II, Jiangsu Xinghua Chushui Electric Thermal Appliance Factory, Taizhou, China), ultrapure water machine (EPED-10TH, Nanjing Yipu Yida Technology Development Co., Ltd., Nanjing, China), X-ray fluorescence spectrometer (ZSX PRIMUS III+, Rigaku Electric Co., Ltd., Beijing, China), X-ray diffractometer (D8 Advance, Bruker Technology Co., Ltd., Beijing, China) and field emission scanning electron microscope system (ZEISS sigma HD, Carl Zeiss Management Co., Ltd., Shanghai, China).

XRD: the maximum output power is 3 kW, the voltage is 20–60 kV, the current is 2–60 mA, and the radius of the goniometer is 185 mm. MDI jade 9 was used for XRD analysis. XRD database is PDF 2009.

SEM-EDS: the acceleration voltage is 10 kV, the magnification is 100–200 k, the secondary electron resolution is 1 nm.

2.2. Experimental Method

The experiment mainly studies the effects of mixture ratio, roasting temperature and roasting time process parameters on the roasting denitrification effect of aluminum dross. The specific experimental steps are as follows:

(1) The additive and fully dried aluminum dross are ground according to the mixture ratio ($m_{additive}$:$m_{aluminum\ dross}$). 50 g homogenized mixture is extracted and placed in the corundum crucible.
(2) The crucible is placed in a batch-type furnace for firing at a certain temperature.
(3) The crucible is removed from the batch-type furnace after firing for a certain period of time and then cooled to room temperature in air.
(4) The cooled aluminum dross is ground into small pieces to complete the calcined sample. The samples are stored in a desiccator for subsequent detection of the denitrification rate, microscopic morphology and material composition of aluminum dross.

2.3. Detection Method

The acid-base constant volume method is used to determine the denitrification rate of aluminum dross. The specific operations are as follows [18]:

(1) First, 2 g aluminum dross is weighed out by balance and then poured into a conical flask containing 150 mL 20% NaOH solution. The stopper needs to be capped tightly.
(2) The conical flask is placed on a universal electric stove and heated until the solution boils. The solution is kept boiling for 2 h for distillation. The distilled ammonia gas is absorbed with 200 mL 40 g/L boric acid solution.
(3) After distillation, 0.05 mol/L dilute hydrochloric acid solution is used for titration, and standard methyl red-methylene blue is used as an indicator. During the titration, the endpoint of the titration is that the solution changes from blue to purple.

The calculation of AlN content and denitrification rate in aluminum dross is as follows:

$$W_1 = \frac{2.05C(V_2 - V_1)}{bM} \times 100\% \tag{1}$$

$$X_b = \frac{W_2 - W_1}{W_2} \times 100\% \tag{2}$$

where, W_1 is the aluminum nitride content in the sample (100%); V_1 is the volume of dilute hydrochloric acid consumed by the test sample (mL); V_2 is the volume of dilute hydrochloric acid consumed by the blank control group (mL); C is the concentration of dilute hydrochloric acid (mol/L); M is the mass of the test sample taken during measurement (g); b is the total mass of aluminum dross added to the test sample under the current mixture ratio (g); X_b is the denitrification rate (100%); W_2 is the aluminum nitride content (100%) in the original aluminum dross.

3. Results and Discussion

3.1. Organization and Composition of Aluminum Dross

XRD, XRF and SEM-EDS were used to analyze aluminum dross. The XRF test results are shown in Table 1, and the XRD test results are shown in Figure 2. It can be seen from Table 1 and Figure 2 that the composition of aluminum dross is relatively complex, mainly containing Al, Si, Cl and other elements, of which the content of Al element accounts for 61.69% at most, and the main phases are Al_2O_3, AlN and NaCl, while the single substance Al phase is relatively small, indicating that the Al element in aluminum dross mostly exists in aluminum-containing compounds, not in the form of an aluminum single substance, so the regenerated aluminum dross is not suitable for remelting and recovering metal aluminum. The SEM-EDS detection results are shown in Figure 3. Aluminum dross is composed of relatively independent particles of different sizes and shapes. The spherical particles in the red frame area in Figure 3a are mainly composed of elements Al, O and N. SEM, micro-area element surface distribution and EDS energy spectrum analysis are shown in Figure 3b,c, respectively.

Table 1. XRF analysis results of original aluminum dross (Unit: wt%).

Element	Al	Si	Cl	Na	Mg	Ca	Fe	K	Ti	Other
Content	61.69	8.58	7.70	4.30	3.32	3.13	2.77	2.47	1.26	4.78

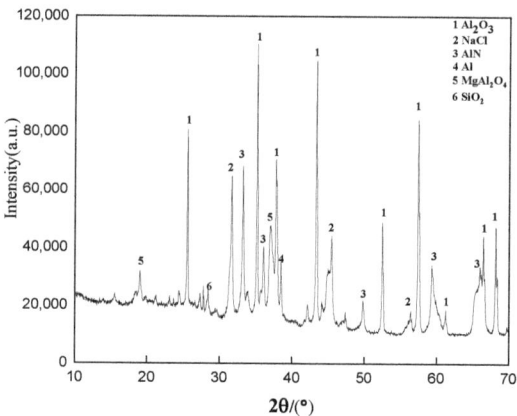

Figure 2. XRD pattern of original aluminum dross.

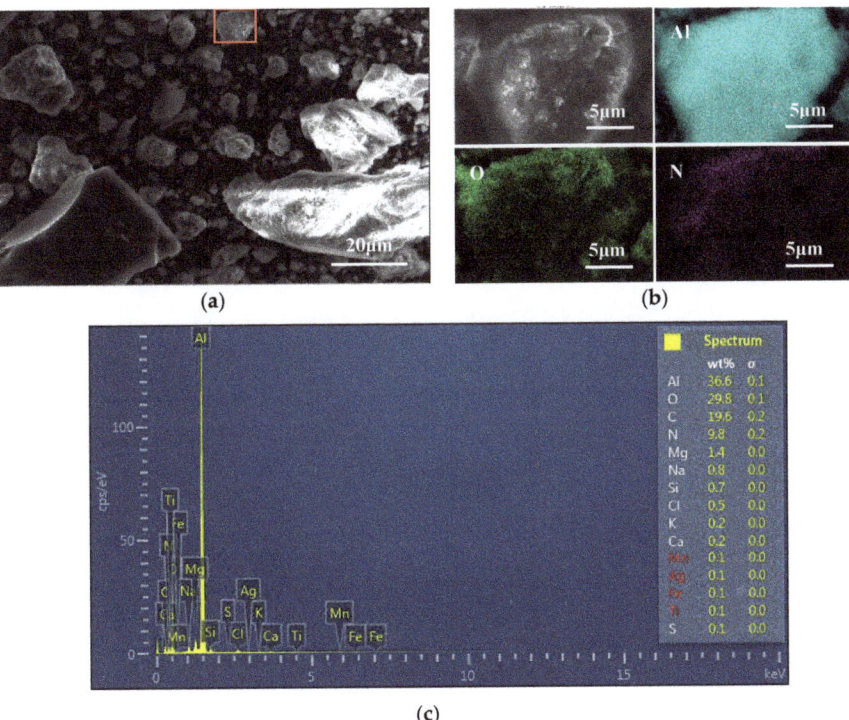

(c)

Figure 3. SEM-EDS image of original aluminum dross. (**a**) SEM of original aluminum dross; (**b**) SEM and micro-area element surface distribution of the red frame area; (**c**) EDS spectrum analysis of the red frame area.

SEM-EDS scanning results combined with XRD detection results show that Al, O and n elements in aluminum dross are mixed and doped with each other. AlN and Al_2O_3 in aluminum dross do not exist alone but are mixed with each other to form a complex mixture. Due to the relatively stable performance of Al_2O_3, the direct roasting and denitrification effect of aluminum dross is poor when Al_2O_3 wraps AlN. Therefore, additives need to be added to break the shell wrapped in AlN to improve the roasting and denitrification efficiency.

The acid-base constant volume method is used to determine the AlN content in the original recycled aluminum dross. With alumina as the blank control group, the results are shown in Table 2.

Table 2. Test results of AlN content in original recycled aluminum dross.

	Raw Material Aluminum Dross	Alumina
Hydrochloric acid consumption/mL	149.6 148.1 148.6	13.4 13.0 12.9
Average hydrochloric acid consumption/mL	148.76	13.1
AlN content/%	13.9%	/

3.2. Effect of Sodium Carbonate and Calcium Oxide on Roasting Denitrification of Aluminum Dross with High Nitrogen Concentration

3.2.1. Effect of Mixture Ratio on Denitrification of Aluminum Dross

The mixture ratio is the most important parameter used as an additive to speed up the denitrification rate of aluminum dross. An excessive mixture ratio will cause the additive to fail to react with the substances in the aluminum dross, resulting in a waste of resources. Too small a mixture ratio will result in poor nitrogen removal. Under constant roasting temperature and roasting time, the effect of the change of additive mass ratio to aluminum dross on the denitrification rate of aluminum dross by the mixture ratio is studied. Aluminum dross with a mixture ratio of 0.2, 0.4, 0.6, 0.8 or 1 is calcined at 900 °C for 4 h. The variation of the denitrification rate of aluminum dross with different mixture ratios is shown in Figure 4.

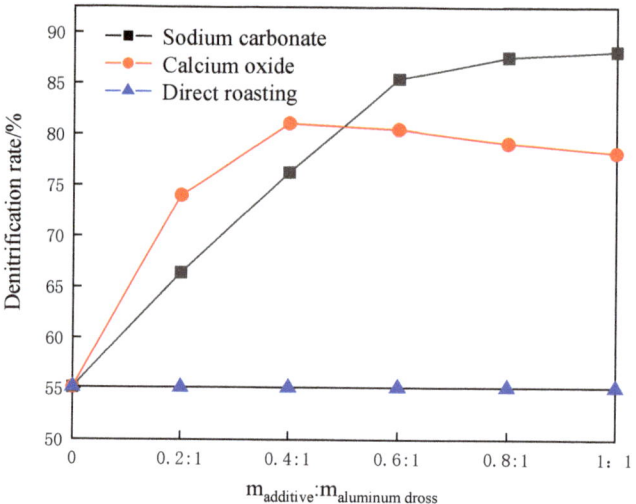

Figure 4. Denitrification rate of aluminum dross varies with mixture ratio.

It can be seen from Figure 4 that the denitrification rate of aluminum dross increases with the increase of the added amount of sodium carbonate. When $m_{sodium\ carbonate}:m_{aluminum\ dross}$ = 0.6, the denitrification rate of aluminum dross reaches 85.48%. The denitrification rate of aluminum dross does not increase significantly when sodium carbonate continues to increase. Under the same process conditions, the denitrification rate of direct roasting aluminum dross is only 55.12%. Na_2CO_3 can chemically react with Al_2O_3 at a certain temperature, as shown in formula (3).

$$Al_2O_3 + Na_2CO_3 = 2NaAlO_2 + CO_2 \qquad (3)$$

Thermodynamic software HSC6.0 is used for the calculation of formula (3). Na_2CO_3 and Al_2O_3 can react spontaneously to form $NaAlO_2$ at 800 °C. The generated $NaAlO_2$ is an important solid phase for dissolving Al_2O_3, the Al_2O_3 shell wrapped with AlN can be broken open and more AlN can be directly exposed to the air, so the denitrification rate is improved [19]. Therefore, $m_{sodium\ carbonate}:m_{aluminum\ dross}$ = 0.6 is the optimal mixture ratio for roasting denitrification with sodium carbonate added.

For the samples added with calcium oxide, the denitrification rate of aluminum dross increases first and then decreases with the increase in the mixture ratio. At the same time, through the experimental phenomenon, it is observed that the agglomeration hardening degree of the treated sample is opposite to the change in denitrification rate, which first decreases and then increases. When the proportioning ratio was 0.2, the denitrification rate was 74.03%. When the blending ratio was 0.4, the agglomeration and hardening degree of the sample was the lowest, and the maximum denitrification rate was 81.08%. The ratio of ingredients continued to increase, and the denitrification rate began to decrease. When the proportioning ratio was 0.6 and 0.8, the denitrification rates were 80.5% and 79.14% respectively. Considering that CaO cannot react directly with AlN, the above phenomena show that the denitrification rate of aluminum dross is affected by the content of metal aluminum and calcium oxide in the material.

The oxidation rate of AlN in aluminum dross can be affected by the environmental atmosphere. The more fully AlN contacts with O_2, the faster its oxidation rate. When the mixture quality is constant, the greater the batching ratio, the lower the proportion of metal aluminum in the mixture, and the smaller the adhesion and wrapping of metal aluminum to the flowing aluminum dross particles after melting under this condition, so as to increase the contact opportunity between AlN in aluminum dross and O_2 in the air and improve the denitrification rate. Continuing to increase the mixture ratio leads to an increase in the proportion of CaO in the mixture. At this time, the particle size of calcium oxide powder is significantly lower than that of aluminum dross. Small calcium oxide powder fills the gaps in the aluminum dross, causing the material to become denser. The reduced AlN contact with O_2 reduces the denitrification rate [20]. Therefore, the optimal mixture ratio of aluminum dross treated with calcium oxide is $m_{calcium\ oxide} : m_{aluminum\ dross} = 0.4$.

To sum up, both Na_2CO_3 and CaO can speed up the roasting denitrification rate of aluminum dross. Among them, Na_2CO_3 reacts with Al_2O_3, so that more AlN is exposed to the air, which improves the denitrification rate. And CaO is to reduce the degree of encapsulation of aluminum dross particles after melting by changing the metal aluminum content in the mixture, thereby improving the denitrification rate. In contrast, adding Na_2CO_3 denitrification is more direct and the effect is better.

3.2.2. Effect of Roasting Temperature on Denitrification of Aluminum Dross

In the case of roasting time = 4 h and additive ratio = 0.6(sodium carbonate)/0.4(calcium oxide), the temperature is set to 600, 700, 800, 900 and 1000 °C to study the denitrification effect of aluminum dross by roasting temperature, as shown in Figure 5.

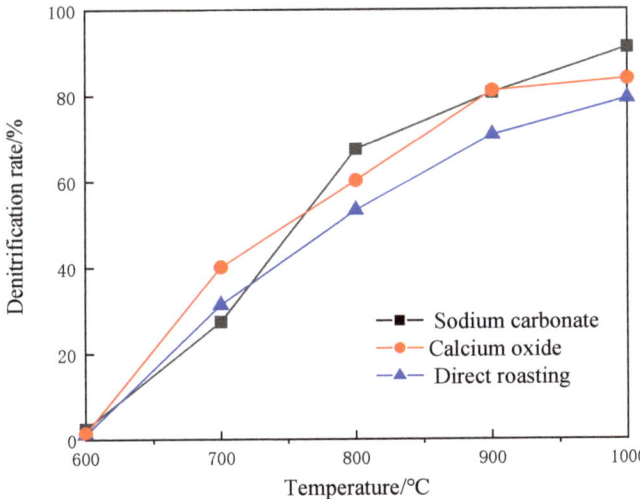

Figure 5. Denitrification rate of aluminum dross varies with roasting temperature.

According to Figure 5, the denitrification rate of the calcined samples with added sodium carbonate gradually increased with increasing temperature. According to the thermodynamic analysis, when the temperature is low, sodium carbonate and Al_2O_3 cannot spontaneously produce stress and do not damage the dense oxide film of Al_2O_3. Therefore, the denitrification effect is similar to that of direct roasting denitrification without substantial change. When the temperature exceeds 700 °C, sodium carbonate and Al_2O_3 react, the dense Al_2O_3 oxide film wrapped in AlN is broken, and the denitrification rate increases sharply. At this time, the denitrification effect is better than that of direct roasting. When the temperature rises to 1000 °C, the denitrification rate can reach 91.15%. In contrast, the direct roasting denitrification rate at this time is only 79.28%. Therefore, the optimal roasting temperature is 1000 °C.

For calcium oxide-added samples, the denitrification rate of aluminum dross increases with temperature. At 300 °C, the denitrification rate of aluminum dross is only 0.98%, which shows that AlN in aluminum dross oxidizes slowly under this condition. At 500 °C, the denitrification rate of the sample is not much different from that of direct roasting, with a denitrification rate of 1.59%. At 700 °C, the metal aluminum reaches the melting point and begins to melt under this condition. After melting, the degree of adhesion and wrapping of the aluminum dross particles decreases, thereby increasing the contact opportunity between AlN in the aluminum dross and O_2 in the air. At this time, the denitrification rate begins to appear significantly. increased, reaching 40.14%. At 900 °C, the difference between the denitrification rate (81.03%) of the roasted samples and that of direct roasting with calcium oxide addition is 10.24% at most. When the temperature continues to increase, the denitrification rate increases significantly slower [21]. Considering the effect of nitrogen removal and processing cost, the optimal roasting temperature is 900 °C. To sum up, the roasting temperature has a significant effect on the roasting denitrification rate of aluminum dross. The effect of both additives in the low-temperature stage was not obvious. As the temperature gradually increases the denitrification rate increases sharply. Through the analysis of the test results, the denitrification effect of Na_2CO_3 is far better than the denitrification effect of direct roasting and slightly better than that of CaO.

3.2.3. Effect of Roasting Time on Denitrification of Aluminum Dross

When the mixture ratio = 0.6 (sodium carbonate) and the roasting temperature = 1000 °C/ mixture ratio = 0.4 (calcium oxide) and the roasting temperature = 900 °C, the roasting time is set as 1 h, 2 h, 3 h, 4 h or 5 h to study the effect of roasting time on the denitrification of aluminum dross, as shown in Figure 6.

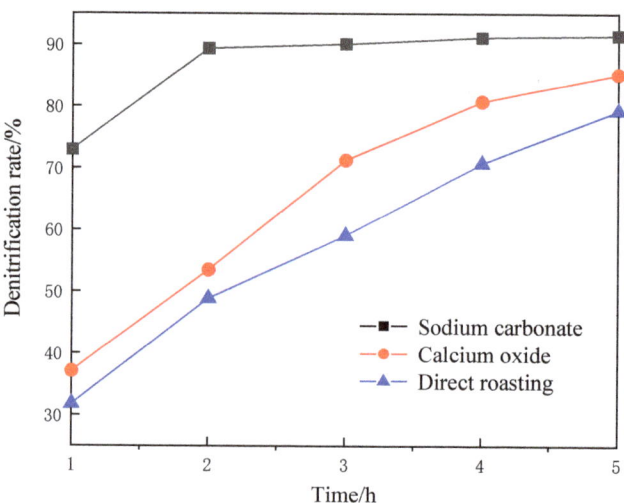

Figure 6. Denitrification rate of aluminum dross as a function of roasting time.

It can be seen from Figure 6 that at the roasting temperature of 1000 °C, sodium carbonate can accelerate the roasting denitrification of aluminum dross. With increasing temperature, the denitrification rate of both samples roasted with sodium carbonate and direct roasting increases. When roasting for 1 h, the denitrification rate of the sample added with sodium carbonate reaches more than 70%, while the sample of direct roasting is only about 30%, with an obvious gap. After calcination for 2 h, the denitrification rate of the sample added with sodium carbonate does not change significantly, which is stable at around 90%. At this time, its denitrification effect is still better than that of direct roasting. The treated samples are found to have increased whiteness with increasing roasting time. Considering that aluminum dross is black, it can be concluded that the reason for this phenomenon is the volatilization of certain impurities in aluminum dross. Therefore, the optimal denitrification process parameters of sodium carbonate-added aluminum dross are: $m_{sodium\ carbonate} : m_{aluminum\ dross}$ = 0.6, roasting temperature = 1000 °C and roasting time = 3 h.

For calcium oxide-added samples, the denitrification rate of aluminum dross increases with time, with a gradually slowing rate. When the roasting time is 1, 2 and 3 h, the denitrification rates of aluminum dross are 37.12%, 53.53% and 71.29%, respectively. When the roasting time is 5 h, the denitrification rate of aluminum dross reaches the highest (85.25%). Therefore, the optimal denitrification process parameters of calcium oxide-added aluminum dross are: $m_{calcium\ oxide} : m_{aluminum\ dross}$ = 0.4, roasting temperature = 900 °C and roasting time = 5 h.

3.3. Microstructure and Composition of Calcined Samples

Under $m_{sodium\ carbonate} : m_{aluminum\ dross}$ = 0.6, roasting temperature = 1000 °C and roasting time = 3 h, XRD detection of the roasted samples with sodium carbonate added is carried out, as shown in Figure 7a. The roasted samples with calcium oxide added under $m_{calcium\ oxide} : m_{aluminum\ dross}$ = 0.4, roasting temperature = 900 °C, roasting time = 5 h are taken for XRD detection, as shown in Figure 7b.

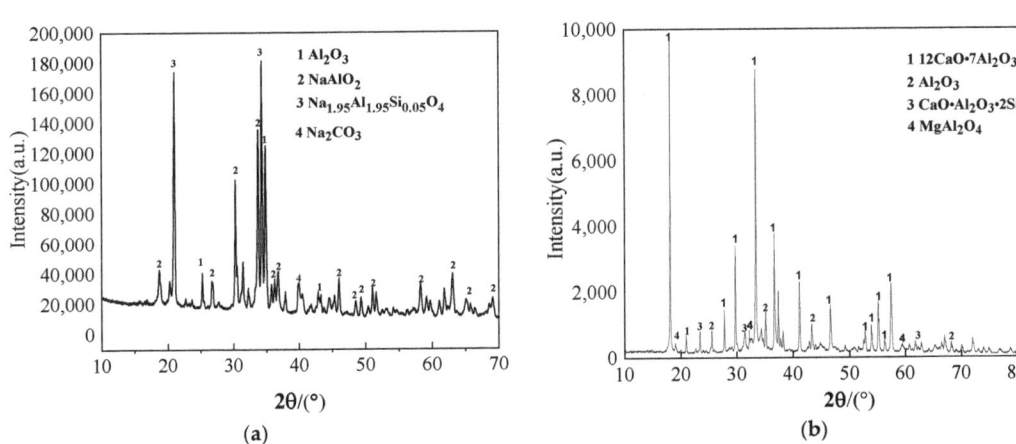

Figure 7. XRD of roasted samples with sodium carbonate and calcium oxide. (a) Add sodium carbonate; (b) Add calcium oxide.

No Al, AlN and NaCl phases were found in the samples, indicating that under this condition, Al in the aluminum dross slag had all reacted with O_2 and N_2 to form Al_2O_3 and AlN, see reaction formulas (4) and (5). AlN is oxidized to Al_2O_3 and shows the disappearance of the phase. The disappearance of the NaCl phase is the reduction of the volatilization of the substance at high temperature, which corresponds to the observed increase in the whiteness of the sample, indicating that the high-temperature waste gas generated during the roasting process not only contains N_2 and various nitrogen oxides but also contains a variety of aluminum dross slag. Due to the impurities volatilized during roasting, the waste gas needs to be treated during the treatment process to prevent environmental pollution.

$$4Al + 3O_2 = 2Al_2O_3 \tag{4}$$

$$2Al + N_2 = 2AlN \tag{5}$$

For the new phases $NaAlO_2$, $Na_{1.95}Al_{1.95}Si_{0.05}O_4$, $12CaO \cdot 7Al_2O_3$ and $CaO \cdot Al_2O_3 \cdot 2SiO_2$, the samples added with sodium carbonate and calcium oxide were further analyzed by SEM-EDS.

SEM-EDS analysis of the sodium carbonate-added sample is shown in Figure 8. The main elements contained in the scanning area are O, Al, Na and Si, of which O content is the highest (52.26%), followed by Al and Na (13.62% and 13.41%) and Si (2.36%). According to Figure 7a, $NaAlO_2$ and $Na_{1.95}Al_{1.95}Si_{0.05}O_4$ are mixed together. Through the analysis of EDS scanning results, the Si content is much lower than the Al content. Therefore, the formation of $NaAlO_2$ from sodium carbonate and alumina is the main reaction during calcination. SiO_2 react with sodium carbonate and alumina to generate $Na_{1.95}Al_{1.95}Si_{0.05}O_4$, see reaction formula (6) [22].

$$1.95NaCO_3 + 1.95Al_2O_3 + SiO_2 = 2Na_{1.95}Al_{1.95}Si_{0.05}O_4 + 1.95CO_2 \tag{6}$$

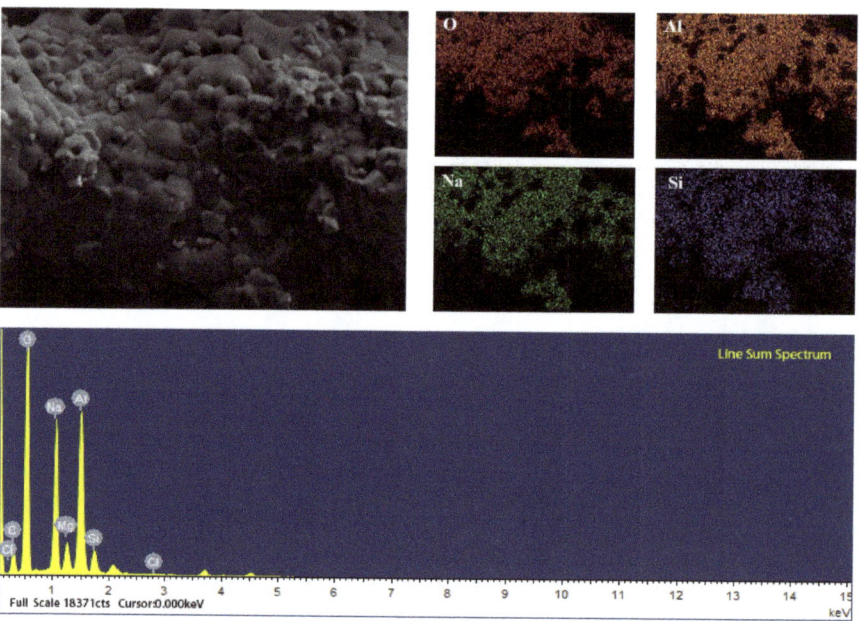

Figure 8. SEM-EDS image of the roasted sample with sodium carbonate added at $m_{sodium\ carbonate}:m_{aluminum\ dross}$ = 0.6, roasting temperature = 1000 °C, roasting time = 3 h.

SEM-EDS analysis of the calcium oxide-added sample is shown in Figure 9. The aluminum dross particles appeared fine and aggregated after adding calcium oxide and calcined. The red frame area in Figure 9 is further subjected to SEM and micro-area element surface distribution, as well as EDS energy spectrum detection. According to the results, the main elements contained in the red box area are Al, O and Ca. Combined with the analysis of Figure 7b, it is believed that the addition of calcium oxide calcination can convert the difficult leaching Al_2O_3 in the aluminum dross into easily leaching $12CaO \cdot 7Al_2O_3$, and at the same time can generate by-product CaO·Al$_2$O$_3$·2SiO$_2$, as shown in the reaction formula (7) and (8).

$$12CaO + 7Al_2O_3 = 12CaO·7Al_2O_3 \tag{7}$$

$$CaO + Al_2O_3 + 2SiO_2 = CaO·Al_2O_3·2SiO_2 \tag{8}$$

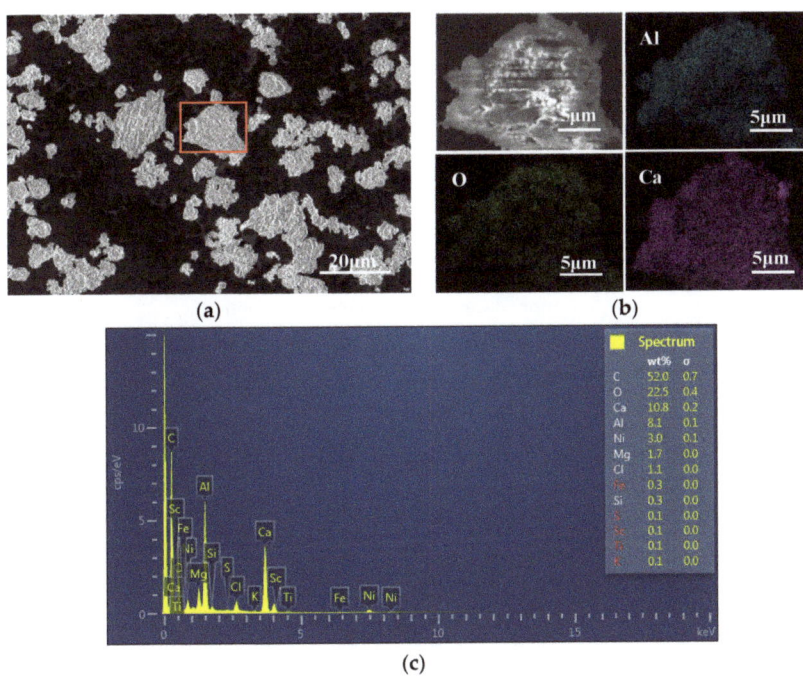

Figure 9. (a–c) SEM-EDS image of calcium oxide-added roasted sample under $m_{calcium\ oxide}$:$m_{aluminum\ dross}$ = 0.4, roasting temperature = 900 °C, roasting time = 5 h.

4. Conclusions

(1) The addition of sodium carbonate and calcium oxide can speed up the roasting denitrification rate of aluminum dross, which can be converted into valuable substances such as NaAlO$_2$ and 12CaO·7Al$_2$O$_3$ after roasting. In the process of treatment, both methods will produce high-temperature exhaust gas containing a variety of nitrogen oxides and impurities in aluminum dross. Therefore, it is necessary to collect and treat the gas in the treatment process to prevent environmental pollution.

(2) Adding sodium carbonate can directly destroy the oxide film wrapped on the surface of AlN, improve the contact area between ain and air, and improve the denitrification rate of aluminum dross.

(3) Calcium oxide mainly depends on reducing the wrapping degree of molten aluminum in the sample to make AlN more fully in contact with O$_2$ in the air, thus increasing the denitrification rate of aluminum dross. It further shows that the lower the content of metal aluminum in aluminum dross, the better the denitrification effect of aluminum dross treated by the calcium oxide roasting method and the lower the roasting temperature and less energy consumption.

(4) The optimal denitrification process parameters of the two additives are obtained by the single factor optimization experiment method. Under $m_{sodium\ carbonate}$:$m_{aluminum\ dross}$ = 0.6, roasting temperature = 1000 °C and roasting time = 4 h, the denitrification rate can

reach 91.32%. Under $m_{calcium\ oxide}:m_{aluminum\ dross}$ = 0.4, roasting temperature = 900 °C and roasting time = 5 h, the denitrification rate can reach 85.25%.

(5) The research results have certain guiding significance for aluminum dross harmless treatment and have reference value for the research and development of new aluminum dross harmless treatment equipment and resource utilization equipment.

Author Contributions: Conceptualization, H.N. and X.W.; methodology, C.L.; validation, C.L.; formal analysis, H.N. and S.L.; investigation, Y.Z. (Yu Zhang) and X.W.; resources, H.N. and S.L.; writing—original draft preparation, Y.Z. (Yu Zhang); writing—review and editing, H.N. and Y.Z. (Yu Zhu); visualization, C.L. and J.Z.; supervision, H.N. All authors have read and agreed to the published version of the manuscript.

Funding: This research was supported by the Priority Academic Program Development of Jiangsu Higher Education Institutions, grant number PAPD. Jiangsu Province Policy Guidance Program (International Science and Technology Cooperation) Project, grant number BZ2021045; Nantong Applied Research Project, grant number JCZ21066, JCZ21043, JCZ21013.

Institutional Review Board Statement: Not applicable.

Informed Consent Statement: Not applicable.

Data Availability Statement: Data is contained within the article.

Conflicts of Interest: The authors declare no conflict of interest.

References

1. Hong, J.-P.; Wang, J.; Chen, H.-Y.; Sun, B.-D.; Li, J.-J.; Chen, C. Process of aluminum dross recycling and life cycle assessment for Al-Si alloys and brown fused alumina. *Trans. Nonferrous Met. Soc. China* **2010**, *20*, 2155–2161. [CrossRef]
2. Yang, C.; Feng, N. Current situation of resource recovery and utilization of industrial secondary aluminum ash. *Mod. Chem. Ind.* **2022**, *42*, 73–77. (In Chinese)
3. Zhou, F. Present situation and Development Countermeasures of China's recycled aluminum industry. *Period. Hebei* **2022**, 48–50. (In Chinese) [CrossRef]
4. Cai, B.; Deng, J.; Tan, X.; Ma, R.; Ren, T.; Xi, R. Characteristics of aluminum ash and slag in reclaimed aluminum industry and its storage environment risk prevention and control. *Inorg. Chem. Ind.* **2021**, 117–121. (In Chinese) [CrossRef]
5. Abdulkadir, A.; Ajayi, A.; Hassan, M.I. Evaluating the chemical composition and the molar heat capacities of a white aluminum dross. *Energy Procedia* **2015**, *75*, 2099–2105. [CrossRef]
6. Adeosun, S.O.; Sekunowo, O.I.; Taiwo, O.O.; Ayoola, W.A.; Machado, A. Physical and mechanical properties of aluminum dross. *Adv. Mater.* **2014**, *3*, 6–10. [CrossRef]
7. Gopienko, V.G.; Kiselev, V.P.; Zobnina, N.S. Methods of manufacture of aluminum powders and their fields of application. *Sov. Powder Metall. Met. Ceram.* **1985**, *23*, 926–930. [CrossRef]
8. Fukumoto, S.; Hookabe, T.; Tsubakino, H. Hydrolysis behavior of aluminum nitride in various solutions. *J. Mater. Sci.* **2000**, *35*, 2743–2748. [CrossRef]
9. Calder, G.V.; Stark, T.D. Aluminum Reactions and Problems in Municipal Solid Waste Landfills. *Pract. Period. Hazard. Toxic Radioact. Waste Manag.* **2010**, *14*, 258–265. [CrossRef]
10. Zhang, Y.; Guo, Z.; Han, Z.; Xiao, X.; Peng, C. Feasibility of aluminium recovery and $MgAl_2O_4$ spinel synthesis from secondary aluminium dross. *Int. J. Minerals. Metall. Mater.* **2019**, *26*, 309–318. (In Chinese) [CrossRef]
11. Shinzato, M.C.; Hypolito, R. Effect of disposal of aluminum recycling waste in soil and water bodies. *Environ. Earth Sci.* **2016**, *75*, 628. [CrossRef]
12. Ministry of Ecology and Environment of the People's Republic of China. *National List of Hazardous Wastes*, 2021 ed.; Order No. 15 of the Ministry of Ecology and Environment; Ministry of Ecology and Environment of the People's Republic of China: Beijing, China, 2021. (In Chinese)
13. Mahinroosta, M.; Allahverdi, A. Enhanced alumina recovery from secondary aluminum dross for high purity nanostructured γ-alumina powder production: Kinetic study. *J. Environ. Manag.* **2018**, *212*, 278–291. [CrossRef]
14. Murayama, N.; Baba, M.; Hayashi, J.-I.; Shibata, J.; Valix, M. Adsorption property of water vapor on $AlPO_4$-5 synthesized from aluminum dross. *Mater. Trans.* **2013**, *54*, 2265–2270. [CrossRef]
15. Lv, S.; Ni, H.; Wang, X.; Ni, W.; Wu, W. Effects of Hydrolysis Parameters on AlN Content in Aluminum Dross and Multivariate Nonlinear Regression Analysis. *Coatings* **2022**, *12*, 552. [CrossRef]
16. Hu, S.; Wang, D.; Hou, D.; Zhao, W.; Li, X.; Qu, T.; Zhu, Q. Research on the Preparation Parameters and Basic Properties of Premelted Calcium Aluminate Slag Prepared from Secondary Aluminum Dross. *Materials* **2021**, *14*, 5855. [CrossRef]
17. Shi, M.; Li, Y. Extraction of Aluminum Based on NH_4HSO_4 Roasting and Water Leaching from Secondary Aluminum Dross. *JOM* **2022**, 1–9. [CrossRef]

18. Shi, M.; Li, Y.; Ni, P. Recycling valuable elements from aluminum dross. *Int. J. Environ. Sci. Technol.* **2022**, 1–10. [CrossRef]
19. Kim, J.; Lee, H. Thermal and carbothermic decomposition of Na_2CO_3 and Li_2CO_3. *Metall. Mater. Trans. B* **2001**, *32*, 17–24. [CrossRef]
20. Huan, S.; Wang, Y.; Di, Y.; You, J.; Peng, J.; Hong, Y. Experimental Study on Alumina Extraction by Calcination of Secondary Aluminum Dross. *Conserv. Util. Miner. Resour.* **2020**, *40*, 34–39.
21. Heo, J.H.; Kim, T.S.; Sahajwalla, V.; Park, J.H. Observations of FeO Reduction in Electric Arc Furnace Slag by Aluminum Black Dross: Effect of CaO Fluxing on Slag Morphology. *Met. Mater. Trans. A* **2020**, *51*, 1201–1210. [CrossRef]
22. Li, Y.; Chen, X.; Liu, B. Experimental Study on Denitrification of Black Aluminum Dross. *JOM* **2021**, *73*, 2635–2642. [CrossRef]

Article

Influence of Contact Interface Friction on Plastic Deformation of Stretch-Bend Forming

Shengfang Zhang, Guangming Lv, Fujian Ma, Ziguang Wang and Yu Liu *

School of Mechanical Engineering, Dalian Jiaotong University, Dalian 116028, China; zsf@djtu.edu.cn (S.Z.); lv0527@163.com (G.L.); mafujianyx@163.com (F.M.); wangzg@djtu.edu.cn (Z.W.)
* Correspondence: liuyu_ly12@126.com

Abstract: The contact interface friction between the specimen and the mold during the stretch-bend is a complex interactive process. Friction causes the uneven distribution of tensile stress on the specimen, which affects the plastic flow of the forming material and the spring-back after forming. In this paper, the analytical model of frictional shear stress and tensile stress distribution in the contact mold segment of the stretch and bend synchronous loading stage was established. The influence law of friction coefficient and contact mold angle on the stress–strain distribution of the specimen contact mold segment was discussed. The effect of key factors affecting the friction state of the contact interface (mold surface roughness and contact mold angle) on the shrinkage deformation of the cross-section and the tensile deformation gradient of the specimen was analyzed by equivalent stretch-bend forming experiments. The results showed that the smaller the surface roughness of the mold was, the better the friction state of the contact interface was, the plastic deformation of the specimen was more uniform, and the difference between the section shrinkage and elongation of the contact mold segment and the suspension segment was smaller. Reducing the contact mold angle of the stretch-bend can bring down the tensile stress difference at both ends of the contact mold segment of the specimen so that the section shrinkage and tensile elongation of the contact mold segment and the suspension segment tend to be consistent.

Keywords: aluminum alloy profiles; stretch-bend forming; friction coefficient; plastic deformation; deformation gradient

Citation: Zhang, S.; Lv, G.; Ma, F.; Wang, Z.; Liu, Y. Influence of Contact Interface Friction on Plastic Deformation of Stretch-Bend Forming. *Coatings* **2022**, *12*, 1043. https://doi.org/10.3390/coatings12081043

Academic Editors: Matic Jovičević-Klug, Patricia Jovičević-Klug and László Tóth

Received: 30 June 2022
Accepted: 21 July 2022
Published: 23 July 2022

Publisher's Note: MDPI stays neutral with regard to jurisdictional claims in published maps and institutional affiliations.

Copyright: © 2022 by the authors. Licensee MDPI, Basel, Switzerland. This article is an open access article distributed under the terms and conditions of the Creative Commons Attribution (CC BY) license (https://creativecommons.org/licenses/by/4.0/).

1. Introduction

With the speed of high-speed railroad operations constantly refreshing, the development of the high-speed railway is entering the post high-speed railway era [1]. The structural strength of the high-speed train body has become one of the bottlenecks restricting the further development of high-speed railways [2]. The high-speed train body is mainly welded with high-strength, complex, cross-sectional aluminum alloy profiles stretch-bend-formed structural parts, which put forward higher requirements for the stretch-bend forming accuracy of aluminum alloy profiles structural parts [3].

Stretch-bend forming applies tangential tension to the ends of the profile during bending contact with the mold to reduce the stress difference between the inner and outer layers, which changes the internal stress distribution state of the cross-section and realizes the plastic forming processing for aluminum alloy profiles with large curvature and complex cross-sections [4]. In the process of stretch-bend forming, the friction of the contact interface between the stretch-bend mold and the profile is inevitable. With the increase of the contact mold angle, the contact interface changes continuously, and the frictional shear stress distribution is not uniform [5]. The tensile stress along the thickness and length of the specimen is distributed in a gradient, resulting in uneven plastic deformation of the stretch-bend specimen. Consequently, internal stress is also non-uniformly distributed after unloading, which makes it difficult to control spring-back and, therefore, dramatically decreases the forming accuracy. The contact interface friction of stretch-bend forming is a cumulative and superimposed dynamic process, which is closely related to some factors, such as material properties, surface roughness, and friction conditions [6,7].

With the wide application of the profile stretch-bend forming process in the processing of large-scale curved structural parts, a lot of research work has been carried out on this process by researchers from different countries. Liang et al. [8] studied the influence of the bending angle of multi-point stretch-bend on the section deformation of the parts by using the ABAQUS finite element analysis method and predicted the forming quality control curve by the support vector regression method. Chen et al. [9] analyzed the spring-back of multi-point flexible stretch-bend forming of Y-section profiles under different process parameters. The effects of process parameters such as horizontal bending radius, vertical bending radius, transition length, rolling mold radius, and wall thickness of the profile on spring-back deviation were discussed. Liu et al. [10] investigated the spring-back characteristics of Z and T-sections aluminum lithium alloy stretch-bend forming using plastic deformation theory, explicit and implicit finite element theory, and experimental methods. The spring-back after unloading is caused by the high strength and modulus of elasticity of the aluminum-based alloy. In addition, the bending radius is closely related to the spring-back radius; the larger the bending radius, the larger the spring-back radius. Liu et al. [11] conducted a finite element simulation analysis of the stretch-bend process of aluminum alloy profiles, showing that increasing the complementary elongation reduces contour error and spring-back amount of the profile, but increases the section distortion and spatial distortion. A backtracking method was proposed to optimize the supplemental stretch rate. Liu et al. [12] studied the hot stretch-bending process of U-shaped titanium by applying numerical and experimental approaches. The electro–thermal–mechanical multi-fields coupling finite element model for U-shaped titanium extrusion hot stretch-bending was established by sequential numerical simulation between electro-thermal coupling and thermal-mechanical coupling. The optimal process parameters for hot stretch-bending of U-shaped extrusion were optimized and the spring-back after hot stretch-bend was evaluated. Welo and Ma [7,13] proposed a new flexible 3D stretch-bending process. Using experimental and numerical approaches, the capabilities of the forming method and machine were verified, and the characteristics and mechanisms were thoroughly explored. It was found that bending radius, friction, local deformation, and clamping have an obvious influence on forming quality. Maati et al. [14], through numerical simulations, demonstrated the influence of constitutive modeling on the prediction of the degree of spring-back in the case of a stretch-bending test. The spring-back depends closely on the material anisotropy. The final spring-back is greater in the direction of 0° for a low value of clamping force. Uemori et al. [15] proposed an explicitly theoretical analysis method based on the maximum load criterion that can easily and rapidly allow us to predict the fracture and spring-back of high-tensile-strength steel sheets by using two no-dimensional parameters (ultimate tensile ratio and ratio of plate thickness to bending radius). Gu et al. [16] conducted a finite element simulation analysis of the stretch-bend process of variable curvature L-section aluminum alloy profiles. The forming accuracy of complex curvature stretch-bend forming structural parts is improved by spring-back compensation for the mold surface curve.

The material properties, cross-sectional shape, and the amount of stretching all affect the amount of spring-back in the stretch-bend forming process. However, the increase of tensile stress will lead to uneven deformation of the specimen and increase the risk of specimen fracture since spring-back is caused by internal stress redistribution during unloading. The uneven plastic deformation will produce local stress concentration, which will increase the amount of spring-back and also intensify cross-sectional distortion. Han [17] obtained the effect law of contact pressure on the friction coefficient through draw-bending experiments of U-section specimens. When the contact pressure is small, the friction coefficient of the contact interface does not change much, and when the contact pressure is large, the friction coefficient of the contact interface increases with the increase of the contact pressure.

The change of the friction coefficient caused by the contact pressure affects the contact stress distribution of the specimen, which leads to the difference in the amount of spring-back. Fox et al. [18] conducted a test on the friction coefficient of sheet drawing and forming and established a mathematical model to describe the friction coefficient under different contact conditions. Liu et al. [19–21] found an analytical model for rotational tensile bending section deformation of thin-walled rectangular tubes and studied the effects of die constraint and interfacial friction on section deformation. The best combination of friction coefficient and minimum cross-sectional distortion was obtained by the rotary draw bending process tests. Guan et al. [22] analyzed the drawing and bending stress state of the sheet and established the relationship between friction coefficient and spring-back angle. The results showed that the frictional force is beneficial for reducing spring-back, but it is difficult to transfer the tensile force uniformly to all sections of the workpiece as the effect of friction force, which makes the specimen deformation uneven. Yang et al. [23], using explicit finite element simulation combined with physical experiments, explored the underlying effects of friction on bending behaviors from multiple aspects such as wrinkling, wall thickness variation, and cross-section deformation. Zhang et al. [24] studied the residual stress distribution of a 2026-T3511 aluminum alloy asymmetric T-section beam under displacement-controlled stretch-bend. A two-step numerical simulation was performed to predict the residual stress superposition in the quenching and subsequent mechanical processes. The spatial variation of the residual stresses in a stretch-bend beam can be mainly attributed to nonuniform plastic deformation through the thickness and longitudinal direction, which is caused by the combined load. Muranaka et al. [25] developed a new "rubber-assisted stretch bending method" by which elastic rubber can be partially applied to the surface of a die for the uniform bending with a constant curvature radius. The spring-back decreased by 21% in comparison with the crank motion simple bending by using ordinary metal dies. Xiang et al. [5] stated that tangential friction occurred at the rubber/metal interface during friction-assisted stretch-bend. Tangential friction assists in modifying stress distribution in the loading process. Liang et al. [26] studied the deformation difference between the contact zone and non-contact zone of profile and roller dies. The results showed that the pre-stretching amount has no obvious effect on the profile web heights in the contact and non-contact areas. However, the profile web thickness becomes thinner at each contact area away from the fixture, while the non-contact area thickness does not change significantly.

The stretch-bend forming process is described and discussed in the above research, including the influence of the contact interface friction between the stretch-bend specimen and the mold on the overall spring-back of the forming specimen after unloading. However, the influence of specific parameters of contact interface friction on the plastic flow of materials in the forming process was not concerned. Since the variation of the contact interface friction coefficient in stretch-bend forming is complex, it is necessary to explore the law of the key parameters affecting the friction state of the contact interface on the plastic flow of the formed specimen.

In this paper, the influence of the contact interface friction state (surface roughness of the mold and the angle of the contact mold) on the plastic deformation of a forming specimen was studied during the stretch-bend forming process. The stress state of the contact interface of the stretch-bend forming specimen was analyzed, and the analytical formula of the cross-section stress-strain distribution of the contact mold segment of the stretch-bend forming specimen was established. An equivalent stretch-bend forming experiment was designed based on the discrete step-by-step loading characteristics of the stretch-bend forming trajectory. By measuring the cross-section width and strain grid of the stretch-bend specimen, the influence law of the specimen plastic deformation with different die surface roughness and die-attaching angle was obtained, which indirectly verifies the stress–strain distribution of the specimen's cross-section. It provides a necessary reference for establishing variable increment (variable tensile increment and variable wrapping angle) trajectory optimization method.

2. Theoretical Analysis of Plastic Deformation in Stretch-Bend Forming

In the process of stretch-bend forming, the seemingly smooth surface of the stretch-bend specimen and the mold is an uneven rough surface. The high contact pressure of the contact interface between the profile and the mold is due to the tangential tension of the stretch-bend. According to Coulomb's law of friction, there will be friction between the contact interface of the profile and the mold, which hinders the plastic deformation of the profile. Since stretch-bend forming is the process of the specimen bending and sticking mold, the specimen is continuously bent toward the mold to fit the mold. During this time, the contact interface and contact pressure change complexly in real-time.

2.1. Force Analysis of Stretch-Bend Forming

The aluminum profile stretch-bend forming process can be divided into three stages: pre-stretch, stretch-bend, and complemental stretch. In the stage of stretch and bend synchronous loading, the specimen plastic deformation, and elongation under tensile load take place. The plastic sliding friction occurs due to the relative sliding between the specimen and the mold contact interface. With the increase of bend contact mold angle, the contact surface between the specimen and the mold keeps increasing continuously. The frictional shear stress on the surface of the specimen contact mold segment is opposite to the direction of tangential tension, which impedes the plastic flow of the specimen surface material and also changes the distribution of the shear stress along the thickness direction of the specimen, affecting the uniformity of plastic deformation.

The following basic assumptions are made for force analysis during stretch and bend synchronous loading.

(1) Section assumption: It is assumed that the sections before and after stretch-bend are flat, the section before stretch-bend is perpendicular to the axis of the profile, and the section after stretch-bend is perpendicular to the tangent of the neutral axis;
(2) Stress assumption: It is assumed that each element of the specimen is in a state of uniaxial tension or uniaxial compression during the stretch-bend forming process;
(3) Material elastic-plasticity assumption: It is assumed that the material is a homogeneous, continuous, and isotropic elastic–plastic deformation body. The elastic–plastic deformation conforms to the loading and unloading deformation law of classical elastic–plastic theory.

The force in the stretch-bend forming process is shown in Figure 1. In the figure, F is the axial tension force; R is the radius of the mold; ρ is the bending radius of the neutral layer of the profile; M is the total bending moment; F_f is the frictional force on the contact mold segment, and the direction is along the tangent direction of the mold arc; F_N is the positive pressure of the profile on the mold, and the direction is perpendicular to the contact surface.

Figure 1 shows the bending and tensile deformation under the combined action of bending moment and tensile force. The contact pressure between the specimen and the mold causes a frictional shear force on the contact surface opposite to the sliding direction of plastic deformation. Due to the continuity of the material, the frictional shear force produces a shearing effect on the plastic deformation section of the specimen. The Coulomb friction coefficient is used to characterize the shear friction force of the contact surface, as shown in Equation (1) [27]. The shear friction coefficient is used to characterize the shear action inside the workpiece, as shown in Equation (2) [28].

$$f = \frac{F_f}{F_N} \tag{1}$$

$$m = \frac{\tau_m}{\sigma_n} \tag{2}$$

where, the F_f is the friction of the contact surface; F_N is the positive pressure of the contact surface; τ_m is the frictional shear stress; σ_n is the shear strength of the material.

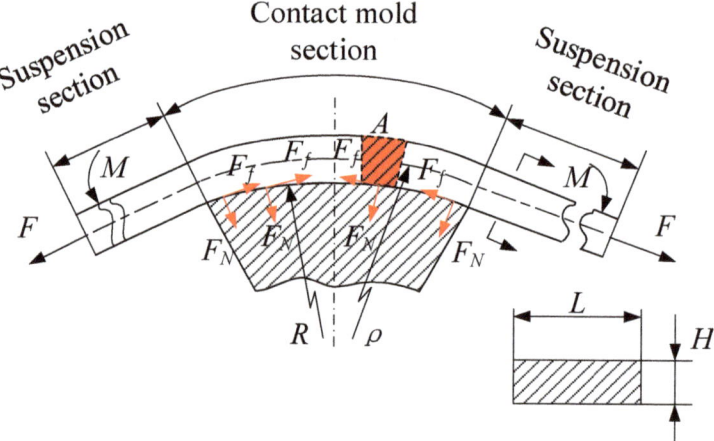

Figure 1. Force analysis of the stretch and bend synchronous loading process.

Due to the axial symmetry of the contact mold segment, 1/2 was taken for analysis. The contact mold segment was divided into innumerable micro-units. The force analysis of the micro-unit at section A point shown in Figure 1 was taken as shown in Figure 2. The arc angle corresponding to the micro-unit is $\Delta\theta$. The tension on the left and right sides of the micro-unit are F_{i+1} and F_i, respectively. The positive pressure of the micro-unit on the arc surface of the mold is F_{N_i}. According to the decomposition of forces and Coulomb's law of friction, the positive pressure F_{N_i} and the frictional force F_{f_i} on the surface of the micro-unit are calculated as:

$$F_{N_i} = F_i \sin\frac{\Delta\theta}{2} + F_{i+1}\sin\frac{\Delta\theta}{2} \tag{3}$$

$$F_{f_i} = \mu F_{N_i} = \mu(F_i \sin\frac{\Delta\theta}{2} + F_{i+1}\sin\frac{\Delta\theta}{2}) \tag{4}$$

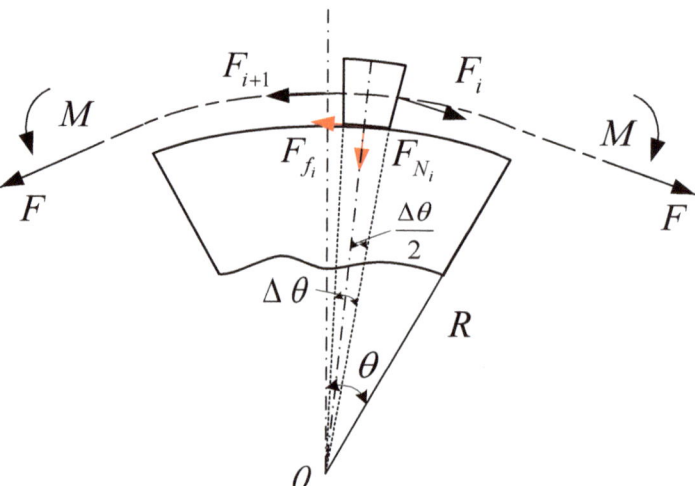

Figure 2. Force analysis of micro-unit at point A.

According to Newton's first law, the equilibrium equation of the tangential force of the i-th micro-unit in the direction of the length of the specimen contact mold segment is as:

$$F_{f_i} = F_i - F_{i+1} \tag{5}$$

Since $\Delta\theta$ is an infinitesimal quantity, there is $\sin\frac{\Delta\theta}{2} \approx \frac{\Delta\theta}{2}$. Take $F_1 = F$, and divide θ into n parts, $n \to \infty$. Equation (5) is brought into Equations (3) and (4), and the positive pressure F_{N_i}, friction force F_{f_i}, and tension force F_i of the i-th micro-unit are obtained by iterative collation, respectively.

$$F_{N_i} = 2\Delta\theta F \frac{(2-\mu\Delta\theta)^{i-1}}{(2+\mu\Delta\theta)^i} \tag{6}$$

$$F_{f_i} = 2\mu\Delta\theta F \frac{(2-\mu\Delta\theta)^{(i-1)}}{(2+\mu\Delta\theta)^i} \tag{7}$$

$$F_i = F \frac{(2-\mu\Delta\theta)^{(i-1)}}{(2+\mu\Delta\theta)^{(i-1)}} \tag{8}$$

As shown in Figure 2, the contact area between the micro-unit and the arc surface of the mold is $S = R\theta L$. According to the formula of shear stress $\sigma = F/S$, the frictional shear stress τ_{f_i} and tensile stress τ_i of the i-th micro-unit can be obtained as follows:

$$\tau_{f_i} = 2\mu\Delta\theta F \frac{(2-\mu\Delta\theta)^{(i-1)}}{(2+\mu\Delta\theta)^i R\Delta\theta L} \tag{9}$$

$$\tau_i = F \frac{(2-\mu\Delta\theta)^{(i-1)}}{(2+\mu\Delta\theta)^{(i-1)} R\Delta\theta L} \tag{10}$$

The parameters of the stretch-bend as shown in Table 1 are brought into Equations (9) and (10). The variation curves of the frictional shear stress and the total tensile stress in the cross-section for each micro-unit in the contact mold segment of the specimen were obtained by numerical calculation, as shown in Figures 3 and 4.

Table 1. Stretch-bend parameters.

Tension F/kN	Bend Angle θ/°	Friction Coefficients μ	Bending Radius R/mm	Section Width L/mm
30	30	0.15, 0.2, 0.25	150	20
30	30, 45, 60	0.2	150	20

It can be seen from Figure 3 that the rightmost micro-unit is subjected to frictional shear stresses of 0.599, 0.799, and 0.99 MPa, respectively, when the friction coefficients of contact sections are 0.15, 0.2, and 0.25, respectively. The normal pressure of the micro-unit from right to left gradually decreases, resulting in a lower frictional force on the micro-unit at the same friction coefficient. As the friction coefficient increases, the frictional shear stress on the micro-unit at the same position becomes larger. It is known that the magnitude of the frictional shear stress of the micro-unit is related to the friction coefficient and the position of the micro-unit. That is, the smaller the coefficient of friction, the farther the location of the micro-unit from the point of tension is, and the smaller the frictional shear stress on the micro-unit is.

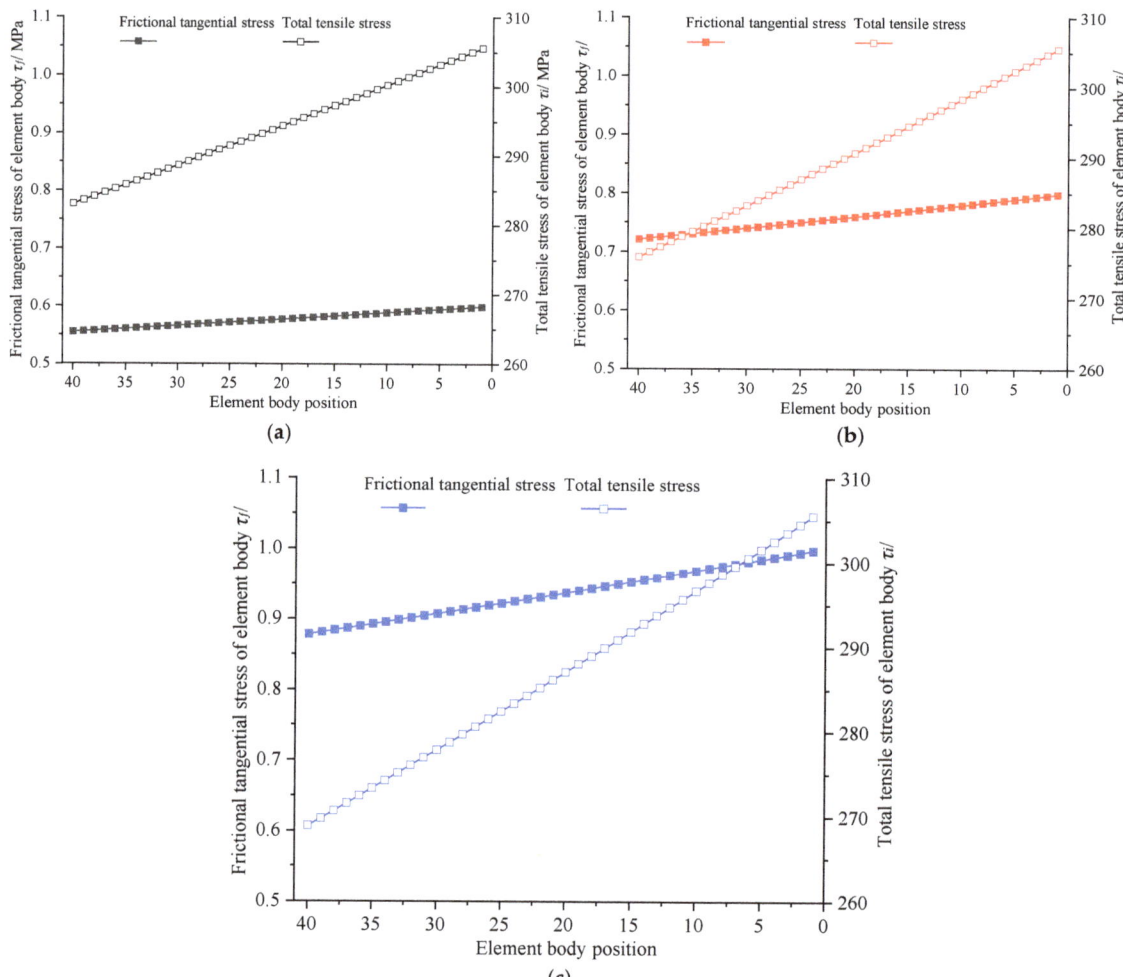

Figure 3. The frictional tangential stress and total tensile stress of each micro-unit under different friction coefficients. (**a**) $\mu = 0.15$; (**b**) $\mu = 0.2$; (**c**) $\mu = 0.2$.

It can be seen from Figure 3 that the total tensile stress on the rightmost micro-unit for all three friction coefficients was 305.577 MPa. From right to left along the direction of the bending arc, the total tensile stress experienced by the micro-unit showed a significant decrease. Since the order of frictional shear stress is much smaller than the tensile stress in the micro-unit cross-section, it is known from Equation (5) that the tensile stress in the micro-unit was $F_{i+1} = F_i - F_{f_i}$. The superimposed accumulation of frictional shear stress makes the tensile stress in the bending specimen unevenly distributed.

It can be seen from Figure 4 that under different bending angles, the varying trends of the frictional shear stress and total tensile stress of the micro-unit were the same, decreasing in turn along the arc direction of the specimen from right to left. The larger the bending angle of the specimen, the smaller the frictional shear stress and the total tensile stress of the micro-units farthest from the point of action of the tensile force are, and the more obvious the uneven stress distribution on the left and right sides of the specimen.

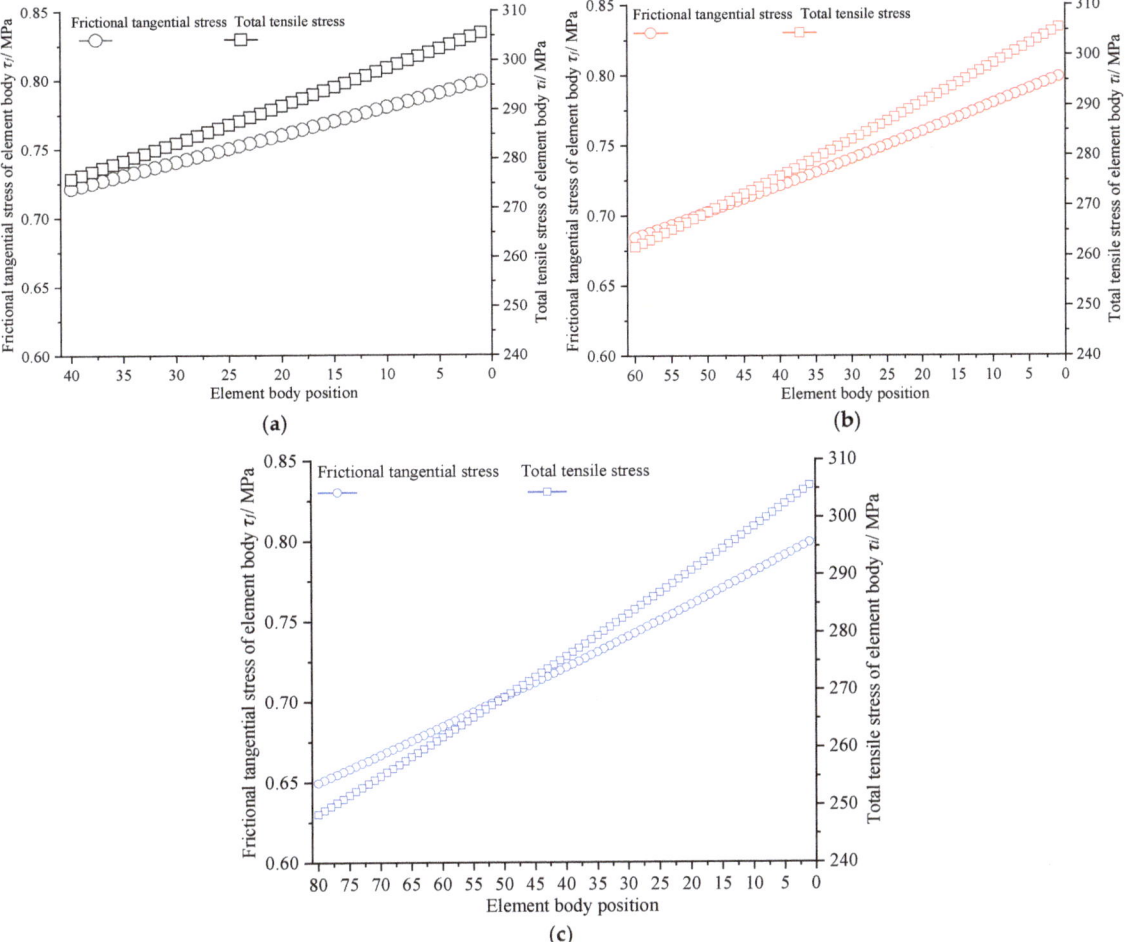

Figure 4. The frictional tangential stress and total tensile stress of each element under different bend angles. (**a**) 30° contact mold; (**b**) 45° contact mold; (**c**) 60° contact mold.

2.2. Strain Distribution of Stretch-Bend Forming

It is assumed that the material enters the plastic deformation stage completely during the stretch-bend forming process. A bilinear hardening power exponential strengthening model [29] was used to depict the stress–strain relationship during bending tensile loading.

$$\sigma = A + B\varepsilon_p{}^n \tag{11}$$

where, A is the initial yield strength of the material; B is the process-hardening modulus, which is the slope of the stress–strain curve in the plastic deformation stage; ε_p is the equivalent plastic strain; n is the strain-hardening index, which reflects the ability of the metal material to resist uniform plastic deformation.

Substituting Equation (11) into Equation (10), the plastic strain at the contact surface of the i-th micro-unit and the mold was obtained as follows.

$$\varepsilon_i = \left(\frac{F(2 - \mu\Delta\theta)^{(i-1)}}{(2 + \mu\Delta\theta)^{(i-1)} R\Delta\theta LB} - \frac{A}{B} \right)^{\frac{1}{n}} \tag{12}$$

The inner material fibers are compressive elastic–plastic deformation, while the outer material fibers are tensile elastic–plastic deformation under the action of the bending moment. The friction produces a shearing effect on the contact surface of the inner layer of the specimen, changing the stress–strain distribution of the cross-section. The stress–strain distribution of the profile cross-section under the action of frictional shear is shown in Figure 5. It can be seen from Figure 5 that the frictional shear stress on the surface of the profile in contact with the mold was the largest, and the frictional shear stress decreased in a gradient along the thickness direction of the profile on the section until the frictional shear stress on the upper surface of the profile approached zero. The tensile stress and strain on the upper surface of the profile were much greater than those on the lower surface.

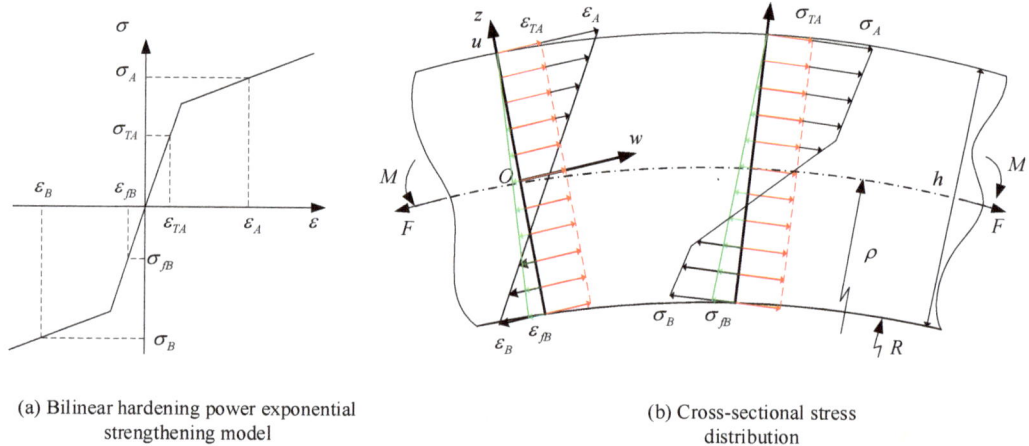

(a) Bilinear hardening power exponential strengthening model

(b) Cross-sectional stress distribution

Figure 5. Stress–strain distribution of the stretch-bend section. (**a**) Bilinear hardening power exponential strengthening model; (**b**) cross-sectional stress distribution; where, wou is the neutral layer coordinate system; σ and ε are the total stress and total strain of the section, respectively; σ_A and ε_A are the total stress and total strain of the outermost layer of the section, respectively; σ_{fB} and ε_{fB} are the frictional shear stress and strain of the contact interface between the inner layer of the section and the mold, respectively; σ_B and ε_B are the total stress and total strain of the innermost fiber, respectively.

3. Aluminum Alloy Profile Simulated Stretch-Bend Experiment

The friction changes the stress and strain distribution within the stretch-bend forming specimen, which affects the forming load as well as the plastic deformation. In order to research the effect of friction of the contact interface on the plastic deformation of the stretch-bend forming specimen, the stretch and bend synchronous loading experiment was used to simulate the stretch-bend forming process of aluminum alloy profiles, and the influence of friction coefficient and contact mold angle on the section deformation and tensile deformation of the specimen was analyzed.

3.1. Experiment Specimen

The experimental material in this study is AL6005A-T6 aluminum alloy profile with the chemical composition shown in Table 2.

Table 2. Chemical composition of Al6005A-T6 aluminum alloy profiles (%, mass fraction).

Si	Mg	Fe	Cu	Mn	Cr	Ti	Zn
0.5–0.9	0.4–0.7	0.35	0.3	0.5	0.3	0.1	0.2

According to the international standard ISO 6892-1-2019 [30], metal material tensile experiment standard at room temperature, in the INSTRON 5982 electronic universal material testing machine by INSTRON CORPORATION of located in Boston, MA, USA for the uniaxial tensile experiment, the load measurement accuracy of the testing machine is not less than ±0.4%, and the movement accuracy error is ±0.0001 mm. The loading rate is 3 mm/ min, and the specimen specifications are 20 mm × 5 mm × 200 mm. We repeated the uniaxial tensile experiment three times and took the average value. Finally, the mechanical property parameters of the experiment aluminum alloy profile specimens are shown in Table 3. The static tensile material engineering stress–strain curves of the specimens are shown in Figure 6.

Table 3. Tensile mechanical properties of Al6005A-T6 aluminum alloy profiles.

Extension Percentage/%	Yield Strength/MPa	Tensile Strength/MPa	Elastic Modulus/GPa	Poisson Ratio	Tension Rupture/kN
17	282.2 ± 1.1288	300.5 ± 1.202	66.77 ± 0.26708	0.3	29.53 ± 0.11812

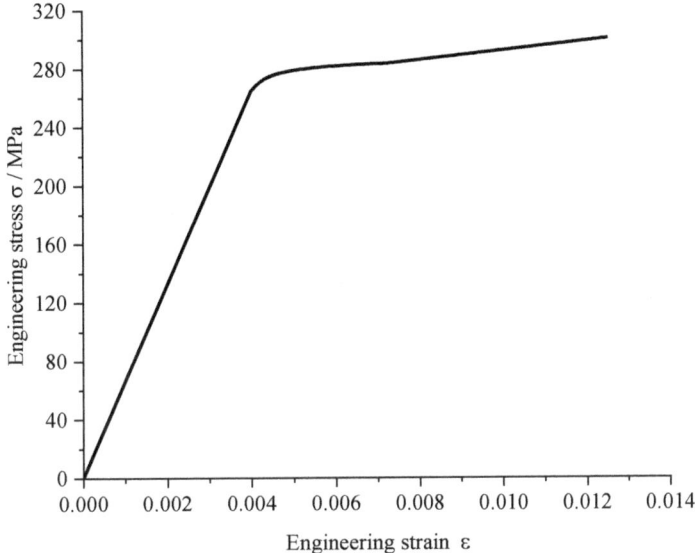

Figure 6. Engineering stress–strain curve of AL6005A-T6 aluminum alloy profile.

To explore the plastic deformation during the simultaneous tensile-bending loading phase, combined with the mechanical properties of the experiment specimen, we ensured that the specimen entered the plastic deformation stage and was not pulled off. Three types of stretch-bend specimens were made according to the different contact mold angles, as shown in Table 4. To analyze the strain variation of the specimen in the length direction, the specimen needs to be pre-treated as follows: (1) a flexible film is sprayed on the upper surface of the specimen, and no flexible film is sprayed on the down surface; (2) the strain measurement grid lines with 2 mm spacing were laser engraved in the deformation area of the specimen to form a strained grid as shown in Figure 7. The laser-engraved strain grid lines are of high accuracy and consistency while not damaging to the specimen itself. Since the flexible film can plastically deform and elongate synchronously with the specimen, the deformation amount of the specimen strain element can be obtained by measuring the grid width of the flexible film after laser engraving.

Table 4. Specimen sizes at different contact mold angles.

Contact Mold Angle $\theta/°$	Specimen Length L/mm	Contact Mold Length L_1/mm	Suspension Length L_2/mm	Clamping Length L_4/mm	Section Width B/mm	Section Thickness H/mm
30	280 ± 0.01	80	100	100	20 ± 0.01	5 ± 0.01
45	320 ± 0.01	120	100	100	20 ± 0.01	5 ± 0.01
60	360 ± 0.01	160	100	100	20 ± 0.01	5 ± 0.01

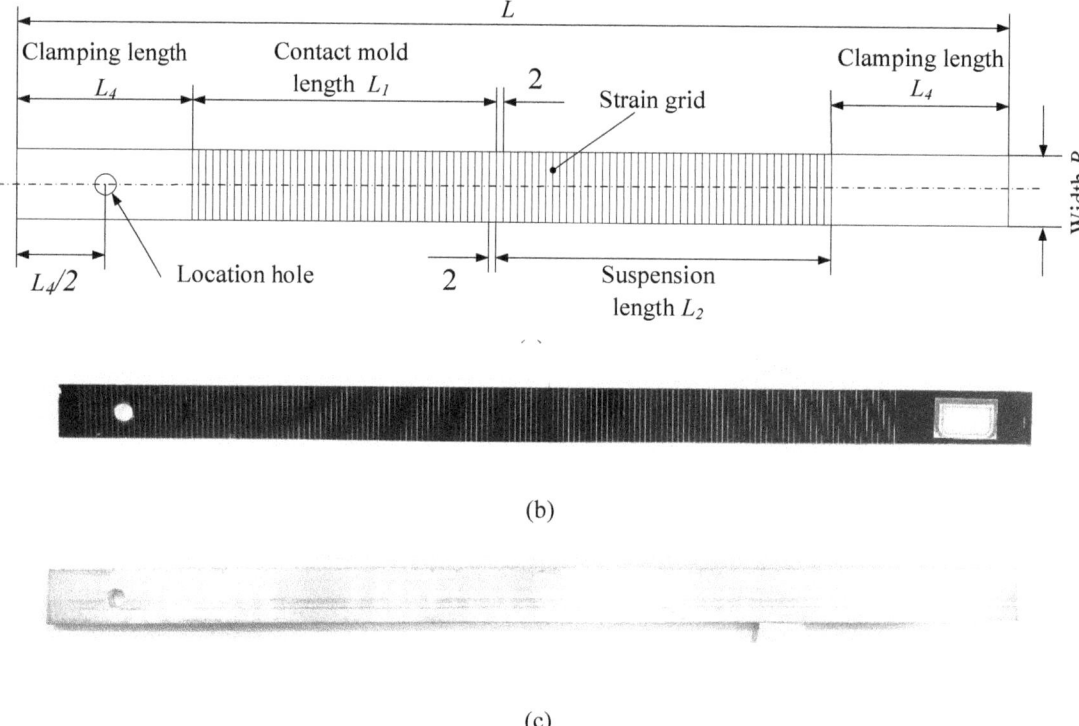

Figure 7. Structure size of the specimen. (**a**) Specimen size; (**b**) Upper surface of test sample; (**c**) Down surface of test sample.

3.2. Equipment and Methods of the Equivalent Stretch-Bend Experiment

To further analyze the effects of contact friction and contact mold angle on plastic deformation of stretch-bend specimens, the equivalent stretch-bend experiment was carried out on the tensile testing machine by auxiliary tooling. The equipment of the equivalent stretch-bend experiment is mainly composed of the main body of the stretch-bend mold, replaceable upper clamp, bending limit roller, fixed block, and other parts, as shown in Figure 8. The replaceable clamp and bending limit rollers can be used at the same time to achieve three types of contact mold angle of bending experiments as shown in Figure 8. The suspension section and the tensile axis of the specimen were ensured to be in the same straight line.

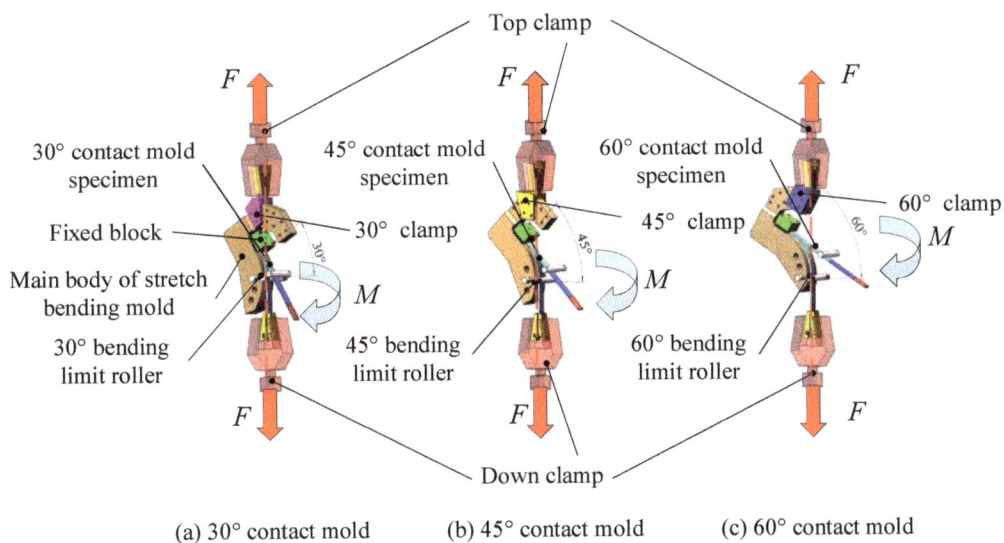

Figure 8. Equipment and methods of the equivalent stretch-bend experiment.

The specific method of the equivalent stretch-bend experiment is as follows: (1) The upper end of the specimen is fixed to the main body of the stretch-bend mold by the fixing block and limit pin. The bending moment M is applied to the specimen as shown in Figure 8 to make the specimen fit on the stretch-bend mold. (2) Lock the bending limit roller to press the specimen onto the stretch-bend mold. (3) Adjust the upper and lower clamps of the INSTRON 5982 electronic universal material test machine to intersect at 90°. The replaceable upper clamp of the stretch-bend experiment equipment and the lower end of the tensile specimen are respectively installed in the upper and lower clamps of the tensile experimental machine and locked. (4) Apply 2 kN pre-tension to the specimen, and then open the bending limit roller. The specimen is kept on bending contact with the mold under pre-tension. (5) The displacement control method is used for tensioning with a loading speed of 3 mm/min and a strain rate of 0.115 s^{-1}. Figure 9 shows the equivalent stretch-bend experiment of 30° contact mold. Repeat steps (1)–(5) for 30°, 45°, and 60° equivalent stretch-bending experiments, respectively. According to the mechanical properties of the material, the tensile-bending specimens are ensured to enter the plastic deformation stage. The tensile amounts of the specimens with different contact mold angles are determined by the equal tensile proportion, and the tensile amount is rounded off. Finally, the tensile rates of 30°, 45°, and 60° contact mold were 4.29%, 4.37%, and 4.44%, respectively. The specific experiment plan is shown in Table 5. As Bhanudas Bachchhav et al. [31] carried out sliding friction experiments with different surface roughness, the research results showed that the rougher the surface, the greater the friction coefficient. Zhang et al. [32] pointed out that when the surface roughness Ra of the mold was 1.6 um, the friction coefficient between the test specimen and the mold was 0.21 in the study of the influence of low-frequency vibration on the friction coefficient of the stretch-bend. Based on the abovementioned relationship between surface roughness and friction coefficient, in this experiment, equivalent stretch-bend experiments were carried out with molds with roughness Ra of 6.3 and 12.5, respectively, to simulate the equivalent stretch-bend forming under two different contact interface friction coefficients.

The specimen after stretch-bend is shown in Figure 10. The micrometer was used to measure the width of the upper and lower edges of the three equal cross-sections $AA'DD'$, $BB'EE'$, and $CC'FF'$ of the specimen contact mold segment and the three equal cross-sections GG', HH', and II' of the specimen suspension segment. The measurement accuracy of micrometer was ±0.001 mm. The strain grids of the contact mold segment and the suspension segment of the specimen were measured one by one in the tensile direction with a 40× light microscope, and the plastic deformation of the strain mesh was calculated. The measurement accuracy of optical microscope was ±0.001 mm. The measurement accuracy was less than an order of magnitude of plastic deformation, so the measurement error did not affect the analysis of plastic deformation results, and had certain credibility.

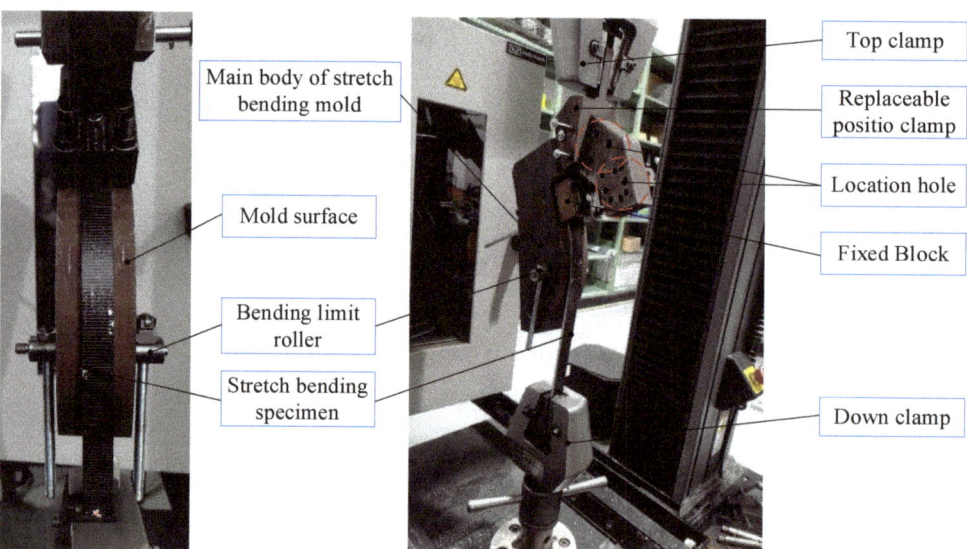

Figure 9. Equivalent stretch-bend experiment.

Table 5. The equivalent of the stretch-bend experiment parameters.

Condition	Mold Surface Roughness Ra/um	Contact Angle θ/°	Specimen Length L/mm	Stretching Value ΔL/mm
1	6.3	30	280 ± 0.01	12 ± 0.0001
2	12.5	30	280 ± 0.01	12 ± 0.0001
3	6.3	45	320 ± 0.01	14 ± 0.0001
4	12.5	45	320 ± 0.01	14 ± 0.0001
5	6.3	60	360 ± 0.01	16 ± 0.0001
6	12.5	60	360 ± 0.01	16 ± 0.0001

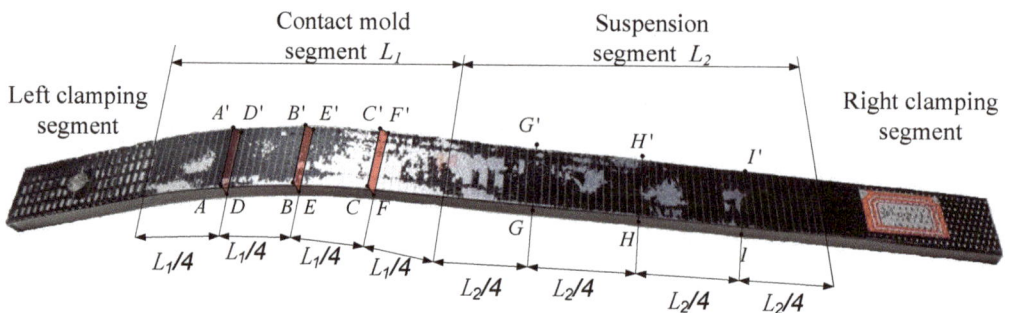

Figure 10. The specimen after stretch-bending.

4. Experiment Results and Analysis

4.1. Effect of Contact Interface Friction and Contact Mold Angle on Shrinkage Deformation of the Cross-Section

4.1.1. Cross-Section Shrinkage Deformation Analysis

Figure 11 shows the width change curve of the upper and lower edges of the specimen cross-section with a smooth mold stretch-bend. It can be seen from Figure 11 that with the 30° contact mold stretch-bend, the upper surface deformed widths of the contact mold segment $AA'DD'$, $BB'EE'$, and $CC'FF'$ cross-sections were 19.87, 19.64, and 19.62 mm, respectively, while the lower surface deformed widths were 19.95, 19.90, and 19.66 mm, respectively. In Figure 11, the deformed widths of the upper and lower surfaces of the left $AA'DD'$ cross-section were between 19.87 and 19.98 mm, and the deformed widths of the upper and lower surfaces of the rightmost $CC'FF'$ cross-section were between 19.58 and 19.2 mm. That is, the width of the upper surface of the specimen was smaller than the width of the lower surface at the same cross-section position of the contact mold segment, and the width of the section from left to right decreased gradually. It should be noted that the deformation of the upper surface was much larger than that of the lower surface in the stretch-bend forming process. The closer to the action point of the tensile force, the greater the overall deformation of the section is. There was no width difference between the upper and lower surfaces in the cross-section of the suspension segment. The width of the neutral layer cross-section can be used for characterization. However, due to the inhomogeneity of the material, there was a slight fluctuation in the width of each cross-section of the suspension segment.

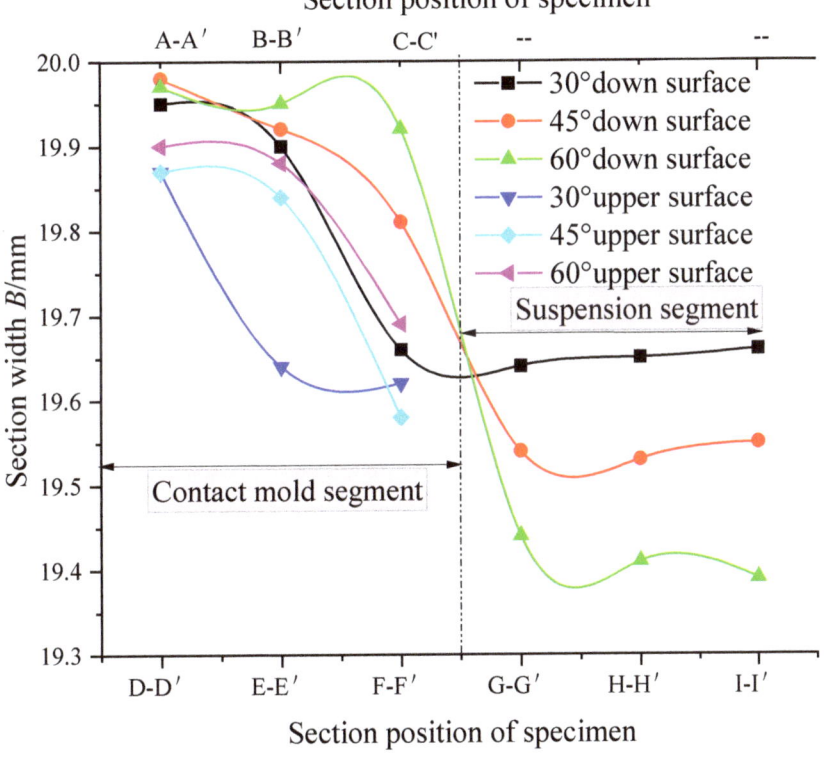

Figure 11. Variation curves of the upper and lower edges of the specimen cross-section with smooth die stretch-bending.

It can also be seen from Figure 11 that the cross-sectional width of the contact mold segment was between 19.58 and 19.98 mm, while the cross-sectional width of the suspension segment was between 19.39 and 19.66 mm. By taking the average of the three sampled section widths of the contact mold segment and analyzing the six sets of laboratory data, it can be seen that the cross-sectional deformation of the suspension segment was greater than that of the contact mold segment by more than 44.1%. With the increase of the contact mold angle, the plastic deformation of each cross-section of the contact mold segment decreased, while the plastic deformation of each cross-section of the suspension segment both had the opposite trend.

Figure 12 shows the average cross-sectional width of the contact mold segment and the suspension segment under the 45° contact mold and stretch-bend. The cross-section of the contact mold segment is a trapezoidal section with a width of 19.76 mm at the top bottom and 19.90 mm at the bottom. The cross-section of the suspension is a rectangular section with a width of 19.54 mm. It can be seen that the cross-sectional shapes of the contact mold segment and the suspension segment of the specimen are the scaled-down trapezoidal section and rectangular section, respectively, after stretch-bend plastic deformation with the rectangular sectional specimen according to the stress–strain distribution of the cross-section during the stretch-bend forming process shown in Figure 5. It is known that the tensile stresses of the contact mold segment are reduced due to the frictional forces of the contact interface and bending compressive stresses, which are distributed in a gradient in the thickness direction. The final result is that the plastic deformation of the upper surface is greater than the lower surface, and the cross-section of the contact mold segment is trapezoidal.

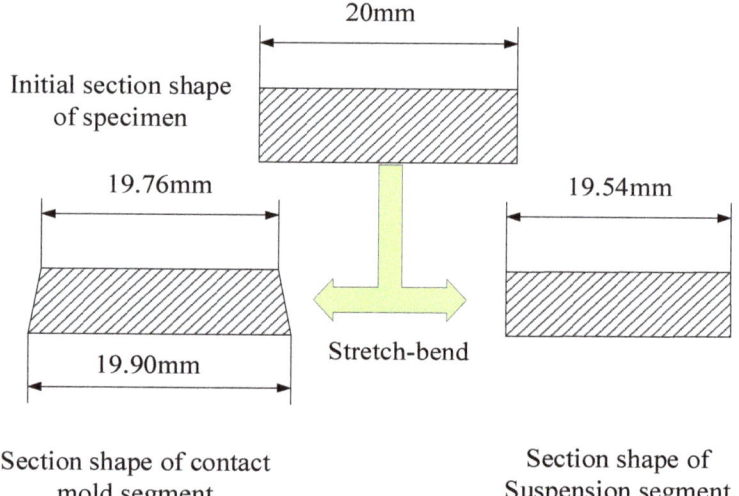

Figure 12. Deformation diagram of 45° contact mold stretch-bend sections.

4.1.2. Effect of Contact Interface Friction on Cross-Sectional Shrinkage Rate

The cross-section shrinkage rate is the ratio of the area of the cross-section in a reduction in the original cross-section area of the specimen under tensile plastic deformation, which is one of the plasticity indicators of the material. To further analyze the influence of contact interface friction on cross-section deformation of the contact mold segment, we calculated the shrinkage rate of each section according to the deformation shape of the specimen cross-section, as shown in Figure 12. Figure 13 shows the variation curve of the cross-section shrinkage rate of each section of the specimens with different roughness of the mold. It can be seen from Figure 13 that in the same contact mold angle, the smoother the contact surface, the smaller the friction coefficient is; then, the higher the section shrinkage of the contact mold segment, the the lower the section shrinkages of the suspension segments are. In addition, the cross-section shrinkage of the contact mold segment was at most 4.45% of that of the suspension segment. This is mainly due to the improvement of the friction state and the reduction of the friction shear stress, which reduces the uneven distribution of tensile stress in the contact mold segment of the specimen.

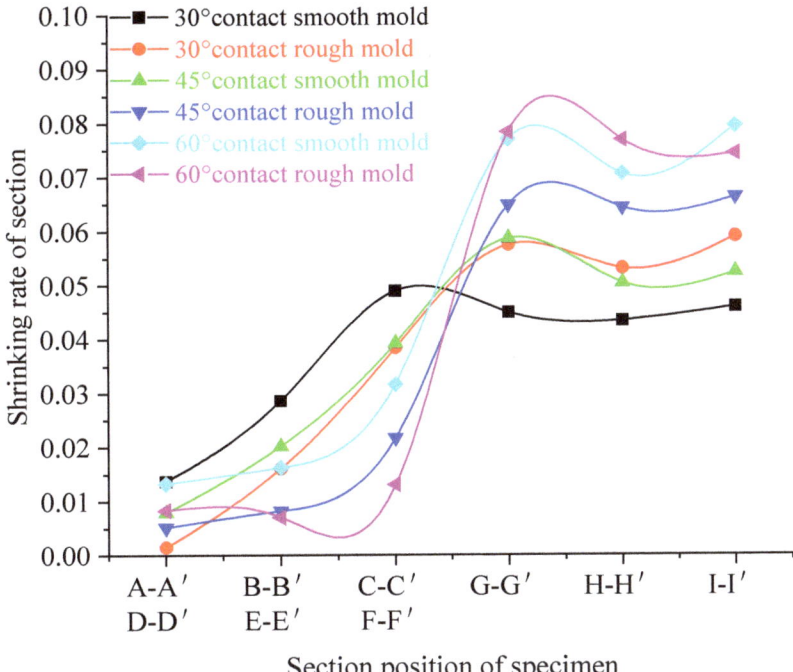

Figure 13. The variation curve of the cross-section shrinkage rate with different roughnesses of the mold.

4.1.3. Effect of Contact Mold Angle on Cross-Sectional Shrinkage Rate

The cross-sectional shrinkage rates of each section of the specimen contact mold segment and the suspension segment were averaged. The distribution of the cross-section shrinkage changes of both the specimen contact mold segment and the suspension segment under different contact mold angles is plotted as shown in Figure 14. It can be found from Figure 14 that with the contact mold angle of the specimen increased from 30° to 60°, the cross-section shrinkage rate in the contact mold segment of the smooth contact surface and rough contact surface decreased by 32.8% and 49.2%, respectively, and the cross-sectional shrinkage rate in the suspension segment increased by 60.7% and 35.3%, respectively. It can also be seen from Figure 14 that at 30°, 45°, and 60° contact mold angles, the smooth mold had an increase of 38.3%, 47.7%, and 53.4%, respectively, in the cross-sectional shrinkage rate of the contact mold segment compared with the rough mold. This is due to the increase of the contact mold angle of the specimen, making the total amount of friction resistance superposition in the contact mold segment become larger. As a result, the plastic deformation of the specimen contact mold segment was reduced, and the corresponding deformation of the suspension segment increased. The angle of the contact mold is inversely proportional to the shrinkage of the cross-section of the contact mold segment and positively proportional to the shrinkage of the suspension segment.

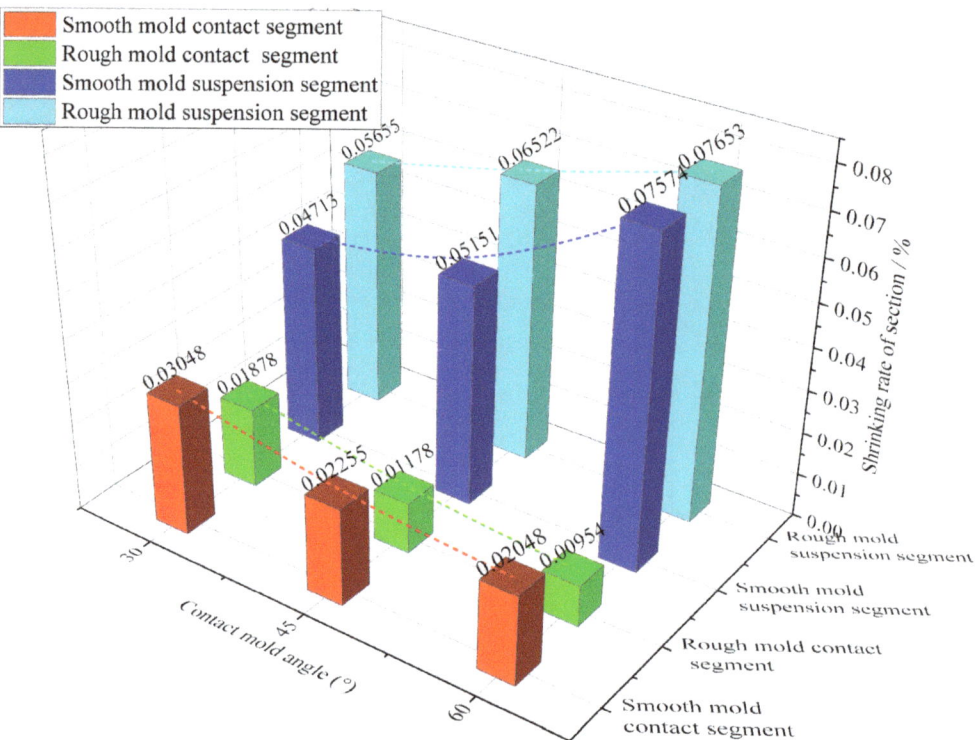

Figure 14. Cross-section shrinkage distribution of the contact mold segment and the suspension segment with different contact mold angles.

4.2. Effect of Contact Interface Friction and Contact Mold Angle on Tensile Deformation of the Cross-Section

4.2.1. Effect of Contact Interface Friction on Tensile Deformation of Contact Mold Segment

Figure 15 shows the strain grid length and deformation distributions of contact mold segments for 30°, 45°, and 60° of contact mold angle specimens under two types of mold surfaces. It can be seen from Figure 15b that the mesh deformation of the leftmost end micro-unit of the 45° contact mold specimen with smooth and rough contact surfaces was 0.025 and 0.02 mm, respectively. The micro-unit mesh deformation amounts were 0.097 and 0.06 mm, respectively. It can also be seen from Figure 15a–c that the mesh deformations of the micro-unit at both ends of the contact mold segment were not much different for stretch-bend the under different contact surfaces. However, the mesh deformation of the corresponding micro-unit in the middle of the contact mold segment for the smooth mold stretch-bend was much larger than that for the rough mold stretch-bend. The mesh deformation of the micro-unit in the contact mold segment increased in a gradient from left to right.

From Equations (9) and (10) and Figure 3, it can be seen that the frictional shear stress was gradually accumulated and superimposed along with the grid mesh from right to left, resulting in a gradient distribution of tensile stress, and the tensile stress of the leftmost element was the smallest. Therefore, the deformation elongation of the micro-unit at the left end measured by the experiment was much smaller than that at the right end. It can be seen from Figure 3 that the larger the friction coefficient of the contact interface, the faster the tensile stress of the micro-unit decreases, which leads to the greater the tensile deformation gradient of the micro-unit of the contact mold segment.

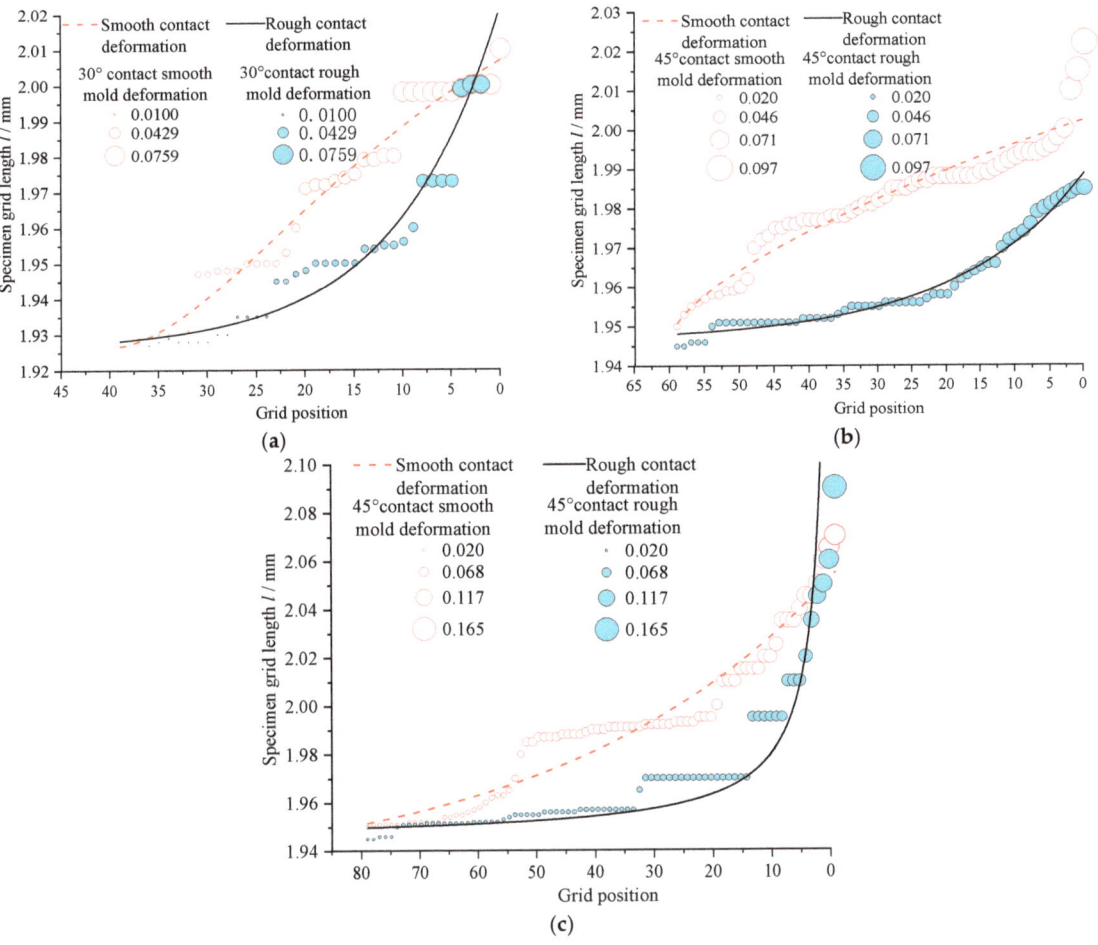

Figure 15. Strain grid length and deformation distribution of the contact mold segment. (**a**) 30° contact mold angle; (**b**) 45° contact mold angle; (**c**) 60° contact mold angle.

There is a certain degree of fluctuation in the deformation of the strain mesh due to measurement errors and the non-uniformity of the material. The strain mesh length in Figure 15 was fitted by the least squares method. It could be found that the deformation gradient of the smooth mold stretch-bend micro-unit mesh was smaller than that of the rough mold by the fitting curve. That is, the rougher the stretch-bend mold is, the greater the gradient of deformation of the contact mold segment is, and the worse the uniformity of the formed specimen is.

The deformation of the strain grid of the 30-degree contact mold segment with roughness Ra = 12.5 um in Figure 15a was extracted. According to the strain calculation formula, the linear strain $\varepsilon_i = (l_i - l_0)/l_i$ of each strain element was calculated, where, l_i is the length of strain grid after stretching, and l_0 is the length of strain grid before stretching. The linear strain of each strain element of the contact mold segment was brought into the static tensile stress–strain curve of the specimen material, and the tensile stress of each element was obtained. The tensile stress of the micro-unit variation curve is plotted as shown in Figure 16.

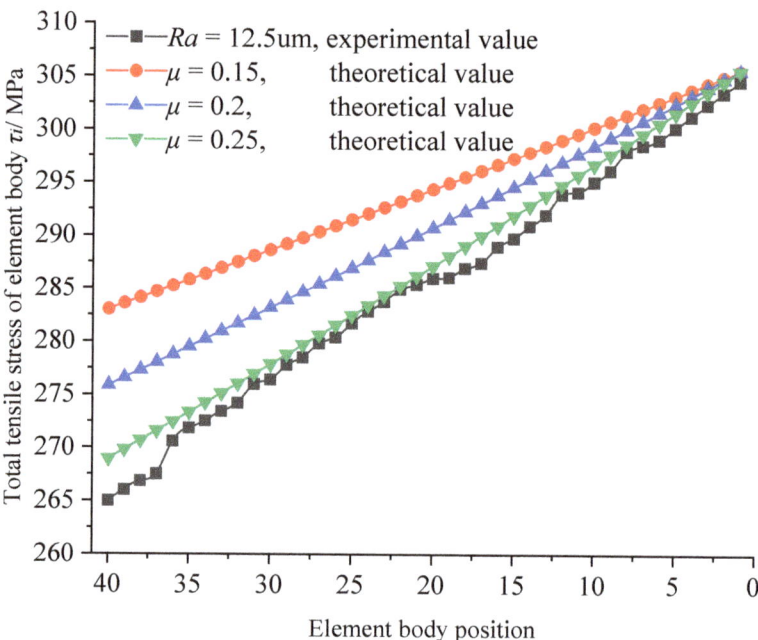

Figure 16. Variation curve of tensile stress.

It can be seen from Figure 16 that the tensile stress of the micro-unit by the experiment decreased from right to left along the specimen direction. When the mold surface roughness Ra was 12.5 um in the experiment and the contact interface friction coefficient μ was 0.25 in the theoretical derivation, the slope of the tensile stress gradient on the micro-unit was close to that. There were numerical fluctuations at individual measurement points, and the maximum error was 6.95 MPa. It can be seen that the experiment verified the gradient distribution law of the tensile stress on the micro-unit along the length of the specimen in the theoretical analysis. At the same time, it also indirectly indicated that the friction coefficient of the contact interface was close to 0.25 when the mold surface roughness was 12.5 um. The above provides a new method to obtain the friction coefficient of stretch-bend forming.

4.2.2. Effect of Contact Interface Friction on Tensile Deformation of the Suspension Segment

Figure 17 shows the strain grid length and deformation distribution of the suspension segment for 30°, 45°, and 60° contact mold angle specimens under two types of mold surfaces of the stretch-bend forming specimen. It can be seen from Figure 17a–c that there was a very small fluctuation in the mesh deformation of the micro-unit in the suspension segment of the specimen. Thereinto, the average deformation of the suspension segment micro-unit mesh for the 30° contact mold stretch-bend-formed specimen under two types of mold surfaces was 0.097 and 0.126 mm, respectively; the average deformation of the suspension segment micro-unit mesh for the 45° contact mold specimen was 0.116 and 0.149 mm, respectively; the average deformation for the 60° contact mold specimen was 0.150 and 0.178 mm, respectively.

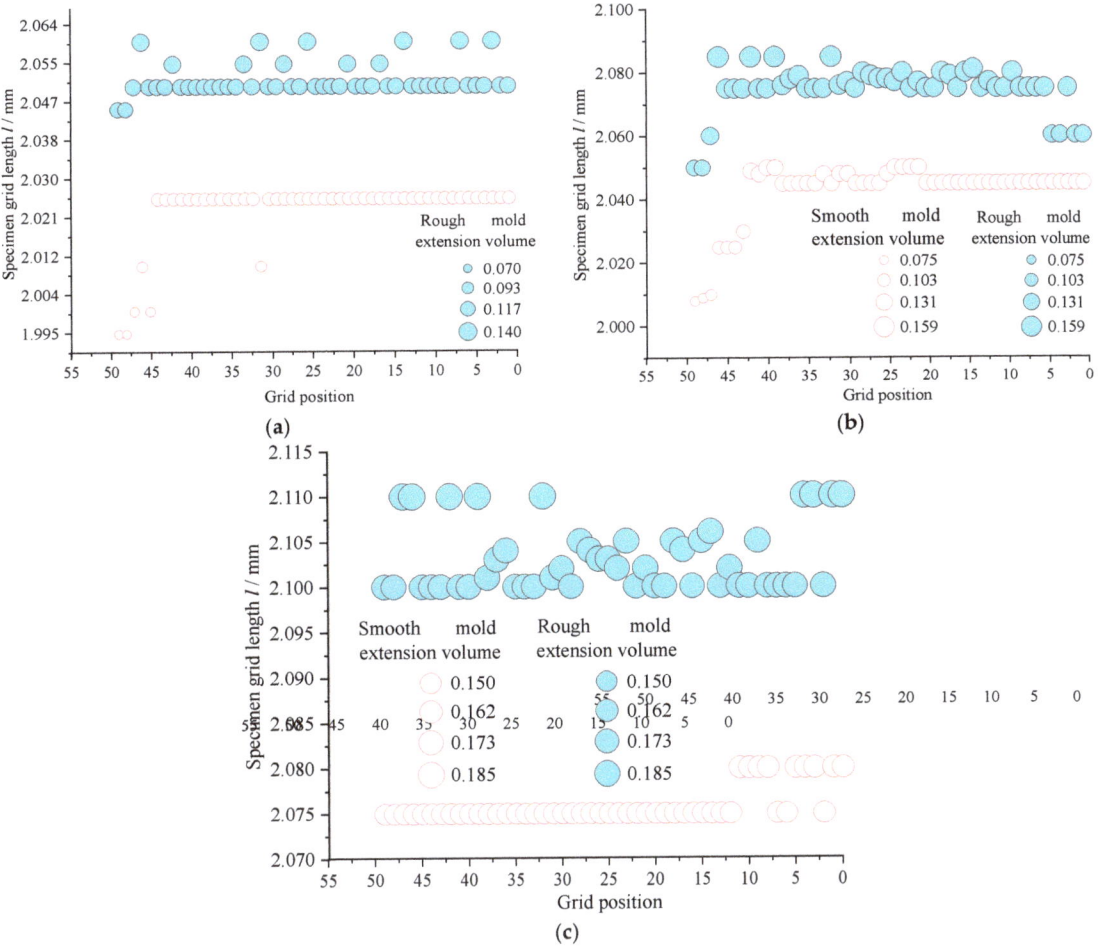

Figure 17. Strain grid length and deformation distribution of suspension segment. (**a**) 30° contact mold; (**b**) 45° contact mold; (**c**) 60° contact mold.

It can be found that the smoother the contact surface, the smaller the tensile deformation of the suspension segment is, which is opposite to the tensile deformation of the contact mold segment in Figure 15. This is because the total amount of tensile displacement of the specimen as a continuous deformed body remains constant during the stretch-bend forming process. The friction force hinders the tensile plastic deformation of the contact mold segment, and the deformation of the suspension segment must increase to compensate for the reduced deformation of the contact mold segment.

4.2.3. Effect of Contact Mold Angle on Tensile Elongation

The elongation of the strain mesh of the micro-unit in the contact segment and the suspension of the specimen were counted, respectively. The tensile elongation distribution of the contact mold segment and the suspension segment with different contact mold angles was plotted as shown in Figure 18. It can be seen from Figure 18 that with the contact mold angle of the specimen increasing from 30° to 60°, the elongation of the contact mold segment of the smooth contact surface and rough contact surface decreased by 39.9% and 45.3%, respectively, and the elongation of the suspension segment increased by 55.2% and

40.8%, respectively. It can also be seen from Figure 14 that at 30°, 45°, and 60° contact mold angles, the smooth mold had an increase of 32.3%, 36.1%, and 38.3%, respectively in the elongation of the contact mold segment compared with the rough mold.

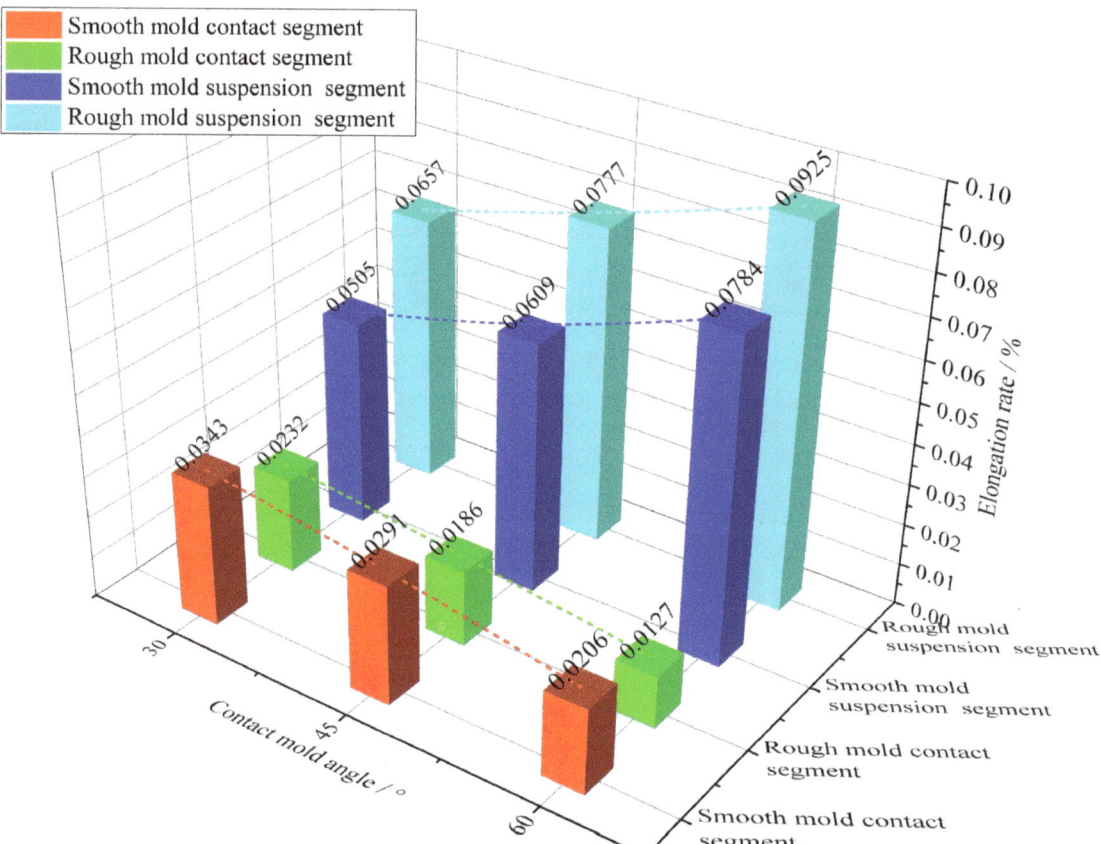

Figure 18. The tensile elongation distribution of the contact mold segment and the suspension segment with different contact mold angle.

According to the changes of frictional tangential stress and total tensile stress of each element under different bend angles, shown in Figure 4, the change gradient of the tensile stress of the micro-unit at the same position of the specimens with different contact mold lengths is the same. Due to the increase of the contact mold segment angle, the contact mold length between the specimen and the mold increased. The tensile stress of the more micro-units continued to decrease, which would lead to a decrease in the overall elongation of the die section. Therefore, with the increase of the contact mold angle, the elongation of the specimen contact mold segment decreases, and the elongation rate of the suspension segment increases. To improve the uniformity of the stretch-bend specimen, the bending contact mold angle of each loading step can be reduced to control the cross-section shrinkage and tensile elongation.

5. Conclusions

In this paper, the frictional plastic deformation behavior of the contact cross-section in the stretch-bend forming process was investigated. The analytical model of stress–strain distribution in the contact mold segment of stretch-bend forming was established by element discretization. The effects of contact section friction state and contact interface angle on the cross-sectional shrinkage deformation, tensile deformation gradient of stretch-bend specimens, and deformation distribution were discussed by an equivalent stretch-bend forming experiment. Through theoretical derivation and equivalent experiments, some conclusions can be summarized as follows:

(1) The frictional shear stress of the discrete units in the contact mold segment was opposite to the tensile stress, and the cumulative superposition in the length direction of the specimen caused the uneven distribution of tensile stress. Generally, the larger the friction coefficient, the farther the micro-unit is away from the action point and the smaller the tensile stress on the micro-unit.

(2) The tensile stress of the contact mold segment gradually decreased from top to bottom along the thickness direction. The plastic deformation of the upper surface was greater than that of the lower surface, and the cross-section shape of the contact mold segment after stretch-bend forming was a trapezoid. The tensile stress and strain in the cross-section of the contact mold segment of the stretch-bend specimen were distributed in a trapezoid shape the cross-section, and the plastic deformation of the upper surface of the section was larger than that on the lower surface. As a result, the cross-section shape of the contact mold segment was trapezoidal after stretch-bend forming.

(3) In terms of the stretch-bend of the smooth mold compared with the rough mold, the cross-section shrinkage of the contact mold segment increased by more than 38.2%, and the elongation of the contact mold segment increased by more than 32.3%. The greatest effective factor of the specimen plastic deformation is the friction state of the contact interface. Generally, the better the friction state of the contact interface, the higher the cross-sectional shrinkage and elongation of the contact mold segment, and the smaller the tensile deformation gradient, and the smaller the cross-section shrinkage and elongation of the suspension segment. In order to improve the plastic deformation uniformity of the contact mold segment, the mold surface can be properly treated to reduce the friction coefficient.

(4) When the contact mold angle increased from 30° to 60°, the cross-section shrinkage of the contact mold segment decreased by more than 32.8%, and the elongation of the contact mold segment decreased by more than 39.9%. With the increase of the contact mold angle, the total amount of frictional shear stress and the tensile stress difference of both sides increased, the cross-section shrinkage and elongation of the contact mold segment decreased, and the cross-section shrinkage and elongation of the suspension segment increased. Therefore, reducing the angle of the contact mold can effectively control the influence of friction shear stress on plastic deformation.

Author Contributions: Conceptualization, S.Z. and G.L.; methodology, F.M.; formal analysis, Z.W.; investigation, Y.L.; data curation, G.L.; writing—original draft preparation, G.L.; writing—review and editing, Y.L.; supervision, S.Z.; project administration, Y.L.; funding acquisition, S.Z. All authors have read and agreed to the published version of the manuscript.

Funding: This research was funded by the National Natural Science Foundation of China, the funder Shengfang Zhang, grant number 52075066, and Open Foundation of Key Laboratory of Fundamental Science for National Defense of Aeronautical Digital Manufacturing Process, the funder Yu Liu, number SHSYS202007.

Institutional Review Board Statement: Not applicable.

Informed Consent Statement: Not applicable.

Data Availability Statement: Not applicable.

Conflicts of Interest: The authors declare no conflict of interest.

References

1. Xiong, J.; Shen, Z. Rise and future development of Chinese high-speed railway. *J. Traffic Transp. Eng.* **2021**, *21*, 6–29. [CrossRef]
2. Ding, S.; Chen, D.; Liu, J. Research development and prospect of China high-speed train. *Chin. J. Theor. Appl. Mech.* **2021**, *53*, 35–50. [CrossRef]
3. Xu, H.; Liu, M.; Guo, Z.; Zou, Y.; Lu, R.; Gu, Z.; Cheng, X. Accuracy control of stretch bending for variable curvature L-section aluminum alloy door column of EMU. *J. Harbin Inst. Technol.* **2021**, *53*, 77–83. [CrossRef]
4. Zhai, R.; Ding, X.; Yu, S.; Wang, C. Stretch bending and springback of profile in the loading method of prebending and tension. *Int. J. Mech. Sci.* **2018**, *144*, 746–764. [CrossRef]
5. Xiang, N.; Shu, Y.; Wang, P.; Huang, T.; Guo, X.; Guo, J.; Chen, X.; Chen, F. Improved forming accuracy through controlling localized sheet metal deformation in the friction-assisted stretch bending process. *Int. J. Adv. Manuf. Technol.* **2021**, *116*, 3635–3650. [CrossRef]
6. Ma, J.; Welo, T.; Blindheim, J.; Ha, T. Effect of Stretching on Springback in Rotary Stretch Bending of Aluminium Alloy Profiles. *Key Eng. Mater.* **2021**, *883*, 175–180. [CrossRef]
7. Welo, T.; Ma, J.; Blindheim, J.; Ha, T.; Ringen, G. Flexible 3D stretch bending of aluminium alloy profiles: An experimental and numerical study. *Procedia Manuf.* **2020**, *50*, 37–44. [CrossRef]
8. Liang, J.; Liao, Y.; Li, Y.; Liang, C. Study on the Influence of Bending Angle of Multipoint Stretch-Bending of Profiles on Section Distortion of Parts. *Math. Probl. Eng.* **2020**, *2020*, 1975805. [CrossRef]
9. Chen, C.; Liang, J.; Li, Y.; Liang, C.; Jin, W. Springback Analysis of Flexible Stretch Bending of Multi-Point Roller Dies Process for Y-Profile under Different Process Parameters. *Metals* **2021**, *11*, 646. [CrossRef]
10. Liu, T.; Wang, Y.; Wu, J.; Xia, X.; Wang, J.; Wang, W.; Wang, S. Springback analysis of Z & T-section 2196-T8511 and 2099-T83 Al–Li alloys extrusions in displacement controlled cold stretch bending. *J. Mater. Process. Technol.* **2015**, *225*, 295–309. [CrossRef]
11. Liu, C.-G.; Zhang, X.-G.; Wu, X.-T.; Zheng, Y. Optimization of post-stretching elongation in stretch bending of aluminum hollow profile. *Int. J. Adv. Manuf. Technol.* **2015**, *82*, 1737–1746. [CrossRef]
12. Liu, B.; Cao, F.; Zeng, Y.; Wu, W. Numerical and experimental study on temperature and springback control of U-shape titanium extrusion hot stretch bending. *Int. J. Light. Mater. Manuf.* **2022**, *5*, 453–469. [CrossRef]
13. Ma, J.; Welo, T. Analytical springback assessment in flexible stretch bending of complex shapes. *Int. J. Mach. Tools Manuf.* **2020**, *160*, 103653. [CrossRef]
14. Maati, A.; Tabourot, L.; Balland, P.; Ouakdi, E.; Vautrot, M.; Ksiksi, N. Constitutive modelling effect on the numerical prediction of springback due to a stretch-bending test applied on titanium T40 alloy. *Arch. Civ. Mech. Eng.* **2015**, *15*, 836–846. [CrossRef]
15. Uemori, T.; Naka, T.; Tada, N.; Yoshimura, H.; Katahira, T.; Yoshida, F. Theoretical predictions of fracture and springback for high tensile strength steel sheets under stretch bending. *Procedia Eng.* **2017**, *207*, 1594–1598. [CrossRef]
16. Gu, Z.; Lv, M.; Li, X.; Xu, H. Stretch bending defects control of L-section aluminum components with variable curvatures. *Int. J. Adv. Manuf. Technol.* **2015**, *85*, 1053–1061. [CrossRef]
17. Han, S. Influence of Frictional Behavior Depending on Contact Pressure on Springback at U Draw Bending. *Trans. Mater. Process.* **2011**, *20*, 344–349. [CrossRef]
18. Fox, R.T.; Maniatty, A.M.; Lee, D. Determination of friction coefficient for sheet materials under stretch-forming conditions. *Met. Mater. Trans. A* **1989**, *20*, 2179–2182. [CrossRef]
19. Liu, K.X.; Liu, Y.L.; Yang, H. An analytical model for the collapsing deformation of thin-walled rectangular tube in rotary draw bending. *Int. J. Adv. Manuf. Technol.* **2013**, *69*, 627–636. [CrossRef]
20. Liu, K.; Liu, Y.; Yang, H. Experimental and FE simulation study on cross-section distortion of rectangular tube under multi-die constraints in rotary draw bending process. *Int. J. Precis. Eng. Manuf.* **2014**, *15*, 633–641. [CrossRef]
21. Liu, K.; Liu, Y.; Yang, H. Experimental study on the effect of dies on wall thickness distribution in NC bending of thin-walled rectangular 3A21 aluminum alloy tube. *Int. J. Adv. Manuf. Technol.* **2013**, *68*, 1867–1874. [CrossRef]
22. Guan, Y.; Zhao, J.; Wang, F.; Ma, L. Influence of friction on springback of plate tension-bending. *J. Plast. Eng.* **2003**, *10*, 43–45. [CrossRef]
23. Yang, H.; Li, H.; Zhan, M. Friction role in bending behaviors of thin-walled tube in rotary-draw-bending under small bending radii. *J. Mater. Process. Technol.* **2010**, *210*, 2273–2284. [CrossRef]
24. Zhang, Z.; Yang, Y.; Li, L.; Yin, J. Distribution of residual stress in an asymmetric T-section beam by stretch-bending. *Int. J. Mech. Sci.* **2019**, *164*, 105184. [CrossRef]
25. Muranaka, T.; Fujita, Y.; Otsu, M.; Haraguchi, O. Development of rubber-assisted stretch bending method for improving shape accuracy. *Procedia Manuf.* **2018**, *15*, 709–715. [CrossRef]
26. Liang, J.; Han, C.; Li, Y.; Yu, K.; Liang, C. Study on deformation difference between the contact zone and the non-contact zone of the flexible 3D stretch bending profile and roller dies based on pre-stretching amount. *Int. J. Adv. Manuf. Technol.* **2020**, *108*, 3579–3589. [CrossRef]
27. Hambleton, J.; Drescher, A. On modeling a rolling wheel in the presence of plastic deformation as a three- or two-dimensional process. *Int. J. Mech. Sci.* **2009**, *51*, 846–855. [CrossRef]
28. Bobrovskij, I.; Khaimovich, A.; Bobrovskij, N.; Travieso-Rodriguez, J.A.; Grechnikov, F. Derivation of the Coefficients in the Coulomb Constant Shear Friction Law from Experimental Data on the Extrusion of a Material into V-Shaped Channels with Different Convergence Angles: New Method and Algorithm. *Metals* **2022**, *12*, 239. [CrossRef]

29. Song, P.; Li, W.; Wang, X. A Study on Dynamic Plastic Deformation Behavior of 5052 Aluminum Alloy. *Key Eng. Mater.* **2019**, *812*, 45–52. [CrossRef]
30. *TS EN ISO 6892-1-2019*; Metallic Materials. Tensile Testing. Part 1: Method of Test at Room Temperature (ISO 6892-1-2019). ISO: Geneva, Switzerland, 2019. Available online: https://www.iso.org/standard/78322.html (accessed on 29 June 2022).
31. Bachchhav, B.; Bagchi, H. Effect of surface roughness on friction and lubrication regimes. *Mater. Today Proc.* **2020**, *38*, 169–173. [CrossRef]
32. Zhang, S.; Lv, G.; Ma, F.; Wang, Z.; Liu, Y. Influence of low-frequency vibration on friction coefficient of contact interface in bend-stretch forming. *Int. J. Light. Mater. Manuf.* **2022**, *5*, 306–314. [CrossRef]

Article

Steel Surface Doped with Nb via Modulated Electron-Beam Irradiation: Structure and Properties

Maxim Sergeevich Vorobyov, Elizaveta Alekseevna Petrikova, Vladislav Igorevich Shin, Pavel Vladimirovich Moskvin, Yurii Fedorovich Ivanov, Nikolay Nikolaevich Koval, Tamara Vasil'evna Koval, Nikita Andreevich Prokopenko, Ruslan Aleksandrovich Kartavtsov and Dmitry Alekseevich Shpanov *

Institute of High Current Electronics, Siberian Branch, Russian Academy of Sciences, 2/3 Akademichesky Ave., Tomsk 634055, Russia; vorobyovms@yandex.ru (M.S.V.); elizmarkova@yahoo.com (E.A.P.); shin.v.i@yandex.ru (V.I.S.); pavelmoskvin@mail.ru (P.V.M.); yufi55@mail.ru (Y.F.I.); koval@opee.hcei.tsc.ru (N.N.K.); tvkoval@tpu.ru (T.V.K.); nick08_phantom@mail.ru (N.A.P.); kartavcov@gmail.com (R.A.K.)
* Correspondence: das138@tpu.ru

Abstract: A niobium film on an AISI 5135 steel substrate was exposed to submillisecond pulsed electron-beam irradiation with controlled energy modulation within a pulse to increase the film–substrate adhesion. This modulated irradiation made it possible to dope the steel-surface layer with Nb through film dissolution in the layer, for which optimum irradiation conditions were chosen from experiments and a mathematical simulation. The irradiated system was tested for surface hardness and wear, and its surface structure and elemental composition were analyzed. The results demonstrate that the microhardness of the irradiated system is much higher and that its wear rate is much lower compared to the initial state.

Keywords: surface modification; doping; grid plasma cathode; electron beam; Nb–steel system; high-rate crystallization

1. Introduction

Currently, structural steels occupy one of the leading positions in industry due to their strong mechanical characteristics, good technological efficiency, and comparatively low cost [1–3]. In particular, they are widely used to manufacture products used in operations with high loads, and because such loads can cause deformation and fracturing (e.g., cracking, spallation [4]), mostly in their surface layers, the surface layers of structural steels are hardened [5]. One of the ways of hardening the surfaces and improving the functional properties of machine parts is to deposit a protective coating on their working surfaces, which can greatly improve their physicomechanical characteristics (strength, wear resistance, etc.) and service life [6]. However, such protectively hardened coatings often suffer from weak adhesion to substrate materials. This problem can be solved by combining several hardening methods so that each shows its advantages and not its disadvantages [7]. The formation of high-strength and thermally stable surface layers through refractory-metal-coating deposition and intense pulsed electron-beam irradiation with amplitude and width modulation makes it possible to greatly improve the service conditions and applicability of structural steels [8–10]. Structural steels doped with Nb, compared to other steels, are lighter and harder, and they have greater strength, higher corrosion resistance, and longer lifetimes; additionally, they are ductile, i.e., they show no brittleness after doping with Nb. The fields in which they are used include mechanical, radio, and nuclear energy engineering, chemical industries, the automobile and building industries, and the aerospace industry in particular [11–22]. Steel doping with V, Ti, Nb, and Zr yields carbides that are poorly soluble in austenite. These elements are efficient (in decreasing the grain size, cold-brittleness threshold, and sensitivity to stress concentration) only if their steel content is low (up to 0.15 wt.%). With higher contents, they decrease the hardenability and brittle-fracture resistance of steel because a large amount of metal carbides precipitates at the grain boundaries. The presence of Nb in proper amounts in steels provides primary structural refinement due to ferrite formation and the filling of its interdendrite spacings with eutectic liquid. With these positive effects, for example, low-alloy structural AISI 5135 steel doped with Nb can be used to manufacture tools for the efficient extrusion of light metals (e.g., Al) and the hot working of bearings; thus, it can be a good alternative to expensive, difficult-to-machine heat-resistant steels doped in their bulk with Mo, W, Ti, and V. The surface modification considered below is a form of finishing and does not require any further treatment [23–27].

2. Test Material and Treatment Methods

The test material was low-carbon AISI 5135 steel ((0.1–0.39) C, (0.17–0.37) Si, (0.5–0.8) Mn, ≤0.3 Ni, ≤0.035 Cr, ≤0.035 P, 0.8–1.1 Cr, ≤0.3 Cu, and balance Fe, wt.%). Its specimens, with dimensions $15 \times 15 \times 5$ mm, were coated with an Nb film 3 µm thick via plasma-assisted arc deposition on the QUINTA vacuum ion-plasma setup [28], which is part of the UNIKUUM complex of unique electrophysical equipment of Russia (https://ckp-rf.ru/catalog/usu/434216/ (accessed on 13 June 2023)). Figure 1 shows a schematic diagram of Nb-film deposition. The deposition was realized using a PINK-P gas plasma generator based on a non-self-sustained low-pressure arc discharge with a hollow and a hot cathode (Figure 1, cathodes 4 and 5, respectively). The generator makes it possible to treat objects up to 40 cm in length and up to 1 kg in weight. It is used for preliminary surface cleaning and heating in argon plasma and for ion-plasma assistance in vacuum-arc deposition, which increases the coating–substrate adhesion and the vacuum-arc stability. The arc evaporator with a 98 wt.% Nb cathode (Figure 1, evaporator 2), compared to its previous version, ensured better cooling of the cathode back's surface and a smaller amount of microdroplets in the coating. Before film deposition, the specimens were washed with petrol in an ultrasonic bath (to remove mechanical and oil contaminants from their surfaces) and, after they were fixed in the specimen holder, they were placed in the vacuum chamber at a distance of 17 cm from the evaporated cathode. Subsequently, the vacuum chamber was pumped to a limiting pressure of 1×10^{-2} Pa, and the specimens were coated with a Nb film. The deposition comprised several stages. First, the specimens were exposed to ion-plasma surface cleaning in argon. Subsequently, their surfaces were bombarded with Nb ions at an argon pressure of 0.15 Pa, arc current of 80 A, substrate-bias voltage of −900 V, and pulse duty factor of 85%, which ensured better film adhesion. Next, the specimens were coated with Nb for 20 min at an argon pressure of 0.3 Pa, arc current of 80 A, and bias voltage of 50 V. The Nb-coated specimens were cooled in the vacuum chamber to room temperature and removed from it for further pulsed electron-beam treatment. The coating thickness determined by the Calotest method was 3 µm.

The Nb film–AISI 5135 steel-substrate system was irradiated with an electron beam on the SOLO setup (also a part of the UNIKUUM complex), whose electron source allows the generation of electron beams with diameters of up to 5 cm, energy of up to 25 keV, and energy density of up to 100 J/cm^2 at a pulse duration of 20–1000 µs [29]. The unique feature of this type of electron source is the possibility of controlling the beam current, which is weakly dependent on the accelerating voltage [30]. Thus, it is possible to control the beam power in the submillisecond range of pulse durations [31] and, hence, the rate of energy delivery to the surface of a treated target within a beam current pulse [32]. The control of energy delivery makes it possible to control the temperature field in the target surface layer and, hence, its structure and phase state.

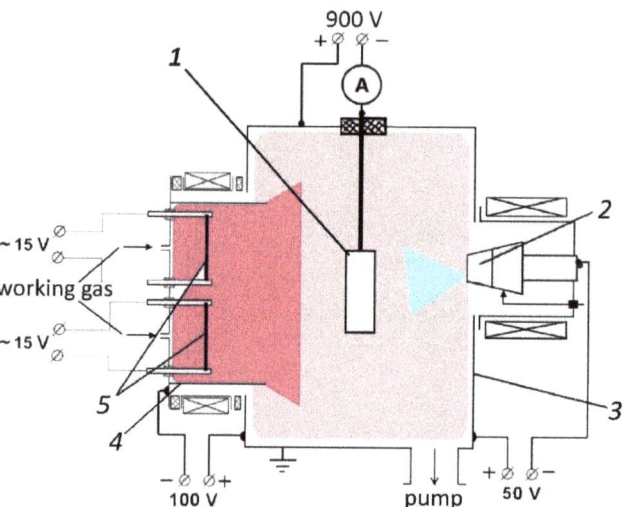

Figure 1. Schematic diagram of Nb-film deposition: 1—specimen holder, 2—arc evaporator with Nb cathode, 3—vacuum chamber (anode), 4—hollow cathode of PINK-P generator, 5—hot W cathode of PINK-P generator.

Figure 2 shows a schematic diagram of electron-beam irradiation. Niobium-coated steel specimen *8* was fixed on two-coordinate manipulator *6* with narrow stainless steel plate *7* with thickness 0.2 mm and height 5–7 mm for thermal-loss reduction. Using the manipulator table, the specimens were moved in vacuum chamber *3* under electron beam *5*. The temperature of the specimen surface was measured with high-speed infrared pyrometer *15* (Kleiber KGA 740-LO), allowing temperature measurements in the range of 300–2300 °C with a response time of 6 μs. The pyrometer used radiation at 2–2.2 μm, which was collected by lens *11* with a surface-spot diameter of about 5 mm at the specimen center. The output signal of the pyrometer was the voltage measuring 0–10 V, which depended linearly on the temperature in the operating range of 2000°. To determine the specimen-surface emissivity, the specimen was heated to 500 °C for 3 min by an electron beam for a duration of 200 μs and repetition frequency of 10 Hz, which did not change the surface properties. After this heating, the heat was distributed over the specimen volume via conduction, such that the specimen was cooled to 300 °C in about 2 min. During the period of cooling, the specimen-surface temperature was measured with the pyrometer, as was the temperature in the specimen bulk with K-type thermocouple *12* built in the specimen on the back side. The results of temperature measurements were compared to determine the surface emissivity.

The modes of irradiation were chosen so that the specimen's surface layer would be heated above the steel's melting temperature (1450–1550 °C) but below the niobium's melting temperature (2500 °C), i.e., to ≈2000 °C. Thus, it was expected that the thin Nb film would dissolve in the molten steel-surface layer kept at 2000 °C. According to the pyrometer specifications, the measured temperature T [°C] is determined by its output signal U_p [V] as follows: $T = 300 + 200\ U_p$. The behavior of U_p with time for different irradiation modes is presented in Figure 3.

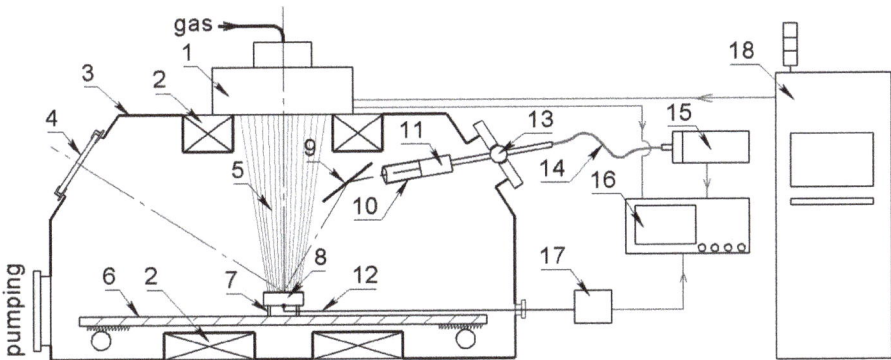

Figure 2. Schematic diagram of electron-beam irradiation: 1—electron source, 2—magnetic coils, 3—vacuum chamber, 4—observation window, 5—pulsed electron beam, 6—two-coordinate manipulator table, 7—fastening plates, 8—irradiated specimen, 9—copper mirror, 10—collimator, 11—lens, 12—thermocouple, 13—vacuum joint, 14—fiber-optic cable, 15—high-speed infrared pyrometer (Kleiber KGA 740-LO), 16—oscilloscope, 17—normalizing converter, 18—power-supply units and automation system.

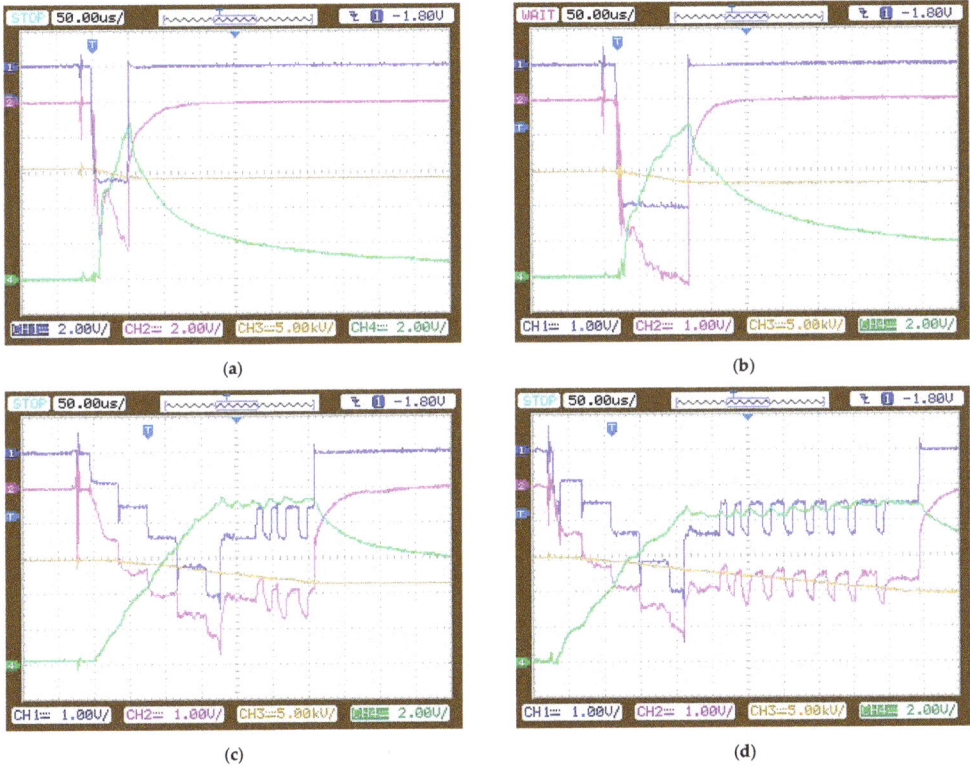

Figure 3. The behavior of Up with time for different irradiation modes: (**a**,**b**)—rectangular current pulses of duration 50 μs and 100 μs; (**c**,**d**)—pulses with varied beam currents at which the surface temperature reached 2000 °C and remained at this level for 150 μs for a total pulse duration of 300 μs and for 350 μs at 500 μs.

To maintain the specimen surface at the same temperature level, the following parameters were chosen for single beam current pulses:
- rectangular current pulses of duration 50 µs (Figure 3a) and 100 µs (Figure 3b);
- pulses with varied beam currents at which the surface temperature reached 2000 °C and remained at this level for 150 µs for a total pulse duration of 300 µs (Figure 3c) and for 350 µs at 500 µs (Figure 3d).

The following color symbols were used on the waveforms (Figure 3): waveforms of discharge current I_d (1, blue, 40 A/div for figure "a" and 20 A/div for others), acceleration-gap current I_0 (2, magenta, 40 A/div for figure "a" and 20 A/div for others), accelerating voltage U_0 (3, brown, 5 kV/div, zero is the same as signal 4), and pyrometer signal U_p (4, green, 2 V/div) for different irradiation modes. The beam was transported to the sample in the plasma that it created. Therefore, current I_0 contained not only the beam current, but also the ion current, which moved from the plasma to the high-voltage electrode.

Figure 4 shows experimental beam-power profiles for the beam-current pulses in Figure 3. It is shown that the beam-current modulation within a submillisecond pulse due to proportional arc-current modulation made it possible to control the beam power and the energy to the target surface within the pulse and, hence, the temperature field in the specimen-surface layer.

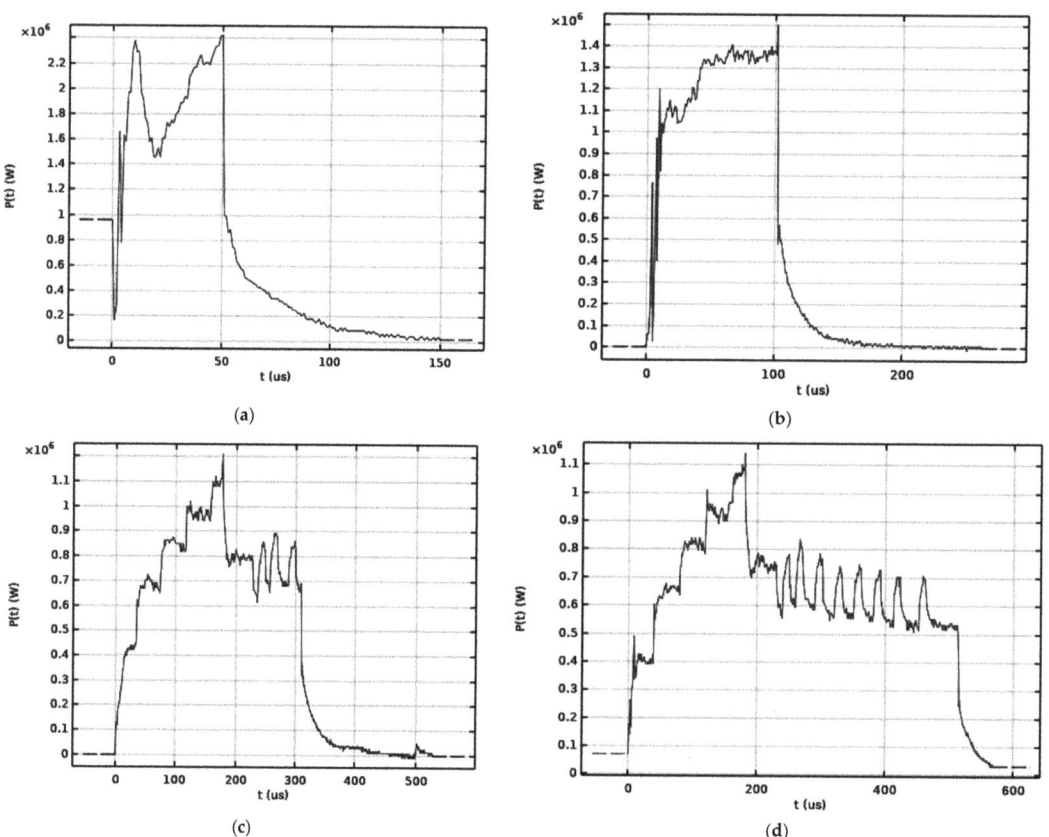

Figure 4. Experimental beam-power profiles within pulses lasting 50 (**a**), 100 (**b**), 300 (**c**), and 500 µs (**d**).

The temperature field formed in the Nb film–AISI 5135 steel-substrate system under intense pulsed electron-beam irradiation (beam energy density 10–50 J/cm^2, current pulse duration 50–500 µs) was estimated using a mathematical simulation.

With an average energy of the beam electrons on the target of ~10 keV, the energy source in the mathematical model can be assumed to act on the surface, and the thermal processes for pulse durations of 50–500 µs can be considered in the one-dimensional approximation because the transverse size of energy exposure is much larger than the depth of thermal-field propagation. With these assumptions, the estimation of the heating is reduced to solving the heat-conduction equation:

$$\rho c \frac{\partial T}{\partial t} = \frac{\partial}{\partial x}\left(\lambda \frac{\partial T}{\partial x}\right) \qquad (1)$$

with a boundary and initial conditions of

$$-\lambda \frac{\partial T(t,0)}{\partial x} = q(t), \quad \frac{\partial T(t,l)}{\partial x} = 0, \quad T(t=0, x) = T_0 \qquad (2)$$

where c is the specific heat capacity, ρ is the density, and λ is the heat conductivity, which is dependent on the temperature and coordinates and corresponds to the coating or substrate, $q(t) = U(t)j(t)$ is the power density of an external heat source, $U(t)$ and $j(t)$ are the accelerating voltage and the beam-current density, respectively, l is the computational domain length, $T_0 = 23$ °C.

Formula (2) is a boundary condition of the second kind (Neumann's condition): a solid is heated from the outside by a stream of high-temperature thermal radiation $q(t)$ and the initial value.

The mathematical model assumes the presence of a two-phase region. In a solid–liquid system, this region is characterized by the average liquid volume fraction, θ. The phase transition falls in the temperature interval, ΔT, at which pint the material state is modeled by a smoothed function of θ, varying from 1 to 0 [33]. The effective heat conductivity of the solid–liquid system λ is related to the solid conductivity λ_s and liquid conductivity λ_l, as

$$\lambda = (1-\theta)\lambda_s + \theta\lambda_l \qquad (3)$$

The density of the two-phase region and its heat capacity were calculated in the same way. The latent heat of fusion L is introduced as an additional term in the heat capacity in the phase transition: $c_i = c_s + L/\Delta T$. The value of the phase-transition interval ΔT is determined from the correspondence between the calculated and experimental surface temperatures.

The problem was solved numerically for table values of thermophysical parameters [34]. For AISI 5135 steel, the melting temperature is $T_l = 1420$ °C, $\Delta T = 8$ °C, and the heat conductivity and the heat capacity at $T < T_l$ measure $\lambda_s = 45.62 - 0.003T - 4.6 \times 10^{-5}T^2 + 3 \times 10^{-8}T^3$ W/(m·°C) and $c_s = 430.75 + 0.3802T - 0.00014T^2$ J/(kg·°C), $\lambda_l = 35$ W/(m·°C), $\rho = 7820$ kg/m^3, $L = 84$ kJ/kg, respectively. For the coating of thickness 3 µm, the Nb's melting temperature is 2741 °C, $\rho_s = 8000$ kg/m^3, $c_s = 268 + 0.048T$ J/(kg·°C), $\lambda_s = 53 + 0.013T$ W/(m·°C). The coating melting temperature is higher than the maximum surface temperature of 2000 °C considered in this study. The coating fails under electron-beam irradiation, changing its properties due to the molten substrate. This is allowed for in the simulation by estimating the thermophysical coefficients as averages for the coating and substrate.

For model pulses close to the experimental pulses (Figure 4), we numerically analyzed the temperature-field dynamics in the Nb film–AISI 5135 steel-substrate system on approaching the surface temperature of 2000 °C. For two rectangular pulses with durations of 50 µs and 100 µs and energy densities of 17 J/cm^2 and 24 J/cm^2, respectively, the surface temperature of 2000 °C was reached at the end of the pulses. The melt depths in these modes were 8 µs and 12 µs, respectively. To increase the thickness of the molten-steel layer, calculations were performed in which the specimens surface was preliminarily heated to ≈200 °C for 200 μs and kept at this temperature for 100 μs and 300 μs (pulses with durations of 300 μs and 500 μs). Figure 5a shows the beam power and the surface temperature for the pulse of duration, 500 μs. The temperature distribution in depth from the surface at different points in time is presented in Figure 5b.

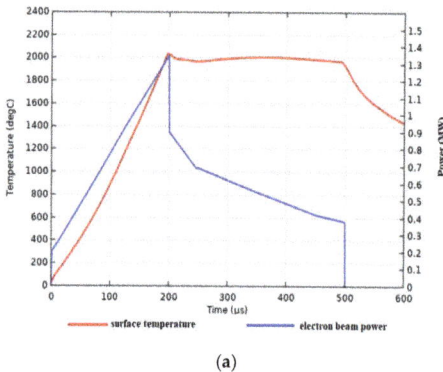 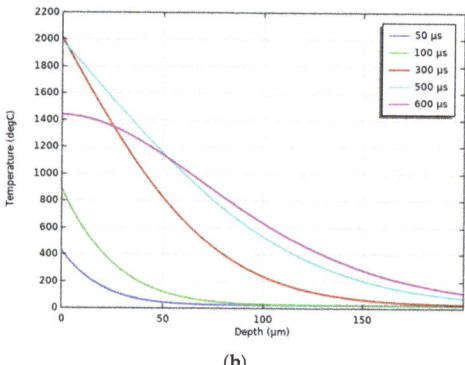

(a) (b)

Figure 5. Surface temperature and beam power vs. time (**a**), and temperature distribution in depth at different time points (**b**) for model pulse of duration 500 μs (for mode in Figure 4d).

In a real electron beam, as numerical calculations show, the following physical processes take place:

(1) energy dissipation (up to 20%) of electrons transported to the target in an extended plasma channel;

(2) a potential drop in the accelerating gap (up to 50%) due to the growth of the spatial charge with an increase in the beam current.

The maximum loss in the target is at a distance of one-third of the depth of the run, so the surface source in the mathematical model is quite correct at $Re \ll 3\sqrt{at}$, and deducted thermal conductivity.

The molten-layer depth depends both on the pulse duration and on the beam power density. Figure 6 shows the melt depths, and the surface melt times for the pulses considered. The melt depth at pulse durations of 50–100 μs was 8–12 μm, and at 500 μs, it increased to 30 μm (Figure 6a), with the surface remaining molten for 480 μs (Figure 6b).

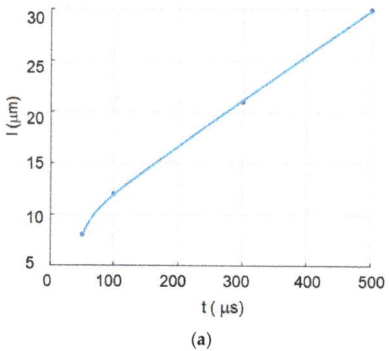 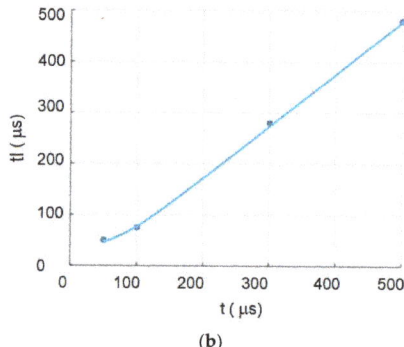

(a) (b)

Figure 6. Melt depths (**a**) and surface melt times (**b**) for model pulses with durations of 50, 100, 300, and 500 μs.

Using the data from numerical calculations, we chose the irradiation modes (Figure 4) that provided surface modifications, with Nb kept in its near-melting state and AISI 5135 steel in its surface-melting state.

Figure 7 shows the surface temperature $T(t)$ according to the calculation and experimental data for different beam powers and durations (Figure 4). The pulsed character of $T(t)$ with the surface temperature remaining constant was due to beam-current fluctuations (Figure 4c,d).

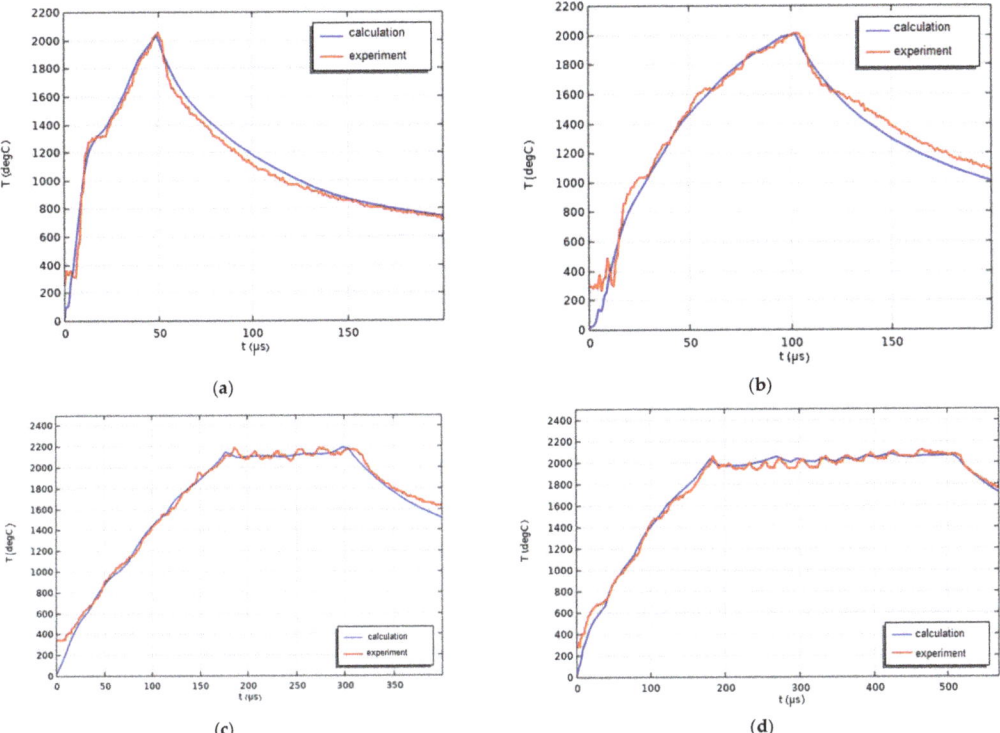

Figure 7. Surface temperatures for current pulse durations of 50 (**a**), 100 (**b**), 300 (**c**), and 500 µs (**d**), according to calculation and experiment.

The phase state of the steel and film–substrate system was examined on an XRD 6000 diffractometer (Shimadzu, Kyoto, Japan) in the Bragg–Brentano geometry with CuKα radiation (λ = 1.5418 nm); the range of diffraction angles was 2θ = 30–80 deg and the scan rate was 2 deg/min. The structure of the irradiated surface was analyzed on a Philips SEM-515 scanning-electron microscope (Amsterdam, The Netherlands). The elemental composition of the surface layer was determined on an EDAX ECON IV micro-analyzer (Philips, Amsterdam, Netherlands, attached to Philips SEM-515). The defect substructure and the phase state of the material were analyzed on a JEOL JEM 2100F transmission-electron microscope (Akishima, Tokyo, Japan). The surface hardness of the steel and film–substrate system at different stages of electron-ion-plasma treatment was determined on a PMT-3 hardness tester at a normal indenter load of 500 mN (LOMO, Saint Petersburg, Russia). The friction and wear coefficients of the surface layer were measured on a Tribotechnic tribometer (Tribotechnic, Clichy, France) in the pin-on-disk geometry at room temperature. The counter body was a SiC ball with a diameter of 6 mm, the track diameter was 4 mm, the rotation rate was 2.5 cm/s, the load was 5 N, and the travel distance was 1000 m. The wear volume of the surface layer was determined after track profilometry. The wear coefficient was estimated by the formula $k = \frac{S \cdot R}{F_n \cdot n}$ mm^3/Nm, where S is the track cross-sectional area, n is the number of circles, R is the circle radius, F_n is the indenter load.

3. Analysis of Strength, Tribological Properties, and Structure

The irradiated specimens of the Nb film–AISI 5135 steel-substrate system (Nb–steel system) were tested for their surface strength (microhardness) and tribological properties (wear resistance and friction coefficient). Figure 8 shows the friction coefficient vs. time and the wear-track cross-section.

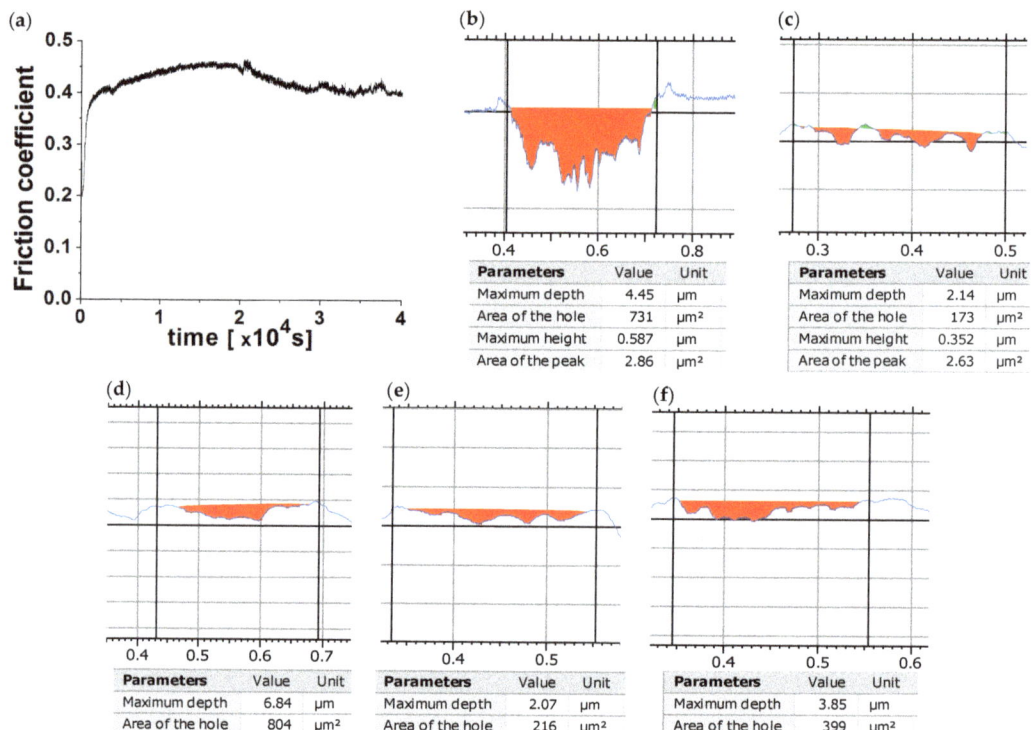

Figure 8. Friction coefficient vs. time (**a**) and characteristic wear-track cross-section: ((**b**) initial, (**c**) 100 µs, (**d**) 50 µs, (**e**) 500 µs, (**f**) 300 µs).

After irradiation in all four modes, the microhardness of the Nb–steel system measured 8–9 GPa, which was higher than the initial value of 3.6 GPa by a factor of 2.2–2.5 (Figure 9). Its wear rate decreased five times when the pulse duration measured 50 µs and 100 µs and three times when the temperature was maintained at about 2000 °C. For the initial sample (without coating), the microhardness was 3.6 GPa and the wear rate was $2.7 \cdot 10^{-6}$ mm^3/(N·m).

The surface structure and the elemental composition of the Nb–steel system before and after irradiation were examined by using scanning-electron microscopy. It can be seen in Figure 10 that microcraters formed on the surface of the system during the irradiation at pulse durations of 50–300 µs. At a pulse duration of 500 µs, no craters were detected (Figure 10e), which was probably due to the evaporation of volatile impurities from the hot specimen's surface [35]. Our X-ray spectral analysis showed that increasing the surface heating time to 300 µs decreased the Nb content in the surface layer from 97.55 at.% to 1.06 at.% (Figure 10a,d), which can be associated both with the film's immersion in the steel substrate and with the Nb's dissolution in the molten steel-surface layer.

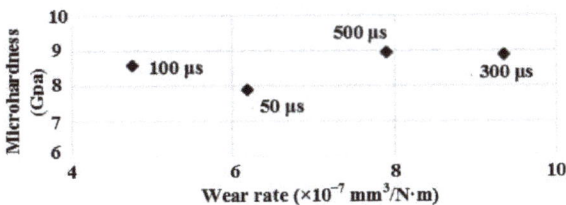

Figure 9. Correlation field for surface microhardness and wear rate of Nb–steel system after electron-beam irradiation.

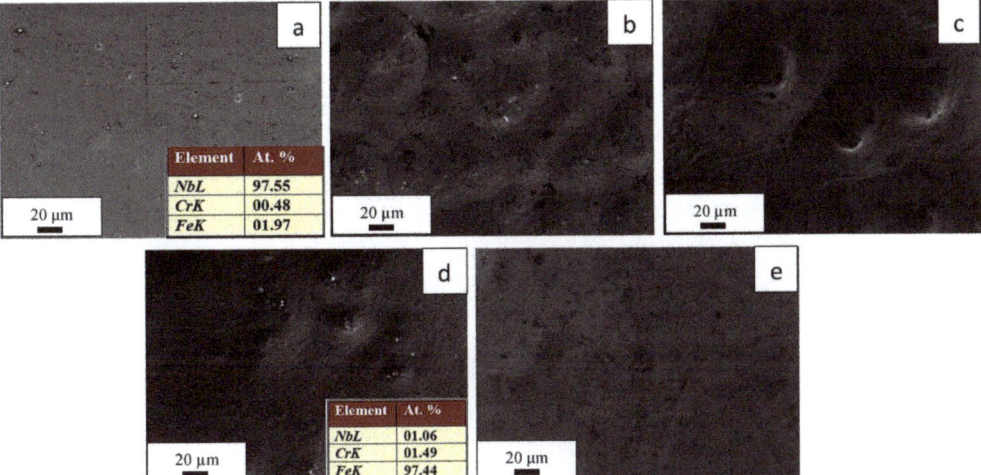

Figure 10. Surface structure of Nb–steel system before (**a**) and after electron-beam irradiation at pulse durations of 50 (**b**), 100 (**c**), 300 (**d**), and 500 µs (**e**).

During the electron-beam irradiation, a cellular structure of high-rate crystallization formed in the surface layer of the Nb–steel system (Figure 11), which was indicative of surface-layer melting. The average cell size increased from 0.2 µm to 0.37 µm when the pulse duration was increased (Figure 12), suggesting that the cooling rate decreased as the pulse became longer, which was confirmed by both the calculations and the real temperature measurements (Figure 7).

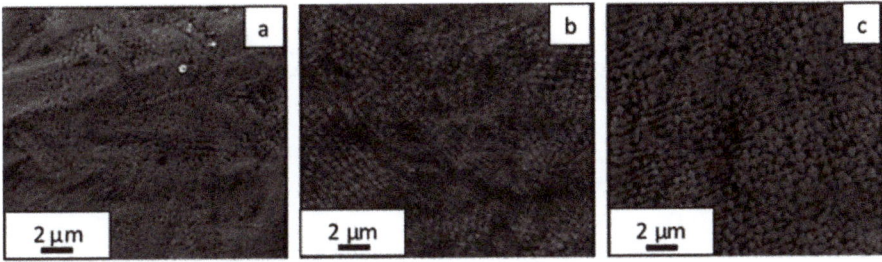

Figure 11. Surface structure of Nb–steel system after electron-beam irradiation for 50 (**a**), 100 (**b**), and 500 µs (**c**).

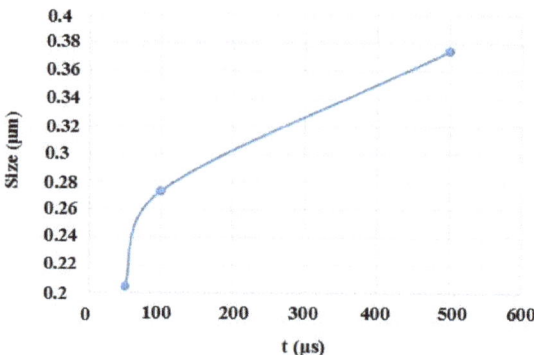

Figure 12. Average size of surface-crystallization cells in Nb–steel system vs. current-pulse duration.

The surface layer formed in the Nb–steel system via high-rate crystallization had a columnar structure (Figure 13). The surface-layer thickness with this columnar structure (high-rate crystallization region) agreed with the calculation data in Figure 5b, measuring about 20–25 μm at t = 300 μs. Our X-ray diffraction analysis showed that the Nb-surface concentration increased from 3.2 mass % (300 μs) to 4.2 mass % (500 μs) (Table 1), probably because the melt filled the surface craters and decreased the surface cooling rate. As a result, some Nb atoms migrated from the crater walls to the surface.

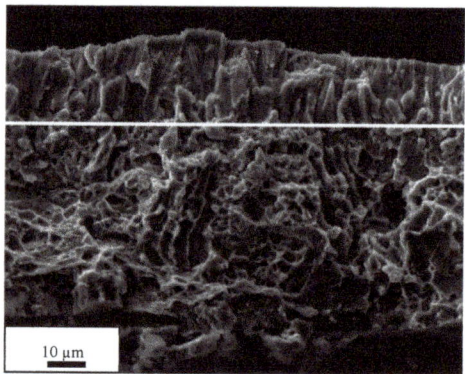

Figure 13. Fracture surface of Nb–steel system irradiated for 300 μs.

Table 1. X-ray diffraction data.

t, μs	Fe, mass %	a(Fe), Å	D, nm	$\Delta d/d$, 10^{-3}	Nb, wt.%	a(Nb), Å	D, nm	$\Delta d/d$, 10^{-3}
0	5.6	2.8763	16.7	0.002	94.4	3.3731	17.4	8.045
50	79.7	2.8475	15.6	1.626	20.3	3.2700	17.9	6.485
100	94.3	2.8491	15.4	1.403	5.7	3.2799	97.2	5.436
300	96.8	2.8527	15.2	1.508	3.2	3.2799	23.4	1.742
500	95.8	2.8668	16.0	1.14	4.2	3.2932	23.6	6.048

The data from the X-ray diffraction analysis showed that the surface layer of the Nb–steel system comprised two main phases: an α-Fe bcc solid solution and niobium (Figure 14, Table 1). During irradiation at t = 300 μs, the relative Nb content in its surface layer decreased from 94.4 to 3.2 wt.%. Simultaneously, the lattice parameters of Nb and

Fe changed, which was probably because the two phases diffused into each other. The formation of solid solutions was also evidenced by changes in the Nb- and Fe-lattice microdistortion ($\Delta d/d$).

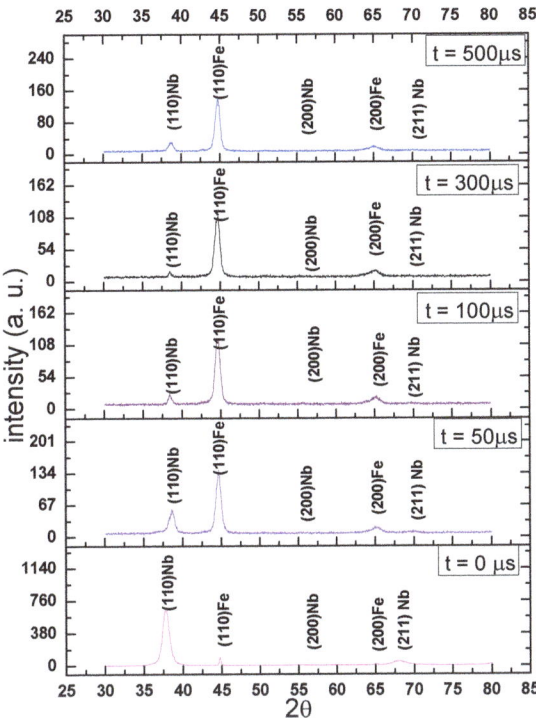

Figure 14. Fragments of X-ray diffraction patterns of Nb–steel system irradiated in different modes.

The data from the transmission-electron microscopy also showed that the initial structure of the AISI 5135 steel was formed by ferrite grains and perlite grains with lamellar morphologies (Figure 15).

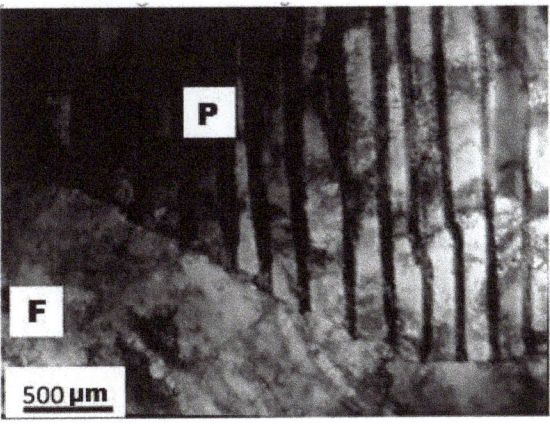

Figure 15. Structure of AISI 5135 steel before modification, with "P" for perlite and "F" for ferrite.

After the electron-beam irradiation of the Nb–steel system, the surface structure of the steel was substantially modified. In a surface layer of 20–30 μm, the structure was typical of high-rate quenching, containing packet martensite (Figure 16a,e) and, more rarely, lamellar martensite (Figure 16b,d). In the volume and at the boundaries of the martensite crystals, there were complex carbide $M_{23}C_6$-like nanosized particles $(Cr, Fe, Nb)_{23}C_6$ (Figure 16f).

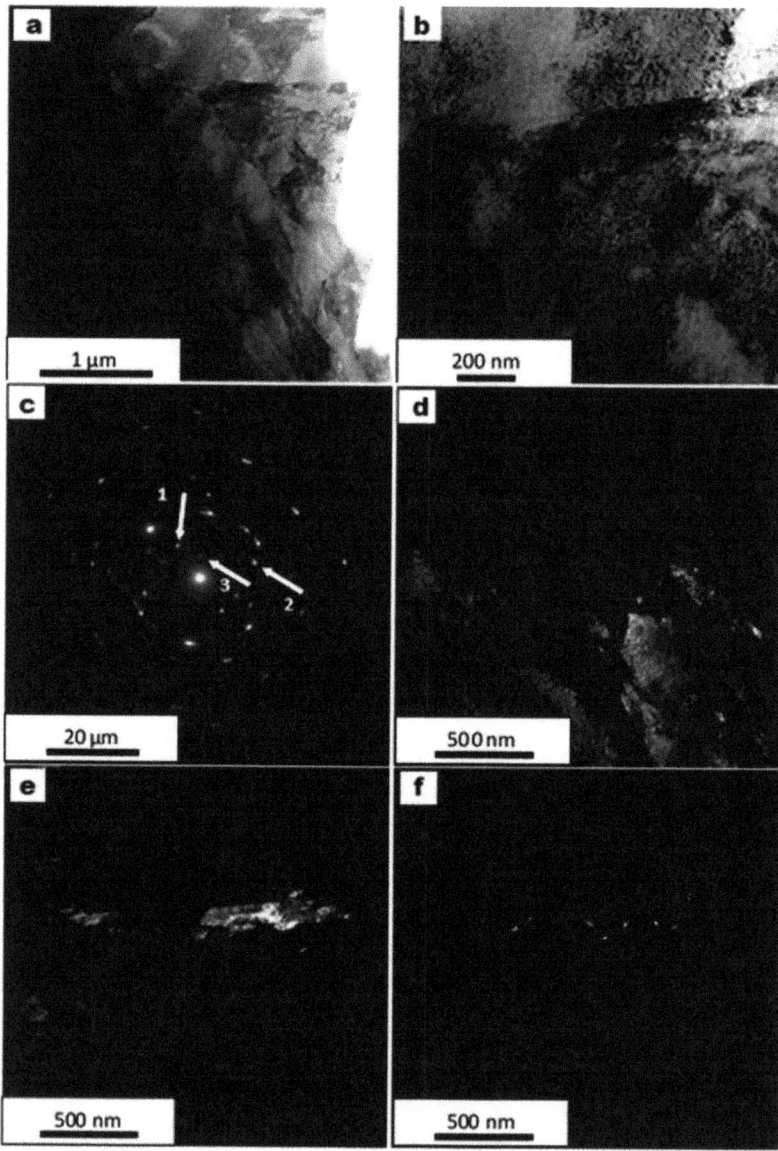

Figure 16. Surface structure of Nb–steel system after electron-beam irradiation at $E = 24$ J/cm^2 for 100 μs: bright fields (**a**,**b**), diffraction pattern (**c**), and dark fields in reflections $[110]\alpha$-Fe (**d**), $[002]\alpha$-Fe (**e**), $[135]M_{23}C_6$ (**f**), shown by arrows 1, 2, 3, respectively, in (**c**).

At a depth of 30–40 μm, the structure corresponded to the high-rate thermal transformation of pearlite (Figure 17). It comprised pearlite grains with two types of cementite particle: extended lamellae (Figure 17a), which is characteristic of lamellar pearlite formed under quasi-equilibrium conditions; and round particles in the form of ferrite lamellae (Figure 17b), which is characteristic of the high-rate thermal transformation of cementite lamellae (dissolution and repeated precipitation).

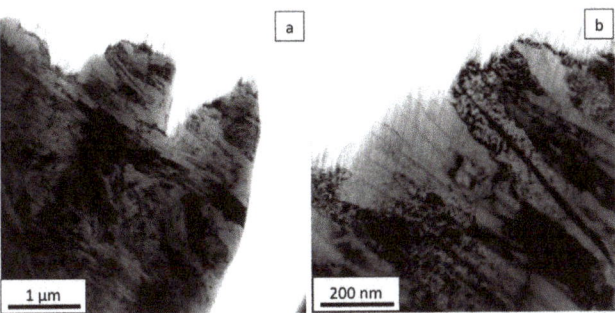

Figure 17. Structure of Nb–steel system 36 µm beneath its surface after irradiation at E = 24 J/cm^2 for 100 µm, extended lamellae (**a**), and round particles in the form of ferrite lamellae (**b**).

From the transformation of the structure and phase state of the surface layer, it was observed that the increase in the microhardness and wear resistance of the modified steel was mainly due to the formation of a hardening (martensite) structure, the precipitation of nanosized $M_{23}C_6$ particles at the martensite crystal boundaries, and pearlite structural transformation via nanosized iron-carbon precipitation in the pearlite lamellae.

At depths of 50–70 µm, the structure was close in phase state and morphology to the initial AISI 5135 steel structure. It contained ferrite and perlite grains (Figure 18).

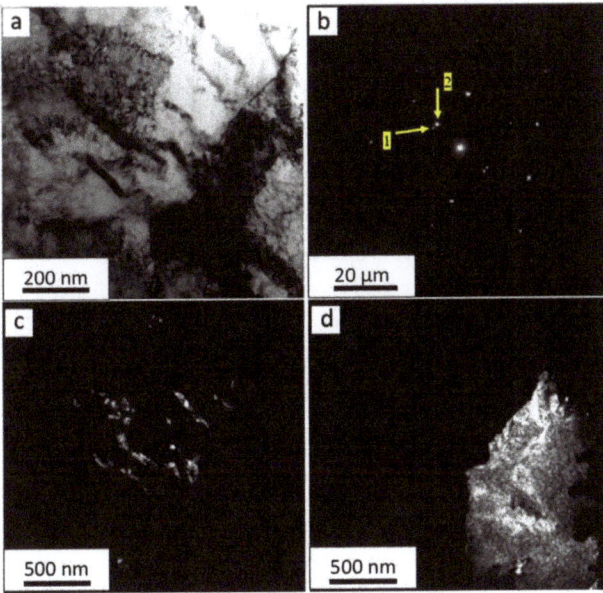

Figure 18. Structure of Nb–steel system 57 µm beneath its surface after irradiation at E = 24 J/cm^2 for 100 µs: (**a**)—bright field, (**b**)—electron-diffraction pattern, (**c**)—dark field in reflection [211]Fe_3C (reflection 1 in (**b**) yellow arrow 1), (**d**)—dark field in reflection [110]α-Fe (reflection 2 in (**b**) yellow arrow 2).

4. Conclusions

Structural AISI 5135 steel was exposed to surface treatment through Nb-film deposition, the surface melting of the steel, the doping of its melt with Nb via the submillisecond modulated electron-beam irradiation of the Nb film–AISI 5135 steel-substrate system (Nb–steel system), and high-rate cooling.

The advantages of steel doping with small amounts of Nb are its simplicity, highly efficient strength enhancement at a Nb content of 0.01–0.04 wt.%, high level of mechanical properties at elevated temperatures, thermal stability up to 1150 °C, and good workability and weldability, which makes it possible to decrease the weight of machines and constructions and increase their reliability and lifetime.

The irradiation was performed at pulse durations of up to 500 µs using the SOLO electron source with a grid-plasma cathode, which made it possible to vary and control the rate of energy delivery to the Nb–steel system's surface within a pulse. The temperature field formed in the Nb–steel system under intense pulsed irradiation was estimated by mathematical simulation methods, and irradiation modes were chosen to provide the near-melting conditions for the niobium and surface-melting conditions for AISI 5135 steel. The results of our study demonstrate that after irradiation in all the modes chosen, the surface microhardness of the Nb–steel system measured 8–9 GPa, which was higher than its initial value of 3.6 GPa by a factor of 2.2–2.5. The wear rate of the Nb–steel system decreased five times when the pulse duration measured 50 µs and 100 µs and three times when the temperature was maintained at about 2000 °C (t = 300 and 500 µs) due to beam-current modulation. The data from the X-ray diffraction analysis using transmission-electron microscopy showed that the increase in the microhardness and wear resistance of the modified steel was mainly due to the transformations of its surface structure and phase state: the formation of a hardening (martensite) structure, the precipitation of nanosized $M_{23}C_6$ particles at the martensite crystal boundaries, and the pearlite's structural transformation via iron-carbide precipitation in the pearlite lamellae.

Thus, our study demonstrates that such combined surface treatments of structural steels hold promise for use in industry.

Author Contributions: Conceptualization, V.I.S., M.S.V. and N.N.K.; methodology, M.S.V., N.N.K., T.V.K. and Y.F.I.; investigation, E.A.P., V.I.S., P.V.M., Y.F.I., N.A.P., R.A.K. and D.A.S.; resources, N.A.P.; data curation, M.S.V.; writing—original draft preparation, V.I.S., M.S.V. and Y.F.I.; writing—review and editing, V.I.S., M.S.V., N.N.K. and D.A.S. All authors have read and agreed to the published version of the manuscript.

Funding: The work was supported by the Russian Science Foundation (project no. 23-29-00998).

Institutional Review Board Statement: Not applicable.

Informed Consent Statement: Not applicable.

Data Availability Statement: Not applicable.

Conflicts of Interest: The authors declare no conflict of interest.

References

1. Beddoes, J. *Introduction to Stainless Steels*, 3rd ed.; ASM International: Novelty, OH, USA, 1999; p. 315.
2. Hosford, W.F. *Iron and Steel*; Cambridge University Press: New York, NY, USA, 2012; p. 298.
3. McGuire, M.F. *Stainless Steels for Design Engineers*; ASM International: Novelty, OH, USA, 2008; p. 304.
4. Baddoo, N.R. Stainless steel in construction: A review of research, applications, challenges and opportunities. *J. Constr. Steel Res.* **2008**, *64*, 1199–1206. [CrossRef]
5. Akhmadeev, Y.H.; Ivanov, Y.F.; Krysina, O.V.; Lopatin, I.V.; Petrikova, E.A.; Rygina, M.E. Electron–ion–plasma modification of carbon steel. *High Temp. Mater. Process.* **2021**, *25*, 47–55. [CrossRef]
6. Pogrebnjak, A.D.; Lisovenko, M.A.; Buranich, V.V.; Turlybekuly, A. Protective coatings with nanoscale multilayer architecture: Current state and main trends. *Phys.-Uspekhi* **2021**, *64*, 253–279. [CrossRef]
7. Budilov, V.V.; Mukhin, V.S.; Yagafarov, I.I. An experimental study of the effect of vacuum ion-plasma coating deposition on the GTE part cylindrical surface accuracy. *Russ. Aeronaut.* **2015**, *58*, 116–119. [CrossRef]

8. Kravchenko, V.N.; Novikov, E.S.; Voropayev, A.I.; Petrov, L.M.; Ivanchuck, S.B. The application of strengthening vacuum ion-plasma treating for the surface layer of constructional metal materials in mechanical engineering and the features of its formation. *J. Phys. Conf. Ser.* **2019**, *1281*, 012040. [CrossRef]
9. Boxman, R.L.; Zhitomirsky, V.N. Vacuum arc deposition devices. *Rev. Sci. Instrum.* **2006**, *77*, 021101. [CrossRef]
10. Ivanov, Y.F.; Kolubaeva, Y.A.; Teresov, A.D.; Koval, N.N.; Lu, F.; Liu, G.; Gao, Y.; Zhang, X.; Tang, Z.; Wang, Q. Electron beam nanostructurization of titanium alloys surface. In Proceedings of the 9th International Conference Modification of Materials with Particle Beams and Plasma Flows, Tomsk, Russia, 21–26 September 2008; pp. 143–146.
11. Morozov, Y.D.; Stepashin, A.M.; Aleksandrov, S.V. Effects of manganese and niobium and rolling conditions on the properties of low-alloy steel. *Metallurgist* **2002**, *5*, 45–47. (In Russian)
12. Odesskii, P.D.; Smirnov, L.A. Vanadium and niobium in microalloyed steel for metal structures. *Steel Transl.* **2005**, *35*, 63–73.
13. Fernandez, A.I.; Uranga, P.; Lopez, B.; Rodrigues-Ibabe, J.M. Dynamic recrystallization behavior covering a wide austenite grain size range in Nb and Nb–Ti microalloyed steels. *Mater. Sci. Eng.* **2003**, *361*, 367–376. [CrossRef]
14. Inoue, K.; Ishikawa, N.; Ohnuma, I.; Ohtani, H.; Ishida, K. Calculation of phase equilibria between austenite and (Nb, Ti, V) (C, N) in microalloyed steels. *ISIJ Int.* **2001**, *41*, 175–182. [CrossRef]
15. De Ardo, J. Fundamental metallurgy of niobium in steel. In Proceedings of the International Symposium Niobium 2001, Orlando, FL, USA, 2–5 December 2001; pp. 571–586.
16. Karjalainen, L.P.; Peura, P.; Porter, D.A. Effects of strain rate changes and strain path on flow stress and recrystallisation kinetics in a Nb bearing microalloyed steels. In *Thermomechanical Processings of Steels, Proceedings of the Church Hous Conference, London, UK, 24–26 May 2000*; IoM Communications Ltd.: London, UK, 2000; Volume 1, pp. 309–321.
17. Khlusova, E.I.; Sych, O.V.; Orlov, V.V. Cold-resistant steels: Structure, properties, and technologies. *Phys. Met. Metallogr.* **2021**, *122*, 579–613. [CrossRef]
18. Posti, A.I. Effect of micro alloying of steel with niobium on the mechanical properties of heat-strengthened rebar. *Litiyoi Metall. Foundry Prod. Metall.* **2021**, *1*, 73–77. (In Russian) [CrossRef]
19. Zhou, Z.; Zhang, Z.; Shan, Q.; Li, Z.; Jiang, Y.; Ge, R. Influence of Heat-Treatment on Enhancement of Yield Strength and Hardness by Ti-V-Nb Alloying in High-Manganese Austenitic Steel. *Metals* **2019**, *9*, 299. [CrossRef]
20. You, K.; Liu, X.; Chen, K.; Wang, W.; He, Z. Microstructure of plasma Nb-alloyed layer on 40Cr13 stainless steel. *Cailiao Kexue Yu Gongyi/Mater. Sci. Technol.* **2015**, *23*, 68–74.
21. Liu, A.M.; Feng, Y.; Zhao, Y.; Ma, M.T.; Lu, H.Z. Effect of niobium and vanadium micro-alloying on microstructure and property of 22MnB5 hot press forming steel. *Mat. Mech. Eng.* **2019**, *43*, 34–37.
22. Liu, S.; Ai, S.; Long, M.; Feng, Y.; Zhao, J.; Zhao, Y.; Gao, X.; Chen, D.; Ma, M. Evolution of Microstructures and Mechanical Properties of Nb-V Alloyed Ultra-High Strength Hot Stamping Steel in Austenitizing Process. *Materials* **2022**, *15*, 8197. [CrossRef]
23. Cherenda, N.N.; Rogovaya, I.S.; Shymanski, V.I.; Uglov, V.V.; Saladukhin, I.A.; Astashynski, V.M.; Kuzmitski, A.M.; Ivanov, Y.F.; Petrikova, E.A. Elemental and phase compositions and mechanical properties of titanium surface layer alloyed by Zr, Nb, and Al under the action of compression plasma flows. *High Temp. Mater. Process.* **2022**, *26*, 1–9. [CrossRef]
24. Goncharenko, I.M.; Grigoriev, S.V.; Lopatin, I.V.; Koval, N.N.; Schanin, P.M.; Tukhfatullin, A.A.; Ivanov, Y.F.; Strumilova, N.V. Surface modification of steels by complex diffusion saturation in low pressure arc discharge. *Surf. Coat. Technol.* **2003**, *169–170*, 419–423. [CrossRef]
25. Shymanski, V.I.; Esipenko, D.V.; Uglov, V.V.; Koval, N.N.; Ivanov, Y.F.; Teresov, A.D. Oxidation behavior of TiCr and TiMo alloys formed by low-energy pulsed electron beam impact. *Surf. Coat. Technol.* **2022**, *434*, 128227. [CrossRef]
26. Ivanov, Y.F.; Shugurov, V.V.; Petrikova, E.A.; Teresov, A.D.; Krysina, O.V.; Tolkachev, O.S. Borating of Steel Initiated by a Pulsed Electron Beam. *AIP Conf. Proc.* **2022**, *2509*, 020090.
27. Gromov, V.E.; Ivanov, Y.F.; Vorobiev, S.V.; Konovalov, S.V. *Fatigue of Steels Modified by High Intensity Electron Beams*; Cambridge International Science Publishing Ltd.: London, UK, 2015; 272 p.
28. Shugurov, V.V.; Koval, N.N.; Krysina, O.V.; Prokopenko, N.A. QUINTA equipment for ion-plasma modification of materials and products surface and vacuum arc plasma-assisted deposition of coatings. *J. Phys. Conf. Ser.* **2019**, *1393*, 012131–012141. [CrossRef]
29. Koval, N.N.; Grigoryev, S.V.; Devyatkov, V.N.; Teresov, A.D.; Schanin, P.M. Effect of Intensified Emission during the Generation of a Submillisecond Low-Energy Electron Beam in a Plasma-Cathode Diode. *IEEE Trans. Plasma Sci.* **2009**, *37*, 1890–1896. [CrossRef]
30. Vorobyov, M.S.; Koval, N.N.; Moskvin, P.V.; Teresov, A.D.; Doroshkevich, S.Y.; Yakovlev, V.V.; Shin, V.I. Electron beam generation with variable current amplitude during its pulse in a source with a grid plasma cathode. *J. Phys. Conf. Ser.* **2019**, *1393*, 012064. [CrossRef]
31. Vorobyov, M.S.; Moskvin, P.V.; Shin, V.I.; Koval, N.N.; Ashurova, K.T.; Doroshkevich, S.Y.; Devyatkov, V.N.; Torba, M.S.; Levanisov, V.A. Dynamic Power Control of a Submillisecond Pulsed Megawatt Electron Beam in a Source with a Plasma Cathode. *Tech. Phys. Lett.* **2021**, *47*, 528–531. [CrossRef]
32. Vorobyov, M.S.; Koval, T.V.; Shin, V.I.; Moskvin, P.V.; Tran, M.K.A.; Koval, N.N.; Ashurova, K.; Doroshkevich, S.Y.; Torba, M.S. Controlling the Specimen Surface Temperature during Irradiation with a Submillisecond Electron Beam Produced by a Plasma-Cathode Electron Source. *IEEE Trans. Plasma Sci.* **2021**, *49*, 2550–2553. [CrossRef]
33. Comsol, A.B. *Comsol Multiphysics*; Heat Transfer Module User's Guide, Version 5; Comsol AB: Stockholm, Sweden, 2018.

34. Babichev, A.P.; Babushkina, N.A.; Bratkovskii, A.M.; Grigor'eva, I.S.; Melikhova, E.Z. *Physical Values: Handbook*; Energoatomizdat: Moscow, Russia, 1991.
35. Ivanov, Y.F.; Gromov, V.E.; Zagulyaev, D.V.; Konovalov, S.V.; Rubannikova, Y.A. Improvement of Functional Properties of Alloys by Electron Beam Treatment. *Steel Transl.* **2022**, *52*, 71–75. [CrossRef]

Disclaimer/Publisher's Note: The statements, opinions and data contained in all publications are solely those of the individual author(s) and contributor(s) and not of MDPI and/or the editor(s). MDPI and/or the editor(s) disclaim responsibility for any injury to people or property resulting from any ideas, methods, instructions or products referred to in the content.

Article

Mathematical Model of Surface Topography of Corroded Steel Foundation in Submarine Soil Environment

Wei Wang [1], Yuan Wang [1,2], Jingqi Huang [1,*] and Lunbo Luo [3]

1 School of Civil and Resource Engineering, University of Science and Technology Beijing, Beijing 100083, China; wangwei_ustb@yeah.net (W.W.); wangyuanhhu@163.com (Y.W.)
2 College of Water Conservancy and Hydropower Engineering, Hohai University, Nanjing 210024, China
3 Institute of Science and Technology, China Three Gorgers Corporation, Beijing 100038, China; lunbo.luo@foxmail.com
* Correspondence: huangjingqi11@163.com

Citation: Wang, W.; Wang, Y.; Huang, J.; Luo, L. Mathematical Model of Surface Topography of Corroded Steel Foundation in Submarine Soil Environment. *Coatings* **2022**, *12*, 1078. https://doi.org/10.3390/coatings12081078

Academic Editors: Matic Jovičević-Klug, Patricia Jovičević-Klug and László Tóth

Received: 27 June 2022
Accepted: 27 July 2022
Published: 30 July 2022

Publisher's Note: MDPI stays neutral with regard to jurisdictional claims in published maps and institutional affiliations.

Copyright: © 2022 by the authors. Licensee MDPI, Basel, Switzerland. This article is an open access article distributed under the terms and conditions of the Creative Commons Attribution (CC BY) license (https://creativecommons.org/licenses/by/4.0/).

Abstract: For the corrosion risk of steel structures in the marine environment, the topography characteristics of corroded steel surfaces were paid little attention to, which has a significant effect on the mechanical properties of the interface between steel foundation and soil medium. An effective mathematical model for reconstructing the topography of corroded steel surface is very helpful for numerically or experimentally studying the soil-corroded steel interaction properties. In this study, an electrolytic accelerated corrosion experiment is conducted first to obtain corroded steel samples, which are exposed to submarine soil and suffer different corrosion degrees. Then, the surface height data of these corroded steel samples are scanned and analyzed. It is found that the height of surface two-dimensional contour curves under different corrosion degrees obeys the Gaussian distribution. Based on the spectral representation method, a mathematical model is developed for the profile height of the corroded steel surface. By comparing the standard deviation, arithmetic mean height and maximum height of reconstructed samples with those of experimental samples, the reliability of the developed mathematical model is proved. The proposed mathematical model can be adapted to reconstruct the surface topography of steel with different corrosion degrees for the following research on the shearing behavior of soil-corroded steel interface.

Keywords: corrosion; surface topography; mathematical model; corroded steel surface; marine soil

1. Introduction

It is a common phenomenon during the service life of steel structures in the marine environment. With the rapid development of exploitations of marine energy, the salt corrosion research of steel structures has been focused on widely [1]. In design, the protective layer and surplus design are often considered to reduce the effect of corrosion [2,3]. However, as time goes on, the protection system may fail. Corrosion will happen and affect the property of the steel structure. Therefore, it is important to study the corrosion characteristics of steel structures in a marine environment.

At present, a large number of studies has been conducted on the corrosion characteristics of steel material in a corrosive environment, with the aid of experiment methods and modern analytical techniques. The relevant research can be found in the work of Kovendhan et al. [4], James and Hattingh [5], Lv et al. [6], Wei et al. [7] and so on. The above studies mainly focus on explaining the occurrence of metal corrosion in terms of lattice and electrochemical mechanisms. However, civil engineers care more about the mechanical properties and residual bearing capacity of steel members or structures after corrosion. There are different methods to consider the effect of corrosion on steel structures. Some researchers simulated the effect of corrosion on the mechanical behavior of steel structures by decreasing the section thickness of steel members. Karagah et al. [8] investigated the effect of corrosion on the axial capacity of short steel columns by reducing the localized thickness in a monotonic axial load experiment. Liu et al. [9] notched in the center zone of steel columns to simulate the loss of section due to corrosion in axial loading tests. Wang et al. [10,11] analyzed the ultimate strength of steel pipe piles in corrosion conditions by converting corrosion effects into thickness loss in numerical simulations. Other similar studies can be found in the literature by Yamamoto and Ikegami [12], Akpan et al. [13], Guo et al. [14] and so on. The above research studies only consider the effect of section thickness loss due to corrosion but ignore the effect of variation of the surface topography of steel structures. To consider the effect of surface topography, Jiang and Soares [15] assumed corrosion pits to be cylindrically shaped with various distributions and intensity or depth in numerical analysis. Ahmmad and Sumi [16] considered the corrosion pit as conical pits with different depth-to-diameter ratios. Nakai et al. [17,18] studied the effect of corrosion on steel plate strength by artificially creating conical corrosion pits in experiments and numerical simulations. Though some researchers have already begun to study the effect of corroded surface topography on the mechanical properties of steel members or structures, the corroded surface topography is always considered simply by setting pits with regular shape and distribution, which is difficult to describe the real surface topography of corroded steel structure. Little research has been devoted to the shearing property of soil-corroded steel surface, which has a significant influence on the bearing capacity of the steel foundation of ocean structures. The main reason can be attributed to the lack of an effective mathematical model of the surface topography of corroded steel.

In this paper, the corrosion characteristics of Q235 steel in a submarine clay soil environment are studied by electrolytic accelerated corrosion experiments. The electrochemical process of corrosion and the law of mass loss are analyzed. To further characterize the corroded steel surface, the surface topography is investigated. After the analysis of two-dimensional profile height on a corroded surface, a mathematical model is constructed by the spectral representation method. The reliability of the mathematical model is verified. The developed model can be employed to generate the surface profile of corroded steel structures as close as possible to the reality in the follow-up research. It is very helpful for studying the corrosion characteristic of the steel surface and the soil-corroded steel interaction properties.

2. Electrolytic Accelerated Corrosion Experiment

Electrolytic accelerated corrosion experiment is a common method for the investigation of the corrosion of metals [19–21]. The acceleration mechanism is to accelerate the electrochemical reaction of the corroded metal by means of an applied current [22]. Compared with natural exposure experiments, the electrolytic accelerated corrosion experiment is more convenient and can save a lot of time. In this study, the electrolytic accelerated corrosion experiment is conducted to investigate the corrosion characteristic of steel structure foundation in submarine soil.

2.1. Experimental Principle

The electrolytic accelerated corrosion is a result of an artificial electrolytic cell reaction. Figure 1a shows the electrochemical reaction process on the surface of the corroded specimen. The corroded electrode loses electrons and undergoes an oxidation reaction, while the electrolyte solution gains electrons for a reduction reaction [19,21,23].

The anodic reaction can be represented as

$$Fe \rightarrow Fe^{2+} + 2e^- \tag{1}$$

$$Fe \rightarrow Fe^{3+} + 3e^- \tag{2}$$

The cathode reaction can be represented as

$$O_2 + 2H_2O + 4e^- \rightarrow 4OH^- \tag{3}$$

The corrosion current is controlled by the applied current during experiments. According to Faraday law [19,21], the theoretical mass loss of electrochemical corrosion can be calculated as

$$\Delta m = \frac{MIT}{zF} \qquad (4)$$

where M represents the molar mass of Fe (56 g·mol^{-1}); I is the experimental current (A); t is the experimental time (s); z is the number of electrons transferred by corrosion of Fe ($z = 2$ or 3); F is Faraday constant which has a value of 96,500 A·s^{-1}.

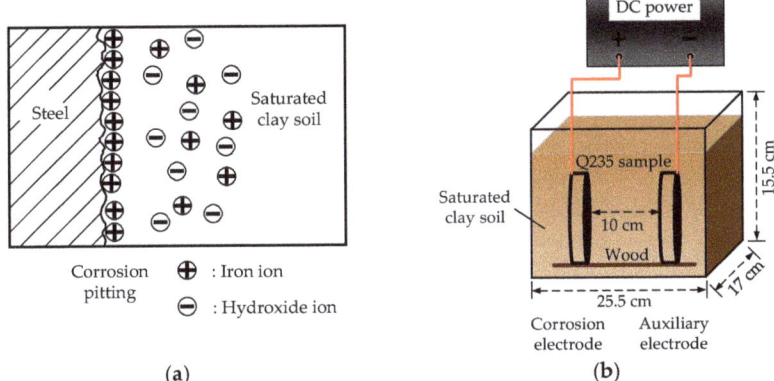

Figure 1. Electrolytic accelerated corrosion experiment: (**a**) electrochemical process; (**b**) experimental system.

2.2. Experimental Process

The experiment is mainly composed of a direct current (DC) power, a glass container, clay soil, an electrolyte solution and a Q235 steel sample. The experiment system is shown in Figure 1b. The Q235 steel is machined to circular samples with a diameter of 61.5 mm and a thickness of 10 mm. In the experimental process, one sample is used as the corrosion electrode connected to the positive pole of DC power; the other is connected to the negative terminal of the power supply as an auxiliary electrode. Throughout the experiment, the two electrodes are fixed in position by a wooden rod and kept 10 cm apart. Waterproof tape is used to wrap the corrosion electrode, leaving only one side. By maintaining a constant current of 3 A throughout the experiment, the corrosion process of the steel sample connected to the positive pole is accelerated. Saturated clay soil is used to simulate the submarine soil environment, which is derived from the coastal region. A 5% mass concentration NaCl solution is added to the clay soil as an electrolyte solution.

After the experiment, the corroded steel sample is removed from the experimental apparatus and cleaned. The whole experiments are set up with ten groups of working conditions with a minimum experimental time of 1 h and a maximum experimental time of 10 h. The duration of each group of experiments differs by one hour. In addition, the uncorroded steel sample is used for a control group.

2.3. Mass Loss per Unit Area

After experiments, the corroded samples are removed from the soil and tore off the waterproof tapes. Then the corrosion products are cleaned. The fragile corrosion products stuck to the surface can be easily removed by flowing water. The rest of the corrosion products are cleaned with ethyl alcohol and distilled water. After drying, the samples are weighed. The mass loss per unit area m_a is employed to describe the corrosion degree of the samples, which is calculated as

$$m_a = \frac{m_0 - m_1}{A} = \frac{m_0 - m_1}{\pi r^2} \quad (5)$$

where m_a is the mass loss per unit area (g/dm^2); m_0 is the original weight of the sample (g); m_1 is the weight of the corroded sample (g); A is the area of the corroded surface (dm^2), which can be given by the radius r of the circular surface as πr^2.

Based on Equation (4), the theoretical values of mass loss per unit area under different corrosion times are calculated by taking z to be 2 and 3, respectively, wherein z represents the number of electrons transferred by corrosion of *Fe*. When z equals 2, the corrosion product is divalent iron, and when z equals 3, the corrosion product is trivalent iron. The comparison between the experimental value and the theoretical value is shown in Figure 2. One finds from Figure 2 that the mass loss per unit area measured by experimental samples is between the theoretical values calculated with z being 2 and 3. It means the experimental corrosion products included both trivalent iron and divalent iron. Another phenomenon can be found in Figure 2 that the mass loss per unit area ma increases linearly with the corrosion time t. It indicates that the rate of corrosion is uniform in the electrolytic accelerated corrosion experiments. The variation of the mass loss per unit area m_a (g/dm^2) with corrosion time t (h) is fitted with a linear function. The fitting formula is given as

$$m_a = 6.6195 \times t \quad (6)$$

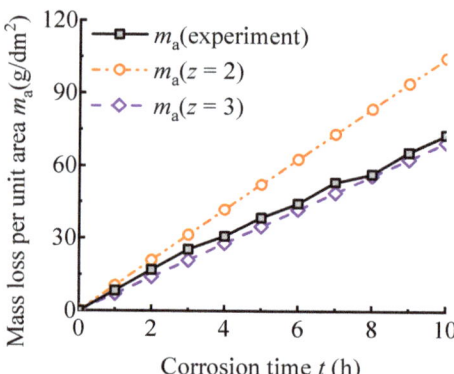

Figure 2. Mass loss per unit area.

By dividing the corrosion process into two electrochemical reaction processes, during which the corrosion product is divalent iron and trivalent iron, respectively, the corrosion mass loss rates of the two electrochemical reaction processes are calculated. The comparison result is shown in Figure 3. In Figure 3, the orange color means the mass loss rate of the electrochemical reaction process during which the corrosion product is divalent iron. The purple color means the mass loss rate of the electrochemical reaction process during which the corrosion product is trivalent iron. It can be found when the corrosion time is less than 3 h, the mass loss rates of the two electrochemical reaction processes are similar. With the increasing corrosion time, the corrosion products are mainly trivalent iron. This phenomenon keeps step with the result of previous studies [10].

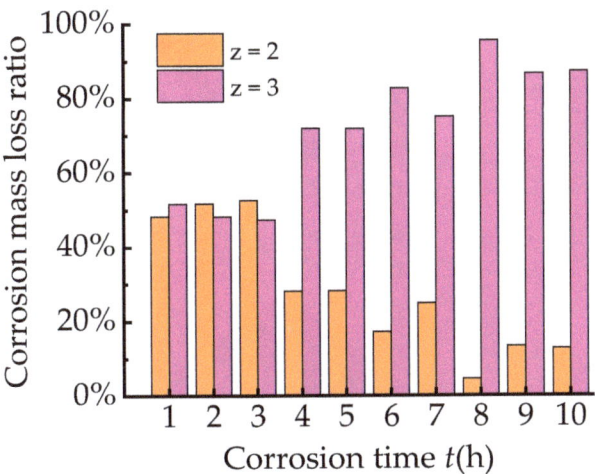

Figure 3. Corrosion mass loss ratio.

3. Mathematical Model of Corroded Steel Surface

3.1. Surface Topography

To investigate the surface characteristic of corroded samples, a binocular laser scanner is employed to measure the height data of the surface topography. In order to reduce the measurement error at the edge of the circular samples, the square area with a 40 mm side length at the center of the specimen is chosen to investigate the surface characteristic of corroded steel. As the scanning interval is set as 0.1 mm, there are 401 contour curves in each sample area. Each scanning curve has 401 data points. The data are stored on the computer. The height coordinate of the data is the distance of the corroded sample surface relative to the laser scanner. For the convenience of analysis, by subtracting the mean value of the scanning height, a series of surface height data $h(x,y)$ is obtained, which fluctuates around 0. The schematic diagram of the surface height data is illustrated in Figure 4.

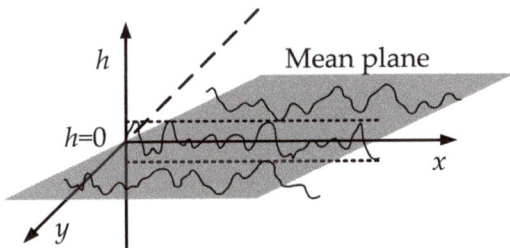

Figure 4. Schematic diagram of surface topography.

The surface curve maps of the corroded surface are drawn in Figure 5. In Figure 5, the corroded surface is composed of a series of uneven two-dimensional contour curves. When the corrosion degree is light, the fluctuation of two-dimensional contour curves is small. With the increasing corrosion degree, the corroded surface has obvious areas of corrosion pits. The number and size of corrosion pits are increasing. Additionally, the fluctuation of the two-dimensional contour curves also increases gradually.

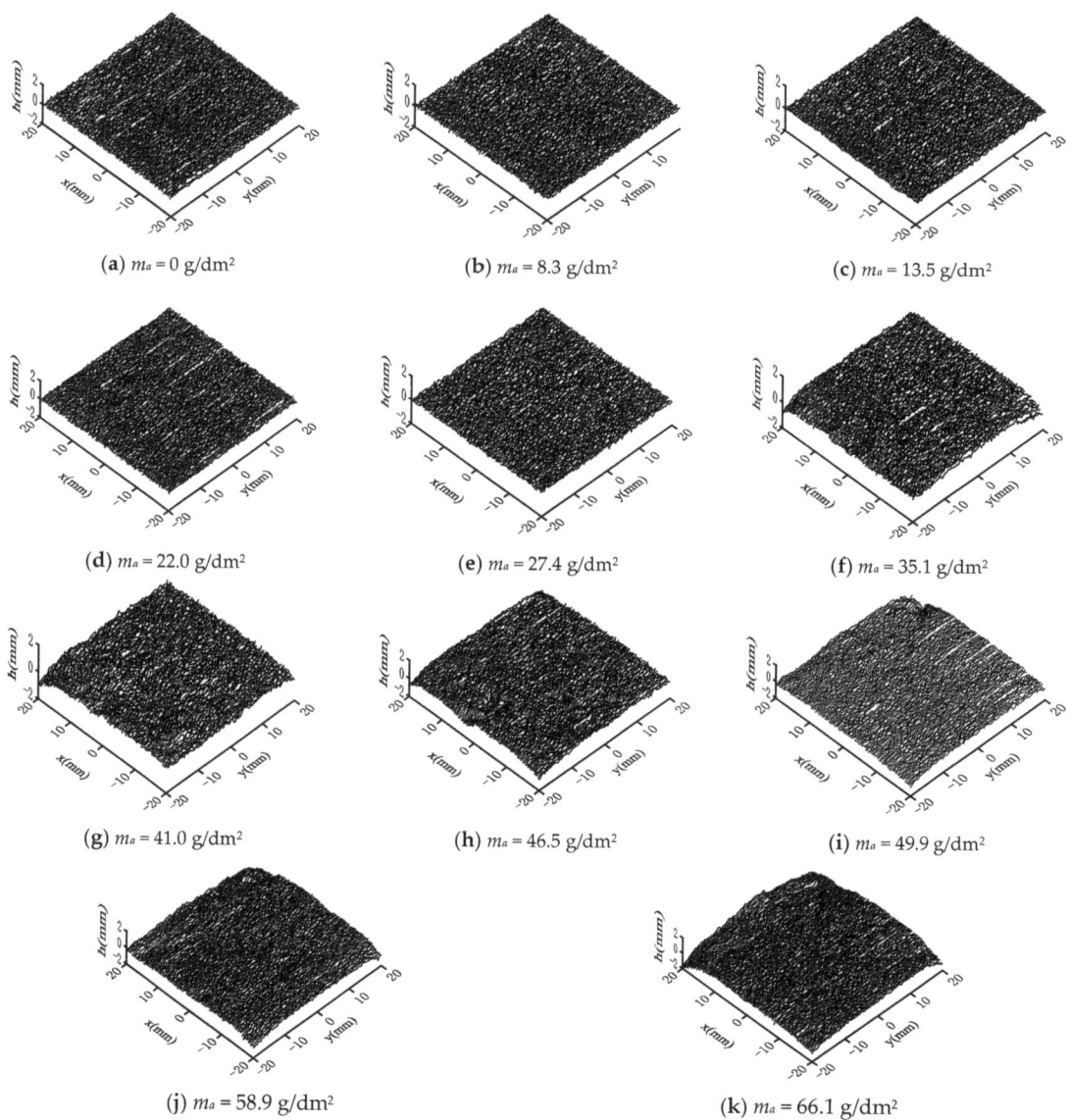

Figure 5. Surface curve maps of corroded steel surface with different corrosion degrees.

3.2. Surface Height Distribution

In order to analyze the distribution of surface height, the surface height probability density of the two-dimensional contour curve is calculated and plotted in Figure 6. Considering that each sample has 401 contour curves, only one representative two-dimensional contour curve is selected for each sample to draw the histogram of surface height, as shown in Figure 6. It can be found from Figure 6 that the probability density of the two-dimensional contour curve of each sample is almost symmetric on both sides of the surface height of 0 mm. The closer the surface height equals 0, the larger the probability density is. The shape of the histogram of surface height probability density is similar to the probability density curve of the Gaussian distribution.

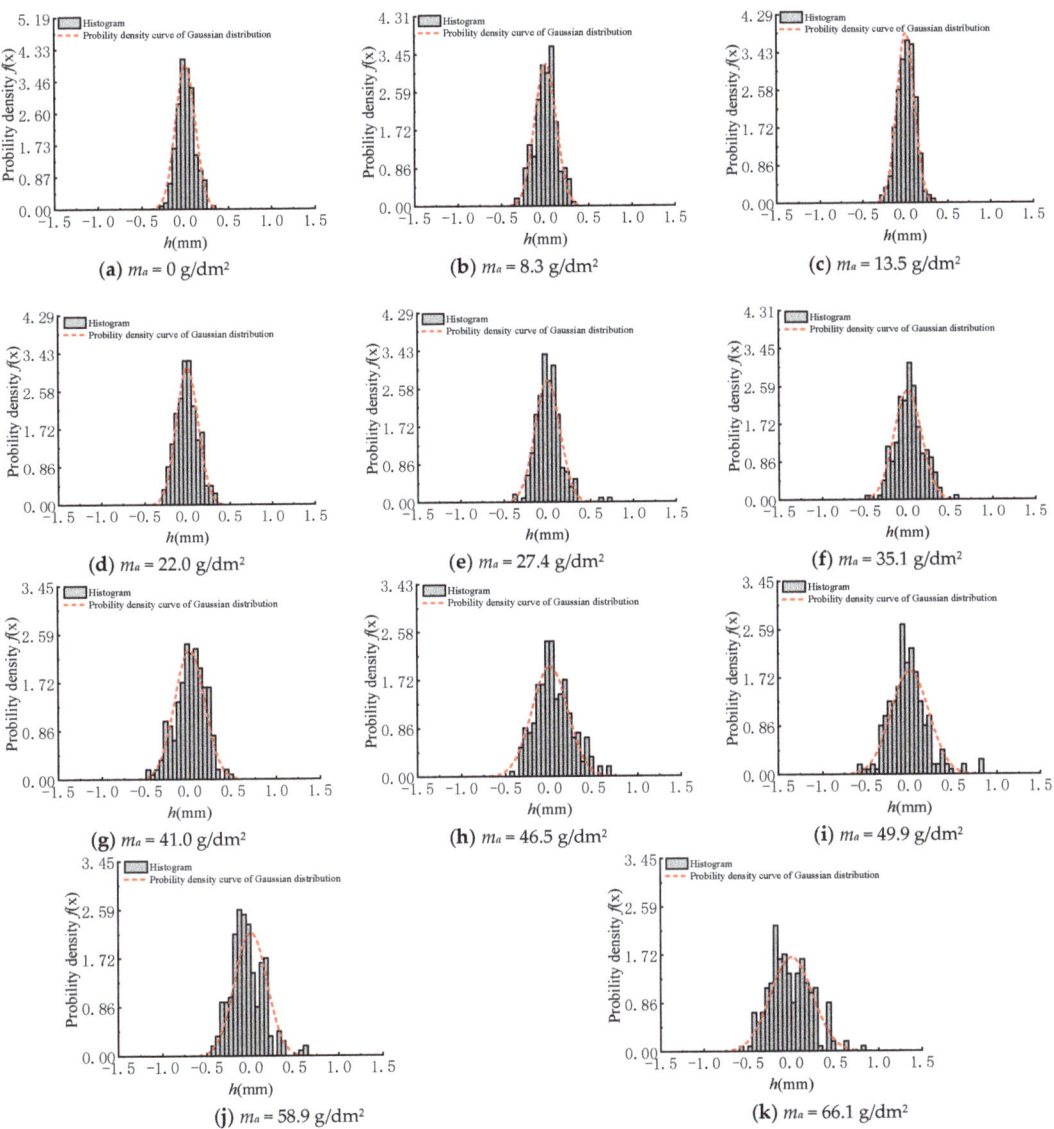

Figure 6. Probability density of surface height of two-dimensional contour curve.

To prove the surface height of the contour curve follows the Gaussian distribution, based on the probability density function of the Gauss distribution, the mean value and standard deviation of the surface height of contour curve are calculated. The probability density curves following the Gaussian distribution are drawn in Figure 6 with calculated mean value and standard deviation to compare with the histogram of probability density. The probability density function of Gauss distribution can be written as

$$f(x; \mu, \sigma) = \frac{1}{\sigma\sqrt{2\pi}} \int_{-\infty}^{x} \exp\left(-\frac{(x-\mu)^2}{2\sigma^2}\right) dx \qquad (7)$$

where μ is the mean value of the surface height of a two-dimensional contour curve, and σ is the standard deviation of the surface height of a two-dimensional contour curve.

The comparison results are shown in Figure 6. It is evident that the height distribution of the two-dimensional contour curve is in accord with the Gaussian distribution form.

The height standard deviation of each two-dimensional contour curve is calculated. The results that corrosion degree ma equals 0 (i.e., no corrosion), 35.1 and 66.1 g/dm² are plotted in Figure 7. It is obvious that the discreteness of the standard deviation is small. Therefore, the mean value of the standard deviation of every two-dimensional contour curve can be adopted as the standard deviation of the surface height of one sample. The variation curve of the standard deviation of surface height of steel samples with different corrosion degrees is drawn in Figure 8. It can be found from this figure that the standard deviation increases with the corrosion degree. It means that the corroded steel surface becomes rougher.

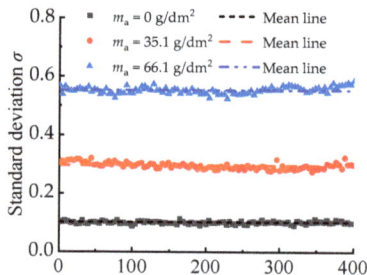

Figure 7. Standard deviation of surface height of two-dimensional contour curves.

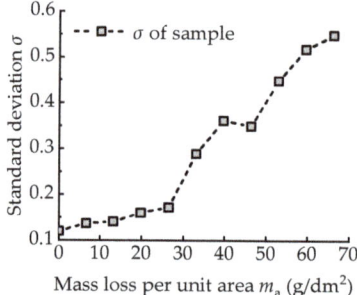

Figure 8. Variation of standard deviation σ with corrosion degree m_a.

3.3. Development of Mathematical Model

The corroded surface is composed of a series of uneven two-dimensional contour curves. These contour curves can be regarded as experimental results of the sample function of a one-dimensional stochastic process. In previous studies, the Monte Carlo method is the universal method to describe stochastic processes. However, it is usually time-consuming [24]. Shinozuka and Deodatis [25] proposed a spectral representation method to generate one-dimensional, uni-variate, stationary, Gaussian stochastic processes. Considering this method is more efficient than the Monte Carlo method in calculation, the spectral representation method is used to develop the mathematical model of corroded surface in the following.

3.3.1. Verification of Stationarity

The corroded surface can be understood as consisting of a series of experimental results of the sample function of the one-dimensional stochastic process. The concept of the stochastic process can be explained in Figure 9. In Figure 9, $h_k(x)$ denotes k-th sample of a one-dimensional stochastic process, where $k = 1, 2, \ldots, N$. x_i means the i-th data point of any sample of one-dimensional stochastic process along x-axis direction, where $I = 1, 2, \ldots, n$. Theoretically, when the numbers of samples (N) and data points (n) both tends to be infinity, the precise corroded surface topography can be obtained.

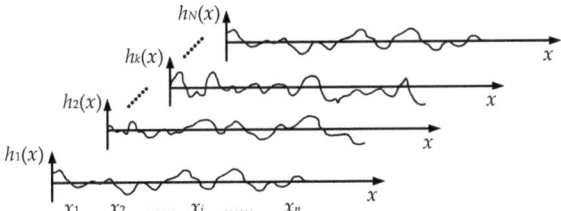

Figure 9. Schematic diagram of stochastic process on corroded surface.

The arbitrary sample of one-dimensional stochastic process consists of a series of stochastic data points along the length x direction, which can be expressed as

$$\{h(x)\} = \{h(x_1), h(x_2), \cdots, h(x_i), \cdots, h(x_n)\} \tag{8}$$

where x is the coordinate along the length direction; $h(x_i)$ is the surface height of i-th data point of the one-dimensional stochastic process.

In order to develop the mathematical model of the stochastic process with the spectral representation method, the stationary of the one-dimensional stochastic process needs to be verified. According to the above analysis in Section 3.2, the stochastic process $\{h(x)\}$ is a Gaussian stochastic process, which is also a secondary moment process. For a secondary moment process $\{h(x)\}$, if the following conditions are satisfied, the process is a stationary stochastic process.

$$(1) \ \forall x \in X, m_h(x) = \text{constant};$$

$$(2) \ \forall \tau \in R, x, x + \tau \in X, R_h(x, x + \tau) = R_h(\tau).$$

where $m_h(x)$ is the mean value of the stochastic process when the position coordinates equal x; τ is the sampling interval along the length direction; $R_h(x,x+\tau)$ is the correlation function when the sampling interval is τ.

The above conditions mean that if the mean value of the stochastic process is a constant independent of position coordinates and the correlation function is a function of sampling interval τ independent of position coordinates, the secondary moment process can be seen as a stationary stochastic process.

Since the two-dimensional contour curves of corroded surface are regarded as a series of samples of a one-dimensional stochastic process, the mean values of surface height of 401 contour curves with different position coordinates are calculated. The variation of mean value with x-coordinate for three corrosion degrees is drawn in Figure 10 as examples. It is obvious that the mean value of the surface height of contour curves is almost equal to zero when the sampling length is greater than 30. When the sampling interval is a unit interval, the correlation function of surface height of 401 contour curves with different position coordinates is calculated. The variation of correlation function with x-coordinate for three corrosion degree are plotted in Figure 11 as an example, too. It can be found that the correlation function tends to be a constant when the sampling length is greater than 30. Figures 10 and 11 show that the mean value and correlation function of surface height of contour curves is independent of position coordinates when the sampling length is greater than 30. Therefore, the contour curves can be regarded as the samples of a one-dimensional stationary Gaussian stochastic process [26,27].

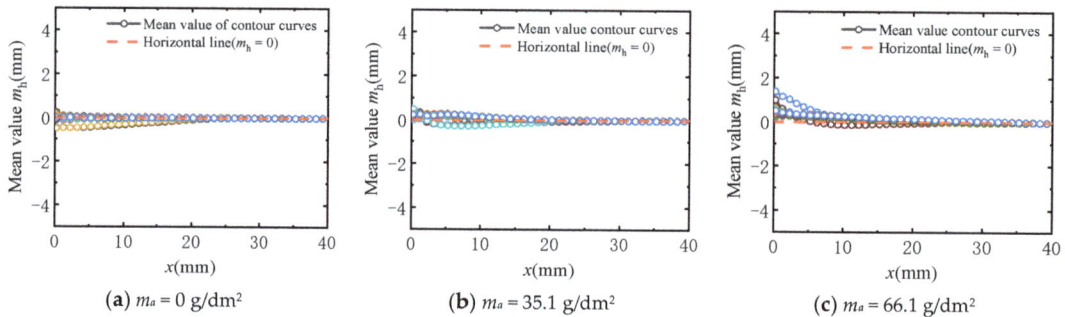

Figure 10. Mean value of surface height of two-dimensional contour curves.

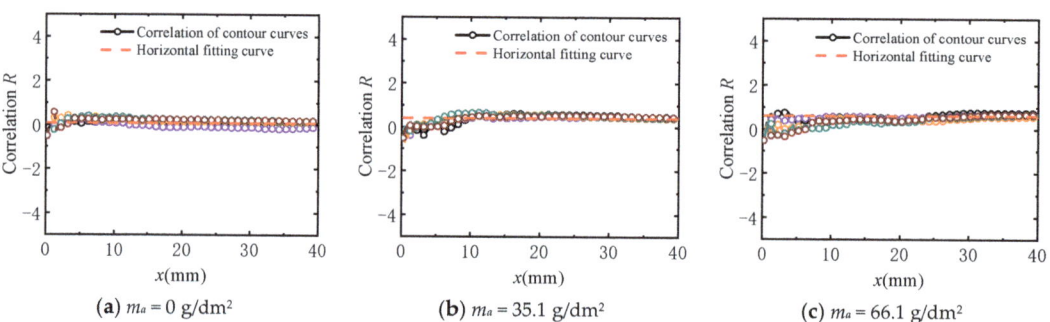

Figure 11. Correlation function of surface height of two-dimensional contour curves.

3.3.2. Mathematical Model

After verification of stationarity, the mathematical model of the corroded surface is developed with the spectral representation method in this subsection. According to the study of Shinozuka and Hu [25,28], a one-dimensional stationary stochastic process $\{h(x)\}$ can be simulated by the following series as $N \to \infty$:

$$h(x) = \sqrt{2} \sum_{n=0}^{N-1} A_n \cos(w_n x + \phi_n) \tag{9}$$

where

$$A_n = (2S_h(w_n)\Delta w)^{1/2}, \, n = 0, 1, 2, \ldots, N-1 \tag{10}$$

$$w_n = n \times \triangle w, \, n = 0, 1, 2, \ldots, N-1 \tag{11}$$

$$\Delta w = \frac{w_n}{N} \tag{12}$$

and

$$A_0 = 0 \text{ or } S_h(w_0 = 0) = 0 \tag{13}$$

In Equation (9), w_n represents an upper cut-off frequency beyond which the power spectral density function $S_h(w_n)$ can be assumed to be zero. Φ_n is an independent random phase angle uniformly distributed in the range $[0, 2\pi]$. The stochastic process $\{h(x)\}$ is periodic with period T_0:

$$T_0 = 2\pi/\Delta w \tag{14}$$

For a one-dimensional stationary stochastic process, the power spectrum function $S_h(w)$ and the autocorrelation function $R(\tau)$ are a pair of Fourier transform pairs. They have the form as

$$\begin{cases} S_h(w) = \int_{-\infty}^{+\infty} R(\tau)e^{-iw\tau}d\tau \\ R(\tau) = \frac{1}{2\pi}\int_{-\infty}^{+\infty} S_h(w)e^{iw\tau}dw \end{cases} \quad (15)$$

Equation (15) is called the Wiener–Khintchine formula, which reveals the connection between the statistical law describing the stationary process from the time perspective and the statistical law describing the stationary process from the frequency perspective.

Since $S_h(w)$ and $R(\tau)$ are even functions, Equation (15) could be rewritten in the form of Equation (16) using Euler's formula as

$$\begin{cases} S_h(w) = 2\int_0^{+\infty} R(\tau)e^{iw\tau}d\tau \\ R(\tau) = \frac{1}{\pi}\int_0^{+\infty} S_h(w)e^{iw\tau}dw \end{cases} \quad (16)$$

The autocorrelation function of surface height is an important parameter to characterize the variation of the parameter space points. The cosine exponential model is always used to fit the autocorrelation function [29,30]. The fitting formula has the form as

$$R(\tau) = exp(-k|\tau|)(cos(w\tau)) \quad (17)$$

where τ is the sampling interval; k and w are the parameters of the fitting formula.

If the unit sampling interval is τ_0, the sampling interval can be expressed as $j \cdot \tau_0$. Therefore, the different values of j are selected, and the length of the sampling interval is different. For two-dimensional contour curves, the maximum sampling interval is the length of the contour curve along the length direction.

According to the definition of autocorrelation function [31], the autocorrelation function of surface height of two-dimensional contour curves could be calculated as

$$R(\tau) = R(i \cdot \tau_0) = E[h(x)h(x+\tau)] \\ = \int_{-\infty}^{+\infty} x_1 x_2 f_2(x_1, x_2; \tau)dx_1 dx_2 = \frac{1}{n-i}\sum_{k=0}^{n-i} h(x_k)h(x_{k+i}) \quad (18)$$

Based on Equations (17) and (18), the autocorrelation function of surface height of two-dimensional contour curves is calculated and fitted. The fitting diagrams for three corrosion degrees are shown in Figure 12. In Figure 12, the black points are the calculated values of the autocorrelation function of contour curves. The red curves are the fitting curves fitted with Equation (17).

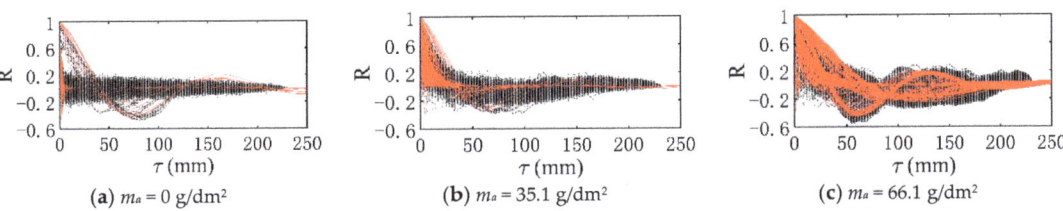

Figure 12. Sample autocorrelation function fitting diagram.

It can be found from Figure 12 that the autocorrelation function varies with the increase in corrosion degree. For the statistical analysis of the parameters k and w of the fitting curves, it is found that the mean values of parameters k and w of all fitting curves for every sample have an S-logistic function form and exponential function form, respectively, with the increasing of corrosion degree. Therefore, the S-logistic function and exponential function are used to fit the variations of parameters k and w, respectively. The fitting formulas are given as

$$k = 0.05665 + \frac{1.57188}{1+(m_a/27.5374)^{8.14368}} \quad (19)$$

$$w = 0.00402 \times 0.3864^{(m_a/6.6195)} \quad (20)$$

The fitting diagrams of parameters k and w are plotted in Figure 13. It can be found that the parameter k decreases slowly when corrosion degree ma is smaller than 20 g/dm². The rate of reduction increases when corrosion degree ma increases from 20 to 40 g/dm². The parameter k tends to be stable when corrosion degree m_a is larger than 40 g/dm². The variation of parameter w differs from that of parameter k. Parameter w decreases when corrosion degree ma is smaller than 20 g/dm². Then with the increasing of corrosion degree m_a, parameter w tends to be stable.

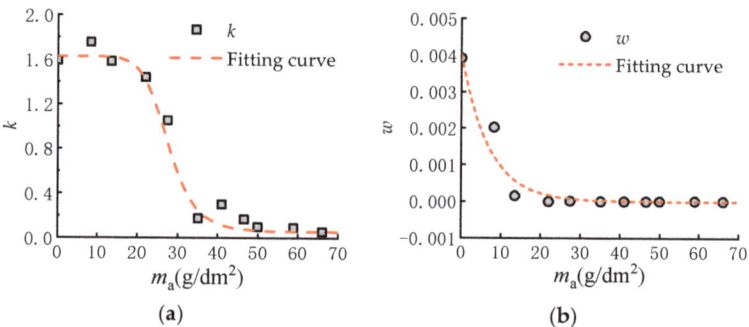

Figure 13. Variation curve of parameters k and w with corrosion degree: (**a**) parameter k; (**b**) parameter w.

Combining Equations (17), (19) and (20), the function for the variation of the autocorrelation function with the corrosion degree m_a can be obtained. According to Equation (16), the Fourier transform of the autocorrelation function is the power spectrum function. Therefore, the function for the variation of the power spectrum function with the corrosion degree ma can be expressed as

$$\begin{cases} S_h(w) = 2\int_0^{+\infty} R(\tau)e^{iw\tau}d\tau \\ R(\tau) = \exp(-k|\tau|)(\cos(w\tau)) \\ k = 0.05665 + \frac{1.57188}{1+(m_a/27.5374)^{8.14368}} \\ w = 0.00402 \times 0.3864^{(m_a/6.6195)} \end{cases} \quad (21)$$

Based on Equations (9)–(10), the stochastic function of one-dimensional stationary stochastic process $\{h(x)\}$ can be written as

$$h(x) = \sqrt{2}\sum_{n=0}^{N-1} \sqrt{2S_h(w_n)\Delta w}\cos(w_n x + \phi_n) \quad (22)$$

Substituting Equation (21) into Equation (22), the stochastic process $\{h(x)\}$ with different corrosion degrees can be simulated.

3.4. Stochastic Result Validation

Based on the above mathematical model of one-dimension stochastic process, the two-dimensional contour curves of corroded surfaces with different corrosion degrees can be generated, as shown in Figure 14. These sample curves fluctuate around 0. With the increase in corrosion degree, the fluctuation amplitude of the sample curve increases. It represents the corroded surface becoming rougher.

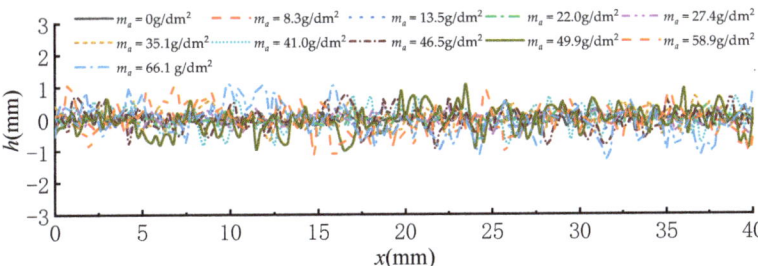

Figure 14. Randomly generated two-dimensional contour curves under different corrosion degree.

In order to prove the reliability of the mathematical model, the standard deviation σ, arithmetic mean height S_a and maximum height S_z of randomly generated two-dimensional contour curves and experimental samples are compared [32,33]. Standard deviation σ represents the dispersion degree of surface height of the two-dimensional contour curve. It can be calculated as

$$\sigma = \sqrt{\frac{\sum_{i=1}^{n} h(x_i)^2}{n}} \quad (23)$$

where n is the number of points of the two-dimensional contour curve; $h(x_i)$ is the surface height when the coordinate is x_i.

The arithmetic mean height S_a is the arithmetic mean of the surface offset within the sampling area, which reflects the fluctuation of surface height. The calculation formula is expressed as

The maximum height S_z is the distance between the maximum surface peak height S_p and the maximum surface valley depth S_v within the sampling range. The calculation formula is written as

$$S_z = S_p - S_v \quad (24)$$

where S_p is the maximum surface peak height; S_v is the maximum surface valley depth.

The comparison results are shown in Figure 15. It can be found that the standard deviation, arithmetic mean height and maximum height of randomly generated two-dimensional contour curves are similar to those of the experiment samples. It proves that the mathematical model of corroded surface topography is reliable. The randomly generated two-dimensional contour curves can be used to represent the corroded surfaces.

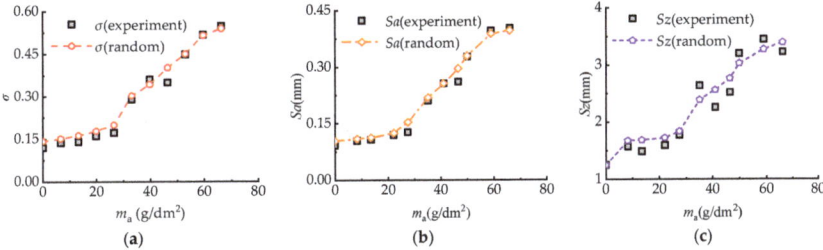

Figure 15. Comparison diagrams: (a) standard deviation σ; (b) arithmetic mean height Sa; (c) maximum height Sz.

4. Discussion

For any point on the corroded surface, it is crossed by a lot of two-dimensional contour curves. For the convenience of computation, any point on the corroded surface is regarded as the mean value of the height of randomly generated two-dimensional contour curves through that point in x-direction and y-direction. Therefore, the surface height of any point of the corroded surface $h(x,y)$ can be expressed as

$$h(x,y) = \frac{h(x) + h(y)}{2} \tag{25}$$

$$h(x) = \sqrt{2} \sum_{n=0}^{N-1} \sqrt{2S_h(w_n)\Delta w} \cos(w_n x + \phi_{nx}) \tag{26}$$

$$h(y) = \sqrt{2} \sum_{n=0}^{N-1} \sqrt{2S_h(w_n)\Delta w} \cos(w_n y + \phi_{ny}) \tag{27}$$

According to the sum to product formula of trigonometric functions, Equation (25) can be written as Equation (29).

$$\cos\alpha + \cos\beta = 2\cos\frac{\alpha+\beta}{2}\cos\frac{\alpha-\beta}{2} \tag{28}$$

$$h(x,y) = \frac{h(x)+h(y)}{2} = \sqrt{2}\left[\sum_{n=0}^{N-1}\sqrt{2S_h(w_n)\Delta w}\cos\left(\frac{(w_n x + w_n y) + (\phi_{nx}+\phi_{ny})}{2}\right)\cos\left(\frac{(w_n x - w_n y) + (\phi_{nx}-\phi_{ny})}{2}\right)\right] \tag{29}$$

Based on the mathematical model of a three-dimensional surface, the stochastic surface topography of different corrosion degrees is generated, as shown in Figure 16. It is evident that the corroded surface is rough, and corrosion pits appear randomly on the surface. In fact, the three-dimensional surface can be regarded as a two-dimensional stochastic field based on the position coordinates. Generating a three-dimensional surface with the mathematical model of a one-dimensional stochastic process is a simplified method. Compared with the simplified method in past works that assumes the corrosion pit is cylindrical shaped pits or conical pits, the reconstructed surface topography is more effective in describing the property of corroded steel surface [15,16].

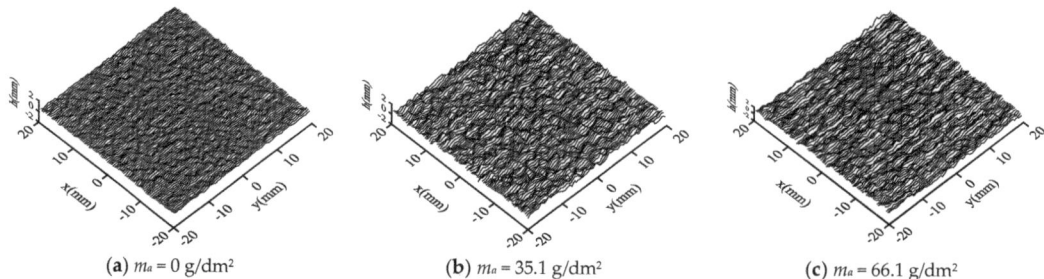

Figure 16. Randomly generated corroded surface with different corrosion degrees.

5. Conclusions

In this paper, the corrosion characteristics in the submarine soil environment of steel structure are studied by electrochemical accelerated corrosion experiments. The corrosion characteristics are summarized through the analysis of electrochemical reactions and mass loss. The variation rules of surface topography of corroded steel samples are investigated by scanning tests. The following conclusion can be drawn:

(1) The mass loss per unit area increases linearly with the corrosion time. Based on Faraday's law, the experimental value and theoretical value of mass loss are compared. The result indicates that when the corrosion time is less than 3 h, the mass loss rates of two electrochemical reaction processes during which the corrosion product is divalent iron and trivalent iron, respectively, are similar. With the increasing corrosion time, the corrosion products of experiments are mainly trivalent iron.

(2) With the increasing corrosion degree, the corroded steel surface becomes rougher, and the number and size of corrosion pits increase. The height of surface two-dimensional contour curves under different corrosion degrees obeys the Gaussian distribution.

Based on the spectral representation method, a mathematical model is developed for a one-dimensional profile of a corroded steel surface. The reliability of the mathematical model is proved by comparing the standard deviation, arithmetic mean height and maximum height of reconstructed samples with those of experimental samples. By assuming that the height of any point on the corroded surface equals the mean value of heights of randomly generated two-dimensional contour curves through that point along x-direction and y-direction, the three-dimensional surface can be reconstructed. The mathematical model can be adapted to reconstruct the surface topography of steel with different corrosion degrees in the following research on the shearing behavior of soil-corroded steel interface.

Author Contributions: Methodology, W.W.; Formal analysis, W.W.; Investigation, W.W.; Writing—original draft, W.W.; Visualization, W.W.; Writing—review and editing, Y.W., J.H. and L.L.; Resources, Y.W. and J.H.; Conceptualization, J.H.; Supervision, J.H. All authors have read and agreed to the published version of the manuscript.

Funding: This research was funded by Beijing Natural Science Foundation Program,(JQ19029), and Fundamental Research Funds for the central universities, (FRF-BD-18-007A).

Institutional Review Board Statement: Not applicable.

Informed Consent Statement: Not applicable.

Data Availability Statement: Not applicable.

Conflicts of Interest: The authors declare no conflict of interest.

References

1. Sathish, T.; Mohanavel, V.; Arunkumar, T.; Raja, T.; Rashedi, A.; Alarifi, I.M.; Badruddin, I.A.; Algahtani, A.; Afzal, A. Investigation of Mechanical Properties and Salt Spray Corrosion Test Parameters Optimization for AA8079 with Reinforcement of TiN + ZrO$_2$. *Materials* **2021**, *14*, 5260. [CrossRef] [PubMed]
2. Momber, A. Corrosion and corrosion protection of support structures for offshore wind energy devices (OWEA). *Mater. Corros.* **2010**, *62*, 391–404. [CrossRef]
3. Kirchgeorg, T.; Weinberg, I.; Hörnig, M.; Baier, R.; Schmid, M.; Brockmeyer, B. Emissions from corrosion protection systems of offshore wind farms: Evaluation of the potential impact on the marine environment. *Mar. Pollut. Bull.* **2018**, *136*, 257–268. [CrossRef] [PubMed]
4. Kovendhan, M.; Kang, H.; Jeong, S.; Youn, J.-S.; Oh, I.; Park, Y.-K.; Jeon, K.-J. Study of stainless steel electrodes after electrochemical analysis in sea water condition. *Environ. Res.* **2019**, *173*, 549–555. [CrossRef]
5. James, M.; Hattingh, D. Case studies in marine concentrated corrosion. *Eng. Fail. Anal.* **2015**, *47*, 1–15. [CrossRef]
6. Lv, K.; Xu, S.; Liu, L.; Wang, X.; Li, C.; Wu, T.; Yin, F. Comparative Study on the Corrosion Behaviours of High-Silicon Chromium Iron and Q235 Steel in a Soil Solution. *Int. J. Electrochem. Sci.* **2020**, 5193–5207. [CrossRef]
7. Wei, B.; Qin, Q.; Bai, Y.; Yu, C.; Xu, J.; Sun, C.; Ke, W. Short-period corrosion of X80 pipeline steel induced by AC current in acidic red soil. *Eng. Fail. Anal.* **2019**, *105*, 156–175. [CrossRef]
8. Karagah, H.; Shi, C.; Dawood, M.; Belarbi, A. Experimental investigation of short steel columns with localized corrosion. *Thin-Walled Struct.* **2015**, *87*, 191–199. [CrossRef]
9. Liu, X.; Nanni, A.; Silva, P.F. Rehabilitation of Compression Steel Members Using FRP Pipes Filled with Non-Expansive and Expansive Light-Weight Concrete. *Adv. Struct. Eng.* **2005**, *8*, 129–142. [CrossRef]
10. Wang, K.; Zhao, M.-J. Mathematical Model of Homogeneous Corrosion of Steel Pipe Pile Foundation for Offshore Wind Turbines and Corrosive Action. *Adv. Mater. Sci. Eng.* **2016**, *2016*, 9014317. [CrossRef]

11. Wang, K.; Li, Z.; Zhao, M. Mechanism of Localized Corrosion of Steel Pipe Pile Foundation for Offshore Wind Turbines and Corrosive Action. *Open Civ. Eng. J.* **2016**, *10*, 685–694. [CrossRef]
12. Yamamoto, N.; Ikegami, K. A Study on the Degradation of Coating and Corrosion of Ship's Hull Based on the Probabilistic Approach. *J. Offshore Mech. Arct. Eng.* **1998**, *120*, 121–128. [CrossRef]
13. Akpan, U.O.; Koko, T.; Ayyub, B.; Dunbar, T. Risk assessment of aging ship hull structures in the presence of corrosion and fatigue. *Mar. Struct.* **2002**, *15*, 211–231. [CrossRef]
14. Guo, J.; Wang, G.; Ivanov, L.; Perakis, A.N. Time-varying ultimate strength of aging tanker deck plate considering corrosion effect. *Mar. Struct.* **2008**, *21*, 402–419. [CrossRef]
15. Jiang, X.; Soares, C.G. Ultimate capacity of rectangular plates with partial depth pits under uniaxial loads. *Mar. Struct.* **2012**, *26*, 27–41. [CrossRef]
16. Ahmmad, M.; Sumi, Y. Strength and deformability of corroded steel plates under quasi-static tensile load. *J. Mar. Sci. Technol.* **2009**, *15*, 1–15. [CrossRef]
17. Nakai, T.; Matsushita, H.; Yamamoto, N. Effect of pitting corrosion on local strength of hold frames of bulk carriers (2nd Report)—Lateral-distortional buckling and local face buckling. *Mar. Struct.* **2004**, *17*, 612–641. [CrossRef]
18. Nakai, T.; Matsushita, H.; Yamamoto, N. Effect of pitting corrosion on the ultimate strength of steel plates subjected to in-plane compression and bending. *J. Mar. Sci. Technol.* **2006**, *11*, 52–64. [CrossRef]
19. Chen, J.; Fu, C.; Ye, H.; Jin, X. Corrosion of steel embedded in mortar and concrete under different electrolytic accelerated corrosion methods. *Constr. Build. Mater.* **2020**, *241*, 117971. [CrossRef]
20. Li, Y.; Xu, C.; Zhang, R.H.; Liu, Q.; Wang, X.H.; Chen, Y.C. Effects of Stray AC Interference on Corrosion Behavior of X70 Pipeline Steel in a Simulated Marine Soil Solution. *Int. J. Electrochem. Sci.* **2017**, 1829–1845. [CrossRef]
21. Xu, Z.; Du, Y.; Qin, R.; Zhang, H. Study of Corrosion Behavior of X80 Steel in Clay Soil with Different Water Contents under HVDC Interference. *Int. J. Electrochem. Sci.* **2020**, 3935–3954. [CrossRef]
22. Wang, X.; Song, X.; Chen, Y.; Wang, Z.; Zhang, L. Corrosion Behavior of X70 and X80 Pipeline Steels in Simulated Soil Solution. *Int. J. Electrochem. Sci.* **2018**, *13*, 6436–6450. [CrossRef]
23. Su, X.; Yin, Z.; Cheng, Y.F. Corrosion of 16Mn Line Pipe Steel in a Simulated Soil Solution and the Implication on Its Long-Term Corrosion Behavior. *J. Mater. Eng. Perform.* **2012**, *22*, 498–504. [CrossRef]
24. Shinozuka, M.; Deodatis, G. Simulation of Multi-Dimensional Gaussian Stochastic Fields by Spectral Representation. *Appl. Mech. Rev.* **1996**, *49*, 29–53. [CrossRef]
25. Shinozuka, M.; Deodatis, G. Simulation of Stochastic Processes by Spectral Representation. *Appl. Mech. Rev.* **1991**, *44*, 191–204. [CrossRef]
26. Liu, Z.; Liu, W.; Peng, Y. Random function based spectral representation of stationary and non-stationary stochastic processes. *Probabilistic Eng. Mech.* **2016**, *45*, 115–126. [CrossRef]
27. Hu, Y.; Tonder, K. Simulation of 3-D random rough surface by 2-D digital filter and fourier analysis. *Int. J. Mach. Tools Manuf.* **1992**, *32*, 83–90. [CrossRef]
28. Hu, B.; Schiehlen, W. On the simulation of stochastic processes by spectral representation. *Probabilistic Eng. Mech.* **1997**, *12*, 105–113. [CrossRef]
29. Uzielli, M.; Vannucchi, G.; Phoon, K.K. Random field characterisation of stress-nomalised cone penetration testing parameters. *Geotechnique* **2005**, *55*, 3–20. [CrossRef]
30. Phoon, K.-K.; Quek, S.T.; An, P. Identification of Statistically Homogeneous Soil Layers Using Modified Bartlett Statistics. *J. Geotech. Geoenvironmental Eng.* **2003**, *129*, 649–659. [CrossRef]
31. Manesh, K.; Ramamoorthy, B.; Singaperumal, M. Numerical generation of anisotropic 3D non-Gaussian engineering surfaces with specified 3D surface roughness parameters. *Wear* **2010**, *268*, 1371–1379. [CrossRef]
32. Gathimba, N.; Kitane, Y.; Yoshida, T.; Itoh, Y. Surface roughness characteristics of corroded steel pipe piles exposed to marine environment. *Constr. Build. Mater.* **2019**, *203*, 267–281. [CrossRef]
33. Xia, M.; Wang, Y.; Xu, S. Study on surface characteristics and stochastic model of corroded steel in neutral salt spray environment. *Constr. Build. Mater.* **2020**, *272*, 121915. [CrossRef]

Article

Corrosion Resistance of Mg/Al Vacuum Diffusion Layers

Shixue Zhang [1], Yunlong Ding [1,*], Zhiguo Zhuang [1] and Dongying Ju [2,3,4]

1 School of Mechanical Engineering, University of Science and Technology Liaoning, Anshan 114051, China
2 School of Materials and Metallurgy, University of Science and Technology Liaoning, Anshan 114051, China
3 Ningbo Haizhi Institute of Materials Industry Innovation, Ningbo 315000, China
4 Department of Information System, Graduate School of Engineering, Saitama Institute of Technology, Fukaya 369-0293, Japan
* Correspondence: dylustl@163.com; Tel.: +86-155-4122-8620

Abstract: This study used a vacuum diffusion welding process to weld magnesium (Mg1) and aluminum (Al1060). The diffusion layers, with different phase compositions, were separated and extracted by grinding. The diffusion layers' microstructures and phase compositions were analyzed using scanning electron microscopy (SEM) and energy dispersive spectroscopy (EDS). Furthermore, the corrosion resistance of each diffusion layer and the substrates were investigated and compared by performing corrosion immersion tests and linear polarization measurements in a 3.5 wt.% NaCl solution. The results showed that diffusion layers consisting of Mg_2Al_3, $Mg_{17}Al_{12}$, and $Mg_{17}Al_{12}$/Mg-based solid solutions were formed at the interface of the Mg1/Al1060 vacuum diffusion joint. Furthermore, each diffusion layer's structure and morphology were of good quality, and the surfaces were free from defects. This result was obtained for a welding temperature of 440 °C and a holding time of 180 min. The corrosion current density of Mg1 was 2.199×10^{-3} A/cm^2, while that of the Al1060, Mg_2Al_3, $Mg_{17}Al_{12}$, and $Mg_{17}Al_{12}$/Mg-based solid solutions increased by order of magnitude, reaching 1.483×10^{-4} A/cm^2, 1.419×10^{-4} A/cm^2, 1.346×10^{-4} A/cm^2, and 3.320×10^{-4} A/cm^2, respectively. The order of corrosion rate was Mg1 > $Mg_{17}Al_{12}$ and Mg-based solid solution > Mg_2Al_3 > $Mg_{17}Al_{12}$ > Al1060. Moreover, all diffusion layers exhibited an improved corrosion resistance compared to Mg1. This was especially the situation for the Mg_2Al_3 layer and $Mg_{17}Al_{12}$ layer, whose corrosion resistances were comparable to that of Al1060.

Keywords: Mg1; Al1060; diffusion welding; intermetallic compounds; diffusion layers; corrosion resistance

1. Introduction

With the continuous and rapid development of modern industrial technology, the sustainable development of lightweight materials for environmental protection and energy savings has attracted more and more attention in today's society [1]. Magnesium and aluminum are two lightweight nonferrous metals of low density and high specific strength compared to common structural materials such as steel. Their alloys have been widely used in global transportation, especially in the automotive and aerospace industries [2–4]. At present, rolling, bonding, and welding are commonly used processes by which it is possible to achieve an effective combination of Mg and Al heterogeneous metals. These processes not only result in an optimization of the structural qualities but also take full advantage of the respective metals' properties [5,6]. The welding methods that are used at present in realizing the combination of Mg/Al heterogeneous metals mainly include laser welding [7], TIG welding [8], stir friction welding [9,10], diffusion welding [11,12], and ultrasonic welding [13]. The chemical reactivities of magnesium and aluminum are relatively high, and defects such as oxidation, cracks, and pores can easily appear in the process of traditional fusion welding. Vacuum diffusion welding is a solid-state welding method that uses a low heat input during the welding process. As a result, the metal base material does not melt; it only undergoes a microscopic plastic deformation at the surface.

The quality of the welded joint is, therefore, relatively stable. This method is suitable for welding metal materials with different physical properties, such as the coefficient of thermal expansion. It is, thus, applicable to magnesium and aluminum.

Corrosion is the physical and chemical reaction between materials and their surroundings, which changes the properties of a material. Corrosion is harmful to the production and development of today's industry. Metal corrosion significantly reduces its service life and easily causes potential safety hazards using precision fields [14,15]. It is well known that magnesium and magnesium alloys experience poor corrosion resistance and are susceptible to corrosion in neutral solutions containing chloride ions. Poor corrosion resistance is a crucial bottleneck that limits their wide application in areas with high safety requirements [16]. However, aluminum and aluminum alloys exhibit an improved corrosion resistance relative to magnesium and magnesium alloys in chlorine-containing environments [17]. Intermetallic compounds (Mg_2Al_3, $Mg_{17}Al_{12}$) are formed in the joint during vacuum diffusion welding of the Mg/Al pair, thereby forming a continuous diffusion layer. Numerous studies have shown that Mg/Al intermetallic compounds are hard and brittle, and their wide distribution at the common interface leads to further deterioration of the mechanical properties of the joint [18–20]. Therefore, in the vacuum diffusion welding of dissimilar metals (e.g., the Mg/Al pair), the primary research has focused on effectively suppressing the formation of brittle intermetallic compounds. Many researchers have extensively investigated the optimization of welding process parameters and the possibility of an implantation of a suitable interlayer to suppress the formation of intermetallic compounds [21–23]. No matter what method is used, the generation of an intermetallic compound cannot be avoided entirely. It is only possible to control its formation, or change its distribution, to a certain extent in the strives to improve the performance of the joint. The generation of intermetallic compounds in joints is inevitable [24–26]. Therefore, under the conditions of the overall controlled corrosion resistance of Mg and Al, it is essential to study the corrosion behavior of the Mg/Al vacuum diffusion composite plates and the intermetallic compounds at the joint. Related workers had made some reports on the corrosion resistance of magnesium-aluminum intermetallic compounds. The research of Zhang et al. [27] showed that the magnesium aluminum intermetallic compound coating could significantly improve the wear resistance and corrosion resistance of magnesium substrate and protect magnesium substrate from wear and corrosion. Bu et al. [28] successfully deposited intermetallic compound $Mg_{17}Al_{12}$ particle reinforced pure Al coatings onto AZ91D magnesium substrate, their measured potentiodynamic polarization curves in 3.5 wt.% NaCl solution showed that the corrosion current density of the magnesium substrate decreased by more than one order of magnitude after the deposition of the coating.

However, to our knowledge, no reports deal with the corrosion resistance evaluation of the Mg/Al vacuum diffusion composite plates, nor with diffusion layers, when immersed in an aggressive NaCl environment. In the present study, the Mg/Al pair welding has been the first one using a vacuum diffusion welding technique. The microstructure and phase composition of the diffusion layers were observed and analyzed by Scanning Electron Microscopy (SEM) and energy dispersive spectroscopy (EDS). Each diffusion layer was separated and extracted. Furthermore, the corrosion mechanism was analyzed using the optical microscope (OM), SEM, and EDS. This was also the situation for the Mg/Al vacuum diffusion composite plates and the different diffusion layers. This paper separated diffusion layers of Mg/Al and their corrosion behavior in a 3.5 wt.% NaCl solution was studied, which was nearly the first try. The purpose of the present study was not only to explore the corrosion resistances of the diffusion layers and the Mg/Al matrix, but also to lay the theoretical foundation for an improvement of the corrosion resistance of magnesium by using alloying aluminum material.

2. Experimental Materials and Methods

2.1. Materials

The base metals selected for the experiments were aluminum (Al1060) and magnesium (Mg1), and the chemical composition of these metals is listed in Tables 1 and 2, respectively. The specimens used for the vacuum diffusion welding test were of size 80 mm × 20 mm × 3 mm.

Table 1. Chemical composition of Mg1 (mass fraction, %).

Mg	Al	Mn	Cu	Si	Fe	Ca	Ni
Bal.	0.2	0.22	0.0008	0.012	0.0021	0.0015	0.0009

Table 2. Chemical composition of Al1060 (mass fraction, %).

Al	Mg	Mn	Cu	Si	Fe	Zn	Ni
Bal.	0.05	0.10	0.007	0.012	0.20	0.25	0.0016

2.2. Preparation of Mg1/Al1060 Vacuum Diffusion Layers

Before welding, the metal surfaces were ground with 400#, 800#, 1200#, 1500#, and 2000# abrasive papers to remove oxide films from these surfaces. After that, it was polished using a polishing cloth and a diamond abrasive paste. Ultrasonic cleaning was performed after the polishing, and absolute ethanol was used to remove any impurities or oil stains from the surface. As the next step, the treated base material was put into a unique mold with a "magnesium on top of aluminum" stacking and placed in a vacuum heating furnace for welding. Based on the phase diagram of the Mg/Al binary alloy, the welding temperature was chosen to be 440 °C. Considering that the Mg/Al vacuum diffusion layers will gradually grow with an extension of the welding time, the holding time during the welding process was selected as 180 min. This time ensures that the diffusion reactions were entirely carried out and that a diffusion layer with sufficient thickness was obtained. This thickness should facilitate the subsequent extraction of diffusion layers with different phase compositions and be sufficiently large for the following corrosion experiments. In order to avoid a thermal shock that would affect the quality of the welded joint, the sample was slowly cooled to room temperature within the furnace after completion of the welding. The heating rate during the welding process was 10 °C/min, and the vacuum pressure was less than 1×10^{-2} Pa. The welding process flow is demonstrated in Figure 1.

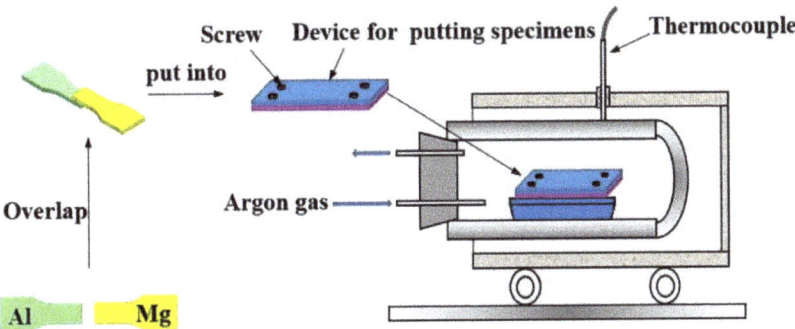

Figure 1. Flow chart of the vacuum diffusion welding process.

2.3. Assessment of Corrosion Resistance

The interface of a Mg1/Al1060 welded joint was sampled by using a wire-cutting technique. The corrosion resistance of these samples (of size 20 mm in length and 8 mm in width) was characterized by corrosion immersion and linear polarization techniques.

The tests were repeated three times under each condition to ensure accuracy and avoid accidental errors. Before these analyses, all samples were polished to 2000 grit and cleaned with anhydrous ethanol. All corrosion resistance tests were carried out in 3.5 wt.% NaCl solution under atmospheric conditions.

2.3.1. Corrosion Immersion Tests

The polished Mg1/Al1060 composite plates were immersed in a 3.5 wt.% NaCl solution and removed every 90 min. During these interruptions, the corrosion products on the surface were removed using ultrasonic cleaning with absolute ethanol, and the plates were then dried with a hair dryer. The evolution of the surface morphologies could, in this way, be followed by using an optical microscope (Thermo Fisher Scientific, Waltham, MA, America).

2.3.2. Linear Polarization

The wire-cut specimens used for linear polarization tests were encapsulated with tooth-powder. The purpose was to expose the diffusion layers both uniformly and continuously. They were, after that, ground from the magnesium substrate side with a metallographic polishing machine. When the grinding was close to the exposure of diffusion layers, they were carefully polished with the polishing cloth and abrasive paste instead of abrasive papers to prevent the removal of abrasive papers being too large to get a single-component diffusion reaction layer; the EDS (Thermo Fisher Scientific, Waltham, MA, America) was used to verify this.

Linear polarization measurements were used for a more quantitative assessment of the corrosion resistance of the diffusion layers. These measurements were carried out for both Mg1 and Al1060 using a CS Instruments Electrochemical Work Station (CS350, Wuhan, China). A comparison between Mg1 and Al1060 was then easily achieved. A conventional three-electrode electrochemical cell setup was employed, which consisted of the test sample as the working electrode (with an exposure area of 1.6 cm^2), a silver chloride electrode as the reference electrode, and a platinum electrode as the counter electrode. The linear polarization measurements were then made by applying a potential in the range of -300 mV to $+300$ mV, with a scan rate of 1 mV/s. Linear polarization curves were obtained, and the electrochemical measurements were completed. Polarization data, including the corrosion potential (E_{corr}), corrosion current density (I_{corr}), polarization resistance (R_p), and corrosion rate (V_{cor}), could be deduced from the linear polarization curves (i.e., log I vs. E plot). Furthermore, the I_{corr} values were obtained from the intersection of the Tafel slope. Moreover, the R_p and Vcor values were calculated using the Corrview 3.10 software, which was provided by the electrochemical workstation (CS350). Furthermore, the samples' surface morphology was studied using SEM and EDS.

3. Results

3.1. Microstructure and Phase Composition of the Mg1/Al1060 Layers

Figure 2 shows the microstructure of the Mg1/Al1060 joint, which was formed at a welding temperature of 440 °C and for a holding time of 180 min. It is clear that the joint is well combined, and there are no defects such as holes, burning, or incomplete fusion. Furthermore, the initial interface between the Mg1 and Al1060 substrates disappeared, and a diffusion layer was formed at the joint position. An enlarged view of the diffusion layer morphology can be seen in Figure 2b, which clearly shows that the diffusion layer is composed of three layers. The organizational structure of layer 1 is relatively uniform, while layer 2 has an irregular columnar structure towards the Mg1 substrate. Moreover, layer 3 consists of a relatively homogeneous eutectic structure and is much thicker than layers 1 and 2. The elemental compositions of these diffusion layers were analyzed using an EDS point scan. The results are shown in Figure 3. The elemental composition of point A in layer 1 was composed of approximately 40% Mg and 60% Al. According to the analysis in Ref. [1] and the Mg/Al alloy phase diagram, the composition of point A was the Mg_2Al_3 phase. Furthermore, point B of layer 2 consisted of nearly 40% Al and 60% Mg, which suggests that layer 2 was composed of the $Mg_{17}Al_{12}$ phase. Figure 3c shows the EDS result from the light point C of layer 3. This point consists of nearly 40% Al and 60% Mg, which was determined to be the $Mg_{17}Al_{12}$ phase. Additionally, Figure 3d shows the EDS result from the dark point D of layer 3, where Mg has increased to nearly 90%, while Al has decreased to nearly 10%. Thus, this point did mainly consist of a Mg-based solid solution. Hence, it was demonstrated that the eutectic structure in layer 3 consisted of both $Mg_{17}Al_{12}$ and a Mg-based solid solution. Based on these results, it can be understood that the order of diffusion layers from the Al side to the Mg side was: Mg_2Al_3 layer, $Mg_{17}Al_{12}$ layer, $Mg_{17}Al_{12}$/Mg-based solid solution layer.

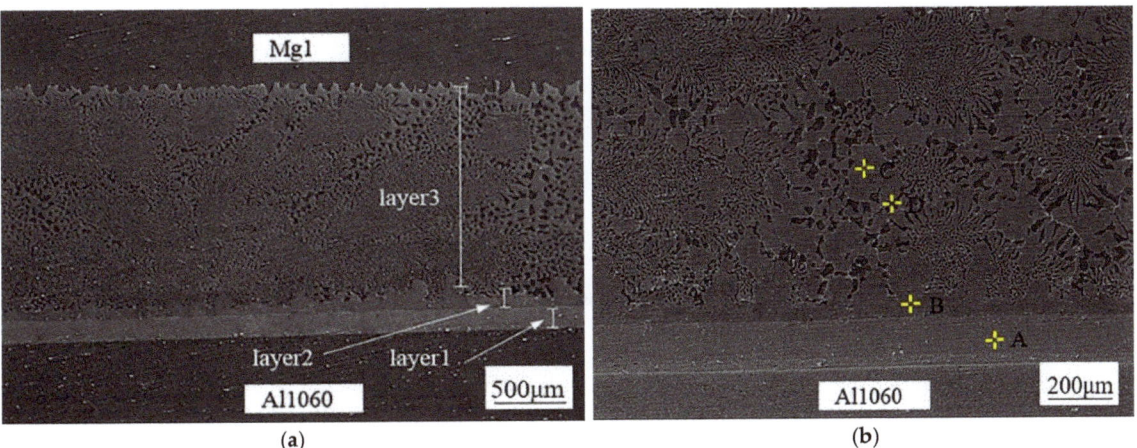

Figure 2. SEM images of the Mg1/Al1060 diffusion layers: (**a**) SEM image at low magnification, (**b**) SEM image at high magnification.

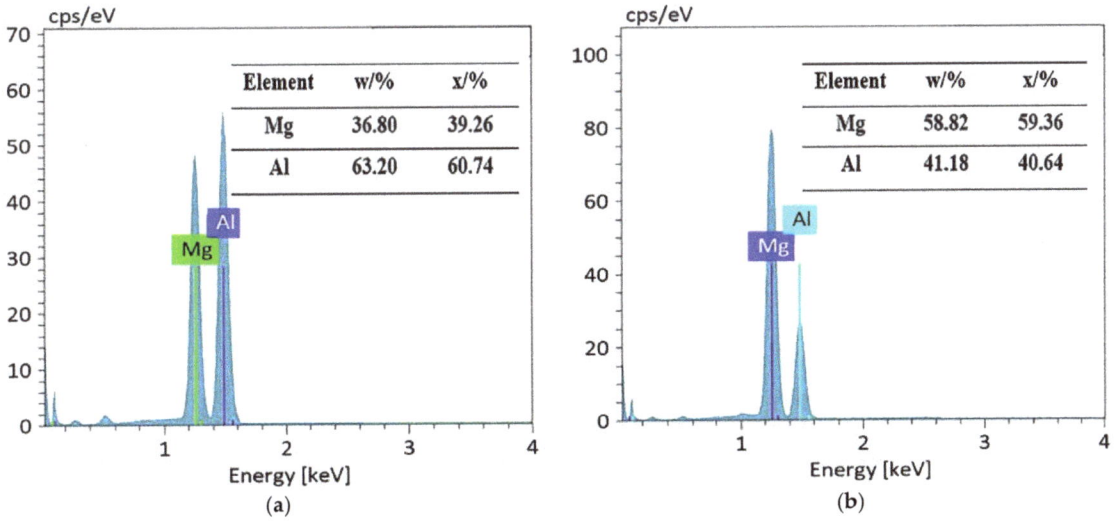

Figure 3. *Cont.*

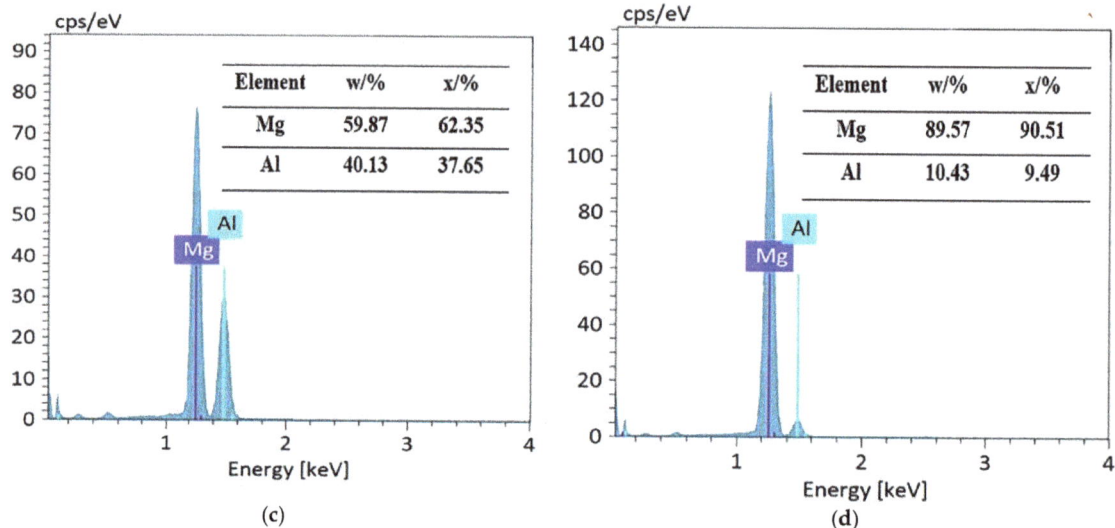

Figure 3. Elemental composition in different regions of the diffusion layer: (**a**) point A, (**b**) point B, (**c**) point C, and (**d**) point D.

3.2. Corrosion Immersion Test Results

Figure 4 shows a significant surface morphology evolution with time for a Mg1/Al1060 composite plate immersed in a 3.5 wt.% NaCl solution. It can be seen that the surface morphology of the diffusion layers and the two substrates was of high quality before any soaking had taken place. Thus, there were no apparent defects. On the contrary, the Mg1 substrate became severely degraded after 90 min of immersion in a 3.5 wt.% NaCl solution. A large piece of the Mg1 metal was corroded away from the composite plate, and the surface of the remaining part turned black (from a metallic luster). However, the surfaces of the diffusion layers and the Al1060 substrate showed no obvious corrosion defects. They had, thus, stronger corrosion resistance than Mg1. Furthermore, the surface of the Mg1 substrate became coarser and darker upon further immersion in the NaCl solution. In addition, typical pitting corrosion occurred on the surface of the Al1060 substrate, and these pits were evenly distributed on the surface. However, there were no noticeable changes on the surfaces of the diffusion layers (see Figure 4c). For even longer immersion times, Mg1 became entirely corroded, and the corrosion of Al1060 worsened. In addition, the pits on the Al1060 surface were gradually enlarged. Interestingly, there were still no evident corrosion-related defects on the surfaces of the diffusion layers (see Figure 4d). When the immersion time reached 360 min, the pitting corrosion on the Al1060 surface became more significant. The pits increased in quantity and gradually appeared as a honeycomb distribution. Some parts of the diffusion layers were also corroded off; traces of pits were left on its surface, with a more pronounced concentration on the $Mg_{17}Al_{12}$ and Mg-based solid solution layer. When the immersion time reached 450 min, the original metal surface of Al1060 was almost completely destroyed. One can also find that the corrosion pits had grown even further and connected in a continuous wave-like pattern. At the same time, the corrosion pits on the diffusion layers had increased in number and significantly enlarged. This result proves that the degree of corrosion became severe with an increase in immersion time.

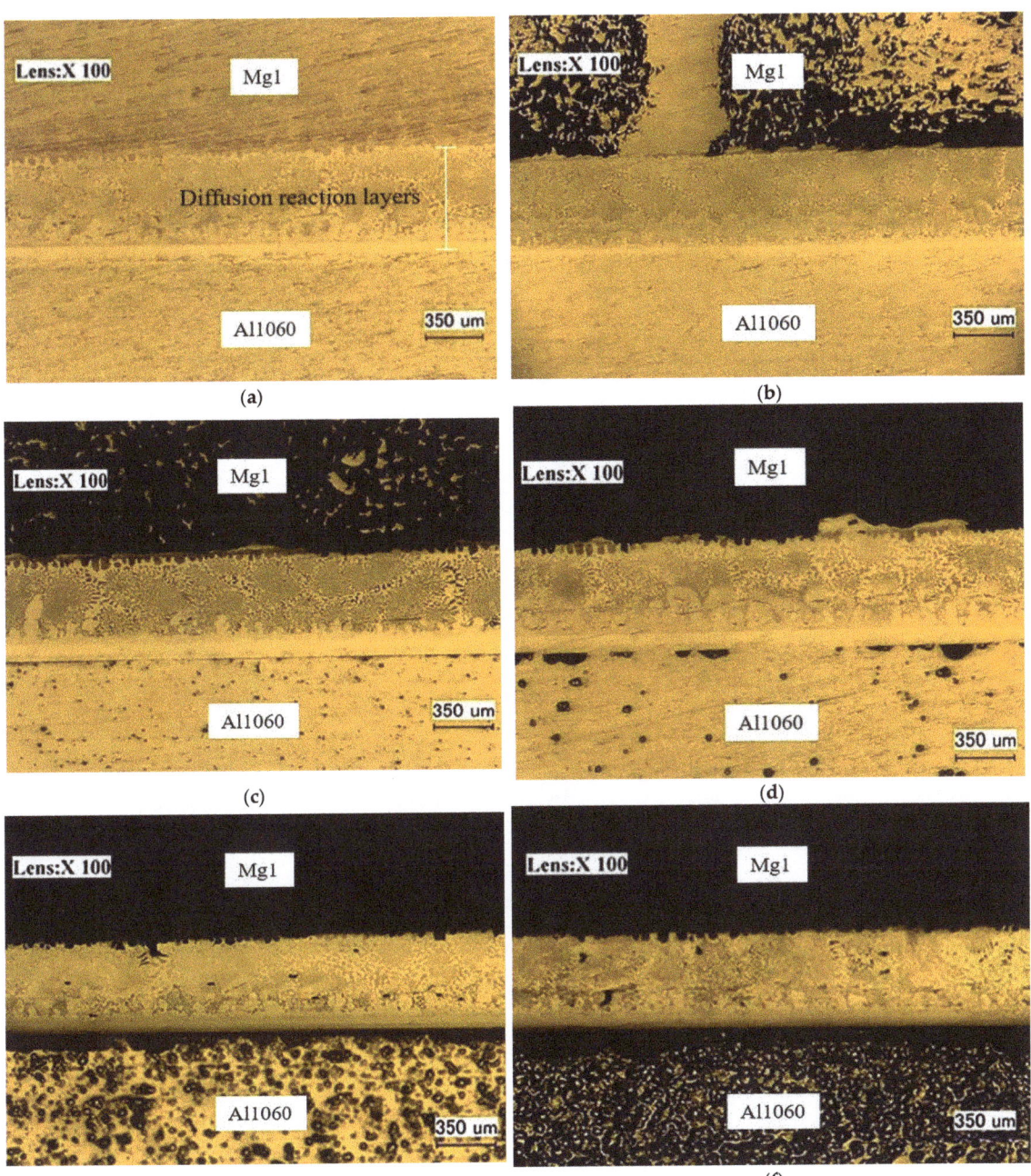

Figure 4. Surface morphology of the substrates and diffusion layers for different immersion times: (**a**) 0 min, (**b**) 90 min, (**c**) 180 min, (**d**) 270 min, (**e**) 360 min, and (**f**) 450 min.

By soaking in a 3.5 wt.% NaCl solution, it could initially be seen that the Mg1 substrate was most easily corroded in the Mg1/Al1060 composite plates. Many studies have confirmed that the oxide films formed on the surfaces of magnesium and magnesium alloys in the air cannot effectively protect the surfaces in a solution containing Cl ions. Rapid corrosion occurs, which is due to the low electrode potential of magnesium. Thus, magnesium will act as an anode when it is in contact with other alloys, causing electrochemical corrosion. This explains the present study's rapid corrosion and degradation of the Mg1 substrate. On the other hand, Al1060 showed a stronger corrosion resistance than Mg1, with typical pitting corrosion occurring after immersion. The continuous pitting corrosion was found to penetrate deep into the metal, causing severe damage to the Al1060 substrate. However, the diffusion layers showed the most negligible corrosion. The explanation is their direct contact with the Mg1 substrate, where Mg1 acts as an anode. Thus, the Mg1 substrate underwent rapid galvanic corrosion, which indirectly protected the diffusion layer. As more and more magnesium substrate was corroded, the protective effect weakened, and the reaction layers gradually corroded. Therefore, the diffusion layers were the least corroded, with corrosion defects gradually appearing on the surface after an immersion time of 360. It was observed that the corrosion defects were mainly concentrated on the $Mg_{17}Al_{12}$ and Mg-based solid solution diffusion layer.

3.3. Cross-Sectional Structure and Energy Spectrum Analysis

Figure 5 shows the cross-sectional structure of each diffusion layer (Mg_2Al_3 layer, $Mg_{17}Al_{12}$ layer, and $Mg_{17}Al_{12}$ and Mg-based solid solution layer) after grinding. It can be seen that the structure and morphology of each diffusion layer are intact, and the surface is free from defects. In order to verify the composition of each diffusion layer, its composition was examined by using EDS. Figure 6 shows the elemental detection results of the EDS surface scan, where red represents the Mg element and green represents the Al element. It can be seen that the Mg and Al elements were uniformly distributed in these samples, which reflects the presence of only one single and homogeneous phase in each of these diffusion layers. The Al element, as represented by green in the Mg_2Al_3 diffusion layer, is dominating in the samples. In comparison, the presence of the Mg element, as represented by red, has significantly increased in the $Mg_{17}Al_{12}$ diffusion layer. While the most apparent Mg element can be observed in $Mg_{17}Al_{12}$ and Mg-based solid solution diffusion reaction layer. These experimental findings are consistent with the theoretical composition of phases in each diffusion layer. The chemical composition of position A in the Mg_2Al_3 layer, position B in the $Mg_{17}Al_{12}$ layer, and position C in the $Mg_{17}Al_{12}$ and Mg-based solid solution layer has been analyzed by performing an EDS area scan. The results are presented in Table 3, where it can be seen that the content of aluminum has gradually decreased, while the content of magnesium has gradually increased when going from box A to box C. This result further verifies the phase composition of the extracted diffusion layers in combination with the Mg/Al alloy phase diagram.

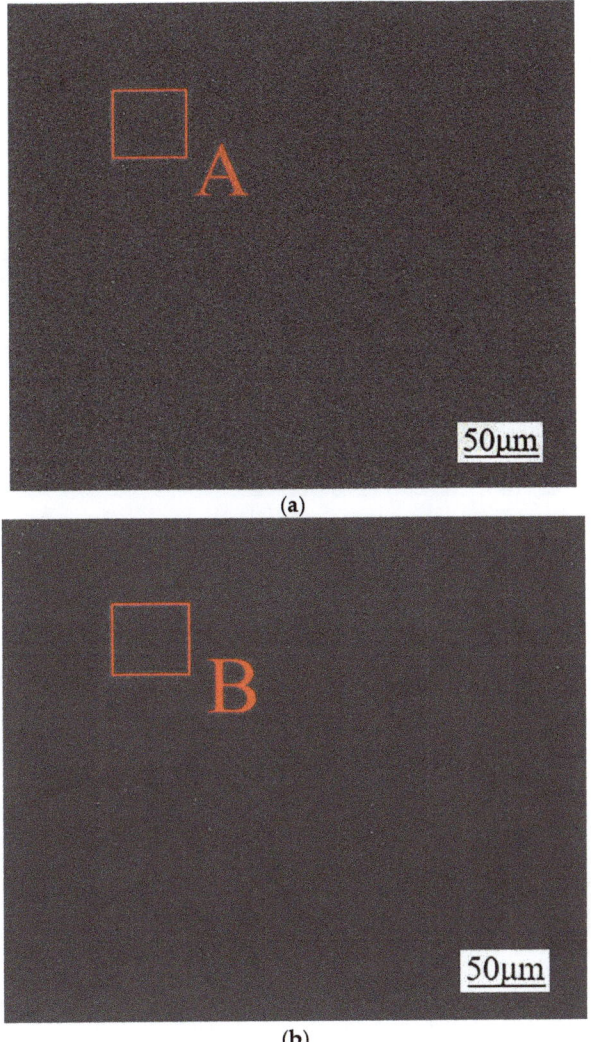

Figure 5. *Cont.*

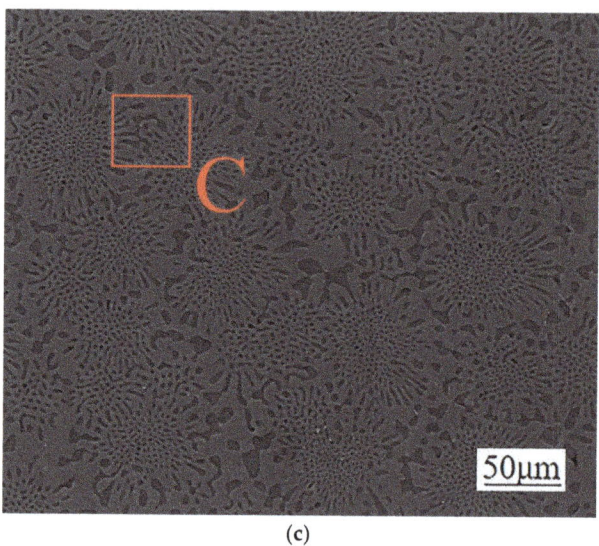

(c)

Figure 5. Cross-sectional structure of different diffusion layers: (**a**) Mg_2Al_3 layer, (**b**) $Mg_{17}Al_{12}$ layer, and (**c**) $Mg_{17}Al_{12}$ and Mg-based solid solution layer.

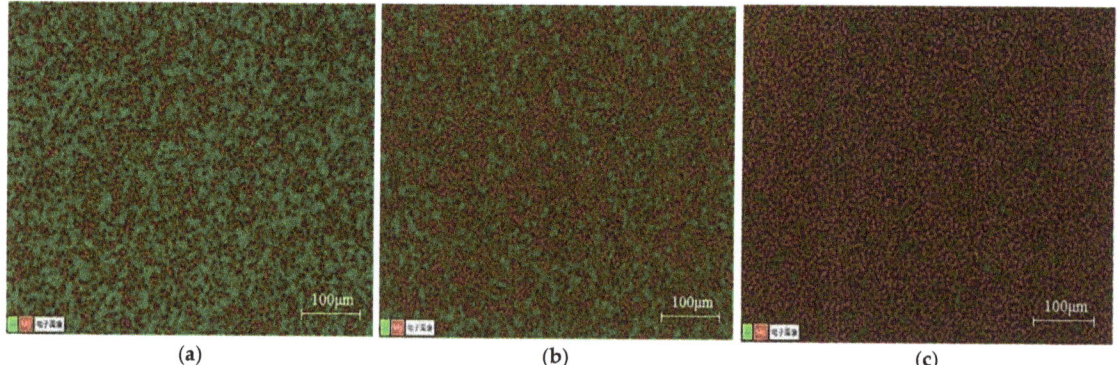

Figure 6. EDS surface scanning analysis of different diffusion layers: (**a**) Mg_2Al_3 layer, (**b**) $Mg_{17}Al_{12}$ layer, and (**c**) $Mg_{17}Al_{12}$ and Mg-based solid solution layer.

Table 3. EDS results of the distinct regions presented in Figure 5.

Position	Mole Fraction/%	
	Al	Mg
A	61.60	38.40
B	39.72	60.28
C	26.14	73.86

3.4. Linear Polarization

3.4.1. Analyses of Polarization Curves

After the immersion in a 3.5 wt.% NaCl solution, Figure 7 shows the polarization curves for the Mg1 and Al1060 substrates. It also shows the polarization curves for the Mg_2Al_3, $Mg_{17}Al_{12}$, $Mg_{17}Al_{12}$, and Mg-based solid solution layers. It is clear that the cathodic reaction is most prominent for the Mg1 substrate, while Al1060, Mg_2Al_3, and

$Mg_{17}Al_{12}$ show a passive and similar electrochemical behavior over a wide potential range. This result suggests that the corrosion resistance of Al1060 and each of the diffusion layers are better than the corrosion resistance of Mg1. Based on the polarization curves, polarization data (including corrosion potential (E_{corr}), corrosion current density (I_{corr}), polarization resistance (R_p), and corrosion rate (V_{cor})) have been calculated and are listed in Table 4.

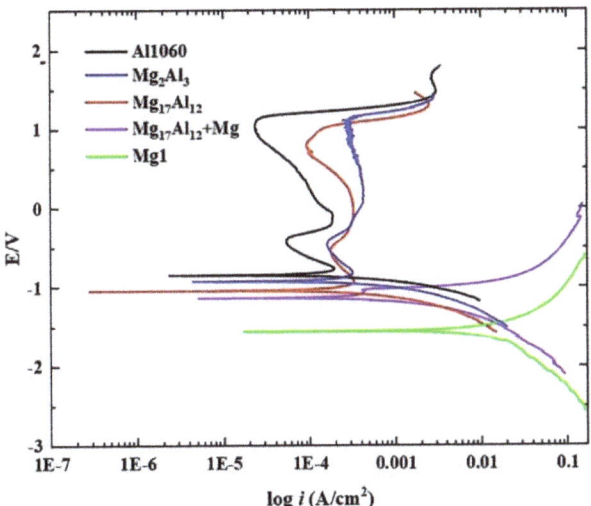

Figure 7. Electrochemical polarization curves of the substrates and different diffusion layers in a 3.5 wt.% NaCl solution.

Table 4. Polarization data of Al1060, Mg1, and diffusion layers in 3.5 wt.% NaCl solution.

Samples	E_{corr} (V)	I_{corr} (A/cm^2)	R_p (Ω cm^2)	V_{cor} (mm/year)
Al1060	−0.84	1.483×10^{-4}	175.96	5.0377
Mg_2Al_3	−0.98	1.419×10^{-4}	183.9	7.2302
$Mg_{17}Al_{12}$	−1.03	1.346×10^{-4}	209.3	6.3529
$Mg_{17}Al_{12}$ + Mg	−1.14	3.320×10^{-4}	81.007	16.414
Mg1	−1.55	2.199×10^{-3}	11.861	115.97

It can be seen that the corrosion potentials of the Al1060 substrate and the Mg_2Al_3 and $Mg_{17}Al_{12}$ layers are very similar, and their absolute values are relatively small. On the contrary, the corrosion potential of the Mg1 substrate is much smaller than those of Al1060 and each diffusion layer. Its absolute value is also the largest, which implies a poor corrosion resistance of Mg1. The order of corrosion current densities for the five tested samples were: Mg1 > $Mg_{17}Al_{12}$ and Mg-based solid solution > Al1060 > Mg_2Al_3 > $Mg_{17}Al_{12}$. The corrosion current density was determined by the dissolution degree of the material. The higher the corrosion current density, the smaller the charge transfer resistance, which means that the material's corrosion resistance is weaker [29]. Furthermore, the Al1060 substrate, and each diffusion layer, showed a reduction of one order for the Icorr value as compared with Mg1 (2.199×10^{-3} A/cm^2). This is a reflection of the much slower degradation of the Al1060 substrate and each diffusion layer as compared with Mg1. Additionally, the polarization resistance (R_p) of the Mg1 substrate was 11.861 Ω cm^2, while $Mg_{17}Al_{12}$ and the Mg-based solid solution layer showed a significant increase (81.007 Ω cm^2). This indicates a marginal improvement in corrosion resistance for $Mg_{17}Al_{12}$ and the Mg-based solid solution layer, while the improvements for the Al1060 substrate and the Mg_2Al_3 and $Mg_{17}Al_{12}$ layers were more obvious. The R_p values increased to 175.96 Ω cm^2, 183.9 Ω cm^2, and 209.3 Ω cm^2, respectively, demonstrating that the corrosion resistance of these three samples increased significantly compared to Mg1. It could also be seen that the corrosion rate of Mg1 had the highest value of 115.97 mm/year. Thus, the quality of the sample would be greatly affected if exposed to a corrosive environment for too long.

The above results show that the Mg1 substrate had the worst corrosion resistance in an aggressive NaCl environment, followed by the $Mg_{17}Al_{12}$ and Mg-based solid solution layer. Furthermore, the Mg_2Al_3 and $Mg_{17}Al_{12}$ layers showed a similar corrosion resistance as the Al1060 substrate, which was significantly better than the corrosion resistance of the Mg1 substrate. In the study of Song et al. [30,31], it was found that the $Mg_{17}Al_{12}$ phase can act as an anode barrier to restrain the overall corrosion of magnesium alloy; the $Mg_{17}Al_{12}$ phase was inert to corrosion in the solution containing cl^-, and a passive regions table over a similar potential range was observed in the polarization curve of $Mg_{17}Al_{12}$ measured by them. Many researchers have used the idea of alloying in the preparation of Mg/Al intermetallic compound coatings on the surface of magnesium alloys. The purpose is to improve the corrosion resistance of magnesium alloys [32–34]. Ji et al. [32] modified the pure aluminum coating deposited on the AZ91D substrate by friction stir-spot-processing. They observed that the Mg_2Al_3 and $Mg_{17}Al_{12}$ phases were irregularly distributed in the coating after modification, and the corrosion current density of the cold-sprayed Al coating was remarkably reduced. Irregularly distributed intermetallic compounds in the coating lead to the significantly enhanced corrosion resistance of the AZ91D substrate. Moreover, the results in the present study strongly agree with the results from the potentiodynamic polarization experiment proposed by Spencer K et al. [34]. This study aimed at Mg/Al vacuum diffusion layers. Each diffusion layer was extracted separately, and the difference in corrosion resistance between different types of Mg/Al intermetallic compounds and the base was investigated in more detail. The results of the corrosion resistance polarization curves also show consistency with the above corrosion immersion tests.

3.4.2. Analysis of Corrosion Morphology

Figure 8 shows the surface morphologies of the Al1060 and Mg1 substrates and of the Mg_2Al_3, $Mg_{17}Al_{12}$, and $Mg_{17}Al_{12}$ and Mg-based solid solution layers after potentiostatic electrochemical measurements. The overall surface compositions were measured by using EDS, and the results are shown in Figure 9. It can be seen that noticeable corrosion pits were formed on the surface of the Al1060 substrate. Some of these corrosion pits appeared to sparkle, indicating that there were corrosion products in the vicinity of the pits (see Figure 8a). However, the original surface of the Mg1 substrate has been severely destroyed (see Figure 8b). The corrosion on this surface was much more severe, with corrosion cracks and pits continuously distributed all over the surface. Moreover, the corrosion products formed on the surface were more compact than those formed on the Al1060 substrate and on each of the diffusion layers. Pitting corrosion occurs in the Mg_2Al_3 layer and $Mg_{17}Al_{12}$ layer likewise (Figure 8c,d). A few pits and cracks could be observed on the surfaces of the Mg_2Al_3 and $Mg_{17}Al_{12}$ layers, even though the whole surfaces were relatively smooth. Corrosion products were distributed over the surface in the form of needles and blocks. While the corrosion was more severe for the $Mg_{17}Al_{12}$ and Mg-based solid solution layer, as compared with Mg_2Al_3 and $Mg_{17}Al_{12}$, the surfaces suffered from corrosion damage with typical pitting and localized corrosion characteristics. Obvious cracks and giant corrosion pits were thus formed on its surfaces (see Figure 8f).

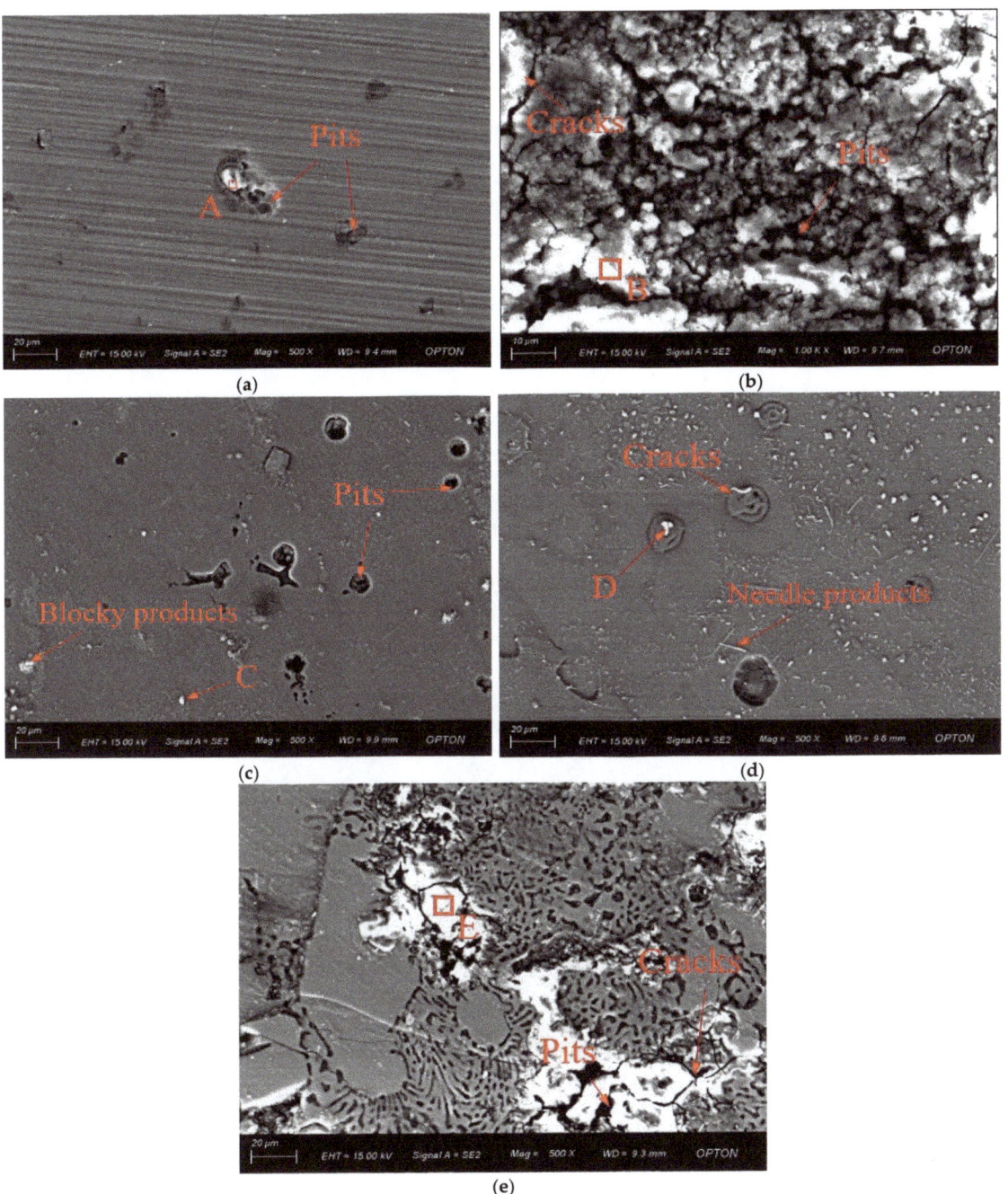

Figure 8. Surface morphologies of the substrates and different diffusion layers after potentiodynamic electrochemical measurements: (**a**) Al1060, (**b**) Mg1, (**c**) Mg_2Al_3 layer, (**d**) $Mg_{17}Al_{12}$ layer, and (**e**) $Mg_{17}Al_{12}$ and Mg-based solid solution layer.

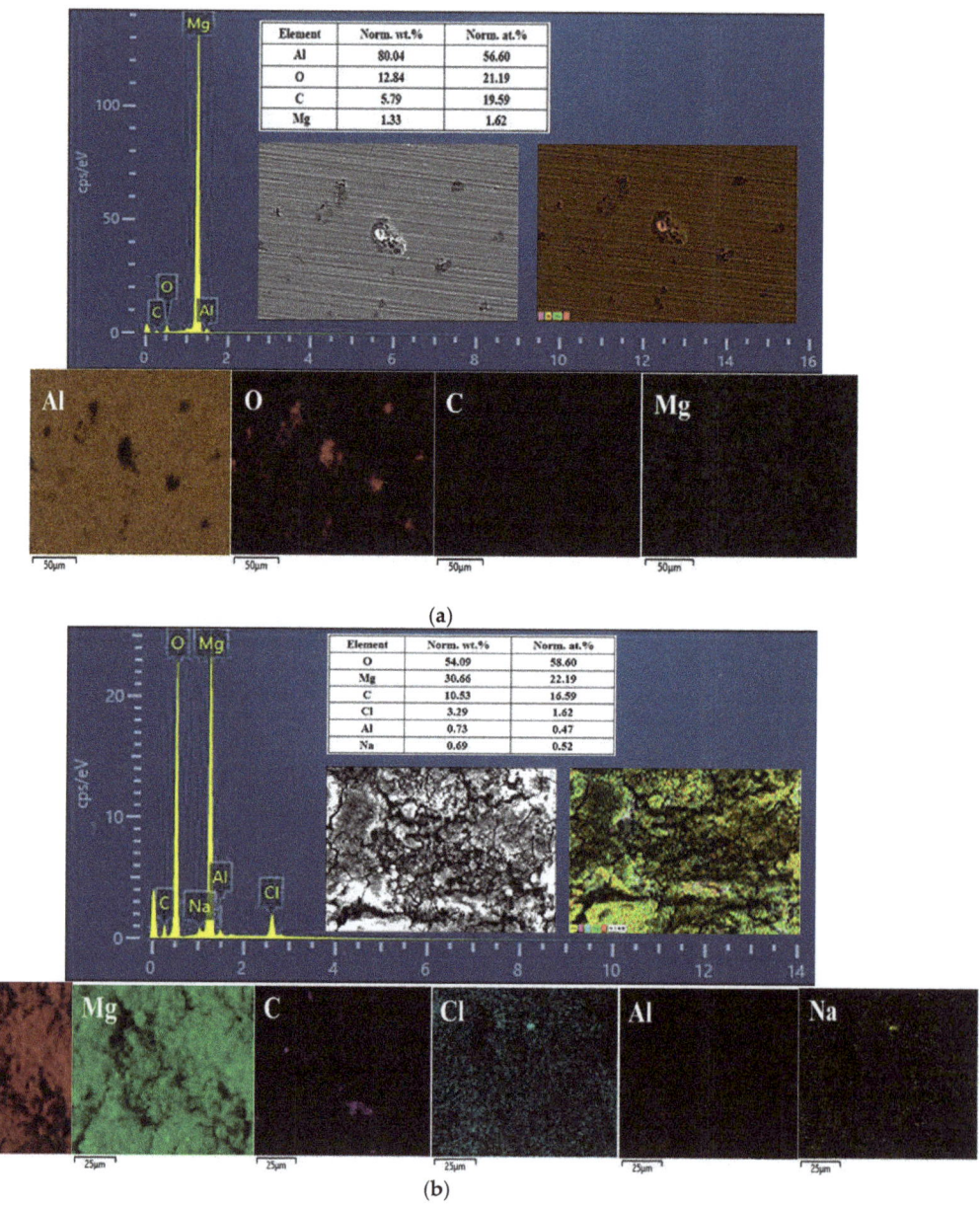

(a)

(b)

Figure 9. *Cont.*

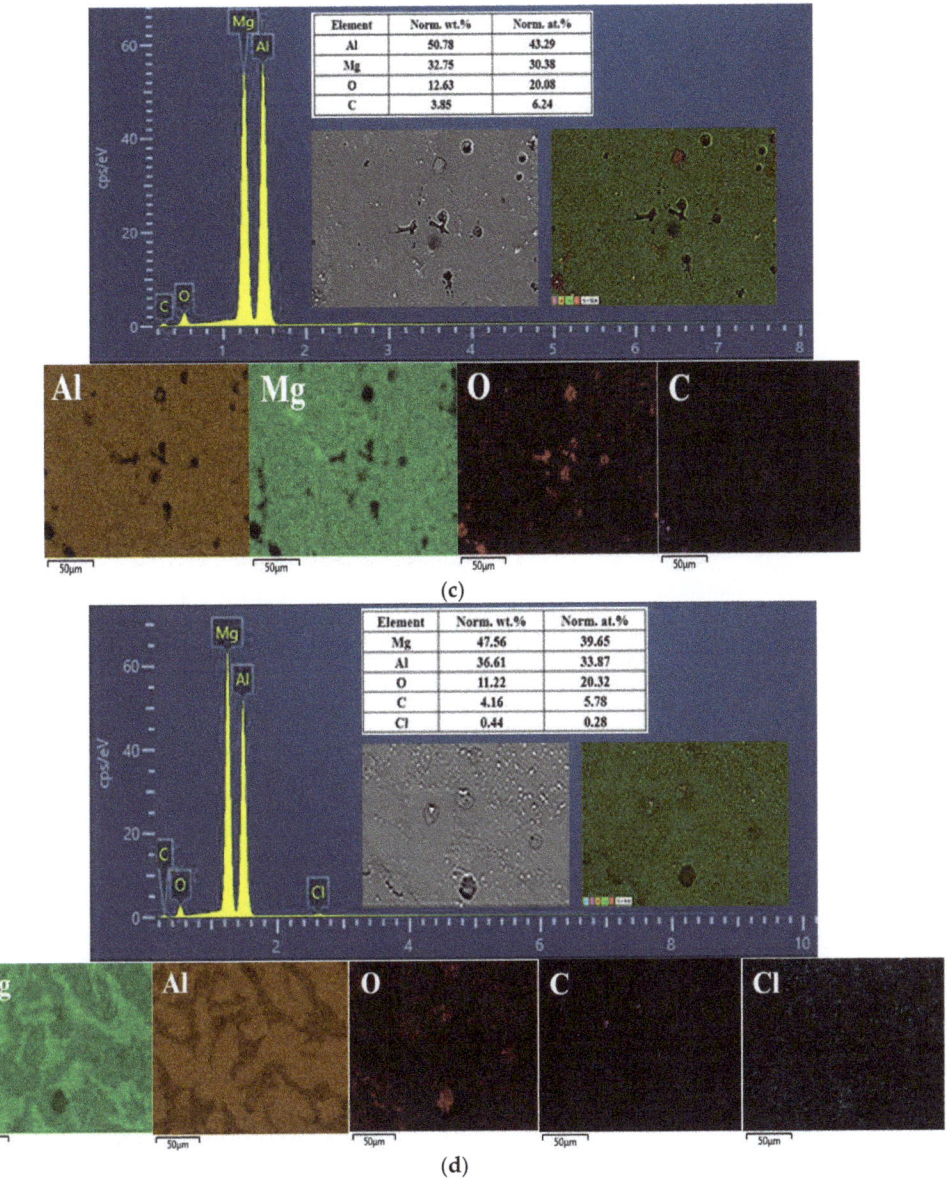

Figure 9. *Cont.*

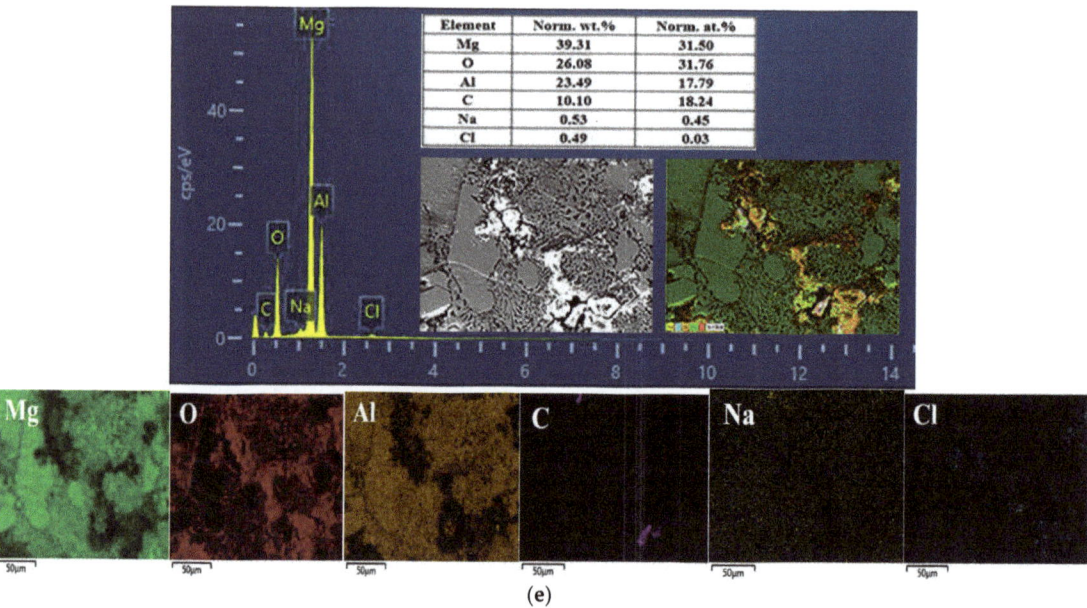

(e)

Figure 9. EDS spectrum and the composition and distribution of elements in the (**a**) Al1060, (**b**) Mg1, (**c**) Mg_2Al_3 layer, (**d**) $Mg_{17}Al_{12}$ layer, and (**e**) $Mg_{17}Al_{12}$ and Mg-based solid solution layer.

EDS determination results provided a more conclusive figure (Figure 9) on which the overall surface composition of the samples was measured. The elemental composition of the sample surface after electrochemical corrosion can be analyzed more intuitively through surface scanning energy spectrum detection [35]. The EDS spectrum and surface composition indicated the presence of aluminum and oxygen only, along with small amounts of carbon and magnesium on the tested Al1060 surface (see Figure 9a). Most areas of its surface have retained their original integrity. The oxygen was mainly distributed in the vicinity of the corrosion pits. Moreover, the presence of magnesium could be attributed to the fact that it is part of the main composition of Al1060, and the presence of carbon could be attributed to the adsorption of CO_2 from the atmosphere. For each sample in Figure 8, Table 5 shows the elemental composition of a surface spot very close to the corrosion products. For example, the EDS analysis of point A shows an Al content of 45.04 wt.% along with an oxygen content of 50.04 wt.%. The corrosion product is, thus, most likely Al_2O_3. In the study by Chakradhar et al. [36], the EDAX studies also confirmed that the hazy white areas on the Al block after electrochemical corrosion relate to aluminum oxide, and its formation can be described by Equations (1)–(3):

$$Al^{3+} + 3Cl^- = AlCl_3 \qquad (1)$$

$$AlCl_3 + 6H_2O = 2Al(OH)_3 + 3HCl \qquad (2)$$

$$2Al(OH)_3 = Al_2O_3 + 3H_2O \qquad (3)$$

The EDS spectrum and the surface composition of Mg1 indicated a very complex composition of the surface (see Figure 9b), in which the magnesium content was relatively small, and the oxygen content was quite large. This indicates that the surface was almost completely covered with a thick layer of compounds. Moreover, the EDS analysis showed that box E was mainly composed of 30.25 wt.% magnesium and 56.42 wt.% oxygen (see Table 5), suggesting that the corrosion product is mainly $Mg(OH)_2$. Since the chemical activity of magnesium is lower than that of sodium, it is impossible for the Mg^{2+} ions to be replaced by Na$^+$ ions in the solution. Thus, it was mainly magnesium that reacted with the aqueous solution, which has taken place as described by Equation (4).

$$Mg + 2H_2O = Mg(OH)_2 + H_2 \qquad (4)$$

Figure 9c,d shows the EDS determination results of the Mg$_2$Al$_3$ layer and Mg$_{17}$Al$_{12}$ layers, respectively; it can be observed that aluminum and magnesium play a dominant role in their surface composition, oxygen is not widely and evenly distributed on their surface, and oxygen was dominating in the vicinity of the corrosion defects (including pits and cracks) as well as in block- and needle-like corrosion products. However, in the surface composition of the tested Mg$_{17}$Al$_{12}$ and Mg-based solid solution layer, magnesium, aluminum, and oxygen predominate (see Figure 9f), and oxygen content reaches 26 wt.%, reflecting the existence of many corrosion products on its surface. Moreover, the elemental composition of the corrosion products of these three diffusion layers was mainly magnesium, aluminum, and oxygen (see Table 5). In conclusion, each diffusion layer's corrosion products were not only single component products. There was probably an occurrence of various oxides and hydroxides of Mg and Al, irregularly attached to the surface.

Table 5. EDS results of the distinct regions are presented in Figure 8.

Position	Mole Fraction/%				
	Mg	Al	O	Cl	C
A	0.31	45.04	50.04	0.93	3.68
B	30.25	0.40	58.42	3.91	7.02
C	21.96	23.48	45.62	2.68	6.26
D	20.47	16.29	48.95	5.62	8.67
E	22.19	12.81	58.57	1.19	4.31

After the potentiodynamic electrochemical measurements, by analyzing the surface morphologies, it was possible to conclude that the corrosion damage of the Mg$_{17}$Al$_{12}$ and Mg-based solid solution layers and of the Mg1 substrate was much more severe than those of the Al1060, Mg$_2$Al$_3$, and Mg$_{17}$Al$_{12}$ layers. As a result, the Mg1 substrate was corroded entirely and destroyed, and a thick layer of corrosion products adhered to its surface. Therefore, these corrosion morphology results have verified the results from the corrosion resistance analysis of the polarization curves presented above.

4. Conclusions

The present study uses a vacuum diffusion welding process to weld Mg1 and Al1060. In addition, several corrosion resistance experiments were conducted on the resulting Al/Mg intermetallic plates, and a series of microstructural observations could also be made. According to the experimental results, the following conclusions could be drawn:

(1) Vacuum diffusion welding could realize the joining of Mg1/Al1060. The microstructure of the joint was excellent, and uniform diffusion layers were formed at the interface after sufficient diffusion of elements in the material structures. The diffusion layers from the Al side to the Mg side were: Mg$_2$Al$_3$, Mg$_{17}$Al$_{12}$, and Mg$_{17}$Al$_{12}$, and a Mg-based solid solution layer.

(2) The results of the corrosion immersion tests have demonstrated that the Mg1 substrate was the first to be corroded in a 3.5 wt.% NaCl solution. Severe corrosion damage occurred on this surface after a short period in the solution. The corrosion rates of the Al1060 substrate and the diffusion layers were, thus, slower. The Mg1 substrate, in direct contact with the diffusion layers, acted as an anode in a galvanic cell. It indirectly protected the diffusion layers, which were the latest to be corroded. Among the diffusion layers, corrosion mainly occurred in the combined Mg$_{17}$Al$_{12}$ and Mg-based solid solution layer.

(3) Linear polarization curves and corrosion morphology analyses also showed that the corrosion resistance of Mg1 was the worst in an aggressive NaCl environment, as compared with the Al1060 substrate and the diffusion layers. It was followed by the combined $Mg_{17}Al_{12}$ and Mg-based solid solution layer. As measured by potential electrochemistry, severe corrosion occurred on the surfaces of these compounds. On the contrary, the Mg_2Al_3 and $Mg_{17}Al_{12}$ layers showed excellent corrosion resistance comparable to that of Al1060. The order of corrosion rate of tested samples was Mg1 > $Mg_{17}Al_{12}$ and Mg-based solid solution > Mg_2Al_3 > $Mg_{17}Al_{12}$ > Al1060.

In this study, the Mg/Al vacuum diffusion layers were extracted separately for the first try, and the corrosion behavior of each diffusion layer and substrate was studied in depth. However, this research is subject to several limitations. The first is the experimental instruments, the electromagnetic interference generated by alternating current in the electrochemical workstation will have a particular impact on the measurement results. Another limitation concerns the characterization method, more research on electrochemical tests (e.g., electrochemical impedance spectroscopy, cyclic polarization, etc.) can be carried out in the future, and the corrosion resistance of samples can be further discussed based on these tests. However, these limitations will not cause significant prejudice to the current research results and will not affect the research.

Author Contributions: Conceptualization, S.Z. and Z.Z.; methodology, Y.D. and D.J.; experiment, S.Z. and Y.D.; data curation, S.Z. and Z.Z; writing—original draft preparation, S.Z.; writing—review and editing, Y.D. and D.J. All authors have read and agreed to the published version of the manuscript.

Funding: This research was funded by Scientific Research Funding Project of the Education Department of Liaoning Province (2019LNQN01) and Doctoral Scientific Research Starting Fund of Department of Science and Technology of Liaoning Province (2021-BS-241).

Institutional Review Board Statement: Not applicable.

Informed Consent Statement: Not applicable.

Data Availability Statement: Not applicable.

Acknowledgments: This research was supported by the "Liaoning Provincial Laboratory for Special Machining of Complex Workpiece Surfaces" and "Magnesium Alloy Rolling Centre", School of Mechanical Engineering, University of Science and Technology, Liaoning. Moreover, the authors would like to express their gratitude to EditSprings (https://www.editsprings.cn (accessed on 22 August 2022)) for the expert linguistic services provided.

Conflicts of Interest: The authors declare no conflict of interest.

References

1. Li, X.R.; Liang, W.; Zhao, X.G.; Zhang, Y.; Fu, X.P. Bonding of Mg and Al with Mg–Al eutectic alloy and its application in aluminum coating on magnesium. *J. Alloy. Compd.* **2009**, *471*, 408–411. [CrossRef]
2. Zeng, X.Y.; Wang, Y.X.; Li, X.Q.; Li, X.J.; Zhao, T.J. Effect of inert gas-shielding on the interface and mechanical properties of Mg/Al explosive welding composite plate. *J. Manuf. Process.* **2019**, *45*, 166–175. [CrossRef]
3. Liu, F.; Wang, H.Y.; Liu, L.M. Characterization of Mg/Al butt joints welded by gas tungsten arc filling with Zn–29.5Al–0.5Ti filler metal. *Mater. Charact.* **2014**, *90*, 1–6. [CrossRef]
4. Maity, S.; Chabri, S.; Chatterjee, S.; Bera, S.; Sinha, A. Micromechanical behavior of β-Al_3Mg_2-dispersed aluminum composite prepared by high-energy ball milling and hot pressing. *J. Compos Mater.* **2016**. [CrossRef]
5. Shan, L.H.; Othman, N.H.; Gerlich, A. Review of research progress on aluminium–magnesium dissimilar friction stir welding. *Sci. Technol. Weld. Joi.* **2017**, *8*, 256–270.
6. Kumar, S.; Wu, G.; Gao, S. Process Parametric Dependency of Axial Downward Force and Macro-and Microstructural Morphologies in Ultrasonically Assisted Friction Stir Welding of Al/Mg Alloys. *Metall. Mater. Trans. A* **2020**, *51*, 2863–2881. [CrossRef]
7. Borrisutthekul, R.; Miyashita, Y.; Mutoh, Y. Dissimilar material laser welding between magnesium alloy AZ31B and aluminum alloy A5052-O. *Sci. Technol. Adv. Mater.* **2005**, *6*, 199–204. [CrossRef]
8. Gao, Q.; Wang, K.H. Influence of Zn Interlayer on Interfacial Microstructure and Mechanical Properties of TIG Lap-Welded Mg/Al Joints. *J. Mater. Eng. Perform.* **2016**, *25*, 756–763. [CrossRef]
9. Zheng, Y.; Pan, X.M.; Ma, Y.L.; Liu, S.M.; Zang, L.B. Microstructure and Corrosion Behavior of Friction Stir-Welded 6061 Al/AZ31 Mg Joints with a Zr Interlayer. *Materials* **2019**, *12*, 1115. [CrossRef]

10. Masoudian, A.; Tahaei, A.; Shakiba, A.; Sharifianjazi, F.; Mouandesi, J.A. Microstructure and mechanical properties of friction stir weld of dissimilar AZ31-O magnesium alloy to 6061-T6 aluminum alloy. *Trans. Nonferrous Met. Soc. China.* **2014**, *24*, 1317–1322. [CrossRef]
11. Afghahi, S.S.S.; Jafarian, M.; Paidar, M.; Jafarian, M. Diffusion bonding of Al 7075 and Mg AZ31 alloys: Process parameters, microstructural analysis and mechanical properties. *Trans. Nonferrous Met. Soc. China.* **2016**, *26*, 1843–1851. [CrossRef]
12. Yin, F.X.; Liu, C.C.; Zhang, Y.G.; Qin, Y.F.; Liu, N. Effect of Ni interlayer on characteristics of diffusion bonded Mg/Al joints. *Mater. Sci. Technol.* **2018**, *34*, 1104–1111. [CrossRef]
13. Gu, X.Y.; Sui, C.L.; Liu, J.; Li, D.L.; Meng, Z.Y.; Zhu, K.X. Microstructure and mechanical properties of Mg/Al joints welded by ultrasonic spot welding with Zn interlayer. *Mater. Des.* **2019**, *181*, 108103. [CrossRef]
14. Tabandeh, L.; Khorramabadi, G.S.; Karami, A.; Atafar, Z.; Sharafi, H.; Dargahi, A.; Amirian, F. Evaluation of heavy metal contamination and scaling and corrosion potential in drinking water resources in Nurabad city of Lorestan. *Int. J. Pharm. Technol.* **2016**, *8*, 13137–13154.
15. Dargahi, A.; Shokri, R.; Mohammadi, M.; Azizi, A.; Tabandeh, L.; Jamshidi, A.; Beidaghi, S. Investigating of the corrosion and deposition potentials of drinking water sources using corrosion index: A case study of Dehloran. *J. Chem. Pharm. Sci.* **2016**, *974*, 2115.
16. Kannan, M.B.; Raman, R. In Vitro Degradation and Mechanical Integrity of Calcium-Containing Magnesium Alloys in Modified-Simulated Body Fluid. *Biomaterials* **2008**, *29*, 2306–2314. [CrossRef]
17. He, M.F.; Hu, W.B.; Zhao, S.; Liu, L.; Wu, Y.T. Novel multilayer Mg–Al intermetallic coating for corrosion protection of magnesium alloy by molten salts treatment. *Trans. Nonferrous Met. Soc. China* **2012**, *22*, 74–78. [CrossRef]
18. Azizi, A.; Alimardan, H. Effect of welding temperature and duration on properties of 7075 Al to AZ31B Mg diffusion bonded joint. *Trans. Nonferrous Met. Soc. China* **2016**, *26*, 85–92. [CrossRef]
19. Zheng, B.; Zhao, L.; Dong, S.J.; Hu, X.B. First Principles Study on Mg-Al Intermetallic Compounds. *Mater. Rep.* **2019**, *14*, 2426–2430.
20. Ma, Y.Z.; Wu, L.; Long, L.P.; Liu, W.S.; Liu, C. Microstructure and mechanic property of Mg/Al joints obtained by vacuum diffusion bonding. *Trans. Nonferrous Met. Soc. China* **2017**, *27*, 1083–1090.
21. Fernandus, M.J.; Senthilkumar, T.; Balasubramanian, V.; Rajakumar, S. Optimizing Diffusion Bonding Parameters in AA6061-T6 Aluminum and AZ80 Magnesium Alloy Dissimilar Joints. *J. Mater. Eng. Perform.* **2012**, *21*, 2303–2315. [CrossRef]
22. Wang, Y.Y.; Luo, G.Q.; Zhang, J.; Sheng, Q.; Zhang, L.M. Microstructure and mechanical properties of diffusion-bonded Mg-Al joints using silver film as interlayer. *Mater. Sci. Eng. A* **2013**, *559*, 868–874. [CrossRef]
23. Shang, J.; Wang, K.H.; Zhou, Q.; Zhang, D.K.; Huang, J. Effect of joining temperature on microstructure and properties of diffusion bonded Mg/Al joints. *Trans. Nonferrous Met. Soc. China* **2012**, *22*, 1961–1966. [CrossRef]
24. Liu, L.; Liu, X.; Liu, S. Microstructure of laser-TIG hybrid welds of dissimilar Mg alloy and Al alloy with Ce as interlayer. *Scripta. Mater.* **2006**, *55*, 383–386. [CrossRef]
25. Peng, L.; Li, Y.J.; Ge, H.R.; Wang, J. Investigation of interfacial structure of Mg/Al vacuum diffusion-bonded joint. *Vacuum* **2006**, *80*, 395–399. [CrossRef]
26. Sato, Y.S.; Park, S.; Michiuchi, M.; Kokawa, H. Constitutional liquation during dissimilar friction stir welding of Al and Mg alloys. *Scripta. Mater.* **2004**, *50*, 1233–1236. [CrossRef]
27. Zhang, M.X.; Huang, H.; Spencer, K. Nanomechanics of Mg-Al intermetallic compounds. *Surf. Coat. Technol.* **2010**, *204*, 2118–2122. [CrossRef]
28. Bu, H.Y.; Yandouzi, M.; Lu, C.; MacDonald, D.; Jodoin, B. Cold spray blended Al + $Mg_{17}Al_{12}$ coating for corrosion protection of AZ91D magnesium alloy. *Surf. Coat. Technol.* **2012**, *207*, 155–162. [CrossRef]
29. Burduhos-Nergis, D.P.; Vizureanu, P.; Sandu, A.V. Phosphate Surface Treatment for Improving the Corrosion Resistance of the C45 Carbon Steel Used in Carabiners Manufacturing. *Materials* **2020**, *13*, 3410. [CrossRef]
30. Song, G.L.; Atrens, A.; Wu, X.L.; Zhang, B. Corrosion behaviour of AZ21, AZ501 and AZ91 in Sodium Chloride. *Corros. Sci.* **1998**, *40*, 1769–1791. [CrossRef]
31. Song, G.L.; Atrens, A. corrosion mechanisms of magnesium alloys. *Adv. Eng. Mater.* **1999**, *1*, 11–33. [CrossRef]
32. Ji, G.; Liu, H.; Yang, G.J.; Luo, X.T.; Li, C.X.; He, G.Y.; Zhou, L.; Liang, T. Formation of Intermetallic Compounds in a Cold-Sprayed Aluminum Coating on Magnesium Alloy Substrate after Friction Stir-Spot-Processing. *J. Therm. Spray Technol.* **2021**, *30*, 1464–1481. [CrossRef]
33. Bu, H.Y.; Yanddouzi, M.; Lu, C.; Jodoin, B. Post-heat Treatment Effects on Cold-Sprayed Aluminum Coatings on AZ91D Magnesium Substrates. *J. Therm. Spray Technol.* **2012**, *21*, 731–739. [CrossRef]
34. Spencer, K.; Zhang, M.X. Heat treatment of cold spray coatings to form protective intermetallic layers. *Scr. Mater.* **2009**, *61*, 44–47. [CrossRef]
35. Bejinariu, C.; Burduhos-Nergis, D.P.; Cimpoesu, N. Immersion Behavior of Carbon Steel, Phosphate Carbon Steel and Phosphate and Painted Carbon Steel in Saltwater. *Materials* **2021**, *14*, 188. [CrossRef]
36. Chakradhar, R.P.S.; Mouli, G.C.; Barshulia, H.; Srivastava, M. Improved Corrosion Protection of Magnesium Alloys AZ31B and AZ91 by Cold-Sprayed Aluminum Coatings. *J. Therm. Spray Technol.* **2021**, *30*, 371–384. [CrossRef]

Review

Sustainable New Technology for the Improvement of Metallic Materials for Future Energy Applications

Patricia Jovičević-Klug [1,2,*] and Michael Rohwerder [1]

[1] Group of Corrosion, Department of Interface Chemistry and Surface Engineering, Max-Planck-Institute for Iron Research, Max-Planck-Str. 1, 40237 Düsseldorf, Germany; rohwerder@mpie.de
[2] Alexander von Humboldt PostDoc Research Fellow, Jean-Paul-Str. 12, 53173 Bonn, Germany
* Correspondence: p.jovicevic-klug@mpie.de

Abstract: The need for a more sustainable and accessible source of energy is increasing as human society advances. The use of different metallic materials and their challenges in current and future energy sectors are the primary focus of the first part of this review. Cryogenic treatment (CT), one of the possible solutions for an environmentally friendly, sustainable, and cost-effective technology for tailoring the properties of these materials, is the focus of second part of the review. CT was found to have great potential for the improvement of the properties of metallic materials and the extension of their service life. The focus of the review is on selected surface properties and corrosion resistance, which are under-researched and have great potential for future research and application of CT in the energy sector. Most research reports that CT improves corrosion resistance by up to 90%. This is based on the unique oxide formation that can provide corrosion protection and extend the life of metallic materials by up to three times. However, more research should be conducted on the surface resistance and corrosion resistance of metallic materials in future studies to provide standards for the application of CT in the energy sector.

Keywords: energy sector; renewable energy; fusion; metallic materials; cryogenic treatment; surface; interface; corrosion

Citation: Jovičević-Klug, P.; Rohwerder, M. Sustainable New Technology for the Improvement of Metallic Materials for Future Energy Applications. *Coatings* **2023**, *13*, 1822. https://doi.org/10.3390/coatings13111822

Academic Editor: Alessandro Latini

Received: 3 October 2023
Revised: 20 October 2023
Accepted: 23 October 2023
Published: 24 October 2023

Copyright: © 2023 by the authors. Licensee MDPI, Basel, Switzerland. This article is an open access article distributed under the terms and conditions of the Creative Commons Attribution (CC BY) license (https://creativecommons.org/licenses/by/4.0/).

1. Introduction

With the growth of the human population, there is an increasing need for a more sustainable and more easily accessible source of energy [1], bringing prosperity, economic development, security, better health care, welfare, and the overall better social and environmental development of mankind [2]. In recent years, many challenges, such as the distribution of natural resources, growth of the population and its needs, economic instability, new war zones, etc. [3,4] have emerged in energy sources based on oil, gas, and coal. These challenges combined with geo-political challenges and environmental challenges such as greenhouse gases, environmental impact, sustainable development, etc. [2,5], are leading to increased efforts in research and the development of new solutions and options for new and more sustainable energy sources. The energy sources can be classified into natural fossil-based (oil, gas, and coal) and renewable types [6]. It is important to note that nuclear energy can be grouped on its own or as part of one of the previously mentioned groups. This is a highly controversial topic, mainly based on whether conventional or advanced nuclear power is discussed [6–9]. In this review, nuclear energy will be grouped on its own.

The current prediction of energy sources for the next 20 years (2030–2040) in the European Union (EU), predicted in the year 2020 [10,11], is shown in Figure 1. Currently, the production of energy is still dominated by fossil-fuel-based sources (70%). The nuclear-based sources have consistently maintained a similar ratio, while the renewable sources are constantly gaining an increasing share. For the next two decades, an increasing use of renewable sources for energy production is predicted to increase by 300% in the EU alone by 2030 and by up to 450% by 2040 compared to the current state [10,11]. The nuclear energy source is predicted to remain a stable energy source throughout this period, especially if fusion is included [6–8]. Figure 1 and Table 1 also show the development of renewable energy sources in the EU over the next 20 years and the expected changes in the redistribution of energy sources within different sectors.

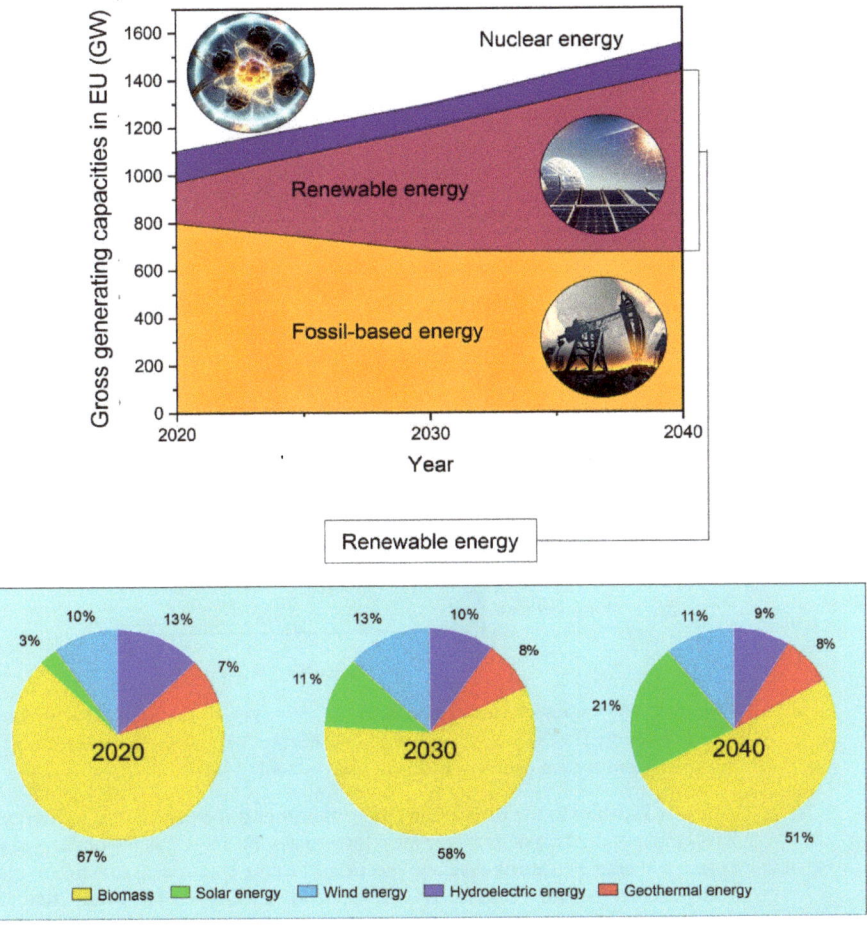

Figure 1. The square graph presents the prediction of gross energy-generating capacities within the EU from current state of 2020 up to the next 20 years (2030–2040) by each sector. The orange color stands for fossil fuels (oil, gas, and coal), red for renewables, and purple for nuclear. The lower pie charts represent the current (year 2020) and predicted (years 2030 and 2040) fractions of the different categories of renewable energy sources.

Production of Electricity per Year for Different Energy Sector

The production of electricity varies within the different categories of the energy sector, and also, the production costs can vary significantly depending on the energy source (Figure 2 and Table 2). Pricing is highly dependent on a variety of external factors, such as subsidies, various taxes, etc., which also vary depending on geopolitical locations and natural resources.

Table 1. Renewable source preconditions for next 20 years for the EU.

	Current State 2020	2030	2040
Hydroelectric energy			
• Land hydroelectric energy (rivers and lakes) • Marine hydroelectric energy (ocean currents, tidal stream, and waves)	13%	10%	9%
Geothermal energy	7%	8%	8%
Biomass	67%	58%	51%
Solar energy			
• Solar thermal • Photovoltaic	3%	11%	21%
Wind energy	10%	13%	11%
Combined contribution of the renewable energy and the total energy production	~25%	~35%	~48%

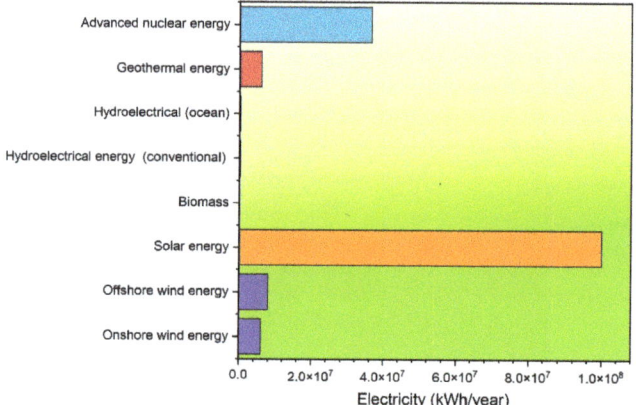

Figure 2. The highest electricity production in kWh/year is possible with solar energy [12,13] and nuclear energy [14]. This is followed by wind energy [15] and geothermal energy [16]. The lowest electricity production comes from the biomass sector [17,18].

In addition to the price of electricity, another important factor in energy production is also the so-called capacity factor (also known as CF). The CF is the unitless ratio of the actual electrical energy output over a given period to the theoretical maximum electrical energy output over the same period. The CF can also be thought of as production efficiency. CF is usually calculated over a year in order to average out temporal variations and to represent the realistic values of energy/electricity produced per maximum capacity of an energy source (Table 2). From Table 2, the cost of electricity for renewable energy sources varies due to different economic perspectives and also the source of production and maintenance and repair costs that need to be considered. The next capacity factors show that the most promising renewable energy sources are geothermal [16], hydroelectric [19–21], and wind [15,22], based on the lowest possible electricity cost and the highest capacity factor.

To cope with the increasing energy demand and consumption, new energy production pathways are expected to evolve and develop to provide greater energy security, reduce global carbon emissions, and lower the financial cost of energy production. The advancement of production processes in the energy sector will require the development and utilization of new materials. This is where metallic materials (metals and alloys) come into play, many of which have been scarcely used or even unused in energy production and will become the most important players in sustainable energy production (see Figure 1 and Table 1).

Table 2. Energy sectors presented in this manuscript and their cost of production and capacity factor [23].

Type of Energy Sector	Cost (EUR/MWh)	Capacity Factor (%)
Hydroelectric energy		
• Land hydroelectric energy	53–326	31–66
• Ocean hydroelectric energy	130–280	39–45
Geothermal energy	49–353	80–90
Biomass	128	64
Solar energy	27–130	12–30
Wind energy		
• Onshore wind energy	24–67	29–52
• Offshore wind energy	60–130	12–48
Advanced nuclear energy	73	94

The successful implementation of such materials will be particularly crucial in applications where high strength and dimensional flexibility combined with high temperature resistance are required. Today, more than 60 different metallic materials are used in one way or another in energy production (as base materials for reactors, storage and accumulation systems, and supporting infrastructure) [24]. Future energy security (Figure 3) requires a critical awareness of the availability, functionality, substitutability, recyclability, and production of metals and alloys [25–27]. Metallic materials are the type of materials that can be newly produced or reused and recycled. Additionally, their properties can be tailored through postprocessing, increasing their flexibility and versatility for various applications. Therefore, the adaptation and development of heat treatment and further processing steps of metallic materials for future applications must be under constant research [25–27]. The factors that influence the value of metallic materials are market availability, substitutability, recyclability, and socio-cultural and environmental impacts [26]. In addition, the development and consideration of new materials such as high-entropy alloys and materials for catalysis, energy generation, and storage applications will bring new challenges and benefits to the energy sector [28,29].

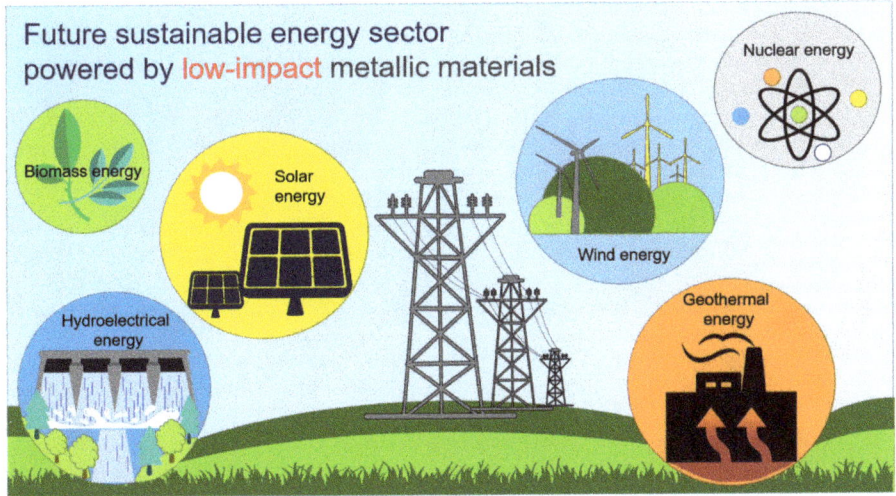

Figure 3. The future sustainable energy sector powered by low-impact metallic materials.

This review aims to provide a comprehensive overview of the metallic materials used in the advanced nuclear and renewable energy sectors of the future, including the challenges (environments) to which these materials are and planned to be exposed on a daily basis. Particularly, this review aims to present a possible novel way of processing metallic materials in a more environmentally friendly way, opening the possibility of improving material properties, extending the life cycle of components, and generating lower CO_2 emissions compared to conventional pathways.

The paper is structured in the following order: in Section 2, the future energy sector is presented, where the renewable energy and advanced nuclear energy sources are introduced and discussed. In the same chapter, the challenges and requirements of the metallic materials used in the given environments of the different energy sources are also presented. Section 3 introduces the emerging green technology of cryogenic heat treatment, which has a great potential to reduce the impact of the energy sector on the environment by improving material properties and extending the life cycle of components. Section 4 discusses the outlook for future technology and the energy sector and provides individual guidelines for future materials implementation in the energy sector.

2. Future Energy Sector

The future energy sector is divided into renewable energy sources and advanced nuclear energy based on fusion. The renewable energy sector is divided into hydroelectric energy, geothermal energy, and biomass, solar, and wind energy, with some subdivisions (see Figure 3 and Table 1).

2.1. Energy Sector and Selection of the Right Material

2.1.1. Advanced Nuclear Energy

For future advanced nuclear reactors, three major subtypes are considered: non-water-cooled reactors, advanced water-cooled reactors, and fusion reactors. The latter's design will be based on the current fission reactors; therefore, the use of similar metallic materials is expected. However, the research into the development of new, more resistant metallic materials and their protection is ongoing [30,31].

The non-water-cooled reactor subtype includes reactors and systems that are still based on fission reaction, but the coolants are either molten salts and are designated as Th-based reactors or high-temperature gases (or cooled with helium, using graphite as a moderator) [30,31]. This category also includes small modular reactors (SMRs), which are fast-cooled reactors based on Na, Pb, and gas cooling [32].

The next category is advanced water-cooled reactors, also based on the fission reaction and using water as a coolant and moderator, i.e., SMRs [30,31]. These reactors are cleaner, fundamentally safer, more fuel efficient, more reliable, and more sustainable than the current generation of reactors [33,34].

The last category is fusion reactors. Fusion reactors are based on fusion plasmas, which are still in the early stages of research and development. The working conditions of these reactors are much more intense compared to those of the fission reactors, reaching temperatures of several thousand degrees during the plasma generation. Therefore, complex testing and development of metallic materials is required for the construction and operation of such reactors. In fact, there are two types of fusion plasmas being considered for the development of future fusion reactors. The first type is based on strong magnetic fields and is known as magnetic confinement fusion (MFC). The second type is based on compressing the deuterium (DT) fuel and heating it rapidly that fusion occurs before the fuel expands; this method is also known as inertial confinement fusion (ICF). Due to the different primary principle compared to fission, the material for fusion application must have a specific non-equilibrium thermodynamic state, two or more main phases, complex grain boundaries; and dislocation systems. Compared to the traditional fission system/environment, the new challenges for metallic materials used in fusion will mainly focus on adaptation to the higher resistance to neutron irradiation, cladding, and higher temperatures and stresses [30,31].

The benefits of a new, advanced generation of nuclear reactors can bring society lower energy costs, increased production, decarbonization of industry, increased efficiency, and significantly reduced environmental and waste hazards. The development and safety of future nuclear power systems depends not only on the type of fuel but also on the material design. In current and future advanced nuclear power, materials are exposed to high-energy particles, high temperatures, pressure changes, and highly corrosive environments. However, the degradation of metallic materials in this environment is complex due to the different materials used for the plant, the complex and highly variable environmental conditions, and the different loading conditions of the material in various applications [30].

The first factor to consider is the thermal ageing and fatigue of metallic materials that occurs in metals exposed to elevated temperatures. This is a critical aspect for metallic materials used in nuclear energy, as it can lead to a short life cycle of the metallic component. The reason for this is the altered microstructure, resulting from the diffusion activated process, which causes changes in mechanical properties and fatigue (including creep fatigue) [30,31].

The next factor (second) is irradiation, which causes changes in the dimensional stability of metallic materials and thus influences the final properties of the metallic component. This is caused by the following five radiation processes: (i) phase (microstructural) instability induced by neutron irradiation, forming increased precipitation and segregation of alloying elements; (ii) radiation-induced hardening and embrittlement; (iii) volume swelling due to void formation; (iv) high-temperature helium embrittlement caused by helium movement towards grain boundaries; and (v) irradiation creep caused by changes in the crystal lattice due to migration of interstitial atoms and dislocations [30,31].

The third factor is the water environment, where water is the primary reactor coolant. Exposure of metallic materials to water, especially with elevated temperatures, can lead to corrosion of metallic materials, which can cause degradation of properties and lead to component failures. The extent of the corrosion is a product of several factors, such as water pH, water purity, material composition, temperature, gas concentrations, etc.). The type of corrosion mechanisms can be divided into general and localized corrosion. General corrosion mechanisms include uniform corrosion, boric acid corrosion, erosion corrosion, and flow-accelerated corrosion. Localized corrosion mechanisms include crevice corrosion, galvanic corrosion, pitting, environmental assisted cracking, and biological corrosion. In addition, stress corrosion cracking (intergranular and transgranular stress corrosion cracking and low-temperature cracking) can also be present under high-loading conditions. In addition to water, molten salts and liquid metals can also cause corrosion and electrochemical reactions that affect the degradation of the metallic material (see Table 3) [30,31].

2.1.2. Hydroelectric Renewable Energy

Hydroelectric renewable energy currently represents 13% of all renewable energy sources in the EU [21]. Hydropower is based on the movement of water through a turbine, which in turn drives a generator to produce electricity. Hydroelectric power plants can be installed in oceans or on rivers and lakes, which are sources of continental hydroelectric power. Ocean electricity sources can be divided into wave energy, tidal current energy, tidal barrage, ocean thermal air conditioning, and ocean thermal energy conversion (also known as OTEC), where the last two options additionally produce heat.

OTEC relies on the steam from the warm surface water to interact with the turbines. However, OTEC is still at an early stage of development, and the application relies on the cold, deep ocean water, which condenses the steam back into water for reuse. For this application to be viable, there must be a temperature difference of at least 20 K between the two layers of the ocean water (surface/deep layer), which is mostly limited to the tropical regions. OTEC can also be combined with ocean thermal air conditioning [35].

Ocean thermal air conditioning can be used to control the air conditioning of buildings by using cold, deep water to cool the fresh water circulating through a building. The cold water must be between 277 K and 284 K. The same technique can also be used in lakes [36,37].

The next option for harvesting electricity from the ocean is tidal barrages. Tidal barrages are based on the normal hydroelectric concept, where instead of the typical river flow, the tide drives the turbines in the barrage and then generates electricity. However, the tidal difference between high and low tide must be at least 3 m for this technique to be viable [38,39].

The next option for generating electricity is tidal stream (current) energy, which uses similar technology as tidal barrages but relies on tidal currents [40]. In this type of installation, the turbines are anchored to the seabed or can be suspended from a buoy to generate electricity. The challenge of this technique is the preservation of the marine environment when deploying this type of energy solution [41–43].

The next type of the energy source is produced by wave power. This type of energy harvesting is based on the prediction of constant wave direction. It is estimated that wave power of solely 2100 TWh per year could be generated by harvesting the naturally occurring waves. To harvest wave energy, cells based on the pressure and then movement of hydraulic pumps are built, which then drive the generator into motion. However, the application of this technology is limited to the areas with constant waves [44–47]. In development are also hybrid wave and wind energy farms [44,48] that use a combination of the aforementioned techniques. The last type of the ocean-based hydroelectric renewable energy is related to the use of osmotic power, which is based on the salt concentration difference between seawater and fresh river water [49–51]. There are two known methods: one is reverse electrodialysis (known as RED) [50,52], and the other is pressure-retarded osmosis (PRO) [53,54].

The continental type of hydroelectric renewable energy can be divided into three parts. The best known and that with the longest tradition is hydroelectric power generation with dams on large rivers or lakes [55]. However, due to its environmental impact, the future of small hydro (also known as SHP) is a promising source of renewable and clean energy that provides significantly lower changes to the environment and disruption of the local ecosystem [49,55–57].

The next option for generating electricity from rivers is run-of-the-river (ROR) hydropower, where the natural flow of the river generates electricity, and the flow of the river determines the amount of electricity generated [58–60]. This type of source is ideal for streams and rivers that can sustain a minimum flow compared to other types. This type delivers cleaner power and generates less greenhouse gas for equivalent energy production compared to other types. Additionally, because these types do not require a reservoir, there is a reduced influence on the environment and flooding [61].

The last option is a special type of hydroelectric power generation using turbines that are completely submerged and, in some cases, anchored to the river bed [62,63]. The advantages of this type are its high efficiency, low maintenance, and high reliability compared to other hydroelectric types [64]. However, the environmental impact can be controversial, especially on the flora and fauna of the marine/lake/river environments [65].

The metallic materials used in the hydroelectric sector must be able to withstand the different environments (low temperature, high pressure, and highly corrosive environment (including chemical agents)) and also have low density, high strength, high toughness and high resistance to wear (especially abrasion), high resistance to corrosion, and high fatigue resistance [66]. The metallic materials used for the turbines are austenitic stainless steel (over 12% Cr content as an alloy), but the turbine blades can also be made of martensitic stainless steel due to the higher strength of the steel [66,67]. Low-head machine parts are made of weathering steel (Corten steel) and various types of stainless steel [68,69], high-strength steel [56,70], ACSR steel [71], cast steel [68], carbon steel [72–74], and ferritic steel [69] (see Section 2.2 for detailed description of metallic materials). Resistance to erosion and cavitation are also important factors to consider for metallic materials used in the hydroelectric sector [66,75,76]. In addition, due to the biologically active environment of hydroelectric plants, biofouling occurs where invasive species grow and accumulate various bacteria [66]. As a result, the metallic materials need to either have self-cleaning capabilities or can be treated with surface specific process to provide resistance to biofouling. Alterative metallic materials that can also be used for turbines and components are also different Al alloys (including Al-Si alloys) [66,71,77], Ti alloys [66,78], Cu alloys [76,79,80], and Ni alloys [81] (see Table 3).

2.1.3. Biomass Renewable Energy

Biomass is a type of energy based on organic sources (animals and plants), the most common sources being plants, waste, and wood. Biomass energy can be used for electricity, heating or even biofuel [82–84].

One type of energy production is thermal conversion, where raw materials (paper or waste) are burned (pyrolysis, gasification, anaerobic decomposition, torrefaction, and co-firing) [83–87]. Most of the research and its emphasis has been on the pyrolysis technique, where the combustion of organic material is carried out in the absence of oxygen under temperatures up to 1173 K [85,88]. Different types of catalysts are used for the pyrolysis of different types of source based biomasses (waste, plants, etc.) [89,90].

Biomass also has great potential to be used in the production of steel for reducing the CO_2 impact of the steel industry, but this is still being researched [84].

The main challenges that are facing metallic materials in the biomass sector are corrosion (pitting corrosion, intergranular corrosion, etc.) [89], high-temperature corrosion, and microbial-assisted corrosion [89,91]. The factors that strongly influence all types of the corrosion are fluid dynamics (including different solutions), gas composition (N_2, CO_2, H_2O, Cl_2, H_2, etc. [90]), deposit composition, and temperature (573–1173 K) [89,90].

Other challenges that metallic materials face in biomass energy are abrasive wear [92], material degradation due to small dusty particles [93], and thermomechanical fatigue (high-temperature fatigue) [94–96].

The most common metals used in the biomass energy sector are carbon low-alloyed steels [97,98], austenitic and martensitic stainless steels [97], Al and Ni alloys [97], and some specialized non-ferrous alloys, such as nickel-cobalt-aluminum alloys (NCA). All these metallic materials are used for the main structure and various components in the biomass sector [82–84] (see Table 3).

2.1.4. Onshore and Off-Shore Renewable Energy

The highest demand and largest source of renewable energy in the global market is currently onshore and offshore wind [99]. Wind energy is based on the conversion of kinetic energy into rotational energy, which is then converted into electrical energy by means of a shaft [100]. It is important to note that some turbine designs can produce more energy compared to others (height, size, etc.). The other way to produce more energy is the so-called yawing technique, where the wind turbine is shifted to face directly into the wind, which can be freely manipulated on demand [101].

Wind energy can be produced both onshore and offshore. Onshore wind energy encompasses the energy produced by the wind on land that comes from the natural movement of air [102]. The advantages of onshore wind energy are cost-effective energy, faster installation and easier maintenance, and a reduced environmental impact compared to offshore energy [102]. However, there are disadvantages to onshore wind energy, such as lower power generation, inconsistent wind, varying wind speeds, and a greater impact on nature [102].

Offshore wind energy, i.e., offshore wind farms, are located out at sea where the wind (sea breeze) has higher speeds and greater consistency [102]. The advantages of this type or, more accurately, placement of wind farms include more space for placement, reduced environmental impact (although this is debatable due to the more complicated and invasive supporting infrastructure), and their greater efficiency compared to onshore turbines. The disadvantages are their higher maintenance and repair costs and higher construction costs [101]. Offshore wind turbines can be anchored to the seabed, known as fixed-bottom turbines, or they can be placed on floating platforms, known as floating turbines [101,102].

Metallic materials are used in wind energy for the main foundation of the structure, tower, gearbox, turbines, bearings, bolts, controllers, casings, and many other components that require high corrosion and wear resistance, resistance to higher loads and dynamic forces (including fatigue), and higher contact pressures [103–105]. The material must also be highly resistant to solid particle erosion caused by dust particles [106].

For these reasons, the most common metallic materials used in the wind renewable energy sector (see Section 2.2) are structural steels, stainless steels (austenitic, duplex, and martensitic steels), electrical steels, cast iron, bearing steels, high Cr steels, low C nitrogen steels, and high Si nodular cast iron [103–105]. The most common non-ferrous alloys used in this sector are Cu-based alloys and Al-based alloys [103–105] (see Table 3).

2.1.5. Solar Renewable Energy

The next renewable energy sector is solar energy, which is based on the system of harvesting solar radiation (photovoltaic) for electricity or collecting solar thermal energy for heating [107]. The photovoltaic (PV) system for harvesting solar radiation is primarily based on solar panels, where solar radiation is absorbed by PV cells in the solar panel. This then creates electrical charges that move and respond to an internal electrical field in the PV cell, causing electricity to flow [108,109]. In addition, photocatalysis can provide additional support for solar energy production and storage by inhibiting the conversion of collected light after exposure when there is insufficient light incoming to the solar panel (night time, low radiance angle, or weather-related obstruction) [110,111].

Solar energy can also be harnessed using concentrating solar thermal power (also known as CSP), which uses a system of mirrors to reflect, concentrate, and convert solar energy into heat or even electricity [108,109].

The advantage of solar energy is that it is the most abundant natural energy source in the world, and solar energy can provide a solid and increasing output efficiency compared to other sources. It has minimal harmful effects on the environment, although this can also be highly controversial [108].

The future of solar energy is also being explored in the context of hydrogen production, which can later be used as a clean energy carrier [112]. H_2 production from sustainable solar energy is a possible environmentally friendly solution for the increasing demand for energy and fuel as well as energy storage and transportation. The production of H_2 from solar energy would be achieved by solar thermolysis and then by electrolysis from solar–thermally produced H_2 and photovoltaic-based hydrogen production [112]. Such systems will also require new adaptations of metallic materials to adapt to new challenges in terms of maintenance and repair in correlation to the high temperatures, hydrogen presence that can cause hydrogen embrittlement, and corrosive environments [112].

The challenges that metallic materials face in the solar energy sector include the corrosion effect between molten salts and thermal storage materials [113], high-temperature corrosion [114], mechanically assisted corrosion [114], localized corrosion (stress corrosion cracking and flow-accelerated corrosion) [114], creep fatigue [115,116], erosion [117,118], oxidation [117], and mechanical properties [119,120].

The most commonly used ferrous alloys in the solar energy sector are austenitic and martensitic stainless steels, carbon steels, Cr-Mo steels, duplex steels, FeCrAl steel, and ferritic-martensitic steels (see Section 2.2). The most-used non-ferrous alloys are Ni alloys, which represent more than 60% of all non-ferrous alloys used in this sector. The other non-ferrous alloys are Al-based alloys, high-entropy alloys (HEA), and Mg-based alloys [121] (see Section 2.2) (see Table 3).

2.1.6. Geothermal Renewable Energy

Geothermal renewable energy is the last renewable energy herein presented. Geothermal energy is a type of thermal energy that originates from the formation of the planet and from radioactive decay of elements [122–124]. The Earth's internal thermal energy flows to the surface by conduction at the rate of 44.2 TW [125] and by radioactive decay at the rate of 30 TW [123]. The output of geothermal energy can currently meet twice the current energy demand from all energy sources (including non-renewable sources), but the challenge lies in the non-renewable energy flow and its extraction [123]. Harvested geothermal energy can be used for electricity or for heating/cooling.

There are several ways to produce electricity directly from geothermal energy. The first and the oldest type is the dry-steam power plant, which is based on the underground steam source [124,126]. The next type is the flash-steam power plant, where the source is underground water (>180 °C) and steam. This is the most common type of electricity source based on geothermal energy [127,128]. The next type is an enhanced geothermal system, which uses fracturing, drilling, and injection to extract fluid from the subsurface, which is then used for heating and electricity generation [129,130]. The next type is the binary cycle power plant, where water is heated underground (100–180 °C), and then, the hot water circulates above ground and heats a liquid organic compound that has a lower boiling point than water. This compound then produces steam, which flows into the turbine and powers the generator to produce electricity [122,124,131].

In addition to electricity, geothermal energy can be used for heating and cooling. Thermal energy can be extracted from low-temperature geothermal plants, co-produced geothermal energy, or by geothermal heat pump [124]. The first type, low-temperature geothermal energy, is based on the extracting energy from the low-temperature pockets (around 150 °C) located a few meters below the surface [132]. The next type is the so-called co-produced geothermal energy, where heat is produced by water that has been heated [133,134]. The last type is the geothermal heat pump, which is installed at a depth of 3 to 90 m. In this system, the temperature difference between both ends of the system is used to transfer energy by either heating or cooling the upper part of the system [135,136].

One of the advantages of geothermal energy compared to other sources is that it can be harvested almost anywhere in the world. Additionally, the power plants can last for decades with proper maintenance, and because there is no seasonal variation in workload, the system can be adapted to different conditions depending on the application and environment [122,124].

The challenges facing metallic materials used for geothermal energy are corrosion (uniform corrosion, pitting corrosion, crack corrosion, stress corrosion cracking, sulfur-assisted corrosion cracking, and galvanic corrosion) [137,138], hydrogen bubbling [139,140], corrosion fatigue [141,142], fatigue [137,138], erosion [140,142–144], wear [145], high pressure [144], high temperature [144], and cavitation and decomposition of alloy structure [138]. The most common metallic materials used in the geothermal energy sector are duplex steels, austenitic and superaustenitic stainless steels, martensitic stainless steels, low-alloyed steels, carbon steels, superferritic steels, Cu-based alloys, Ni-based alloys, and Ti-based alloys (see references in Section 2.2) (see Table 3).

2.2. Cost of Maintenance and Repairs in Future Energy Sector and Search of the Solutions for Lowering the Costs

Cost of Maintenance and Repairs of Future Energy Sector

Maintenance costs vary for each of the described energy sectors due to the different technologies used to produce electricity or heat (see Figure 4a). For advanced nuclear power, projections are based on current nuclear power sources and can be up to 20% of the initial investment [146,147]. For biomass energy, maintenance costs are estimated to be up to 35% due to the unique environment. However, different technologies require different levels of maintenance and servicing of the components, so maintenance costs can also be as low as 15% of the investment over time [148,149].

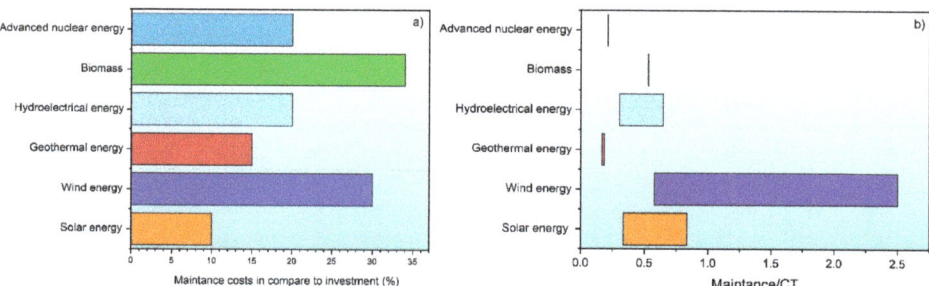

Figure 4. (**a**) Maintenance costs for each energy source type over time. Note that the figures are the maximum value that maintenance can cost in terms of investment over time, but this can vary due to different techniques. (**b**) Maintenance costs for each energy sector normalized by their corresponding maximum and minimum CF. The range of CF was taken from Table 2.

Maintenance costs for hydropower can vary from 1.5 to 20% depending on the type of technology used and the environment in which the plant is located [150,151]. The next sector is geothermal sector, where the maintenance costs can reach up to 15% of investment over the course of the life span of a typical power plant [152,153]. In the wind energy sector, the onshore or offshore location of the wind farms plays an important role in maintenance and repair costs, which can vary the costs somewhere between 20% and 30% of the total investment [103,154]. The last sector is solar energy, which has the one of the lowest maintenance costs (up to 10%) compared to all sectors. However, it is important to note that this does not reflect how much energy is actually produced by it (see Table 2) [155].

In order to put the maintenance cost in relation to the actual output of the individual energy source, the maintenance costs are normalized by the CF of the individual energy source, which is presented in Figure 4b. As can be seen, the wind energy has a very wide range due to the high maintenance costs that can be associated with specific maintenance issues of a wind farm. The solar, hydroelectric, and biomass sources show an intermediate influence of the maintenance costs, while geothermal and advanced nuclear energy have the lowest influence of maintenance costs in relation to their effective production capacity. This Figure 4 clearly shows that the improvement of materials will play an important role in the development and improved cost reduction of renewable energy power plants.

2.3. Metallic Mateirals Used in the Energy Sector

Table 3 summarizes the various metallic materials used in the advanced nuclear and renewable energy sectors.

Table 3. The list of metallic materials used in advanced nuclear energy and renewable energy sectors.

Energy Sector	Metallic Materials		Application
	Ferrous Alloys	Non-Ferrous Alloys	
Advanced Nuclear Energy		Zr-based alloys (Zircaloy-4 (Zr-Sn-Fe-Cr), Zirlo, and M4 (Nb-based alloy)) [30,31]	Water reactors [30,31]
Advanced Nuclear Energy	Austenitic stainless steel (AISI 316 [31], AISI 316 SS [31], AISI 316L [156], AISI 316 LN [157], AISI 304 [31,157–160], AISI 304L [161–164], AISI 304 N [161], AISI 304 SS [31], AISI 347 [31], AISI 308L [157], AISI 308 SS [31], AISI 310 SS [156,165], AISI 309L [31], AISI 309 SS [31], AISI 321 SS [31], AISI 403 [31], AISI 410 [31], AISI 347 SS [31], AISI 630 [31], AISI D9 [31], HT-UPS [31], AISI 4340)		Water reactors, piping, pressurizer, steam generator, pump, valve casing, plunger, control rod drive mechanism, and core internal structure [30,31]
Advanced Nuclear Energy	Cast-austenitic stainless steel (CF3, CF3A, CF3M, CF8, CF8A, CF8M, AISI 304 SS, AISI 304L SS, AISI 316 SS, AISI 316L SS, AISI 321 SS, AISI 347 SS) [31]		Primary cooling piping system, reactor coolant, auxiliary system, pump casing, valve bodies, and cooling circuit [30,31]
Advanced Nuclear Energy		Ni-based alloys (600 [161,166,167], 690 [161,168], 625 [31], 718 [31], X-750 [31], 800 [31], 800 H [165], 182 [31], 82 [31])	Piping, steam generators, tubes, and working component in high corrosive environments [30,31]
Advanced Nuclear Energy	Low-alloyed steel (Ferritic steels: A105 [169], A106 GrB [31], A182 [169], A216 GrWCB [31], A302 GrB [169,170], A333 Gr6 [31], SA212 B [169]; A508 Gr3 [169,171–174], A516 Gr70 [31], A533 A [31], A555 B [31], 15Kh2NMFA [175,176], 08Kh18N10T [177]; bainitic steels: 1Cr1Mo0.25v [31], 2Cr1MoGr 22 [31], NiCrMoV [31]; duplex steel: 2507 [159], DSS [178], Fe20Cr9Ni [179]; carbon steel: AISI 1018 [159])		Steamless piping, gorging, casting, bolting, plate, pressure vessels, piping, and feedwater lines; internal stainless steel cladding; steam generator channel heads [30,31]
Advanced Nuclear Energy	*Fusion RAFM steel (Eurofer97 [180–183], CLAM [31], Infrafm [31], FB2h [31]; Rusfer [31]; 9Cr-2WVTa [31]) Ferritic steel [31]		First wall at reactor, blanket, shield, vacuum vessel, and divertor [30,31]
Advanced Nuclear Energy		Fusion ODS alloy [31] ODS ferritic alloy [31]	
Advanced Nuclear Energy		Fusion other alloys SiC composites [31] W and W-based alloys [184–186] Cu-based alloys [186] Pb-Al alloy [187,188] Mo-based alloys [184] Nb-based alloy [184] V-based alloy [184] C-fiber components [190] HEA [189]	Structural and insulating application, joints, and filaments [30,31]

Table 3. Cont.

Energy Sector	Metallic Materials		Application
	Ferrous Alloys	Non-Ferrous Alloys	
Hydroelectrical Renewable Energy	Austenitic stainless steel (AISI 316/ AISI 316L [71,191–193], AISI 304/ AISI 304L [66,79,194,195], AISI 325 [71], ASTM A743 [79,196], ASTM CF20 [71])		
Hydroelectrical Renewable Energy		Non-ferrous alloys Al-based alloys [66,71,77] Ti-based alloys [66,78] Cu-based alloys [76,79,80] Ni-based alloys ([81,89]I)	Turbines and other components [66]
Hydroelectrical Renewable Energy	Martensitic stainless steel (AISI 410 [197–200], ASTM F6NM [201,202], 13Cr4Ni [202–204], AISI 410T [79], AISI 410 [71], AISI 430 [69], ASTM FV520B [205], ASTM CA6NM [196])		Turbines, shear pins, and other components [66]
Hydroelectrical Renewable Energy	Other steels High-strength steel [56,70,206] ACSR [71,77] Stainless steel [72,77,79,199,206] Cast steel [71] Nitronic steel [202] Carbon steel [72] High-speed steel [207] Electric steel [69] Constructional steel [69]		Supporting systems and other components [66]
Biomass Renewable Energy	Carbon steels and low-alloyed steels (2.25Cr-1Mo [97,98], 5Cr-1Mo [98], 9Cr-1Mo [98,208])		Construction of the plant, pumps, pipes, valves, fittings, and digester tanks [82–84,105,209]
Biomass Renewable Energy		Non-ferrous alloys Al-based alloys [97,210,211] Ni-based alloys [97,212] NCA [213]	Specialized components [82–84]
Biomass Renewable Energy	Stainless steels Austenitic (AISI 304 L [97,214], AISI 316 L [74]) Martensitic (AISI 409 [74], AISI 410 [74], AISI 416 [74])		Construction of the plant, pumps, pipes, valves, fittings, and digester tanks [82–84,105,209]
Biomass Renewable Energy	Cr-steels (12-13Cr [208], 13Cr [208], 14.5Cr [208], 16Cr [208], 12Cr-5Ni-2Mo [208], 11.5Cr-2Mo [208])		

Table 3. Cont.

Energy Sector	Metallic Materials		Application
	Ferrous Alloys	Non-Ferrous Alloys	
Wind and Offshore Wind Renewable Energy	Structural steel (S235J2 [215,216], S355J2 [215,216], S35G10 [217], S460 [216], S690 [216], S355 [216], S420M3Z [218], S500M3Z [219])		
Wind and Offshore Wind Renewable Energy	Stainless steel Duplex stainless steel (mostly AISI 2205 [220]) Austenitic stainless steel (22Cr25NiWCoCu [221], AISI 304L [222–224], AISI 904L [225,226]) Martensitic stainless steel (mainly from 4xx series, such as AISI 440 C [227,228])		Foundation, tower, gear, and casing of the wind turbines [105]
Wind and Off-shore Wind Renewable Energy		Non-ferrous alloys Cu-based alloys [229,230] Al-based alloys mostly from 2xxx and 6xxx series [218,231,232]	
Wind and Off-shore Wind Renewable Energy	Other types of steel Electric steels [233] Cast iron [234,235] High-Si nodular cast iron (EN GJS500-14 [235], EN GJS450-18 [235], EN GJS600-10 [235]) Bearing steels (mainly AISI 52100 [227,228]) High-Cr steel [227,228] Low-C nitrogen steel [227,228]		
Solar Renewable Energy	Austenitic stainless steel (AISI 304 [236,237], AISI 304L [238], AISI 316 [236,239], AISI 316L [237,240,241], AISI 321 [242,243], AISI 347 [242], AISI 347H [236])		
Solar Renewable Energy	Martensitic stainless steel (AISI 420 [244], EN 1.4903 [236], EN 1.4923 [236], AISI T91 [245], VM12 [246])		
Solar Renewable Energy		Ni-based alloy (IN 230 [247], IN 600 [248,249], IN 617 [247,249], IN 625 [236,239,247], IN HT700 [250], IN 800H [251], C-276 [252], XH [249], H230 [249], HR120 [249])	Are used for base in solar-thermal panels, pumps, tanks, and heat exchangers [105]
Solar Renewable Energy	Other steels Carbon steels [236,253] Cr-Mo steels [254] Duplex steels [255] FeCrAl steels [241] Ferritic-martensitic P91 [248]		
Solar Renewable Energy		Other non-ferrous alloys Al-based alloys, mostly from series 7xxx [256–259] HEA [260,261] Mg-based alloys, mostly Ti-Y combination [262,263]	

Table 3. Cont.

Energy Sector	Metallic Materials		Application
	Ferrous Alloys	Non-Ferrous Alloys	
Geothermal Renewable Energy	Duplex steels (2205 [264], 2507 [264], 2707 [264])		Heat exchangers, filters, pumps, valves, piping, and condensers [209]
Geothermal Renewable Energy	Austenitic and superaustenitic steels (AISI 304 [265], AISI 304L [264], AISI 310S [264], AISI 316L [264,266], AISI 321 [264], UNS N08031 [266], N08020 [267], N08026 [267], N08825 [267], N08330 [267], S31254 [267])		
Geothermal Renewable Energy	Martensitic stainless steels (mostly from 4xx series, AISI 400 [268,269], AISI 430 [269], AISI 431 [269])		
Geothermal Renewable Energy		Non-ferrous alloys Ti-based alloys [138] Ni-based alloys [138,264] Cu-based alloys [269]	Construction of the plant [267]
Geothermal Renewable Energy	Other steels Low-alloyed steels [266] Carbon steel [270] Superferritic steels (S44627 [267], S44700 [267], S44800 [267])		

2.4. Solutions for Lowering the Costs of Maintenance and Prolonging the Component Durability

As mentioned before, the major maintenance costs are the repair and replacement of materials used in the power plants. It is important to extend the service life of materials and to design materials and components that can be easily and cost-effectively replaced or repaired. A common solution to address these challenges for metallic materials used in different energy sectors is mostly by cathodic corrosion protection in combination with coatings [271,272]. Cathodic protection of the metallic materials is an electrochemical technique to protect and control corrosion of the material [273,274].

Coatings, especially organic coatings, can also be applied without the combination of cathodic protection, which is mainly used for materials that are immersed in water [275–277]; often, a combined use is chosen. Cathodic protection can be also achieved by some metallic coatings, such as zinc alloy coatings or by a combination of metallic and organic coatings (see, for example, [278,279]). However, coatings and linings can also be applied alone as a passive corrosion protection or in a so-called duplex system, where both coatings and linings are used simultaneously as a multilayer system [218,280].

Other options for surface treatment to improve resistance to environmental factors and prolong component life include surface treatments such as laser treatment, electron beam, induction heating, plasma nitriding, and selection of the appropriate heat treatment to achieve the desired microstructure [281–284]. While coatings and surface treatments can be a good technique to overcome many challenges, certain applications that require specialized metallic materials can make this technique very limited. This is particularly an issue when the application is under harsh conditions such as simultaneous high temperatures and high loads, which require either metallic materials that are difficult to coat or specialized coatings and surface treatments that can be very expensive and have limited-service life due to combined wear, erosion, and corrosion effects [285,286].

This requires a holistic approach to material treatment that is not limited to the surface of the material. A common approach for metallic materials is to use conventional heat treatments to tailor individual properties. However, conventional heat treatments typically involve a trade-off where certain properties are improved at the expense of others, typically resulting in metallic materials with high strength but low fatigue and corrosion properties and vice versa [287,288].

As a result, more sophisticated and complex processing and treatment of metallic materials are being explored to overcome such trade-offs. One of the new options, which has also been tested in the steel industry, is the use of cryogenic treatment, which can improve various properties of metallic materials, including corrosion performance, without adding a coating to the surface [289–292]. A more detailed presentation and explanation of cryogenic treatment and its application to surface and corrosion properties is described in Section 3.

3. Cryogenic Treatments in Energy Sector

The technology of cryogenic treatment (CT) has made tremendous progress in the last 10 years in its application on metallic materials in various sectors ranging from medicine, aerospace, robotics, materials science (including the steel industry), nanotechnology, and mining to even more specialized disciplines [289,293]. The technique has evolved from the first attempts to treat materials at cryogenic temperatures in the 19th century by James Dewar and Karol Olszewski using liquefied gases (nitrogen and hydrogen). Later, the first real scientific observation and documentation of CT was made by NASA (National Aeronautics and Space Administration) in the mid-20th century, when they observed changes in the properties of materials used in space shuttles returning from space [289,293]. The selected aluminum components were harder and more wear-resistant after returning to Earth than they were before the space mission [289,293]. Since then, CT has been slowly adapted with different techniques and applications to metallic materials in order to improve macroscopic and microscopic properties. In the literature, CT can also be called sub-zero treatment, ultra-low temperature processing, or cryo-processing [289,293–296].

The application of CT in the energy sector can be of particular interest due to the variety of metallic materials that are used in extreme conditions (high-temperature and high-pressure environments, highly corrosive environments, highly abrasive environments, etc.), as discussed in Section 2. However, the application of CT in the energy sector is still in its infancy, mainly due to the slow introduction and development of this treatment scheme and the limited research focus on applications in the energy sector.

3.1. Mechanisms of Cryogenic Treatments

The mechanisms of CTs are based on the type, which is defined by the selected temperature regime for the CT (Figure 5). CT is usually applied after the material has been hardened and quenched and before being tempered, usually for 24 h at a predetermined temperature [289–293,297,298]. The most common and the one with the longest tradition is the conventional cryogenic treatment (CCT), where temperatures as low as 193 K are used [299].

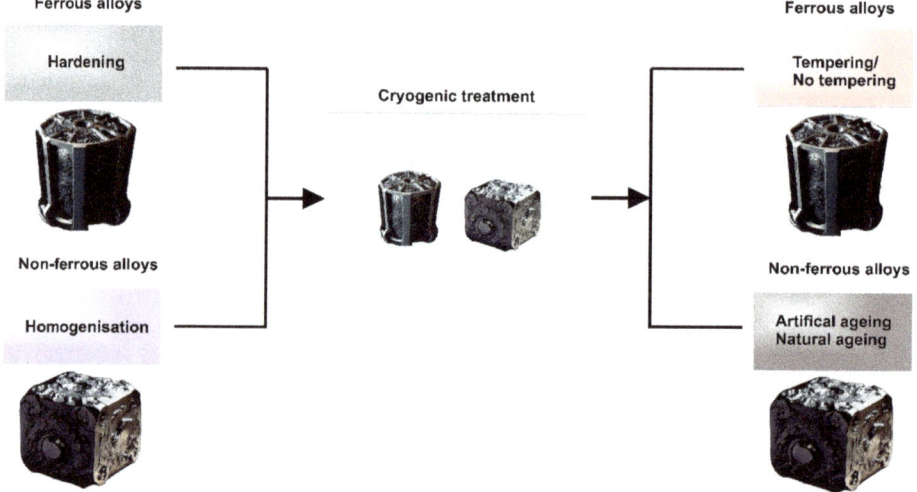

Figure 5. The heat treatment route(s) for both ferrous and non-ferrous alloys when CT is applied.

The reason for CCT being the most-used type in the past was the easy availability of media to which the material is exposed, namely dry ice (solid CO_2) [293,299]. In the past, it was also believed that temperatures as low as 193 K were sufficient to transform all the retained austenite (RA) in ferrous alloys to martensite, thereby increasing wear resistance and fatigue strength [293,299]. The transformation of RA to martensite, particularly in steels, was one of the key properties for which CTs were commonly applied, which also propagated the initial research on CT [293,299]. Unfortunately, the negative results of the first experiments with CCT led many companies to abandon the application and development of this treatment (1940s–1950s) [293,299]. This was mainly due to a misunderstanding of the martensitic transformation and its temperatures as well as simplistic and inconsistent treatment procedures [293,299]. It was not until years later, after NASA observations and detailed documentation of the changes at lower temperatures, that the next two types of CT were developed and tested for materials science applications: shallow (SCT) and deep cryogenic treatment (DCT) [293,299].

Shallow cryogenic treatment is defined between 193 K and 113 K. During SCT, more than 50% of the RA is converted to martensite for generally any ferrous alloy that has instable austenite formation during quenching, causing a change in mechanical properties (increased hardness), size reduction of carbides, and increased precipitation of carbides [300,301]. With the positive results of SCT, the research on CT blossomed and led to further research at even lower temperatures, resulting in the development of deep cryogenic treatment.

Temperatures for deep cryogenic treatment are below 113 K and typically go as low as 4 K, which is the temperature of liquid helium. However, the most-used temperature is 77 K, the temperature of liquid nitrogen, which is the most-used medium in DCT due to abundancy of the media and economic reasons. With DCT, for ferrous alloys, most of the RA is converted to martensite (>90%), the precipitation of carbides is increased, grain refinement and precipitation of nanocarbides occurs, and changes to residual stresses are formed [302]. Special mention should be made to a specific type of DCT, the multi-stage deep cryogenic treatment (MCT), where the DCT treatment of the material consists of rapid changes between SCT and DCT temperatures for a predefined time and number of cycles to manipulate predefined properties [303]. DCT performance is influenced by the selected cooling temperature, cooling–warming rate, time the material is exposed to DCT, type of metallic material (ferrous/non-ferrous alloy or type of steel), chemical composition of the metallic material, hardening process, tempering temperature, and also the microstructural phenomena present within the microstructure (such as transformation-induced plasticity (TRIP), austenite reversion transformation (ART), and twinning-induced plasticity (TWIP)) [304–310].

All types of CT alter the bulk and surface properties of metallic materials. The bulk properties affected by CT are mechanical properties ((micro)hardness [311–314], toughness [311,315–317], strength [318–320], and fatigue [307,321,322]) and magnetism [304,323]. The surface properties affected by CT are corrosion resistance [324–332], wear resistance [321,333–335], roughness [336], and oxide formation [324–326,336].

Bulk properties and, to some extent, selected surface properties have been studied in more detail than others. There are still many unknowns and great potential in surface properties and corrosion resistance, which is also the focus of the following section of this review.

3.2. Energy Sector and Position of Cryogenic Treatments

Cryogenic processing has a great potential in the energy sector due to the use of different materials, from metallic to non-metallic. The application of CT, especially for metallic materials, has a great potential because it improves the properties of metallic materials needed in different energy sectors, from corrosion and wear resistance to mechanical properties and surface modifications [337,338]. At the same time, it does not require the additional application of any other coating treatment to improve the properties (see Section 3.3.1).

However, the application of CT in this sector has not been widespread due to the lack of known test methods and quantification and qualification methods. Only a few attempts have been made to provide systematic guidelines for standards and application of CT for metallic materials [293,306,309,339–341]. An additional obstacle was that in the past, there were no large capacity tanks, and no providers of these services or systems were available on an industrial scale, but this is now changing and, in some cases, improving with the establishment of CT-specialized companies, communities, and even patents [294–296,332,342–347]. CT was also not well transferred to other disciplines, as CT was mainly reserved and developed for improving tools. The research was (and still is) mainly focused on tool steels, such as high-speed steels, hot work tool steels, and cold work tool steels, where the emphasis is on mechanical and wear properties [334,348–351].

As a result, the majority of other types of steels and alloys have been left out of the focus. There is some limited research on non-ferrous alloys, but even these are mostly related to aluminum alloys used or related to the tooling industry. The study of non-ferrous alloys (Al-, Ni-, and Ti-based alloys) showed the improvement of mechanical properties [352–358] such as microhardness [352,354,359–361], fatigue [352], fracture toughness [362,363], impact toughness [352], and tensile strength [352,354].

The following sections present metallic materials that have been tested by cryogenic treatment, the results of which have the potential to be used in the energy sector.

3.3. Effect of Cryogenic Treatments on Surface, Interface, and Corrosion Properties of Metallic Mateirals Used in the Energy Sector

3.3.1. Metallic Materials Being Tested for the Use in the Energy Sector

There are many metallic materials (ferrous and non-ferrous alloys) that are suitable for use in the energy sector that have already been tested through various cryogenic treatments, and studies have resulted in changes in the microstructure of metallic materials, resulting in changes in the properties of the material. The following ferrous and non-ferrous alloys are used in the following sectors (Table 4).

Table 4. The list of ferrous alloys that were cryogenically treated and have the potential for use in the energy sector.

Ferrous Alloys	Grades of Steel	Tested Properties	Possibilities of Application in Selected Energy Sector
Austenitic stainless steel	AISI 304 [364–370], AISI 304L [308,319,371–374], AISI 304LN [374], AISI 316 [374–380], AISI 316L [192,341,348,381–384], AISI 316LN [374,385], AISI 321 [386,387], AISI 347 [388,389]	Hardness, microhardness, wear (abrasive wear), fracture toughness, impact toughness, compressive strength, tensile strength, yield strength, elongation, friction, erosion, strain-hardening exponent, surface roughness, machining of steel, fatigue, residual stress, surface chemistry, and oxidation	In all energy sectors
Martensitic stainless steel	AISI 410 [390], AISI 420 [349,390–392], AISI 420 MOD [392], AISI 430 [393,394], AISI 431 [304,309,395], AISI 440C [366], AISI P91 [396], 10Cr13Co13Mo5NiW1VE [397], 13Cr4NiMo [315], 10Cr [398].	Yield strength, elongation, tensile strength, wear, hardness, impact toughness, fracture toughness, magnetism tribocorrosion, electrochemistry, and corrosion resistance (also stress corrosion cracking)	In all energy sectors.
Duplex steels	AISI 2205 [399,400], AISI 2507 [401–404]	Hardness, wear, machinability, residual stress, and corrosion resistance	Mostly in wind and solar energy
Carbon steels	IS 2062 [405], AISI 1045 [406–412], AISI 1018 [413]	Hardness, wear, surface roughness, tensile strength, yield strength, ultimate tensile strength, elongation, and residual stress	Steels can be used in hydroelectrical, biomass, solar, and geothermal energy
Other steels	Nitronic steels 40 [414], 50 [415] High-strength steels ASTM A36 [416] Cast steels ASTM A743 [417], SAE J431 G10 [418] ACSR [419] Bearing steel AISI 52100 [326,328,420–422] Low-alloyed steels SAE 1008 [423], AISI 4340 [424], AISI 4140 [424] Structural steel S235 [425], S355 [426,427], S460 [428]	Residual tress, hardness, friction, wear, fatigue, impact toughness, corrosion resistance, and machinability	In all energy sectors

Table 5 presents the non-ferrous alloys that have been CT-treated and have potential in the current and future energy sectors.

3.4. Effect of Cryogenic Treatments on Metallic Materials Potenitally Used in the Energy Sector

The surface properties that are the focus of this review and that also need more attention in order to carry out more research on them are corrosion resistance and oxide formation, while wear resistance and roughness have been observed and researched by many studies in the cryogenic community (see Section 3).

Table 5. The list of non-ferrous alloys that were cryogenically treated and have potential for use in the energy sector.

Non-Ferrous Alloys	Grades of Alloys	Tested Properties	Possibilities of Application in Selected Energy Sector
Al-based alloy	2xxx series: 2024 [354,429] 3xxx series: A356 [430,431], A390 [432] 6xxx series: 6026 [352,433–435], 6061 [353,436,437], 6063 [362] 7xxx series: 7075 [330,331,361,438–442]	Hardness, wear (abrasion), corrosion resistance, tensile strength, machinability, fatigue, strain-hardening coefficient, residual stress, fracture toughness, and corrosion resistance	Mostly in hydroelectrical, biomass, wind, and solar energy
Ni-based alloy	Inconel: 200 [443], 600 [444,445], 617 [446], 625 [447–449], 690 [450], 800 [451], 800H [452–454] Hastelloy C276 [455], C22 [456,457], X [458]	Fatigue, surface roughness, machinability, durability, impact toughness, microhardness, and tensile strength	In all energy sectors
Other alloys	HEA [459] W-based alloys [460] Cu-based alloys [461] Ti-based alloys Ti6Al4V [451,462]	Microhardness, compressive strength, and plasticity	Mostly in advanced nuclear power (fusion), geothermal, and solar energy

3.4.1. Oxide Formation

Oxide formation is one of the properties that is seldomly researched and not fully understood in CT. The fact is that most of the studies focus on the corrosion resistance and its improvement by CT, and not many studies strive for deeper understanding of the origin of altered corrosion resistance by CT. A major contribution is provided by passive layers and oxide formation (corrosion products) that can be manipulated by CT and CT-induced changes to the bulk properties of the treated material. The influence of CT on oxide formation has been demonstrated for bearing, high-speed, and cold work tool steels [324–326,336]. The oxidation dynamics after the application of CT was mainly studied by Jovičević-Klug et al., where the observations showed a different development of oxides compared to conventional heat treatment (CHT).

Jovičević-Klug et al. 2021 [336] suggested that the chemical composition of the oxide formation directly corresponds to the higher number of precipitates and the higher surface-to-volume ratio of the carbides. Furthermore, the study indicates that the reduced amount of carbide clusters after CT could be directly correlated with the passivation layer and the oxidation state of the surface and the corresponding corrosion products.

In the next study, Jovičević-Klug et al. 2021 [325] suggested that the Cr oxide layer is thicker on the cryogenically treated samples compared to the CHT samples. These observations also suggest that due to the formation of the Cr-oxide-passivation layer on the CT sample, there is no microscopic-related stress corrosion cracking of the matrix, which in turn, combined with the thicker passivation layer, reduces corrosion propagation.

The next factor observed in relation to CT was the formation of Fe oxides. The study by Jovičević-Klug et al. [326] suggested that Fe-oxides form different layers compared to the CT sample, which is attributed to the local excessive corrosion damage in the CHT sample.

The same researchers, Jovičević-Klug et al. 2022 [324], also observed the different layering of the oxides in the samples. The results of ToF-SIMS provided the novel insight that nitrogen from CT is present in greater amounts in the CT samples, which then influences the complex oxide formations (corrosion products), which ultimately influence the corrosion resistance. Nitrogen acts as an exalter for the formation of green rust, which then acts as a precursor for the formation of the next layer (magnetite). As a result, corrosion propagation is greatly retarded due to the higher density and stability of magnetite. The same study also confirmed that the CT-induced passive film is more stable than its CHT counterpart. As a result, the CT-treated sample showed lower corrosion and wear loss, which was also confirmed in extreme environments (elevated temperatures and vibrations).

The above examples show that there is a need for research on oxide formation as the basis for successful tailoring of corrosion resistance and prolonged component life of treated materials. The studies only focused on tool and bearing steels, which means that other steels such as high-Cr steels, stainless steels, duplex steels, and non-ferrous alloys are still potential research areas with great opportunities for the application of CT to manipulate oxide formation and modify corrosion resistance. To date, no similar studies or research have been conducted or found for non-ferrous alloys. Section 3.4.2 discusses corrosion resistance.

3.4.2. Corrosion Resistance

The influence of CT on corrosion resistance has not only been investigated in relation to tool steels, but many studies have also tested other ferrous (bearing steels and stainless steels) and non-ferrous (mostly Al-based alloys) alloys. The first part focuses on the corrosion testing of ferrous alloys, while the second part focuses on the non-ferrous alloys in relation to CT.

Corrosion Resistance of Ferrous Alloys

The studies on tool steels showed that corrosion resistance is influenced by CT [329,392,463–466]. The corrosion resistance of bearing and tool steels can be improved by up to 65% in an alkaline environment, with the improvement depending on the steel type and heat treatment strategy [326]. This was also observed by Senthilkumar 2014 [327], who found that in alkali conditions, CT improves corrosion resistance, which was postulated to be a result of formation of more stable passive film. Furthermore, in extreme alkaline environments, such as elevated temperatures and vibrations, the CT-treated samples (tool steels) suggested improvement of corrosion resistance by 90% in the study of Jovičević-Klug et al. 2022 [324]. Also, the study by Jovičević-Klug et al. 2021 [326] showed that in an alkaline environment, the formation of pits is modified by CT (for tool and bearing steels). The study showed that pits in CT specimens expand only in the exposed upper part and decrease continuously deeper into the material. It was suggested that this is due to the confinement of the corrosion attack to the grain boundaries and the exposure of the pit opening to the oxidative media, which is limited by the change in orientation of the crack with respect to the sample surface. In addition, the 2021 study by Jovičević-Klug et al. [325] also showed that in the alkaline environment, the CT samples did not show any stress corrosion cracking of the passivation layer, and the presence of Mo in the steel allowed the continuous growth of the protective Cr oxide layer, which reduced the formation and growth of pits. The results show that CT samples have a 3× slower corrosion rate of pitting corrosion, which can be directly correlated to the slower material degradation and prolonged functionality of the metallic material.

Only a few studies have been conducted on stainless steels and a few other types of steels that are more commonly used in energy sector applications. The studies showed different results of CT on the corrosion resistance of steels used in the energy sector. A study by Wang et al. 2020 [467] showed that there is an increase in corrosion resistance for high-strength stainless steel. On contrary, a study by Baldissera and Delprete 2010 [468] postulated that CT has no effect on austenitic stainless steel. Another study by Cai et al. 2016 [469] indicated that for austenitic stainless steel, CT could improve corrosion resistance, which is suggested through Cr-carbide precipitation at the austenite grain boundary, which then reduces the intergranular corrosion. For martensitic stainless steels, CT has been shown to improve corrosion resistance in correlation with both the general and pitting corrosion, as was shown by Ramos et al. 2017 [366]. Another explanation for the higher pitting corrosion potential was proposed by He et al. 2021 [470], in which pitting corrosion was reduced by increased carbide precipitation and Si segregation at the interface boundaries between $M_{23}C_6$ and martensite in the matrix. For structural steels, a 95% improvement in corrosion resistance was determined by Ramesh et al. 2019 [392], which is suggested to be a consequence of uniform and homogenous carbide precipitation and microstructure modification.

The above literature review shows that there has been some research on corrosion enhancement with CT but only on a limited selection of ferrous alloys. Furthermore, the review shows that there is a great need for research on the corrosion resistance of ferrous alloys used in the energy sector in combination with CT. Such research could open up new avenues and applications for CT to improve corrosion resistance alone or in combination with coatings, which could further expand the energy sector from both an economic and sustainable point of view.

Corrosion Resistance of Non-Ferrous Alloys

The corrosion resistance of CT-treated non-ferrous alloys has been mainly focused on the Al-based alloys of the 2xxx [471], 5xxx [355], and 7xxx [330,331,440] series. The study by Cabeza et al. 2015 [471] on Al-based alloys from the 2xxx series suggested that CT improves the resistance to stress corrosion cracking due to changes in compressive residual stresses. Another study by Aamir et al. 2016 [355] showed that for the 5xxx Al-alloy, the corrosion resistance is increased due to the minimization of dislocation densities and noncontinuous distribution of the β-phase. From the 7xxx series, the tested representative was the 7075 Al-alloy. A study by Ma et al. 2021 [440] showed an improvement in corrosion resistance after the application of CT, which was attributed to the increased precipitation of the η' phase. They postulated that the grain boundary from the η' resulted in short chains of carbides, which then blocked corrosion channeling, thus enhancing the corrosion resistance of the alloy. Similar observations were also made by Su et al. 2021 [331]. Ma et al., from their study in 2022 [330], additionally showed that the optimized combination of aging and CT can influence the rate of the corrosion improvement when CT is applied.

Compared to ferrous alloys, research on non-ferrous alloys is also considered to be lacking and is mostly focused on specific alloys, mainly aluminum alloys. The review clearly confirms the lack of research on non-ferrous alloys, which have a great potential for use in the future energy sector. The lack of research can be particularly evident in the case of Ni alloys and corrosion resistance in combination with CT, which are one of the main non-ferrous alloys used in different energy sectors due to their versatility. Other non-ferrous alloys such as Cu-based, Mg-based, V-based, W-based, etc., are also completely excluded from the studies, and therefore, this could be another potentially interesting niche to study in more depth the influence of CT on these alloys, which could be applied to the future energy sector. Furthermore, in most cases, the reasons for improved or sometimes reduced corrosion performance are based on speculation. Fundamental research is needed to elucidate the reasons for the effects of CT on corrosion performance.

4. Economic and Ecological Aspects and Future Role of Cryogenic Treatment in Future Energy Sector

While it is clear that the application of CT to ferrous and non-ferrous alloys has great potential due to its versatile effect on bulk and surface properties, the next question that comes to mind is the economic and environmental aspects of its application. CT uses mostly liquid N_2 as a coolant media, which is highly considered as a viable option for the conventional heat treatment of metallic materials. After CT treatment, LN_2 evaporates to become nitrogen gas (N_2) and becomes part of the air (78% of air consists of N_2). It leaves no harmful residue to the industry and the environment and no health hazards compared to other processing/machining techniques [369]. Therefore, it is considered as a recycling and environmentally friendly approach to improve the materials. Furthermore, it is suggested in some works, see, e.g., Hong and Broomer [369] and Dosset [472], showing a cost reduction of about 50%; however, more insight into the economic advantages is expected in the near future, which is expected to cause an increasing interest in the energy sector. In conclusion, CT shows great potential to improve the corrosion performance of materials, and the process is environmentally unharmful.

Based on all these facts, CT has a bright future in the future energy sector, where advanced knowledge of more economical and ecological impacts on the environment is being considered. Not only that, but with the trend of diminishing natural resources and more recycling options, CT also has an answer, as no additional treatment is required. CT is also expected to play an important role in emerging materials for energy applications and storage (HEA and catalytic materials), which have not yet been explored and applied.

5. Conclusions

This review first provides an overview of the metallic materials used in the energy sector and then of the application of CT on metallic materials used in different energy sectors. In addition, this review also presents a synopsis of the current work and results on surface properties and corrosion, with critical comments to provide a future possibility for metallic materials in relation to CT and oxide formation and corrosion resistance. The review also highlights which materials should be prioritized for CT testing due to lack of research but are of high importance for applications (or are already in use) in different energy sectors.

The main conclusions of the study can be summarized as follows:

- The energy sector has a great demand for the improvement of metallic materials;
- Available green and cost-effective CT technology has been proven to effectively improve the bulk and surface properties of metallic materials;
- CT improves corrosion resistance by up to 90% depending on metallic materials and environmental conditions;
- CT also produces a unique sequence of oxide formation that effectively influences the improved corrosion resistance of cryogenically treated metallic materials;
- The result of CT is a reduction in material degradation and a possible 3-fold increase in the service life of the treated metallic material;
- Further detailed and systematic investigation of the effectiveness of CT is required, using both experiments and modeling of both ferrous and non-ferrous alloys. Combined with detailed microstructural investigations, the mechanisms responsible for changes in metallic material properties can be clearly identified, and standards for the application of CT in the energy sector can be established.

Author Contributions: Conceptualization, P.J.-K.; investigation, P.J.-K.; resources, P.J.-K. and M.R.; data curation, P.J.-K.; writing—original draft preparation, P.J.-K.; writing—review and editing, P.J.-K. and M.R.; visualization, P.J.-K.; supervision, M.R.; project administration, P.J.-K.; funding acquisition, P.J.-K. All authors have read and agreed to the published version of the manuscript.

Funding: The financial support was provided by Alexander von Humboldt Foundation throughout the Humboldt Research Fellowship Programme for PostDocs.

Institutional Review Board Statement: Not applicable.

Informed Consent Statement: Not applicable.

Data Availability Statement: Data will be available upon request.

Conflicts of Interest: The authors declare no conflict of interest.

References

1. Candra, O.; Chammam, A.; Alvarez, J.R.N.; Muda, I.; Aybar, H. The Impact of Renewable Energy Sources on the Sustainable Development of the Economy and Greenhouse Gas Emissions. *Sustainability* **2023**, *15*, 2104. [CrossRef]
2. Owusu, P.A.; Asumadu-Sarkodie, S. A Review of Renewable Energy Sources, Sustainability Issues and Climate Change Mitigation. *Cogent Eng.* **2016**, *3*, 1167990. [CrossRef]
3. Sinha, A.; Sengupta, T. Impact of Natural Resource Rents on Human Development: What Is the Role of Globalization in Asia Pacific Countries? *Resour. Policy* **2019**, *63*, 101413. [CrossRef]
4. Zallé, O. Natural Resources and Economic Growth in Africa: The Role of Institutional Quality and Human Capital. *Resour. Policy* **2019**, *62*, 616–624. [CrossRef]
5. Kumar, M. Social, Economic, and Environemtal of Renewable Energy Resources. In *Wind Solar Hybrid Renewable Energy System*; DoB-Books on Demand: Norderstedt, Germany, 2020; pp. 237–247.
6. Zohuri, B.; McDaniel, P. Energy Insight: An Energy Essential Guide. In *Introduction to Energy Essentials*; Academic Press: Cambridge, MA, USA, 2021; pp. 321–370. [CrossRef]
7. Smith, C.L.; Cowley, S. The Path to Fusion Power. *Philos. Trans. A Math. Phys. Eng. Sci.* **2010**, *368*, 1091. [CrossRef] [PubMed]

8. Dream of Unlimited, Clean Nuclear Fusion Energy within Reach | Research and Innovation. Available online: https://ec.europa.eu/research-and-innovation/en/horizon-magazine/dream-unlimited-clean-nuclear-fusion-energy-within-reach (accessed on 23 July 2023).
9. Carter, T.; Baalrud, S.; Betti, R.; Ellis, T.; Foster, J.; Geddes, C.; Gleason, A.; Holland, C.; Humrickhouse, P.; Kessel, C.; et al. *Powering the Future: Fusion and Plasmas*; US Department of Energy (USDOE): Loa Alamos, NM, USA, 2018.
10. EU Energy Outlook 2050—How Will Europe Evolve over the Next 30 Years?—Energy BrainBlog. Available online: https://blog.energybrainpool.com/en/eu-energy-outlook-2050-how-will-europe-evolve-over-the-next-30-years-3/ (accessed on 28 June 2023).
11. Renewable Energy on the Rise: 37% of EU's Electricity—Products Eurostat News—Eurostat. Available online: https://ec.europa.eu/eurostat/web/products-eurostat-news/-/ddn-20220126-1 (accessed on 28 June 2023).
12. Forbes Home. Solar. Available online: https://www.forbes.com/home-improvement/solar/ (accessed on 23 July 2023).
13. How Much Energy Does a Solar Farm Produce? [Solar Farms Explained]. Available online: https://solarenergyhackers.com/how-much-energy-does-a-solar-farm-produce/ (accessed on 23 July 2023).
14. U.S. Energy Information Administration (EIA). Frequently Asked Questions (FAQs). Available online: https://www.eia.gov/tools/faqs/faq.php?id=104&t=3 (accessed on 23 July 2023).
15. Pearce-Higgins, J.W.; Stephen, L.; Douse, A.; Langston, R.H.W. Greater Impacts of Wind Farms on Bird Populations during Construction than Subsequent Operation: Results of a Multi-Site and Multi-Species Analysis. *J. Appl. Ecol.* **2012**, *49*, 386–394. [CrossRef]
16. Kjeld, A.; Bjarnadottir, H.J.; Olafsdottir, R. Life Cycle Assessment of the Theistareykir Geothermal Power Plant in Iceland. *Geothermics* **2022**, *105*, 102530. [CrossRef]
17. Whole Building Design Guide. Biomass for Electricity Generation | WBDG. Available online: https://www.wbdg.org/resources/biomass-electricity-generation (accessed on 23 July 2023).
18. Fachagentur Nachwachsende Rohstoffe eV Agency for Renewable Resources. *Bioenergy in Germany Facts and Figures 2020*; Fachagentur Nachwachsende Rohstoffe eV Agency for Renewable Resources: Gülzow-Prüzen, Germany, 2020.
19. Tidal Energy: Can It Be Used to Generate Electricity. Available online: https://justenergy.com/blog/tidal-energy-electricity/ (accessed on 23 July 2023).
20. Hallidays. How Much Energy Does Hydropower Produce Each Year. Available online: https://www.hallidayshydropower.com/how-much-energy-hydropower/ (accessed on 23 July 2023).
21. IEA. Hydropower. Available online: https://www.iea.org/energy-system/renewables/hydropower (accessed on 23 July 2023).
22. Global Offshore Wind Capacity Factor 2021 | Statista. Available online: https://www.statista.com/statistics/1368679/global-offshore-wind-capacity-factor/ (accessed on 23 July 2023).
23. Agency, I.R.E. *Renewable Power Generation Costs in 2019*; International Renewable Energy Agency (IRENA): Abu Dhabi, United Arab Emirates, 2020; ISBN 978-92-9260-244-4.
24. IEA. Executive Summary—The Role of Critical Minerals in Clean Energy Transitions—Analysis. Available online: https://www.iea.org/reports/the-role-of-critical-minerals-in-clean-energy-transitions/executive-summary (accessed on 23 July 2023).
25. Watari, T.; Nansai, K.; Nakajima, K. Review of Critical Metal Dynamics to 2050 for 48 Elements. *Resour. Conserv. Recycl.* **2020**, *155*, 104669. [CrossRef]
26. Zepf, V.; Simmons, J.; Reller, A.; Ashfield, M.; Rennie, C. *Materials Critical to the Energy Industry An Introduction*, 2nd ed.; BP p.l.c.: London, UK, 2014.
27. McKinsey. The Raw-Materials Challenge: How the Metals and Mining Sector Will Be at the Core of Enabling the Energy Transition. Available online: https://www.mckinsey.com/industries/metals-and-mining/our-insights/the-raw-materials-challenge-how-the-metals-and-mining-sector-will-be-at-the-core-of-enabling-the-energy-transition (accessed on 23 July 2023).
28. Fu, M.; Ma, X.; Zhao, K.; Li, X.; Su, D. High-Entropy Materials for Energy-Related Applications. *iScience* **2021**, *24*, 102177. [CrossRef]
29. Xin, Y.; Li, S.; Qian, Y.; Zhu, W.; Yuan, H.; Jiang, P.; Guo, R.; Wang, L. High-Entropy Alloys as a Platform for Catalysis: Progress, Challenges, and Opportunities. *ACS Catal.* **2020**, *10*, 11280–11306. [CrossRef]
30. Allen, T.; Busby, J.; Meyer, M.; Petti, D. Materials Challenges for Nuclear Systems. *Mater. Today* **2010**, *13*, 14–23. [CrossRef]
31. Odette, R.G.; Zinkle, S.J. (Eds.) *Structural Alloys for Nuclear Energy Applications*; Elsevier: Amsterdam, The Netherlands, 2019. Available online: https://books.google.de/books?hl=en&lr=&id=KAF0AwAAQBAJ&oi=fnd&pg=PP1&dq=alloys+in+nuclear+energy&ots=kPDtxJjGm_&sig=0_gOE-NqXXv-PV27O1LfuvFqJBM&redir_esc=y#v=onepage&q=alloys%20in%20nuclear%20energy&f=false (accessed on 3 July 2023).
32. Advanced Nuclear Reactors 101. Available online: https://www.rff.org/publications/explainers/advanced-nuclear-reactors-101/ (accessed on 23 July 2023).
33. IAEA. *Advanced Large Water Cooled Reactors. A Supplement to: IAEA Advanced Reactors Information System (ARIS) 2020 Edition*; IAEA: Vienna, Austria, 2020.
34. Tian, W. Grand Challenges in Advanced Nuclear Reactor Design. *Front. Nucl. Eng.* **2022**, *1*, 1000754. [CrossRef]
35. Xiao, C.; Gulfam, R. Opinion on Ocean Thermal Energy Conversion (OTEC). *Front. Energy Res.* **2023**, *11*, 1115695. [CrossRef]
36. Elahee, K.; Jugoo, S. Ocean Thermal Energy for Air-Conditioning: Case Study of a Green Data Center. *Energy Sources Part A Recovery Util. Environ. Eff.* **2013**, *35*, 679–684. [CrossRef]

37. Hunt, J.D.; Zakeri, B.; Nascimento, A.; Garnier, B.; Pereira, M.G.; Bellezoni, R.A.; de Assis Brasil Weber, N.; Schneider, P.S.; Machado, P.P.B.; Ramos, D.S. High Velocity Seawater Air-Conditioning with Thermal Energy Storage and Its Operation with Intermittent Renewable Energies. *Energy Effic.* **2020**, *13*, 1825–1840. [CrossRef]
38. Etemadi, A.; Emami, Y.; AsefAfshar, O.; Emdadi, A. Electricity Generation by the Tidal Barrages. *Energy Procedia* **2011**, *12*, 928–935. [CrossRef]
39. Ringwood, J.V.; Faedo, N. Tidal Barrage Operational Optimisation Using Wave Energy Control Techniques. *IFAC-PapersOnLine* **2022**, *55*, 148–153. [CrossRef]
40. Walker, S.R.J.; Thies, P.R. A Life Cycle Assessment Comparison of Materials for a Tidal Stream Turbine Blade. *Appl. Energy* **2022**, *309*, 118353. [CrossRef]
41. Zheng, J.; Dai, P.; Zhang, J. Tidal Stream Energy in China. *Procedia Eng.* **2015**, *116*, 880–887. [CrossRef]
42. Radfar, S.; Panahi, R.; Javaherchi, T.; Filom, S.; Mazyaki, A.R. A Comprehensive Insight into Tidal Stream Energy Farms in Iran. *Renew. Sustain. Energy Rev.* **2017**, *79*, 323–338. [CrossRef]
43. Blunden, L.S.; Bahaj, A.S. Tidal Energy Resource Assessment for Tidal Stream Generators. *Proc. Inst. Mech. Eng. Part A J. Power Energy* **2007**, *221*, 137–146. [CrossRef]
44. Watson, S.; Moro, A.; Reis, V.; Baniotopoulos, C.; Barth, S.; Bartoli, G.; Bauer, F.; Boelman, E.; Bosse, D.; Cherubini, A.; et al. Future Emerging Technologies in the Wind Power Sector: A European Perspective. *Renew. Sustain. Energy Rev.* **2019**, *113*, 109270. [CrossRef]
45. Majidi, A.G.; Bingölbali, B.; Akpınar, A.; Rusu, E. Wave Power Performance of Wave Energy Converters at High-Energy Areas of a Semi-Enclosed Sea. *Energy* **2021**, *220*, 119705. [CrossRef]
46. Langhamer, O.; Haikonen, K.; Sundberg, J. Wave Power—Sustainable Energy or Environmentally Costly? A Review with Special Emphasis on Linear Wave Energy Converters. *Renew. Sustain. Energy Rev.* **2010**, *14*, 1329–1335. [CrossRef]
47. Clément, A.; McCullen, P.; Falcão, A.; Fiorentino, A.; Gardner, F.; Hammarlund, K.; Lemonis, G.; Lewis, T.; Nielsen, K.; Petroncini, S.; et al. Wave Energy in Europe: Current Status and Perspectives. *Renew. Sustain. Energy Rev.* **2002**, *6*, 405–431. [CrossRef]
48. Rusu, L. The Wave and Wind Power Potential in the Western Black Sea. *Renew. Energy* **2019**, *139*, 1146–1158. [CrossRef]
49. Rahman, A.; Farrok, O.; Haque, M.M. Environmental Impact of Renewable Energy Source Based Electrical Power Plants: Solar, Wind, Hydroelectric, Biomass, Geothermal, Tidal, Ocean, and Osmotic. *Renew. Sustain. Energy Rev.* **2022**, *161*, 112279. [CrossRef]
50. Zhang, Z.; Wen, L.; Jiang, L. Nanofluidics for Osmotic Energy Conversion. *Nat. Rev. Mater.* **2021**, *6*, 622–639. [CrossRef]
51. Xin, W.; Zhang, Z.; Huang, X.; Hu, Y.; Zhou, T.; Zhu, C.; Kong, X.Y.; Jiang, L.; Wen, L. High-Performance Silk-Based Hybrid Membranes Employed for Osmotic Energy Conversion. *Nat. Commun.* **2019**, *10*, 3876. [CrossRef]
52. Wang, Q.; Gao, X.; Zhang, Y.; He, Z.; Ji, Z.; Wang, X.; Gao, C. Hybrid RED/ED System: Simultaneous Osmotic Energy Recovery and Desalination of High-Salinity Wastewater. *Desalination* **2017**, *405*, 59–67. [CrossRef]
53. Han, G.; Zhang, S.; Li, X.; Chung, T.S. Progress in Pressure Retarded Osmosis (PRO) Membranes for Osmotic Power Generation. *Prog. Polym. Sci.* **2015**, *51*, 1–27. [CrossRef]
54. Kim, J.; Jeong, K.; Park, M.J.; Shon, H.K.; Kim, J.H. Recent Advances in Osmotic Energy Generation via Pressure-Retarded Osmosis (PRO): A Review. *Energies* **2015**, *8*, 11821–11845. [CrossRef]
55. Llamosas, C.; Sovacool, B.K. The Future of Hydropower? A Systematic Review of the Drivers, Benefits and Governance Dynamics of Transboundary Dams. *Renew. Sustain. Energy Rev.* **2021**, *137*, 110495. [CrossRef]
56. Kumar, R.; Singal, S.K. Penstock Material Selection in Small Hydropower Plants Using MADM Methods. *Renew. Sustain. Energy Rev.* **2015**, *52*, 240–255. [CrossRef]
57. Gemechu, E.; Kumar, A. A Review of How Life Cycle Assessment Has Been Used to Assess the Environmental Impacts of Hydropower Energy. *Renew. Sustain. Energy Rev.* **2022**, *167*, 112684. [CrossRef]
58. Borkowski, D.; Cholewa, D.; Korzeń, A. Run-of-the-River Hydro-PV Battery Hybrid System as an Energy Supplier for Local Loads. *Energies* **2021**, *14*, 5160. [CrossRef]
59. Venus, T.E.; Hinzmann, M.; Bakken, T.H.; Gerdes, H.; Godinho, F.N.; Hansen, B.; Pinheiro, A.; Sauer, J. The Public's Perception of Run-of-the-River Hydropower across Europe. *Energy Policy* **2020**, *140*, 111422. [CrossRef]
60. François, B.; Hingray, B.; Raynaud, D.; Borga, M.; Creutin, J.D. Increasing Climate-Related-Energy Penetration by Integrating Run-of-the River Hydropower to Wind/Solar Mix. *Renew. Energy* **2016**, *87*, 686–696. [CrossRef]
61. Douglas, T.; Broomhall, P.; Orr, C. *Run-of-River Hydropower in BC A Citizen's Guide to Understanding Approvals, Impacts and Sustainability of Independent Power Projects*; British Columbia Ministry of Environment: Coquitlam, BC, Canada, 2007.
62. Ramadan, A.; Nawar, M.A.A.; Mohamed, M.H. Performance Evaluation of a Drag Hydro Kinetic Turbine for Rivers Current Energy Extraction—A Case Study. *Ocean Eng.* **2020**, *195*, 106699. [CrossRef]
63. Badrul Salleh, M.; Kamaruddin, N.M.; Mohamed-Kassim, Z. Savonius Hydrokinetic Turbines for a Sustainable River-Based Energy Extraction: A Review of the Technology and Potential Applications in Malaysia. *Sustain. Energy Technol. Assess.* **2019**, *36*, 100554. [CrossRef]
64. EduRev. Which One of the Following Turbines Is Used in under Water Powerstations? (a) Pelton Turbine (b) Deriaz Turbine (c) Tubular Turbined (d) Turgo-Impulse Turbine. Correct Answer Is Option "C". Can You Explain This Answer? | EduRev Mechanical Engineering Question. Available online: https://edurev.in/question/1587343/Which-one-of-the-following-turbines-is-used-in-under-water-powerstations-a-Pelton-turbineb-Deriaz-tu (accessed on 24 July 2023).

65. Farr, H.; Ruttenberg, B.; Walter, R.K.; Wang, Y.H.; White, C. Potential Environmental Effects of Deepwater Floating Offshore Wind Energy Facilities. *Ocean Coast. Manag.* **2021**, *207*, 105611. [CrossRef]
66. Quaranta, E.; Davies, P. Emerging and Innovative Materials for Hydropower Engineering Applications: Turbines, Bearings, Sealing, Dams and Waterways, and Ocean Power. *Engineering* **2022**, *8*, 148–158. [CrossRef]
67. Tong, C. Introduction to Materials for Advanced Energy Systems. *MRS Bull.* **2020**, *45*, 317. [CrossRef]
68. Quaranta, E. Estimation of the Permanent Weight Load of Water Wheels for Civil Engineering and Hydropower Applications and Dataset Collection. *Sustain. Energy Technol. Assess.* **2020**, *40*, 100776. [CrossRef]
69. Hoffer, A.E.; Tapia, J.A.; Petrov, I.; Pyrhönen, J. Design of a Stainless Core Submersible Permanent Magnet Generator for Tidal Energy. In Proceedings of the IECON 2019-45th Annual Conference of the IEEE Industrial Electronics Society, Lisbon, Portugal, 14–17 October 2019; pp. 1010–1015. [CrossRef]
70. Rodiouchkina, M.; Lindsjö, H.; Berglund, K.; Hardell, J. Effect of Stroke Length on Friction and Wear of Self-Lubricating Polymer Composites during Dry Sliding against Stainless Steel at High Contact Pressures. *Wear* **2022**, *502–503*, 204393. [CrossRef]
71. Arun, K.; Kumar, K.M.; Karthikeyan, K.M.B.; Mohanasutan, S. Analysis on Influence of Bucket Angle of Pelton Wheel Turbine for Its Structural Integrity Using Aluminium Alloy (A390), Austenitic Stainless Steel (CF20), Grey Cast Iron (325) and Martensitic Stainless Steel (410). *Mater. Today Proc.* **2022**, *62*, 1045–1053. [CrossRef]
72. Musabikha, S.; Pratikno, H.; Utama, I.K.A.P.; Mukhtasor. Material Selection for Vertical Axis Tidal Current Turbine Using Multiple Attribute Decision Making (MADM). *IOP Conf. Ser. Mater. Sci. Eng.* **2021**, *1158*, 012001. [CrossRef]
73. Salter, S.H.; Taylor, J.R.M.; Caldwell, N.J. Power Conversion Mechanisms for Wave Energy. *Proc. Inst. Mech. Eng. Part M J. Eng. Marit. Environ.* **2002**, *216*, 1–27. [CrossRef]
74. Kofoed, J.P.; Frigaard, P.; Friis-Madsen, E.; Sørensen, H.C. Prototype Testing of the Wave Energy Converter Wave Dragon. *Renew. Energy* **2006**, *31*, 181–189. [CrossRef]
75. Jiang, X.; Overman, N.; Canfield, N.; Ross, K. Friction Stir Processing of Dual Certified 304/304L Austenitic Stainless Steel for Improved Cavitation Erosion Resistance. *Appl. Surf. Sci.* **2019**, *471*, 387–393. [CrossRef]
76. Suwanit, W.; Gheewala, S.H. Life Cycle Assessment of Mini-Hydropower Plants in Thailand. *Int. J. Life Cycle Assess.* **2011**, *16*, 849–858. [CrossRef]
77. Azevedo, C.R.F.; Cescon, T. Failure Analysis of Aluminum Cable Steel Reinforced (ACSR) Conductor of the Transmission Line Crossing the Paraná River. *Eng. Fail. Anal.* **2002**, *9*, 645–664. [CrossRef]
78. Ujwala, M.M.; Chaithanya, M.P.; Chowdary, K.; Naik, M.L.S. Design and Analysis of Low Head, Light Weight Kaplan Turbine Blade. *Int. Ref. J. Eng. Sci. IRJES* **2017**, *6*, 17–25.
79. Marangoni, P.R.D.; Robl, D.; Berton, M.A.C.; Garcia, C.M.; Bozza, A.; Porsani, M.V.; Dalzoto, P.D.R.; Vicente, V.A.; Pimentel, I.C. Occurrence of Sulphate Reducing Bacteria (SRB) Associated with Biocorrosion on Metallic Surfaces in a Hydroelectric Power Station in Ibirama (SC)—Brazil. *Braz. Arch. Biol. Technol.* **2013**, *56*, 801–809. [CrossRef]
80. Linhardt, P. Unusual Corrosion of Nickel-Aluminium Bronze in a Hydroelectric Power Plant. *Mater. Corros.* **2015**, *66*, 1536–1541. [CrossRef]
81. Ebrahimi, A. *Advances in Modelling and Control of Wind and Hydrogenerators Edited*, 1st ed.; Ebrahimi, A., Ed.; IntechOpen: London, UK, 2020.
82. Suopajärvi, H.; Umeki, K.; Mousa, E.; Hedayati, A.; Romar, H.; Kemppainen, A.; Wang, C.; Phounglamcheik, A.; Tuomikoski, S.; Norberg, N.; et al. Use of Biomass in Integrated Steelmaking—Status Quo, Future Needs and Comparison to Other Low-CO_2 Steel Production Technologies. *Appl. Energy* **2018**, *213*, 384–407. [CrossRef]
83. Ropital, F. Current and Future Corrosion Challenges for a Reliable and Sustainable Development of the Chemical, Refinery, and Petrochemical Industries. *Mater. Corros.* **2009**, *60*, 495–500. [CrossRef]
84. Mousa, E.; Wang, C.; Riesbeck, J.; Larsson, M. Biomass Applications in Iron and Steel Industry: An Overview of Challenges and Opportunities. *Renew. Sustain. Energy Rev.* **2016**, *65*, 1247–1266. [CrossRef]
85. ORNL. Degradation of Structural Alloys In Biomass-Derived Pyrolysis Oil. Available online: https://www.ornl.gov/publication/degradation-structural-alloys-biomass-derived-pyrolysis-oil (accessed on 12 July 2023).
86. Giudicianni, P.; Gargiulo, V.; Grottola, C.M.; Alfè, M.; Ferreiro, A.I.; Mendes, M.A.A.; Fagnano, M.; Ragucci, R. Inherent Metal Elements in Biomass Pyrolysis: A Review. *Energy Fuels* **2021**, *35*, 5407–5478. [CrossRef]
87. Biomass Energy. Available online: https://education.nationalgeographic.org/resource/biomass-energy/ (accessed on 12 July 2023).
88. Miandad, R.; Rehan, M.; Barakat, M.A.; Aburiazaiza, A.S.; Khan, H.; Ismail, I.M.I.; Dhavamani, J.; Gardy, J.; Hassanpour, A.; Nizami, A.S. Catalytic Pyrolysis of Plastic Waste: Moving toward Pyrolysis Based Biorefineries. *Front. Energy Res.* **2019**, *7*, 27. [CrossRef]
89. Wei, L.; Wang, S.; Liu, G.; Hao, W.; Liang, K.; Yang, X.; Diao, W. Corrosion Behavior of High-Cr-Ni Materials in Biomass Incineration Atmospheres. *ACS Omega* **2022**, *7*, 21546–21553. [CrossRef] [PubMed]
90. Berlanga, C.; Ruiz, J.A. Study of Corrosion in a Biomass Boiler. *J. Chem.* **2013**, *2013*, 272090. [CrossRef]
91. Perego, P.; Fabiano, B.; Pastorino, R.; Randi, G. Microbiological Corrosion in Aerobic and Anaerobic Waste Purification Plants: Safety and Efficiency Problems. *Bioprocess Eng.* **1997**, *17*, 103–109. [CrossRef]

92. Miryalkar, P.; Chavitlo, S.; Tandekar, N.; Valleti, K. Improving the Abrasive Wear Resistance of Biomass Briquetting Machine Components Using Cathodic Arc Physical Vapor Deposition Coatings: A Comparative Study. *J. Vac. Sci. Technol. A Vac. Surf. Film.* **2021**, *39*, 63404. [CrossRef]
93. Bojanowska, M.; Chmiel, J.; Sozańska, M.; Chmiela, B.; Grudzień, J.; Halska, J. Issues of Corrosion and Degradation under Dusty Deposits of Energy Biomass. *Energies* **2021**, *14*, 534. [CrossRef]
94. Calmunger, M.; Wärner, H.; Chai, G.; Segersäll, M. Thermomechanical Fatigue of Heat Resistant Austenitic Alloys. *Procedia Struct. Integr.* **2023**, *43*, 130–135. [CrossRef]
95. Wärner, H.; Chai, G.; Moverare, J.; Calmunger, M. High Temperature Fatigue of Aged Heavy Section Austenitic Stainless Steels. *Materials* **2021**, *15*, 84. [CrossRef] [PubMed]
96. Wärner, H.; Calmunger, M.; Chai, G.; Johansson, S.; Moverare, J. Thermomechanical Fatigue Behaviour of Aged Heat Resistant Austenitic Alloys. *Int. J. Fatigue* **2019**, *127*, 509–521. [CrossRef]
97. Jun, J.; Warrington, G.L.; Keiser, J.R.; Connatser, R.M.; Sulejmanovic, D.; Brady, M.P.; Kass, M.D. Corrosion of Ferrous Structural Alloys in Biomass Derived Fuels and Organic Acids. *Energy Fuels* **2021**, *35*, 12175–12186. [CrossRef]
98. Jun, J.; Frith, M.G.; Connatser, R.M.; Keiser, J.R.; Brady, M.P.; Lewis, S. Corrosion Susceptibility of Cr-Mo Steels and Ferritic Stainless Steels in Biomass-Derived Pyrolysis Oil Constituents. *Energy Fuels* **2020**, *34*, 6220–6228. [CrossRef]
99. Vinnes, M.K.; Worth, N.A.; Segalini, A.; Hearst, R.J. The Flow in the Induction and Entrance Regions of Lab-Scale Wind Farms. *Wind Energy* **2023**, *26*, 1049–1065. [CrossRef]
100. Wind Energy. Available online: https://www.irena.org/Energy-Transition/Technology/Wind-energy (accessed on 19 July 2023).
101. NREL. Wind Energy Basics. Available online: https://www.nrel.gov/research/re-wind.html (accessed on 19 July 2023).
102. National Grid Group. Onshore vs Offshore Wind Energy: What's the Difference? Available online: https://www.nationalgrid.com/stories/energy-explained/onshore-vs-offshore-wind-energy (accessed on 19 July 2023).
103. Greco, A.; Sheng, S.; Keller, J.; Erdemir, A. Material Wear and Fatigue in Wind Turbine Systems. *Wear* **2013**, *302*, 1583–1591. [CrossRef]
104. Reddy, S.S.P.; Suresh, R.; Hanamantraygouda, M.B.; Shivakumar, B.P. Use of Composite Materials and Hybrid Composites in Wind Turbine Blades. *Mater. Today Proc.* **2021**, *46*, 2827–2830. [CrossRef]
105. Worldsteel.Org. Energy. Available online: https://worldsteel.org/steel-topics/steel-markets/energy/ (accessed on 16 July 2023).
106. Khalfallah, M.G.; Koliub, A.M. Effect of Dust on the Performance of Wind Turbines. *Desalination* **2007**, *209*, 209–220. [CrossRef]
107. German Energy Solutions. Solar Thermal Energy. Available online: https://www.german-energy-solutions.de/GES/Redaktion/EN/Text-Collections/EnergySolutions/EnergyGeneration/solar-thermal-energy.html (accessed on 23 July 2023).
108. Kannan, N.; Vakeesan, D. Solar Energy for Future World:—A Review. *Renew. Sustain. Energy Rev.* **2016**, *62*, 1092–1105. [CrossRef]
109. Department of Energy. How Does Solar Work? Available online: https://www.energy.gov/eere/solar/how-does-solar-work (accessed on 19 July 2023).
110. Negi, C.; Kandwal, P.; Rawat, J.; Sharma, M.; Sharma, H.; Dalapati, G.; Dwivedi, C. Carbon-Doped Titanium Dioxide Nanoparticles for Visible Light Driven Photocatalytic Activity. *Appl. Surf. Sci.* **2021**, *554*, 149553. [CrossRef]
111. Yang, X.; Wang, D. Photocatalysis: From Fundamental Principles to Materials and Applications. *ACS Appl. Energy Mater.* **2018**, *1*, 6657–6693. [CrossRef]
112. Hosseini, S.E.; Wahid, M.A. Hydrogen from Solar Energy, a Clean Energy Carrier from a Sustainable Source of Energy. *Int. J. Energy Res.* **2020**, *44*, 4110–4131. [CrossRef]
113. Guillot, S.; Faik, A.; Rakhmatullin, A.; Lambert, J.; Veron, E.; Echegut, P.; Bessada, C.; Calvet, N.; Py, X. Corrosion Effects between Molten Salts and Thermal Storage Material for Concentrated Solar Power Plants. *Appl. Energy* **2012**, *94*, 174–181. [CrossRef]
114. Walczak, M.; Pineda, F.; Fernández, Á.G.; Mata-Torres, C.; Escobar, R.A. Materials Corrosion for Thermal Energy Storage Systems in Concentrated Solar Power Plants. *Renew. Sustain. Energy Rev.* **2018**, *86*, 22–44. [CrossRef]
115. González-Gómez, P.A.; Rodríguez-Sánchez, M.R.; Laporte-Azcué, M.; Santana, D. Calculating Molten-Salt Central-Receiver Lifetime under Creep-Fatigue Damage. *Sol. Energy* **2021**, *213*, 180–197. [CrossRef]
116. Rabelo, M.; Zahid, M.A.; Khokhar, M.Q.; Sim, K.; Oh, H.; Cho, E.C.; Yi, J. Mechanical Fatigue Life Analysis of Solar Panels under Cyclic Load Conditions for Design Improvement. *J. Braz. Soc. Mech. Sci. Eng.* **2022**, *44*, 87. [CrossRef]
117. Galiullin, T.; Gobereit, B.; Naumenko, D.; Buck, R.; Amsbeck, L.; Neises-von Puttkamer, M.; Quadakkers, W.J. High Temperature Oxidation and Erosion of Candidate Materials for Particle Receivers of Concentrated Solar Power Tower Systems. *Sol. Energy* **2019**, *188*, 883–889. [CrossRef]
118. Ali, H.M. Phase Change Materials Based Thermal Energy Storage for Solar Energy Systems. *J. Build. Eng.* **2022**, *56*, 104731. [CrossRef]
119. Ma, T.; Yang, C.; Guo, W.; Lin, H.; Zhang, F.; Liu, H.; Zhao, L.; Zhang, Y.; Wang, Y.; Cui, Y.; et al. Flexible Pt3Ni-S-Deposited Teflon Membrane with High Surface Mechanical Properties for Efficient Solar-Driven Strong Acidic/Alkaline Water Evaporation. *ACS Appl. Mater. Interfaces* **2020**, *12*, 27140–27149. [CrossRef] [PubMed]
120. Li, F.; Li, N.; Wang, S.; Qiao, L.; Yu, L.; Murto, P.; Xu, X.; Li, F.; Li, N.; Wang, S.; et al. Self-Repairing and Damage-Tolerant Hydrogels for Efficient Solar-Powered Water Purification and Desalination. *Adv. Funct. Mater.* **2021**, *31*, 2104464. [CrossRef]
121. Rojas-Morín, A.; Flores-Salgado, Y.; Alvarez-Brito, O.; Jaramillo-Mora, J.; Barba-Pingarrón, A. Thermal Analysis Using Induction and Concentrated Solar Radiation for the Heating of Metals. *Results Eng.* **2022**, *14*, 1000431. [CrossRef]

122. Department of Energy. Geothermal Basics. Available online: https://www.energy.gov/eere/geothermal/geothermal-basics (accessed on 19 July 2023).
123. Wayback Machine. Available online: https://web.archive.org/web/20120217184740/http://geoheat.oit.edu/bulletin/bull28-3/art2.pdf (accessed on 19 July 2023).
124. Geothermal Energy. Available online: https://education.nationalgeographic.org/resource/geothermal-energy/ (accessed on 19 July 2023).
125. Vacquier, V. A Theory of the Origin of the Earth's Internal Heat. *Tectonophysics* **1998**, *291*, 1–7. [CrossRef]
126. Zarrouk, S.J.; Moon, H. Efficiency of Geothermal Power Plants: A Worldwide Review. *Geothermics* **2014**, *51*, 142–153. [CrossRef]
127. Tomarov, G.V.; Borzenko, V.I.; Shipkov, A.A. Application of Hydrogen–Oxygen Steam Generators for Secondary Flash Steam Superheating at Geothermal Power Plants. *Therm. Eng.* **2021**, *68*, 45–53. [CrossRef]
128. Farsi, A.; Rosen, M.A. Comparison of Thermodynamic Performances in Three Geothermal Power Plants Using Flash Steam. *ASME Open J. Eng.* **2022**, *1*, 011016. [CrossRef]
129. Wu, Y.; Li, P. The Potential of Coupled Carbon Storage and Geothermal Extraction in a CO_2-Enhanced Geothermal System: A Review. *Geotherm. Energy* **2020**, *8*, 19. [CrossRef]
130. Kumari, W.G.P.; Ranjith, P.G. Sustainable Development of Enhanced Geothermal Systems Based on Geotechnical Research—A Review. *Earth Sci. Rev.* **2019**, *199*, 102955. [CrossRef]
131. Nasruddin, N.; Dwi Saputra, I.; Mentari, T.; Bardow, A.; Marcelina, O.; Berlin, S. Exergy, Exergoeconomic, and Exergoenvironmental Optimization of the Geothermal Binary Cycle Power Plant at Ampallas, West Sulawesi, Indonesia. *Therm. Sci. Eng. Prog.* **2020**, *19*, 100625. [CrossRef]
132. Duggal, R.; Rayudu, R.; Hinkley, J.; Burnell, J.; Wieland, C.; Keim, M. A Comprehensive Review of Energy Extraction from Low-Temperature Geothermal Resources in Hydrocarbon Fields. *Renew. Sustain. Energy Rev.* **2022**, *154*, 111865. [CrossRef]
133. Van Nimwegen, D.; Van 't Westende, J.; Shoeibi-Omrani, P.; Dinkelman, D.; Peters, E. Sustainability of Geothermal Energy: Handling Co-Produced Gas. In Proceedings of the 3rd EAGE Global Energy Transition, GET 2022, The Hague, The Netherlands, 7–9 November 2022; Volume 2022, pp. 51–55. [CrossRef]
134. Nash, S.S. Offshore Co-Produced Critical Minerals and Geothermal Energy Generation. In Proceedings of the Offshore Technology Conference, Houston, TX, USA, 2–5 May 2022. [CrossRef]
135. Maddah, S.; Goodarzi, M.; Safaei, M.R. Comparative Study of the Performance of Air and Geothermal Sources of Heat Pumps Cycle Operating with Various Refrigerants and Vapor Injection. *Alex. Eng. J.* **2020**, *59*, 4037–4047. [CrossRef]
136. Farzanehkhameneh, P.; Soltani, M.; Moradi Kashkooli, F.; Ziabasharhagh, M. Optimization and Energy-Economic Assessment of a Geothermal Heat Pump System. *Renew. Sustain. Energy Rev.* **2020**, *133*, 110282. [CrossRef]
137. Mahmoud, M.; Ramadan, M.; Pullen, K.; Abdelkareem, M.A.; Wilberforce, T.; Olabi, A.G.; Naher, S. A Review of Grout Materials in Geothermal Energy Applications. *Int. J. Thermofluids* **2021**, *10*, 100070. [CrossRef]
138. Kaya, T.; Hoşhan, P.; Jeotermal Mühendislik, O.; ve Ticaret AŞAnkara, S.; Araştırma Merkezi, T. Corrosion and Material Selection for Geothermal Systems. In Proceedings of the World Geothermal Congress, Antalya, Turkey, 24–29 April 2005; pp. 24–29.
139. Kittel, J.; Ropital, F.; Grosjean, F.; Joshi, G. Evaluation of the Interactions Between Hydrogen and Steel in Geothermal Conditions with H2S. In Proceedings of the World Geothermal Congress, Antalya, Turkey, 24–29 April 2005.
140. Brownlie, F.; Hodgkiess, T.; Pearson, A.; Galloway, A.M. A Study on the Erosion-Corrosion Behaviour of Engineering Materials Used in the Geothermal Industry. *Wear* **2021**, *477*, 203821. [CrossRef]
141. Steel, Q.; Wang, P.; Qiu, B.; Chen, A.; Fink, C.G.; Turner, W.D.; Paul, G.T.; Wolf, M.; Pfennig, A. Failure of Standard Duplex Stainless Steel X2CrNiMoN22-5-3 under Corrosion Fatigue in Geothermal Environment. *IOP Conf. Ser. Mater. Sci. Eng.* **2020**, *894*, 012015. [CrossRef]
142. Tayactac, R.G.; Basilia, B. Corrosion in the Geothermal Systems: A Review of Corrosion Resistance Alloy (CRA) Weld Overlay Cladding Applications. *IOP Conf. Ser. Earth Environ. Sci.* **2022**, *1008*, 012018. [CrossRef]
143. Jasper, S.; Gradzki, D.P.; Bracke, R.; Hussong, J.; Petermann, M.; Lindken, R. Nozzle Cavitation and Rock Erosion Experiments Reveal Insight into the Jet Drilling Process. *Chem. Ing. Tech.* **2021**, *93*, 1610–1618. [CrossRef]
144. Phi, T.; Elgaddafi, R.; Al Ramadan, M.; Fahd, K.; Ahmed, R.; Teodoriu, C. Well Integrity Issues: Extreme High-Pressure High-Temperature Wells and Geothermal Wells a Review. In Proceedings of the Society of Petroleum Engineers—SPE Thermal Well Integrity and Design Symposium 2019, TWID 2019, Banff, AB, Canada, 19–21 November 2019. [CrossRef]
145. Boakye, G.O.; Kovalov, D.; Thorhallsson, A.I.; Karlsdottir, S.N.; Oppong Boakye, G.; Thórhallsson, A.Í.; Karlsdóttir, S.N.; Motoiu, V. Friction and Wear Behaviour of Surface Coatings for Geothermal Applications. In Proceedings of the World Geothermal Congress, Reykjavik, Iceland, 24–27 October 2021.
146. Bates, M.; Valderrama, B.; Bickford, E.; Chan, V.; Tian, L.; Programs, L.; Shah, J.; Wagner, J. *Pathways to Commercial Liftoff: Advanced Nuclear*; US Department of Energy: Washington, DC, USA, 2023.
147. Energy Options Network. *An Energy Innovation Reform Project Report Prepared by the Energy Options Network What Will Advanced Nuclear Power Plants Cost? A Standardized Cost Analysis of Advanced Nuclear Technologies in Commercial Development*; Energy Options Network: New York, NY, USA, 2018.
148. Claverton Group. Owning and Operating Costs of Waste and Biomass Power Plants. Available online: https://claverton-energy.com/owning-and-operating-costs-of-waste-and-biomass-power-plants.html (accessed on 19 July 2023).

149. International Renewable Energy Agency. *Renewable Energy Technologies: Cost Analysis Series Biomass for Power Generation Acknowledgement*; International Renewable Energy Agency: Abu Dhabi, United Arab Emirates, 2012.
150. Blakers, A.; Stocks, M.; Lu, B.; Cheng, C. A Review of Pumped Hydro Energy Storage. *Prog. Energy* **2021**, *3*, 022003. [CrossRef]
151. ETSAP. *Giorgio Hydropower*; Energy Technology Systems Analysis Programme: Paris, France, 2010.
152. Whole Building Design Guide (WBDG). Geothermal Electric Technology. Available online: https://www.wbdg.org/resources/geothermal-electric-technology (accessed on 19 July 2023).
153. LinkedIn. Cost of Geothermal Energy. Available online: https://www.linkedin.com/pulse/cost-geothermal-energy-abdelilahsoni/ (accessed on 19 July 2023).
154. Carroll, J.; McDonald, A.; McMillan, D. Failure Rate, Repair Time and Unscheduled O&M Cost Analysis of Offshore Wind Turbines. *Wind Energy* **2016**, *19*, 1107–1119. [CrossRef]
155. Deb, D.; Brahmbhatt, N.L. Review of Yield Increase of Solar Panels through Soiling Prevention, and a Proposed Water-Free Automated Cleaning Solution. *Renew. Sustain. Energy Rev.* **2018**, *82*, 3306–3313. [CrossRef]
156. Sabzi, M.; Mousavi Anijdan, S.H.; Eivani, A.R.; Park, N.; Jafarian, H.R. The Effect of Pulse Current Changes in PCGTAW on Microstructural Evolution, Drastic Improvement in Mechanical Properties, and Fracture Mode of Dissimilar Welded Joint of AISI 316L-AISI 310S Stainless Steels. *Mater. Sci. Eng. A* **2021**, *823*, 141700. [CrossRef]
157. Tan, J.; Zhang, Z.; Zheng, H.; Wang, X.; Gao, J.; Wu, X.; Han, E.H.; Yang, S.; Huang, P. Corrosion Fatigue Model of Austenitic Stainless Steels Used in Pressurized Water Reactor Nuclear Power Plants. *J. Nucl. Mater.* **2020**, *541*, 152407. [CrossRef]
158. Jo, B.; Okamoto, K.; Kasahara, N. Creep Buckling of 304 Stainless-Steel Tubes Subjected to External Pressure for Nuclear Power Plant Applications. *Metals* **2019**, *9*, 536. [CrossRef]
159. Sadawy, M.M.; El Shazly, R.M. Nuclear Radiation Shielding Effectiveness and Corrosion Behavior of Some Steel Alloys for Nuclear Reactor Systems. *Def. Technol.* **2019**, *15*, 621–628. [CrossRef]
160. Deng, P.; Peng, Q.; Han, E.H.; Ke, W.; Sun, C. Proton Irradiation Assisted Localized Corrosion and Stress Corrosion Cracking in 304 Nuclear Grade Stainless Steel in Simulated Primary PWR Water. *J. Mater. Sci. Technol.* **2021**, *65*, 61–71. [CrossRef]
161. Macdonald, D.D.; Engelhardt, G.R.; Petrov, A. A Critical Review of Radiolysis Issues in Water-Cooled Fission and Fusion Reactors: Part I, Assessment of Radiolysis Models. *Corros. Mater. Degrad.* **2022**, *3*, 470–535. [CrossRef]
162. Chen, X.; Cheng, G.; Hou, Y.; Li, J. Oxide-Inclusion Evolution in the Steelmaking Process of 304L Stainless Steel for Nuclear Power. *Metals* **2019**, *9*, 257. [CrossRef]
163. Yeh, C.P.; Tsai, K.C.; Huang, J.Y. Effects of Relative Humidity on Crevice Corrosion Behavior of 304L Stainless-Steel Nuclear Material in a Chloride Environment. *Metals* **2019**, *9*, 1185. [CrossRef]
164. Yeom, H.; Dabney, T.; Pocquette, N.; Ross, K.; Pfefferkorn, F.E.; Sridharan, K. Cold Spray Deposition of 304L Stainless Steel to Mitigate Chloride-Induced Stress Corrosion Cracking in Canisters for Used Nuclear Fuel Storage. *J. Nucl. Mater.* **2020**, *538*, 152254. [CrossRef]
165. Guo, X.; Liu, Z.; Li, L.; Cheng, J.; Su, H.; Zhang, L. Revealing the Long-Term Oxidation and Carburization Mechanism of 310S SS and Alloy 800H Exposed to Supercritical Carbon Dioxide. *Mater. Charact.* **2022**, *183*, 111603. [CrossRef]
166. Scott, P.M.; Combrade, P. General Corrosion and Stress Corrosion Cracking of Alloy 600 in Light Water Reactor Primary Coolants. *J. Nucl. Mater.* **2019**, *524*, 340–375. [CrossRef]
167. Ahn, T.M. Long-Term Initiation Time for Stress-Corrosion Cracking of Alloy 600 with Implications in Stainless Steel: Review and Analysis for Nuclear Application. *Prog. Nucl. Energy* **2021**, *137*, 103760. [CrossRef]
168. Chakrabarti, C.K.; Kumar, N.; Mishra, R.K.; Bhattacharya, S.; Sengupta, P.; Kaushik, C.P. Palladium Telluride within Nuclear Waste Containing Borosilicate Glass. *Prog. Nucl. Energy* **2022**, *148*, 104236. [CrossRef]
169. Zhou, L.; Dai, J.; Li, Y.; Dai, X.; Xie, C.; Li, L.; Chen, L. Research Progress of Steels for Nuclear Reactor Pressure Vessels. *Materials* **2022**, *15*, 8761. [CrossRef]
170. Bandaru, S.V.R.; Villanueva, W.; Thakre, S.; Bechta, S. Multi-Nozzle Spray Cooling of a Reactor Pressure Vessel Steel Plate for the Application of Ex-Vessel Cooling. *Nucl. Eng. Des.* **2021**, *375*, 111101. [CrossRef]
171. Gao, Z.; Lu, C.; He, Y.; Liu, R.; He, H.; Wang, W.; Zheng, W.; Yang, J. Influence of Phase Transformation on the Creep Deformation Mechanism of SA508 Gr.3 Steel for Nuclear Reactor Pressure Vessels. *J. Nucl. Mater.* **2019**, *519*, 292–301. [CrossRef]
172. Hong, S.; Hyun, S.M.; Kim, J.M.; Lee, Y.S.; Kim, M.C. Effect of Mo and V Addition on Microstructure and Mechanical Properties of SA508 Gr.1A Steel for Pipeline in Nuclear Power Plants. *Met. Mater. Trans. A Phys. Met. Mater. Sci.* **2022**, *53*, 1499–1511. [CrossRef]
173. Carter, M.; Gasparrini, C.; Douglas, J.O.; Riddle, N.; Edwards, L.; Bagot, P.A.J.; Hardie, C.D.; Wenman, M.R.; Moody, M.P. On the Influence of Microstructure on the Neutron Irradiation Response of HIPed SA508 Steel for Nuclear Applications. *J. Nucl. Mater.* **2022**, *559*, 153435. [CrossRef]
174. Ma, X.; She, M.; Zhang, W.; Song, L.; Qiu, S.; Liu, X.; Zhang, R. Microstructure Characterization of Reactor Pressure Vessel Steel A508-3 Irradiated by Heavy Ion. *J. Phys. Conf. Ser.* **2021**, *2133*, 012015. [CrossRef]
175. Vértesy, G.; Gasparics, A.; Szenthe, I.; Gillemot, F.; Uytdenhouwen, I. Inspection of Reactor Steel Degradation by Magnetic Adaptive Testing. *Materials* **2019**, *12*, 963. [CrossRef] [PubMed]
176. Loktionov, V.; Lyubashevskaya, I.; Terentyev, E. The Regularities of Creep Deformation and Failure of the VVER's Pressure Vessel Steel 15Kh2NMFA-A in Air and Argon at Temperature Range 500–900 °C. *Nucl. Mater. Energy* **2021**, *28*, 101019. [CrossRef]

177. Voevodyn, V.; Mytrofanov, S.; Hozhenko, S.V.; Vasylenko, R.L.; Krainyuk, E.; Bazhukov, V.; Palii; Mel'nyk, P.E. Nonmetallic Inclusions in 08Kh18N10T Steel as the Cause of Initiation of Defects in Heat-Exchange Tubes of Steam Generators of Nuclear Power Plants. *Mater. Sci.* **2019**, *55*, 152–159. [CrossRef]
178. Fan, Y.; Liu, T.G.; Xin, L.; Han, Y.M.; Lu, Y.H.; Shoji, T. Thermal Aging Behaviors of Duplex Stainless Steels Used in Nuclear Power Plant: A Review. *J. Nucl. Mater.* **2021**, *544*, 152693. [CrossRef]
179. Chen, Y.; Yang, B.; Zhou, Y.; Wu, Y.; Zhu, H. Evaluation of Pitting Corrosion in Duplex Stainless Steel Fe20Cr9Ni for Nuclear Power Application. *Acta Mater.* **2020**, *197*, 172–183. [CrossRef]
180. Rieth, M.; Dürrschnabel, M.; Bonk, S.; Jäntsch, U.; Bergfeldt, T.; Hoffmann, J.; Antusch, S.; Simondon, E.; Klimenkov, M.; Bonnekoh, C.; et al. Technological Processes for Steel Applications in Nuclear Fusion. *Appl. Sci.* **2021**, *11*, 11653. [CrossRef]
181. Duerrschnabel, M.; Jäntsch, U.; Gaisin, R.; Rieth, M. Microstructural Insights into EUROFER97 Batch 3 Steels. *Nucl. Mater. Energy* **2023**, *35*, 101445. [CrossRef]
182. Klimenkov, M.; Jäntsch, U.; Rieth, M.; Möslang, A. Correlation of Microstructural and Mechanical Properties of Neutron Irradiated EUROFER97 Steel. *J. Nucl. Mater.* **2020**, *538*, 152231. [CrossRef]
183. Cabet, C.; Dalle, F.; Gaganidze, E.; Henry, J.; Tanigawa, H. Ferritic-Martensitic Steels for Fission and Fusion Applications. *J. Nucl. Mater.* **2019**, *523*, 510–537. [CrossRef]
184. Snead, L.L.; Hoelzer, D.T.; Rieth, M.; Nemith, A.A.N. Refractory Alloys: Vanadium, Niobium, Molybdenum, Tungsten. In *Structural Alloys for Nuclear Energy Applications*; Elsevier: Amsterdam, The Netherlands, 2019; pp. 585–640. [CrossRef]
185. Fu, T.; Cui, K.; Zhang, Y.; Wang, J.; Shen, F.; Yu, L.; Qie, J.; Zhang, X. Oxidation Protection of Tungsten Alloys for Nuclear Fusion Applications: A Comprehensive Review. *J. Alloys Compd.* **2021**, *884*, 161057. [CrossRef]
186. Terentyev, D.; Rieth, M.; Pintsuk, G.; Riesch, J.; Von Müller, A.; Antusch, S.; Mergia, K.; Gaganidze, E.; Schneider, H.C.; Wirtz, M.; et al. Recent Progress in the Assessment of Irradiation Effects for In-Vessel Fusion Materials: Tungsten and Copper Alloys. *Nucl. Fusion* **2022**, *62*, 026045. [CrossRef]
187. Alzahrani, J.S.; Alrowaili, Z.A.; Eke, C.; Mahmoud, Z.M.M.; Mutuwong, C.; Al-Buriahi, M.S. Nuclear Shielding Properties of Ni-, Fe-, Pb-, and W-Based Alloys. *Radiat. Phys. Chem.* **2022**, *195*, 110090. [CrossRef]
188. Alzahrani, J.S.; Alrowaili, Z.A.; Saleh, H.H.; Hammoud, A.; Alomairy, S.; Sriwunkum, C.; Al-Buriahi, M.S. Synthesis, Physical and Nuclear Shielding Properties of Novel Pb–Al Alloys. *Prog. Nucl. Energy* **2021**, *142*, 103992. [CrossRef]
189. Pickering, E.J.; Carruthers, A.W.; Barron, P.J.; Middleburgh, S.C.; Armstrong, D.E.J.; Gandy, A.S. High-Entropy Alloys for Advanced Nuclear Applications. *Entropy* **2021**, *23*, 98. [CrossRef]
190. Xin, Z.; Ma, Y.; Chen, Y.; Wang, B.; Xiao, H.; Duan, Y. Fusion-Bonding Performance of Short and Continuous Carbon Fiber Synergistic Reinforced Composites Using Fused Filament Fabrication. *Compos. B Eng.* **2023**, *248*, 110370. [CrossRef]
191. Jiang, X.; Overman, N.; Smith, C.; Ross, K. Microstructure, Hardness and Cavitation Erosion Resistance of Different Cold Spray Coatings on Stainless Steel 316 for Hydropower Applications. *Mater. Today Commun.* **2020**, *25*, 101305. [CrossRef]
192. Kumar, M.; Singh Sidhu, H.; Singh Sidhu, B. Influence of Ultra-Low Temperature Treatments on the Slurry Erosion Performance of Stainless Steel-316L. In *Advances in Mechanical and Materials Technology*; Lecture Notes in Mechanical Engineering; Springer: Singapore, 2022; pp. 1273–1286. [CrossRef]
193. Pitsikoulis, S.A.; Tekumalla, S.; Sharma, A.; Wong, W.L.E.; Turkmen, S.; Liu, P. Cavitation Hydrodynamic Performance of 3-D Printed Highly Skewed Stainless Steel Tidal Turbine Rotors. *Energies* **2023**, *16*, 3675. [CrossRef]
194. Jung, H.J.; Jabbar, H.; Song, Y.; Sung, T.H. Hybrid-Type (D33 and D31) Impact-Based Piezoelectric Hydroelectric Energy Harvester for Watt-Level Electrical Devices. *Sens. Actuators A Phys.* **2016**, *245*, 40–48. [CrossRef]
195. Gavriilidis, I.; Huang, Y. Finite Element Analysis of Tidal Turbine Blade Subjected to Impact Loads from Sea Animals. *Energies* **2021**, *14*, 7208. [CrossRef]
196. Saenz-Betancourt, C.C.; Rodríguez, S.A.; Coronado, J.J. Effect of Boronising on the Cavitation Erosion Resistance of Stainless Steel Used for Hydromachinery Applications. *Wear* **2022**, *498–499*, 204330. [CrossRef]
197. Azevedo, C.R.F.; Magarotto, D.; Araújo, J.A.; Ferreira, J.L.A. Bending Fatigue of Stainless Steel Shear Pins Belonging to a Hydroelectric Plant. *Eng. Fail. Anal.* **2009**, *16*, 1126–1140. [CrossRef]
198. Sankar, S.; Nataraj, M.; Raja, V.P. Failure Analysis of Shear Pins in Wind Turbine Generator. *Eng. Fail. Anal.* **2011**, *18*, 325–339. [CrossRef]
199. Zhang, Q.L.; Hu, L.; Hu, C.; Wu, H.G. Low-Cycle Fatigue Issue of Steel Spiral Cases in Pumped-Storage Power Plants under China's and US's Design Philosophies: A Comparative Numerical Case Study. *Int. J. Press. Vessel. Pip.* **2019**, *172*, 134–144. [CrossRef]
200. Moreira, D.C.; Furtado, H.C.; Buarque, J.S.; Cardoso, B.R.; Merlin, B.; Moreira, D.D.C. Failure Analysis of AISI 410 Stainless-Steel Piston Rod in Spillway Floodgate. *Eng. Fail. Anal.* **2019**, *97*, 506–517. [CrossRef]
201. Maekai, I.A.; Harmain, G.A.; Zehab-ud-Din; Masoodi, J.H. Resistance to Slurry Erosion by WC-10Co-4Cr and Cr3C2 − 25(Ni20Cr) Coatings Deposited by HVOF Stainless Steel F6NM. *Int. J. Refract. Met. Hard Mater.* **2022**, *105*, 105830. [CrossRef]
202. Chauhan, A.K.; Goel, D.B.; Prakash, S. Erosion Behaviour of Hydro Turbine Steels. *Bull. Mater. Sci.* **2008**, *31*, 115–120. [CrossRef]
203. Kumar, S.; Chaudhari, G.P.; Nath, S.K.; Basu, B. Effect of Preheat Temperature on Weldability of Martensitic Stainless Steel. *Mater. Manuf. Process.* **2012**, *27*, 1382–1386. [CrossRef]

204. Rückle, D.; Schellenberg, G.; Ottens, W.; Leibing, B.; Locquenghien, F. Corrosion Fatigue of CrNi13-4 Martensitic Stainless Steel for Francis Runners in Dependency of Water Quality. In Proceedings of the Eurocorr 2022, Berlin, Germany, 28 August–1 September 2022.
205. Kevin, P.S.; Tiwari, A.; Seman, S.; Mohamed, S.A.B.; Jayaganthan, R. Erosion-Corrosion Protection Due to Cr_3C_2-NiCr Cermet Coating on Stainless Steel. *Coatings* **2020**, *10*, 1042. [CrossRef]
206. Titus, J.; Ayalur, B. Design and Fabrication of In-Line Turbine for Pico Hydro Energy Recovery in Treated Sewage Water Distribution Line. *Energy Procedia* **2019**, *156*, 133–138. [CrossRef]
207. Turnock, S.R.; Muller, G.; Nicholls-Lee, R.F.; Denchfield, S.; Hindley, S.; Shelmerdine, R.; Stevens, S. Development of a Floating Tidal Energy System Suitable for Use in Shallow Water. In Proceedings of the 7th European Wave and Tidal Energy Conference, Porto, Portugal, 11–13 September 2007.
208. Keiser, J. *Materials Degradation of Biomass-Derived Oils*; U.S. Department of Energy: Washington, DC, USA, 2021.
209. Special Piping Materials. Stainless Steel and Renewable Energy. Available online: https://specialpipingmaterials.com/stainless-steel-and-renewable-energy/ (accessed on 16 July 2023).
210. Guerrero, G.R.; Sevilla, L.; Soriano, C. Laser and Pyrolysis Removal of Fluorinated Ethylene Propylene Thin Layers Applied on EN AW-5251 Aluminium Substrates. *Appl. Surf. Sci.* **2015**, *353*, 686–692. [CrossRef]
211. Divandari, M.; Jamali, V.; Shabestari, S.G. Effect of Strips Size and Coating Thickness on Fluidity of A356 Aluminium Alloy in Lost Foam Casting Process. *Int. J. Cast Met. Res.* **2013**, *23*, 23–29. [CrossRef]
212. Jun, J.; Su, Y.F.; Keiser, J.R.; Wade, J.E.; Kass, M.D.; Ferrell, J.R.; Christensen, E.; Olarte, M.V.; Sulejmanovic, D. Corrosion Compatibility of Stainless Steels and Nickel in Pyrolysis Biomass-Derived Oil at Elevated Storage Temperatures. *Sustainability* **2022**, *15*, 22. [CrossRef]
213. Diaz, F.; Wang, Y.; Moorthy, T.; Friedrich, B. Degradation Mechanism of Nickel-Cobalt-Aluminum (NCA) Cathode Material from Spent Lithium-Ion Batteries in Microwave-Assisted Pyrolysis. *Metals* **2018**, *8*, 565. [CrossRef]
214. Brady, M.P.; Leonard, D.N.; Keiser, J.R.; Cakmak, E.; Whitmer, L.E. Degradation of Components After Exposure in a Biomass Pyrolysis System. *Corrosion* **2019**, *75*, 1136–1145. [CrossRef]
215. Braun, M. Statistical Analysis of Sub-Zero Temperature Effects on Fatigue Strength of Welded Joints. *Weld. World* **2022**, *66*, 159–172. [CrossRef]
216. Braun, M.; Scheffer, R.; Fricke, W.; Ehlers, S. Fatigue Strength of Fillet-Welded Joints at Subzero Temperatures. *Fatigue Fract. Eng. Mater. Struct.* **2020**, *43*, 403–416. [CrossRef]
217. Igwemezie, V.; Mehmanparast, A.; Kolios, A. Materials Selection for XL Wind Turbine Support Structures: A Corrosion-Fatigue Perspective. *Mar. Struct.* **2018**, *61*, 381–397. [CrossRef]
218. Momber, A. Corrosion and Corrosion Protection of Support Structures for Offshore Wind Energy Devices (OWEA). *Mater. Corros.* **2011**, *62*, 391–404. [CrossRef]
219. Martin, F.; Schröter, F. Stahllösungen Für Offshore-Windkraftanlagen. *Stahlbau* **2005**, *74*, 435–442. [CrossRef]
220. Ferro, P.; Berto, F.; Tang, K. UNS S32205 Duplex Stainless Steel SED-Critical Radius Characterization. *Metall. Ital.* **2020**, 29–38.
221. Sroka, M.; Zielinski, A.; Puszczalo, T.; Sówka, K.; Hadzima, B. Structure of 22Cr25NiWCoCu Austenitic Stainless Steel After Ageing. *Arch. Metall. Mater.* **2022**, *67*, 175–180. [CrossRef]
222. Ermakova, A.; Mehmanparast, A.; Ganguly, S. A Review of Present Status and Challenges of Using Additive Manufacturing Technology for Offshore Wind Applications. *Procedia Struct. Integr.* **2019**, *17*, 29–36. [CrossRef]
223. Nakagawa, T.; Matsushima, H.; Ueda, M.; Ito, H. Corrosion Behavior of SUS 304L Steel in Concentrated K_2CO_3 Solution. *ECS Meet. Abstr.* **2020**, *MA2020-02*, 1228. [CrossRef]
224. Röhsler, A.; Sobol, O.; Hänninen, H.; Böllinghaus, T. In-Situ ToF-SIMS Analyses of Deuterium Re-Distribution in Austenitic Steel AISI 304L under Mechanical Load. *Sci. Rep.* **2020**, *10*, 10545. [CrossRef]
225. Yang, D.; Huang, Y.; Peng, P.; Liu, X.; Zhang, B. Passivation Behavior and Corrosion Resistance of 904L Austenitic Stainless Steels in Static Seawater. *Int. J. Electrochem. Sci.* **2019**, *14*, 6133–6146. [CrossRef]
226. Sanni, O.; Popoola, A.P.I.; Fayomi, O.S.I. Electrochemical Analysis of Austenitic Stainless Steel (Type 904) Corrosion Using Egg Shell Powder in Sulphuric Acid Solution. *Energy Procedia* **2019**, *157*, 619–625. [CrossRef]
227. Evans, M.H. White Structure Flaking (WSF) in Wind Turbine Gearbox Bearings: Effects of 'Butterflies' and White Etching Cracks (WECs). *Mater. Sci. Technol.* **2013**, *28*, 3–22. [CrossRef]
228. Gooch, D.J. Materials Issues in Renewable Energy Power Generation. *Int. Mater. Rev.* **2013**, *45*, 1–14. [CrossRef]
229. Chen, Y.; Cai, G.; Bai, R.; Ke, S.; Wang, W.; Chen, X.; Li, P.; Zhang, Y.; Gao, L.; Nie, S.; et al. Spatiotemporally Explicit Pathway and Material-Energy-Emission Nexus of Offshore Wind Energy Development in China up to the Year 2060. *Resour. Conserv. Recycl.* **2022**, *183*, 106349. [CrossRef]
230. Lloberas-Valls, J.; Benveniste Perez, G.; Gomis-Bellmunt, O. Life-Cycle Assessment Comparison between 15-MW Second-Generation High Temperature Superconductor and Permanent-Magnet Direct-Drive Synchronous Generators for Offshore Wind Energy Applications. *IEEE Trans. Appl. Supercond.* **2015**, *25*, 5204209. [CrossRef]
231. Kumar, A.; Dwivedi, A.; Paliwal, V.; Patil, P.P. Free Vibration Analysis of Al 2024 Wind Turbine Blade Designed for Uttarakhand Region Based on FEA. *Procedia Technol.* **2014**, *14*, 336–347. [CrossRef]
232. Siddiqui, R.A.; Abdullah, H.A.; Al-Belushi, K.R. Influence of Aging Parameters on the Mechanical Properties of 6063 Aluminium Alloy. *J. Mater. Process. Technol.* **2000**, *102*, 234–240. [CrossRef]

233. Wang, H.; Lamichhane, T.N.; Paranthaman, M.P. Review of Additive Manufacturing of Permanent Magnets for Electrical Machines: A Prospective on Wind Turbine. *Mater. Today Phys.* **2022**, *24*, 100675. [CrossRef]
234. Lüddecke, F.; Rücker, W.; Seidel, M.; Assheuer, J. Tragverhalten von Stahlgussbauteilen in Offshore-Windenergie-Anlagen Unter Vorwiegend Ruhender Und Nicht Ruhender Beanspruchung. *Stahlbau* **2008**, *77*, 639–646. [CrossRef]
235. Bleicher, C.; Niewiadomski, J.; Kansy, A.; Kaufmann, H. High-Silicon Nodular Cast Iron for Lightweight Optimized Wind Energy Components. *Int. J. Offshore Polar Eng.* **2022**, *32*, 201–209. [CrossRef]
236. Soleimani Dorcheh, A.; Durham, R.N.; Galetz, M.C. Corrosion Behavior of Stainless and Low-Chromium Steels and IN625 in Molten Nitrate Salts at 600 °C. *Sol. Energy Mater. Sol. Cells* **2016**, *144*, 109–116. [CrossRef]
237. Li, H.; Feng, X.; Wang, X.; Yang, X.; Tang, J.; Gong, J. Impact of Temperature on Corrosion Behavior of Austenitic Stainless Steels in Solar Salt for CSP Application: An Electrochemical Study. *Sol. Energy Mater. Sol. Cells* **2022**, *239*, 111661. [CrossRef]
238. Pantelis, D.I.; Griniari, A.; Sarafoglou, C. Surface Alloying of Pre-Deposited Molybdenum-Based Powder on 304L Stainless Steel Using Concentrated Solar Energy. *Sol. Energy Mater. Sol. Cells* **2005**, *89*, 1–11. [CrossRef]
239. Wang, M.; Zeng, S.; Zhang, H.; Zhu, M.; Lei, C.; Li, B. Corrosion Behaviors of 316 Stainless Steel and Inconel 625 Alloy in Chloride Molten Salts for Solar Energy Storage. *High. Temp. Mater. Process.* **2020**, *39*, 340–350. [CrossRef]
240. Yin, Y.; Rumman, R.; Sarvghad, M.; Bell, S.; Ong, T.C.; Jacob, R.; Liu, M.; Flewell-Smith, R.; Sheoran, S.; Severino, J.; et al. Role of Headspace Environment for Phase Change Carbonates on the Corrosion of Stainless Steel 316L: High Temperature Thermal Storage Cycling in Concentrated Solar Power Plants. *Sol. Energy Mater. Sol. Cells* **2023**, *251*, 112170. [CrossRef]
241. Cionea, C.; Abad, M.D.; Aussat, Y.; Frazer, D.; Gubser, A.J.; Hosemann, P. Oxide Scale Formation on 316L and FeCrAl Steels Exposed to Oxygen Controlled Static LBE at Temperatures up to 800 °C. *Sol. Energy Mater. Sol. Cells* **2016**, *144*, 235–246. [CrossRef]
242. Liu, Q.; Qian, J.; Neville, A.; Pessu, F. Solar Thermal Irradiation Cycles and Their Influence on the Corrosion Behaviour of Stainless Steels with Molten Salt Used in Concentrated Solar Power Plants. *Sol. Energy Mater. Sol. Cells* **2023**, *251*, 112141. [CrossRef]
243. Brown-Shaklee, H.J.; Carty, W.; Edwards, D.D. Spectral Selectivity of Composite Enamel Coatings on 321 Stainless Steel. *Sol. Energy Mater. Sol. Cells* **2009**, *93*, 1404–1410. [CrossRef]
244. Sierra, C.; Vázquez, A.J. High Solar Energy Concentration with a Fresnel Lens. *J. Mater. Sci.* **2005**, *40*, 1339–1343. [CrossRef]
245. Mallco, A.; Pineda, F.; Mendoza, M.; Henriquez, M.; Carrasco, C.; Vergara, V.; Fuentealba, E.; Fernandez, A.G. Evaluation of Flow Accelerated Corrosion and Mechanical Performance of Martensitic Steel T91 for a Ternary Mixture of Molten Salts for CSP Plants. *Sol. Energy Mater. Sol. Cells* **2022**, *238*, 111623. [CrossRef]
246. Pineda, F.; Mallco, A.; De Barbieri, F.; Carrasco, C.; Henriquez, M.; Fuentealba, E.; Fernández, Á.G. Corrosion Evaluation by Electrochemical Real-Time Tracking of VM12 Martensitic Steel in a Ternary Molten Salt Mixture with Lithium Nitrates for CSP Plants. *Sol. Energy Mater. Sol. Cells* **2021**, *231*, 111302. [CrossRef]
247. The Amazing Role of High-Temperature Nickel Alloys and Stainless Steels for Concentrated Solar Power. Available online: https://nickelinstitute.org/en/blog/2021/september/the-amazing-role-of-high-temperature-nickel-alloys-and-stainless-steels-for-concentrated-solar-power/ (accessed on 18 July 2023).
248. Grégoire, B.; Oskay, C.; Meißner, T.M.; Galetz, M.C. Corrosion Mechanisms of Ferritic-Martensitic P91 Steel and Inconel 600 Nickel-Based Alloy in Molten Chlorides. Part II: NaCl-KCl-MgCl$_2$ Ternary System. *Sol. Energy Mater. Sol. Cells* **2020**, *216*, 110675. [CrossRef]
249. Balat-Pichelin, M.; Sans, J.L.; Bêche, E.; Charpentier, L.; Ferrière, A.; Chomette, S. Emissivity at High Temperature of Ni-Based Superalloys for the Design of Solar Receivers for Future Tower Power Plants. *Sol. Energy Mater. Sol. Cells* **2021**, *227*, 111066. [CrossRef]
250. Zhu, M.; Yi, H.; Lu, J.; Huang, C.; Zhang, H.; Bo, P.; Huang, J. Corrosion of Ni–Fe Based Alloy in Chloride Molten Salts for Concentrating Solar Power Containing Aluminum as Corrosion Inhibitor. *Sol. Energy Mater. Sol. Cells* **2022**, *241*, 111737. [CrossRef]
251. Peng, Y.; Shinde, P.S.; Reddy, R.G. High-Temperature Corrosion Rate and Diffusion Modelling of Ni-Coated Incoloy 800H Alloy in MgCl$_2$–KCl Heat Transfer Fluid. *Sol. Energy Mater. Sol. Cells* **2022**, *243*, 111767. [CrossRef]
252. Bell, S.; de Bruyn, M.; Steinberg, T.; Will, G. C-276 Nickel Alloy Corrosion in Eutectic Na$_2$CO$_3$/NaCl Molten Salt under Isothermal and Thermal Cycling Conditions. *Sol. Energy Mater. Sol. Cells* **2022**, *240*, 111695. [CrossRef]
253. García-Martín, G.; Lasanta, M.I.; Encinas-Sánchez, V.; de Miguel, M.T.; Pérez, F.J. Evaluation of Corrosion Resistance of A516 Steel in a Molten Nitrate Salt Mixture Using a Pilot Plant Facility for Application in CSP Plants. *Sol. Energy Mater. Sol. Cells* **2017**, *161*, 226–231. [CrossRef]
254. Cheng, W.J.; Chen, D.J.; Wang, C.J. High-Temperature Corrosion of Cr–Mo Steel in Molten LiNO$_3$–NaNO$_3$–KNO$_3$ Eutectic Salt for Thermal Energy Storage. *Sol. Energy Mater. Sol. Cells* **2015**, *132*, 563–569. [CrossRef]
255. Sarvghad, M.; Will, G.; Steinberg, T.A. Corrosion of Steel Alloys in Molten NaCl + Na$_2$SO$_4$ at 700 °C for Thermal Energy Storage. *Sol. Energy Mater. Sol. Cells* **2018**, *179*, 207–216. [CrossRef]
256. Karalis, D.G.; Pantelis, D.I.; Papazoglou, V.J. On the Investigation of 7075 Aluminum Alloy Welding Using Concentrated Solar Energy. *Sol. Energy Mater. Sol. Cells* **2005**, *86*, 145–163. [CrossRef]
257. Shimizu, Y.; Kawaguchi, T.; Sakai, H.; Dong, K.; Kurniawan, A.; Nomura, T. Al-Ni Alloy-Based Core-Shell Type Microencapsulated Phase Change Material for High Temperature Thermal Energy Utilization. *Sol. Energy Mater. Sol. Cells* **2022**, *246*, 111874. [CrossRef]

258. Sugianto, A.; Tjahjono, B.S.; Mai, L.; Wenham, S.R. Investigation of Unusual Shunting Behavior Due to Phototransistor Effect in N-Type Aluminum-Alloyed Rear Junction Solar Cells. *Sol. Energy Mater. Sol. Cells* **2009**, *93*, 1986–1993. [CrossRef]
259. Krause, J.; Woehl, R.; Rauer, M.; Schmiga, C.; Wilde, J.; Biro, D. Microstructural and Electrical Properties of Different-Sized Aluminum-Alloyed Contacts and Their Layer System on Silicon Surfaces. *Sol. Energy Mater. Sol. Cells* **2011**, *95*, 2151–2160. [CrossRef]
260. Patel, K.; Sadeghilaridjani, M.; Pole, M.; Mukherjee, S. Hot Corrosion Behavior of Refractory High Entropy Alloys in Molten Chloride Salt for Concentrating Solar Power Systems. *Sol. Energy Mater. Sol. Cells* **2021**, *230*, 111222. [CrossRef]
261. He, C.Y.; Gao, X.H.; Dong, M.; Qiu, X.L.; An, J.H.; Guo, H.X.; Liu, G. Further Investigation of a Novel High Entropy Alloy MoNbHfZrTi Based Solar Absorber Coating with Double Antireflective Layers. *Sol. Energy Mater. Sol. Cells* **2020**, *217*, 110709. [CrossRef]
262. Yamada, Y.; Miura, M.; Tajima, K.; Okada, M.; Yoshimura, K. Optical Switching Durability of Switchable Mirrors Based on Magnesium–Yttrium Alloy Thin Films. *Sol. Energy Mater. Sol. Cells* **2013**, *117*, 396–399. [CrossRef]
263. Bao, S.; Tajima, K.; Yamada, Y.; Okada, M.; Yoshimura, K. Magnesium–Titanium Alloy Thin-Film Switchable Mirrors. *Sol. Energy Mater. Sol. Cells* **2008**, *92*, 224–227. [CrossRef]
264. Hot Rocks—Geothermal and the Role of Nickel. Available online: https://nickelinstitute.org/en/blog/2021/august/hot-rocks-geothermal-and-the-role-of-nickel/ (accessed on 18 July 2023).
265. Wang, G.G.; Zhu, L.Q.; Liu, H.C.; Li, W.P. Galvanic Corrosion of Ni–Cu–Al Composite Coating and Its Anti-Fouling Property for Metal Pipeline in Simulated Geothermal Water. *Surf. Coat. Technol.* **2012**, *206*, 3728–3732. [CrossRef]
266. Keserović, A.; Bäßler, R.; Kamah, Y. *Suitability of Alloyed Steels in Highly Acidic Geothermal Environments*; NACE: San Antonio, TX, USA, 2014.
267. Nogara, J.; Zarrouk, S.J. Corrosion in Geothermal Environment Part 2: Metals and Alloys. *Renew. Sustain. Energy Rev.* **2018**, *82*, 1347–1363. [CrossRef]
268. Jalaluddin; Miyara, A.; Tsubaki, K.; Inoue, S.; Yoshida, K. Experimental Study of Several Types of Ground Heat Exchanger Using a Steel Pile Foundation. *Renew. Energy* **2011**, *36*, 764–771. [CrossRef]
269. Faizal, M.; Bouazza, A.; Singh, R.M. Heat Transfer Enhancement of Geothermal Energy Piles. *Renew. Sustain. Energy Rev.* **2016**, *57*, 16–33. [CrossRef]
270. Vitaller, A.V.; Angst, U.M.; Elsener, B. Corrosion Behaviour of L80 Steel Grade in Geothermal Power Plants in Switzerland. *Metals* **2019**, *9*, 331. [CrossRef]
271. AI Technology Inc. Solar Energy Enhancement Protection Coating, Sealant and Adhesive. Available online: https://www.aitechnology.com/products/solar/solar-energy-coating/ (accessed on 23 July 2023).
272. Strom-Forschung De. Protecting Wind Turbines from Corrosion. Available online: https://www.strom-forschung.de/projects/wind-energy/protecting-wind-turbines-from-corrosion/ (accessed on 23 July 2023).
273. Wang, C.; Gao, W.; Liu, N.; Xin, Y.; Liu, X.; Wang, X.; Tian, Y.; Chen, X.; Hou, B. Covalent Organic Framework Decorated TiO$_2$ Nanotube Arrays for Photoelectrochemical Cathodic Protection of Steel. *Corros. Sci.* **2020**, *176*, 108920. [CrossRef]
274. Angst, U.M. A Critical Review of the Science and Engineering of Cathodic Protection of Steel in Soil and Concrete. *Corrosion* **2019**, *75*, 1420–1433. [CrossRef] [PubMed]
275. Brenna, A.; Beretta, S.; Ormellese, M. AC Corrosion of Carbon Steel under Cathodic Protection Condition: Assessment, Criteria and Mechanism. A Review. *Materials* **2020**, *13*, 2158. [CrossRef] [PubMed]
276. Zhang, R.; Yu, X.; Yang, Q.; Cui, G.; Li, Z. The Role of Graphene in Anti-Corrosion Coatings: A Review. *Constr. Build. Mater.* **2021**, *294*, 123613. [CrossRef]
277. Hussain, A.K.; Seetharamaiah, N.; Pichumani, M.; Chakra, C.S. Research Progress in Organic Zinc Rich Primer Coatings for Cathodic Protection of Metals—A Comprehensive Review. *Prog. Org. Coat.* **2021**, *153*, 106040. [CrossRef]
278. Krieg, R.; Vimalanandan, A.; Rohwerder, M. Corrosion of Zinc and Zn-Mg Alloys with Varying Microstructures and Magnesium Contents. *J. Electrochem. Soc.* **2014**, *161*, C156–C161. [CrossRef]
279. Hausbrand, R.; Stratmann, M.; Rohwerder, M. Corrosion of Zinc–Magnesium Coatings: Mechanism of Paint Delamination. *Corros. Sci.* **2009**, *51*, 2107–2114. [CrossRef]
280. Sun, W.; Wang, N.; Li, J.; Xu, S.; Song, L.; Liu, Y.; Wang, D. Humidity-Resistant Triboelectric Nanogenerator and Its Applications in Wind Energy Harvesting and Self-Powered Cathodic Protection. *Electrochim. Acta* **2021**, *391*, 138994. [CrossRef]
281. Hautfenne, C.; Nardone, S.; De Bruycker, E.; Hautfenne, C. Influence of Heat Treatments and Build Orientation on the Creep Strength of Additive Manufactured IN718 Contact Data. In Proceedings of the 4th International ECCC Conference, Düsseldorf, Germany, 10–14 September 2017.
282. Reisgen, U.; Olschok, S.; Jakobs, S. Laser Beam Welding in Vacuum of Thick Plate Structural Steel. In Proceedings of the ICALEO 2013—32nd International Congress on Applications of Lasers and Electro-Optics, Orlando, FL, USA, 23–27 October 2013; pp. 341–350.
283. Vázquez, A.J.; Rodriguez, G.P.; de Damborenea, J. Surface Treatment of Steels by Solar Energy. *Sol. Energy Mater.* **1991**, *24*, 751–759. [CrossRef]
284. Pisarek, J.; Frączek, T.; Popławski, T.; Szota, M. Practical and Economical Effects of the Use of Screen Meshes for Steel Nitriding Processes with Glow Plasma. *Energies* **2021**, *14*, 3808. [CrossRef]

285. Micallef, C.; Zhuk, Y.; Aria, A.I. Recent Progress in Precision Machining and Surface Finishing of Tungsten Carbide Hard Composite Coatings. *Coatings* **2020**, *10*, 731. [CrossRef]
286. Pan, X.; Shi, Z.; Shi, C.; Ling, T.C.; Li, N. A Review on Concrete Surface Treatment Part I: Types and Mechanisms. *Constr. Build. Mater.* **2017**, *132*, 578–590. [CrossRef]
287. Poulain, R.; Amann, F.; Deya, J.; Bourgon, J.; Delannoy, S.; Prima, F. Short-Time Heat Treatments in Oxygen Strengthened Ti-Zr α Titanium Alloys: A Simple Approach for Optimized Strength-Ductility Trade-Off. *Mater. Lett.* **2022**, *317*, 132114. [CrossRef]
288. Osunbunmi, I.S.; Ajide, O.O.; Oluwole, O.O. Effect of Heat Treatment on the Mechanical and Microstructural Properties of a Low Carbon Steel. In Proceedings of the International Conference of Mechanical Engineering, Energy Technology and Management, IMEETMCON 2018, Ibadan, Nigeria, 4–7 September 2018.
289. Jovičević-Klug, P.; Podgornik, B. Review on the Effect of Deep Cryogenic Treatment of Metallic Materials in Automotive Applications. *Metals* **2020**, *10*, 434. [CrossRef]
290. Baldissera, P.; Delprete, C. Deep Cryogenic Treatment: A Bibliographic Review. *Open Mech. Eng. J.* **2008**, *2*, 32–39. [CrossRef]
291. Vengatesh, M.; Srivignesh, R.; Pradeep, T.; Karthik, N.R. Review on Cryogenic Trearment of Steels. *Int. Res. J. Eng. Technol.* **2016**, *3*, 417–422.
292. Sonar, T.; Lomte, S.; Gogte, C. Cryogenic Treatment of Metal—A Review. *Mater. Today Proc.* **2018**, *5*, 25219–25228. [CrossRef]
293. Jovičević-Klug, P. *Mechanisms and Effect of Deep Cryogenic Treatment on Steel Properties*; Institute of Metals and Technology: Ljubljana, Slovenia, 2022.
294. Workman, K.J.; Pitts, D.W. Method of Treating Brake Pads. U.S. Patent US5447035A, 29 September 1994.
295. Leach, J.M.; Harvey, W.L. Method of Cryogenically Hardening an Insert in an Article. U.S. Patent No. 4,336,077, 3 March 1980.
296. Voorhees, J.E. Process for Treating Materials to Improve Their Structural Characteristics. U.S. Patent No. 4,482,005, 13 November 1984.
297. Yildiz, Y.; Nalbant, M. A Review of Cryogenic Cooling in Machining Processes. *Int. J. Mach. Tools Manuf.* **2008**, *48*, 947–964. [CrossRef]
298. Akincioğlu, S.; Gökkaya, H.; Uygur, İ. A Review of Cryogenic Treatment on Cutting Tools. *Int. J. Adv. Manuf. Technol.* **2015**, *78*, 1609–1627. [CrossRef]
299. Barron, R.F. Cryogenic Treatment of Metals to Improve Wear Resistance. *Cryogenics* **1982**, *22*, 409–413. [CrossRef]
300. Senthilkumar, D.; Rajendran, I. Influence of Shallow and Deep Cryogenic Treatment on Tribological Behavior of En 19 Steel. *J. Iron Steel Res. Int.* **2011**, *18*, 53–59. [CrossRef]
301. Senthilkumar, D.; Rajendran, I.; Pellizzari, M.; Siiriainen, J. Influence of Shallow and Deep Cryogenic Treatment on the Residual State of Stress of 4140 Steel. *J. Mater. Process. Technol.* **2011**, *211*, 396–401. [CrossRef]
302. Senthilkumar, D. Cryogenic Treatment: Shallow and Deep. In *Encyclopedia of Iron, Steel, and Their Alloys*; Totten, G.E., Colas, R., Eds.; Taylor and Francis: New York, NY, USA, 2016; pp. 995–1007. ISBN 9781351254496.
303. Ciski, A.; Nawrocki, P.; Babul, T.; Hradil, D. Multistage Cryogenic Treatment of X153CrMoV12 Cold Work Steel. In Proceedings of the 5th International Conference Recent Trends in Structural Materials, Pilsen, Czech Republic, 14–16 November 2018; Volume 461, pp. 0120121–0120126.
304. Jovičević-Klug, P.; Jovičević-Klug, M.; Thormählen, L.; McCord, J.; Rohwerder, M.; Godec, M.; Podgornik, B. Austenite Reversion Suppression with Deep Cryogenic Treatment: A Novel Pathway towards 3rd Generation Advanced High-Strength Steels. *Mater. Sci. Eng. A* **2023**, *873*, 145033. [CrossRef]
305. Jovičević-Klug, P.; Tegg, L.; Jovičević-Klug, M.; Parmar, R.; Amati, M.; Gregoratti, L.; Almasy, L.; Cairney, J.M.; Podgornik, B. Understanding Carbide Evolution and Surface Chemistry during Deep Cryogenic Treatment in High-Alloyed Ferrous Alloy. *Appl. Surf. Sci.* **2023**, *610*, 155497. [CrossRef]
306. Jovičević-Klug, P.; Puš, G.; Jovičević-Klug, M.; Žužek, B.; Podgornik, B. Influence of Heat Treatment Parameters on Effectiveness of Deep Cryogenic Treatment on Properties of High-Speed Steels. *Mater. Sci. Eng. A* **2022**, *829*, 142157. [CrossRef]
307. Jovičević-Klug, P.; Guštin, A.Z.; Jovičević-Klug, M.; Šetina Batič, B.; Lebar, A.; Podgornik, B. Coupled Role of Alloying and Manufacturing on Deep Cryogenic Treatment Performance on High-Alloyed Ferrous Alloys. *J. Mater. Res. Technol.* **2022**, *18*, 3184–3197. [CrossRef]
308. Jovičević-Klug, P.; Lipovšek, N.; Jovičević-Klug, M.; Mrak, M.; Ekar, J.; Ambrožič, B.; Dražić, G.; Kovač, J.; Podgornik, B. Assessment of Deep Cryogenic Heat-Treatment Impact on the Microstructure and Surface Chemistry of Austenitic Stainless Steel. *Surf. Interfaces* **2022**, *35*, 102456. [CrossRef]
309. Jovičević-Klug, P.; Jovičević-Klug, M.; Sever, T.; Feizpour, D.; Podgornik, B. Impact of Steel Type, Composition and Heat Treatment Parameters on Effectiveness of Deep Cryogenic Treatment. *J. Mater. Res. Technol.* **2021**, *14*, 1007–1020. [CrossRef]
310. Pellizzari, M. Influence of Deep Cryogenic Treatment on the Properties of Conventional and PM High Speed Steels. *Metall. Ital.* **2008**, *100*, 17–22.
311. Senthilkumar, D. Influence of Deep Cryogenic Treatment on Hardness and Toughness of En31 Steel. *Adv. Mater. Process. Technol.* **2019**, *5*, 114–122. [CrossRef]
312. Baldissera, P. Deep Cryogenic Treatment of AISI 302 Stainless Steel: Part I—Hardness and Tensile Properties. *Mater. Des.* **2010**, *31*, 4725–4730. [CrossRef]
313. Amini, K.; Negahbani, M.; Ghayour, H. The Effect of Deep Cryogenic Treatment on Hardness and Wear Behavior of the H13 Tool Steel. *Metall. Ital.* **2015**, *107*, 53–58.

314. Elango, G.; Raghunath, B.K.; Thamizhmaran, K. Effect of Cryogenic Treatment on Microstructure and Micro Hardness of Aluminium (LM25)—SiC Metal Matrix Composite. *J. Eng. Res.* **2014**, *10*, 64–68. [CrossRef]
315. Peng, J.; Zhou, B.; Li, Z.; Huo, D.; Xiong, J.; Zhang, S. Effect of Tempering Process on the Cryogenic Impact Toughness of 13Cr4NiMo Martensitic Stainless Steel. *J. Mater. Res. Technol.* **2023**, *23*, 5618–5630. [CrossRef]
316. Çakir, F.H.; Çelik, O.N. The Effects of Cryogenic Treatment on the Toughness and Tribological Behaviors of Eutectoid Steel. *J. Mech. Sci. Technol.* **2017**, *31*, 3233–3239. [CrossRef]
317. Sola, R.; Giovanardi, R.; Parigi, G.; Veronesi, P. A Novel Method for Fracture Toughness Evaluation of Tool Steels with Post-Tempering Cryogenic Treatment. *Metals* **2017**, *7*, 75. [CrossRef]
318. Nanesa, C.H.; Jahazi, M. Simultaneous Enhancement of Strength and Ductility in Cryogenically Treated AISI D2 Tool Steel. *Mater. Sci. Eng. A* **2014**, *598*, 413–419. [CrossRef]
319. Jovičević-Klug, P.; Jovičević-Klug, M.; Rohwerder, M.; Godec, M.; Podgornik, B. Complex Interdependency of Microstructure, Mechanical Properties, Fatigue Resistance and Residual Stress of Austenitic Stainless Steels AISI 304L. *Materials* **2023**, *16*, 2638. [CrossRef]
320. Weng, Z.; Gu, K.; Wang, K.; Liu, X.; Wang, J. The Reinforcement Role of Deep Cryogenic Treatment on the Strength and Toughness of Alloy Structural Steel. *Mater. Sci. Eng. A* **2020**, *772*, 138698. [CrossRef]
321. Korade, D.; Ramana, K.V.; Jagtap, K. Wear and Fatigue Behaviour of Deep Cryogenically Treated H21 Tool Steel. *Trans. Indian Inst. Met.* **2020**, *73*, 843–851. [CrossRef]
322. Bensely, A.; Shyamala, L.; Harish, S.; Mohan Lal, D.; Nagarajan, G.; Junik, K.; Rajadurai, A. Fatigue Behaviour and Fracture Mechanism of Cryogenically Treated En 353 Steel. *Mater. Des.* **2009**, *30*, 2955–2962. [CrossRef]
323. Jovicevic-Klug, M.; Jovicevic-Klug, P.; McCord, J.; Podgornik, B. Investigation of Microstructural Attributes of Steel Surfaces through Magneto-Optical Kerr Effect. *J. Mater. Res. Technol.* **2021**, *11*, 1245–1259. [CrossRef]
324. Jovičević-Klug, P.; Jovičević-Klug, M.; Podgornik, B. Unravelling the Role of Nitrogen in Surface Chemistry and Oxidation Evolution of Deep Cryogenic Treated High-Alloyed Ferrous Alloy. *Coatings* **2022**, *12*, 213. [CrossRef]
325. Jovičević-Klug, M.; Jovičević-Klug, P.; Kranjec, T.; Podgornik, B. Cross-Effect of Surface Finishing and Deep Cryogenic Treatment on Corrosion Resistance of AISI M35 Steel. *J. Mater. Res. Technol.* **2021**, *14*, 2365–2381. [CrossRef]
326. Jovičević-Klug, P.; Kranjec, T.; Jovičević-Klug, M.; Kosec, T.; Podgornik, B. Influence of the Deep Cryogenic Treatment on AISI 52100 and AISI D3 Steel's Corrosion Resistance. *Materials* **2021**, *14*, 6357. [CrossRef]
327. Senthilkumar, D.; Bracke, J.; Slootsman, N. Corrosion and Elastic Behaviour of Cryogenically Treated En 19 Steel. *Corros. Manag.* **2014**, *117*, 16–21.
328. Wang, W.; Srinivasan, V.; Siva, S.; Albert, B.; Lal, M.; Alfantazi, A. Corrosion Behavior of Deep Cryogenically Treated AISI 420 and AISI 52100 Steel. *Corrosion* **2014**, *70*, 708–720. [CrossRef]
329. Ramesh, S.; Bhuvaneswari, B.; Palani, G.S.; Lal, D.M.; Iyer, N.R. Effects on Corrosion Resistance of Rebar Subjected to Deep Cryogenic Treatment. *J. Mech. Sci. Technol.* **2017**, *31*, 123–132. [CrossRef]
330. Ma, S.; Su, R.; Li, G.; Qu, Y.; Li, R. Effect of Deep Cryogenic Treatment on Corrosion Resistance of AA7075-RRA. *J. Phys. Chem. Solids* **2022**, *167*, 110747. [CrossRef]
331. Su, R.; Ma, S.; Wang, K.; Li, G.; Qu, Y.; Li, R. Effect of Cyclic Deep Cryogenic Treatment on Corrosion Resistance of 7075 Alloy. *Met. Mater. Int.* **2022**, *28*, 862–870. [CrossRef]
332. Garrison, T.C. Process for Improving the Corrosion Resistance of Metals. CN1795278A, 24 May 2004.
333. Gunes, I.; Uzun, M.; Cetin, A.; Aslantas, K.; Cicek, A. Evaluation of Wear Performance of Cryogenically Treated Vanadis 4 Extra Tool Steel. *Kov. Mater.* **2016**, *54*, 195–204. [CrossRef]
334. Koneshlou, M.; Meshinchi, A.K.; Khomamizadeh, F. Effect of Cryogenic Tratment on Microstucture, Mechanical and Wear Behaviors of AISI H13 Hot Work Tool Steel. *Cryogenics* **2011**, *51*, 55–61. [CrossRef]
335. Senthilkumar, D.; Rajendran, I. Optimization of Deep Cryogenic Treatment to Reduce Wear Loss of 4140 Steel. *Mater. Manuf. Process.* **2011**, *27*, 567–572. [CrossRef]
336. Jovičević-Klug, P.; Jenko, M.; Jovičević-Klug, M.; Šetina Batič, B.; Kovač, J.; Podgornik, B. Effect of Deep Cryogenic Treatment on Surface Chemistry and Microstructure of Selected High-Speed Steels. *Appl. Surf. Sci.* **2021**, *548*, 149257. [CrossRef]
337. Cryogenic Treatment Applications Find Potential in the Energy Sector. Available online: https://www.cryogenicsociety.org/index.php?option=com_dailyplanetblog&view=entry&year=2023&month=05&day=02&id=188:cryogenic-treatment-applications-find-potential-in-the-energy-sector (accessed on 20 July 2023).
338. Industrial Heating. Deep Cryogenic Treatment for Marine and Oil-and-Gas Applications. 3 December 2018. Available online: https://www.industrialheating.com/articles/94598-deep-cryogenic-treatment-for-marine-and-oil-and-gas-applications (accessed on 20 July 2023).
339. Darwin, J.D.; Mohan Lal, D.; Nagarajan, G. Optimization of Cryogenic Treatment to Maximize the Wear Resistance of 18% Cr Martensitic Stainless Steel by Taguchi Method. *J. Mater. Process Technol.* **2008**, *195*, 241–247. [CrossRef]
340. Kumar, M.C.; Vijayakumar, P.; Narayan, B. Optimization of Cryogenic Treatment Parameters to Maximise the Tool Wear of HSS Tools by Taguchi Method. *Int. J. Mod. Eng. Res.* **2012**, *2*, 3051–3055.
341. Baloji, D.; Anil, K.; Satyanarayana, K.; Ul Haq, A.; Singh, S.K.; Naik, M.T. Evaluation and Optimization of Material Properties of ASS316L at Sub-Zero Temperature Using Taguchi Robust Design. *Mater. Today Proc.* **2019**, *18*, 4475–4481. [CrossRef]
342. Lance, J.W.; Jones, H.M. Material Treatment by Cryogenic Cooling. U.S. Patent No. 3,891,477, 9 September 1971.

343. Jin, F.; Wu, L.; Xiong, C.; Wei, J.; Wen, X. A Kind of Cryogenic Treating Process of Cast Aluminium Alloy Piston Piece. CN103498117B, 29 September 2013.
344. De Carvalho Eduardo, A.; Atem De Carvalho, R. Special Steels; Cryogenic Process for the Production Thereof; Use of Special Steels in a Saline and/or High-Pressure Environment. WO2014008564A1, 9 July 2013.
345. Jin, F.; Zhou, Z.; Wei, J.; Xiong, C.; Wu, L. High-Speed Steel Cryogenic Treatment Process. CN103525997A, 29 September 2013.
346. Pang, B.; Wang, B.; Zhao, X. High-Efficiency Energy-Saving Apparatus Used for Cryogenic Treatment. CN103589840A, 26 November 2013.
347. Kamody, D.J. Process for the Cryogenic Treatment of Metal Containing Materials. US5875636A, 1 October 1997.
348. Li, Z.; Wang, Y.; Wang, J.; Zhang, Y. Effect of Cryogenic Heat Treatment and Heat Treatment on the Influence of Mechanical, Energy, and Wear Properties of 316L Stainless Steel by Selective Laser Melting. *JOM* **2022**, *74*, 3855–3868. [CrossRef]
349. Prieto, G.; Ipiña, J.E.P.; Tuckart, W.R. Cryogenic Treatments on AISI 420 Stainless Steel: Microstructure and Mechanical Properties. *Mater. Sci. Eng. A* **2014**, *605*, 236–243. [CrossRef]
350. Zhirafar, S.; Rezaeian, A.; Pugh, M. Effect of Cryogenic Treatment on the Mechanical Properties of 4340 Steel. *J. Mater. Process. Technol.* **2007**, *186*, 298–303. [CrossRef]
351. Baldissera, P.; Delprete, C. Effects of Deep Cryogenic Treatment on Static Mechanical Properties of 18NiCrMo5 Carburized Steel. *Mater. Des.* **2009**, *30*, 1435–1440. [CrossRef]
352. Jovičević-Klug, M.; Rezar, R.; Jovičević-Klug, P.; Podgornik, B. Influence of Deep Cryogenic Treatment on Natural and Artificial Aging of Al-Mg-Si Alloy EN AW 6026. *J. Alloys Compd.* **2022**, *899*, 163323. [CrossRef]
353. Nageswara Rao, P.; Jayaganthan, R. Effects of Warm Rolling and Ageing after Cryogenic Rolling on Mechanical Properties and Microstructure of Al 6061 Alloy. *Mater. Des.* **2012**, *39*, 226–233. [CrossRef]
354. Shahsavari, A.; Karimzadeh, F.; Rezaeian, A.; Heydari, H. Significant Increase in Tensile Strength and Hardness in 2024 Aluminum Alloy by Cryogenic Rolling. *Procedia Mater. Sci.* **2015**, *11*, 84–88. [CrossRef]
355. Aamir, A.; Lei, P.; Lixiang, D.; Zengmin, Z. Effects of Deep Cryogenic Treatment, Cryogenic and Annealing Temperatures on Mechanical Properties and Corrosion Resistance of AA5083 Aluminium Alloy. *Int. J. Microstruct. Mater. Prop.* **2016**, *11*, 339–358. [CrossRef]
356. Gu, K.; Zhang, H.; Zhao, B.; Wang, J.; Zhou, Y.; Li, Z. Effect of Cryogenic Treatment and Aging Treatment on the Tensile Properties and Microstructure of Ti-6Al-4V Alloy. *Mater. Sci. Eng. A* **2013**, *584*, 170–176. [CrossRef]
357. Vinothkumar, T.S.; Kandaswamy, D.; Prabhakaran, G.; Rajadurai, A. Effect of Dry Cryogenic Treatment on Vickers Hardness and Wear Resistance of New Martensitic Shape Memory Nickel-Titanium Alloy. *Eur. J. Dent.* **2015**, *9*, 513–517. [CrossRef]
358. Anil Kumar, B.K.; Ananthaprasad, M.G.; Gopalakrishna, K. Action of Cryogenic Chill on Mechanical Properties of Nickel Alloy Metal Matrix Composites. In Proceedings of the International Conference on Advances in Materials and Manufacturing Applications, Bangalore, India, 14–16 July 2016; pp. 1–11.
359. Khedekar, D.; Gogte, C.L. Development of the Cryogenic Processing Cycle for Age Hardenable AA7075 Aluminium Alloy and Optimization of the Process for Surface Quality Using Gray Relational Analysis. *Mater. Today Proc.* **2018**, *5*, 4995–5003. [CrossRef]
360. Yuanchun, H.; Li, Y.; Ren, X.; Xiao, Z. Effect of Deep Cryogenic Treatment on Aging Processes of Al–Mg–Si Alloy. *Phys. Met. Metallogr.* **2019**, *120*, 914–918. [CrossRef]
361. Adin, M.Ş. Performances of Cryo-Treated and Untreated Cutting Tools in Machining of AA7075 Aerospace Aluminium Alloy. *Eur. Mech. Sci.* **2023**, *7*, 70–81. [CrossRef]
362. Vendra, S.S.L.; Goel, S.; Kumar, N.; Jayaganthan, R. A Study on Fracture Toughness and Strain Rate Sensitivity of Severely Deformed Al 6063 Alloys Processed by Multiaxial Forging and Rolling at Cryogenic Temperature. *Mater. Sci. Eng. A* **2017**, *686*, 82–92. [CrossRef]
363. Sonia, P.; Verma, V.; Saxena, K.K.; Kishore, N.; Rana, R.S. Effect of Cryogenic Treatment on Mechanical Properties and Microstructure of Aluminium 6082 Alloy. *Mater. Today Proc.* **2020**, *26*, 2248–2253. [CrossRef]
364. Özbek, N.A.; Çİçek, A.; Gülesİn, M.; Özbek, O. Application of Deep Cryogenic Treatment to Uncoated Tungsten Carbide Inserts in the Turning of AISI 304 Stainless Steel. *Met. Mater. Trans. A Phys. Met. Mater. Sci.* **2016**, *47*, 6270–6280. [CrossRef]
365. Çiçek, A.; Kıvak, T.; Ekici, E. Optimization of Drilling Parameters Using Taguchi Technique and Response Surface Methodology (RSM) in Drilling of AISI 304 Steel with Cryogenically Treated HSS Drills. *J. Intell. Manuf.* **2015**, *26*, 295–305. [CrossRef]
366. Ramos, L.B.; Simoni, L.; Mielczarski, R.G.; Vega, M.R.O.; Schroeder, R.M.; De Fraga Malfatti, C. Tribocorrosion and Electrochemical Behavior of DIN 1.4110 Martensitic Stainless Steels After Cryogenic Heat Treatment. *Mater. Res.* **2017**, *20*, 460–468. [CrossRef]
367. Ben Fredj, N.; Sidhom, H. Effects of the Cryogenic Cooling on the Fatigue Strength of the AISI 304 Stainless Steel Ground Components. *Cryogenics* **2006**, *46*, 439–448. [CrossRef]
368. Nalbant, M.; Yildiz, Y. Effect of Cryogenic Cooling in Milling Process of AISI 304 Stainless Steel. *Trans. Nonferrous Met. Soc. China* **2011**, *21*, 72–79. [CrossRef]
369. Hong, S.Y.; Broomer, M.; Hong, S.Y.; Broomer, M. Economical and Ecological Cryogenic Machining of AISI 304 Austenitic Stainless Steel. *Clean Technol. Environ. Policy* **2000**, *2*, 157–166. [CrossRef]
370. Dhananchezian, M.; Pradeep Kumar, M.; Sornakumar, T. Cryogenic Turning of AISI 304 Stainless Steel with Modified Tungsten Carbide Tool Inserts. *Mater. Manuf. Process.* **2011**, *26*, 781–785. [CrossRef]
371. Johan Singh, P.; Guha, B.; Achar, D.R.G. Fatigue Life Improvement of AISI 304L Cruciform Welded Joints by Cryogenic Treatment. *Eng. Fail. Anal.* **2003**, *10*, 1–12. [CrossRef]

372. Singh, P.J.; Mannan, S.L.; Jayakumar, T.; Achar, D.R.G. Fatigue Life Extension of Notches in AISI 304L Weldments Using Deep Cryogenic Treatment. *Eng. Fail. Anal.* **2005**, *12*, 263–271. [CrossRef]
373. Oh, D.; Song, S.; Kim, N.; Kim, M. Effect of Cryogenic Temperature on Low-Cycle Fatigue Behavior of AISI 304L Welded Joint. *Metals* **2018**, *8*, 657. [CrossRef]
374. Read, D.T.; Reed, R.P. Fracture and Strength Properties of Selected Austenitic Stainless Steels at Cryogenic Temperatures. *Cryogenics* **1981**, *21*, 415–417. [CrossRef]
375. Manimaran, G.; Pradeep Kumar, M.; Venkatasamy, R. Influence of Cryogenic Cooling on Surface Grinding of Stainless Steel 316. *Cryogenics* **2014**, *59*, 76–83. [CrossRef]
376. Kennedy, F.E.; Ye, Y.; Baker, I.; White, R.R.; Barry, R.L.; Tang, A.Y.; Song, M. Development of a New Cryogenic Tribotester and Its Application to the Study of Cryogenic Wear of AISI 316 Stainless Steel. *Wear* **2022**, *496–497*, 204309. [CrossRef]
377. Bhaskar, L.; Raj, D.S. Evaluation of the Effect of Cryogenic Treatment of HSS Drills at Different Holding Time in Drilling AISI 316-SS. *Eng. Res. Express* **2020**, *2*, 025005. [CrossRef]
378. Norberto López de Lacalle, L.; Gandarias, A.; López de Lacalle, L.N.; Aizpitarte, X.; Lamikiz, A. High Performance Drilling of Austenitic Stainless Steels. *Int. J. Mach. Mach. Mater.* **2008**, *3*, 1–17. [CrossRef]
379. Uhlář, R.; Hlaváč, L.; Gembalová, L.; Jonšta, P.; Zuchnický, O. Abrasive Water Jet Cutting of the Steels Samples Cooled by Liquid Nitrogen. *Appl. Mech. Mater.* **2013**, *308*, 7–12. [CrossRef]
380. Çiçek, A.; Uygur, I.; Kvak, T.; Altan Zbek, N. Machinability of AISI 316 Austenitic Stainless Steel with Cryogenically Treated M35 High-Speed Steel Twist Drills. *J. Manuf. Sci. Eng.* **2012**, *134*, 061003. [CrossRef]
381. Gao, Q.; Jiang, X.; Sun, H.; Zhang, Y.; Fang, Y.; Mo, D.; Li, X. Performance and Microstructure of TC4/Nb/Cu/316L Welded Joints Subjected to Cryogenic Treatment. *Mater. Lett.* **2022**, *321*, 132453. [CrossRef]
382. Sugavaneswaran, M.; Kulkarni, A. Effect of Cryogenic Treatment on the Wear Behavior of Additive Manufactured 316L Stainless Steel. *Tribol. Ind.* **2019**, *41*, 33–42. [CrossRef]
383. Wang, C.; Lin, X.; Wang, L.; Zhang, S.; Huang, W. Cryogenic Mechanical Properties of 316L Stainless Steel Fabricated by Selective Laser Melting. *Mater. Sci. Eng. A* **2021**, *815*, 141317. [CrossRef]
384. Kosaraju, S.; Singh, S.K.; Buddi, T.; Kalluri, A.; Ul Haq, A. Evaluation and Characterisation of ASS316L at Sub-Zero Temperature. *Adv. Mater. Process. Technol.* **2020**, *6*, 445–455. [CrossRef]
385. Kvackaj, T.; Rozsypalova, A.; Kocisko, R.; Bidulska, J.; Petrousek, P.; Vlado, M.; Pokorny, I.; Sas, J.; Weiss, K.P.; Duchek, M.; et al. Influence of Processing Conditions on Properties of AISI 316LN Steel Grade. *J. Mater. Eng. Perform.* **2020**, *29*, 1509–1514. [CrossRef]
386. Salehi, M.; Eskandari, M.; Yeganeh, M. Characterizations of the Microstructure and Texture of 321 Austenitic Stainless Steel After Cryo-Rolling and Annealing Treatments. *J. Mater. Eng. Perform.* **2023**, *32*, 816–834. [CrossRef]
387. Aletdinov, A.; Mironov, S.; Korznikova, G.; Konkova, T.; Zaripova, R.; Myshlyaev, M.; Semiatin, S.L. EBSD Investigation of Microstructure Evolution during Cryogenic Rolling of Type 321 Metastable Austenitic Steel. *Mater. Sci. Eng. A* **2019**, *745*, 460–473. [CrossRef]
388. Aurich, J.C.; Mayer, P.; Kirsch, B.; Eifler, D.; Smaga, M.; Skorupski, R. Characterization of Deformation Induced Surface Hardening during Cryogenic Turning of AISI 347. *CIRP Ann.* **2014**, *63*, 65–68. [CrossRef]
389. Mayer, P.; Skorupski, R.; Smaga, M.; Eifler, D.; Aurich, J.C. Deformation Induced Surface Hardening When Turning Metastable Austenitic Steel AISI 347 with Different Cryogenic Cooling Strategies. *Procedia CIRP* **2014**, *14*, 101–106. [CrossRef]
390. Laksanasittiphan, S.; Tuchinda, K.; Manonukul, A.; Suranuntchai, S. Use of Deep Cryogenic Treatment to Reduce Particle Contamination Induced Problem in Hard Disk Drive. *Key Eng. Mater.* **2017**, *730*, 265–271. [CrossRef]
391. Prieto, G.; Tuckart, W.R. Influence of Cryogenic Treatments on the Wear Behavior of AISI 420 Martensitic Stainless Steel. *J. Mater. Eng. Perform.* **2017**, *26*, 5262–5271. [CrossRef]
392. Ramesh, S.; Bhuvaneshwari, B.; Palani, G.S.; Mohan Lal, D.; Mondal, K.; Gupta, R.K. Enhancing the Corrosion Resistance Performance of Structural Steel via a Novel Deep Cryogenic Treatment Process. *Vacuum* **2019**, *159*, 468–475. [CrossRef]
393. Makalesi, A.; Şenel, S.; Koçar, O.; Kocaman, E.; Özdamar, O.; Bülent, Z.; Üniversitesi, E.; Fakültesi, M.; Bölümü, M.M. AISI 430 Çeliklerin Derin Kryonejik İşlem Sonrası Mekanik ve Mikroyapısal Özelliklerinin İncelenmesi. *Eur. J. Sci. Technol.* **2021**, *32*, 1000–1005. [CrossRef]
394. ŞİRİN, Ş.; AKINCIOĞLU, S. Investigation of Friction Performance and Surface Integrity of Cryogenically Treated AISI 430 Ferritic Stainless Steel. *Int. Adv. Res. Eng. J.* **2021**, *5*, 194–201. [CrossRef]
395. Yıldız, E.; Altan Özbek, N.; Ankara Hidrolik Mak San Tic Ltd.; Şti, H. Effect of cryogenic treatment and tempering temperature on mechanical and microstructural properties of aisi 431 steel. *Int. J. 3D Print. Technol. Digit. Ind.* **2022**, *6*, 74–82. [CrossRef]
396. Tembwa, E.N. Softening Response of As-Normalized and Cryogenic-Soaked P91 Martensitic Steels p y g P91 Martensitic Steels. Master's Thesis, University of Johannesburg, Johannesburg, South Africa, 2018.
397. Zhao, Z.; Yu, M.; Han, C.; Yang, Z.; Teng, P.; Zhong, J.; Li, S.; Liu, J. Effects of Carbide Evolution on SCC Behaviors of 10Cr13Co13Mo5NiW1VE Martensitic Stainless Steel. Available online: https://papers.ssrn.com/sol3/papers.cfm?abstract_id=4400821 (accessed on 2 October 2023).
398. Zhang, H.; Ji, X.; Ma, D.; Tong, M.; Wang, T.; Xu, B.; Sun, M.; Li, D. Effect of Aging Temperature on the Austenite Reversion and Mechanical Properties of a Fe–10Cr–10Ni Cryogenic Maraging Steel. *J. Mater. Res. Technol.* **2021**, *11*, 98–111. [CrossRef]
399. Dhananchezian, M.; Priyan, M.R.; Rajashekar, G.; Narayanan, S.S. Study the Effect of Cryogenic Cooling On Machinability Characteristics During Turning Duplex Stainless Steel 2205. *Mater. Today Proc.* **2018**, *5*, 12062–12070. [CrossRef]

400. Koppula, S.; Rajkumar, A.; Krishna, S.H.; Prudhvi, R.S.; Aparna, S.; Subbiah, R. Improving the Mechanical Properties of AISI 2205 Duplex Stainless Steel by Cryogenic Treatment Process. *E3S Web Conf.* **2020**, *184*, 01019. [CrossRef]
401. Narayanan, D.; Jagadeesha, T. Process Capability Improvement Using Internally Cooled Cutting Tool Insert in Cryogenic Machining of Super Duplex Stainless Steel 2507. In *Innovative Product Design and Intelligent Manufacturing Systems*; Lecture Notes in Mechanical Engineering; Springer: Singapore, 2020; pp. 323–330. [CrossRef]
402. Narayanan, D.; Salunkhe, V.G.; Dhinakaran, V.; Jagadeesha, T. Experimental Evaluation of Cutting Process Parameters in Cryogenic Machining of Duplex Stainless Steel. In *Advances in Industrial Automation and Smart Manufacturing*; Lecture Notes in Mechanical Engineering; Springer: Singapore, 2021; Volume 23, pp. 505–516. [CrossRef]
403. Kanagaraju, T.; Boopathy, S.R.; Gowthaman, B. Effect of Cryogenic and Wet Coolant Performance on Drilling of Super Duplex Stainless Steel (2507). *Mater. Express* **2020**, *10*, 81–93. [CrossRef]
404. Sastry, C.C.; Abeens, M.; Pradeep, N.; Manickam, M.A.M. Microstructural Analysis, Radiography, Tool Wear Characterization, Induced Residual Stress and Corrosion Behavior of Conventional and Cryogenic Trepanning of DSS 2507. *J. Mech. Sci. Technol.* **2020**, *34*, 2535–2547. [CrossRef]
405. Pradeep Samuel, A.; Arul, S. Effect of Cryogenic Treatment on the Mechanical Properties of Low Carbon Steel IS 2062. *Mater. Today Proc.* **2018**, *5*, 25065–25074. [CrossRef]
406. Thornton, R.; Slatter, T.; Lewis, R. Effects of Deep Cryogenic Treatment on the Wear Development of H13A Tungsten Carbide Inserts When Machining AISI 1045 Steel. *Prod. Eng.* **2014**, *8*, 355–364. [CrossRef]
407. Govindaraju, N.; Shakeel Ahmed, L.; Pradeep Kumar, M. Experimental Investigations on Cryogenic Cooling in the Drilling of AISI 1045 Steel. *Mater. Manuf. Process.* **2014**, *29*, 1417–1421. [CrossRef]
408. Dilip Jerold, B.; Pradeep Kumar, M. Experimental Investigation of Turning AISI 1045 Steel Using Cryogenic Carbon Dioxide as the Cutting Fluid. *J. Manuf. Process.* **2011**, *13*, 113–119. [CrossRef]
409. Mahendran, R.; Rajkumar, P.; Nirmal Raj, L.; Karthikeyan, S.; Rajeshkumar, L. Effect of Deep Cryogenic Treatment on Tool Life of Multilayer Coated Carbide Inserts by Shoulder Milling of EN8 Steel. *J. Braz. Soc. Mech. Sci. Eng.* **2021**, *43*, 378. [CrossRef]
410. Karnan, B.; Kuppusamy, A.; Latchoumi, T.P.; Banerjee, A.; Sinha, A.; Biswas, A.; Subramanian, A.K. Multi-Response Optimization of Turning Parameters for Cryogenically Treated and Tempered WC–Co Inserts. *J. Inst. Eng. India Ser. D* **2022**, *103*, 263–274. [CrossRef]
411. Senthilkumar, D. Deep Cryogenic Treatment of En 31 and En 8 Steel for the Development of Wear Resistance. *Adv. Mater. Process. Technol.* **2021**, *8*, 1769–1776. [CrossRef]
412. Senthilkumar, D. Effect of Deep Cryogenic Treatment on Residual Stress and Mechanical Behaviour of Induction Hardened En 8 Steel. *Adv. Mater. Process. Technol.* **2016**, *2*, 427–436. [CrossRef]
413. Arunkarthikeyan, K.; Balamurugan, K. Performance Improvement of Cryo Treated Insert on Turning Studies of AISI 1018 Steel Using Multi Objective Optimization. In Proceedings of the International Conference on Computational Intelligence for Smart Power System and Sustainable Energy, CISPSSE 2020, Keonjhar, India, 29–31 July 2020. [CrossRef]
414. Wigley, D.A. *The Metallurgical Structure and Mechanical Properties at Low Temperature of Nitronic 40 with Particular Reference to Its Use in the Construction of Models for Cryogenic Wind Tunnels*; National Aeronautics and Space Administration: Hampton, VA, USA, 1982.
415. Gaddam, S.; Haridas, R.S.; Sanabria, C.; Tammana, D.; Berman, D.; Mishra, R.S. Friction Stir Welding of SS 316 LN and Nitronic 50 Jacket Sections for Application in Superconducting Fusion Magnet Systems. *Mater. Des.* **2022**, *221*, 110949. [CrossRef]
416. Sastry, C.C.; Hariharan, P.; Pradeep Kumar, M.; Muthu Manickam, M.A. Experimental Investigation on Boring of HSLA ASTM A36 Steel under Dry, Wet, and Cryogenic Environments. *Mater. Manuf. Process.* **2019**, *34*, 1352–1379. [CrossRef]
417. Y Chow, J.G.; Klamut, C.J. *Properties of Cast C£-8 Stainless-Steel Weldments at Cryogenic Temperatures*; Brookhaven National Lab.: Upton, NY, USA, 1981.
418. Thornton, R.; Slatter, T.; Jones, A.H.; Lewis, R. The Effects of Cryogenic Processing on the Wear Resistance of Grey Cast Iron Brake Discs. *Wear* **2011**, *271*, 2386–2395. [CrossRef]
419. Franco Steier, V.; Kalombo Badibanga, R.; Roberto Moreira Da Silva, C.; Magalhães Nogueira, M.; Araújo, J.A. Effect of Chromium Nitride Coatings and Cryogenic Treatments on Wear and Fretting Fatigue Resistance of Aluminum. *Electr. Power Syst. Res.* **2014**, *116*, 322–329. [CrossRef]
420. Kara, F.; Çiçek, A. Multiple Regression and ANN Models for Surface Quality of Cryogenically-Treated AISI 52100 Bearing Steel Micro Machining of Ti6Al4V and Inconel 718 View Project Improvement of Machining Precesses View Project. *Artic. J. Balk. Tribol. Assoc.* **2013**, *19*, 570–584.
421. Gunes, I.; Cicek, A.; Aslantas, K.; Kara, F. Effect of Deep Cryogenic Treatment on Wear Resistance of AISI 52100 Bearing Steel. *Trans. Indian Inst. Met.* **2014**, *67*, 909–917. [CrossRef]
422. Villa, M.; Pantleon, K.; Somers, M.A.J. Enhanced Carbide Precipitation during Tempering of Sub-Zero Celsius Treated AISI 52100 Bearing Steel. In Proceedings of the Heat Treat & Surface Engineering Conference & Expo, Chennai, India, May 16–18 2013; pp. 1–7.
423. Hong, S.Y.; Ding, Y. Micro-Temperature Manipulation in Cryogenic Machining of Low Carbon Steel. *J. Mater. Process Technol.* **2001**, *116*, 22–30. [CrossRef]
424. Jamali, A.R.; Khan, W.; Chandio, A.D.; Anwer, Z.; Jokhio, M.H.; Karachi, P. Effect of Cryogenic Treatment on Mechanical Properties of AISI 4340 and AISI 4140 Steel. *J. Eng. Technol.* **2019**, *38*, 2413–7219. [CrossRef]

425. Lisiecki, A.; Ślizak, D.; Kukofka, A. Laser Cladding of Co-Based Metallic at Cryogenic Conditions. *J. Achiev. Mater. Manuf. Eng.* **2019**, *95*, 20–31. [CrossRef]
426. Keseler, H.; Westermann, I.; Kandukuri, S.Y.; Nøkleby, J.O.; Holmedal, B. Permanent Effect of a Cryogenic Spill on Fracture Properties of Structural Steels. *IOP Conf. Ser. Mater. Sci. Eng.* **2015**, *102*, 012004. [CrossRef]
427. Abdin, A.; Feyzabi, K.; Hellman, O.; Nordström, H.; Rasa, D.; Thaung Tolförs, G.; Öqvist, P.-O. *Methods to Create Compressive Stress in High Strength Steel Components*; Ångströmlaboratoriet: Uppsala, Sweden, 2018.
428. Walters, C.L.; Alvaro, A.; Maljaars, J. The Effect of Low Temperatures on the Fatigue Crack Growth of S460 Structural Steel. *Int. J. Fatigue* **2016**, *82*, 110–118. [CrossRef]
429. Wang, C.; Yi, Y.; Huang, S.; Dong, F.; He, H.; Huang, K.; Jia, Y. Experimental and Theoretical Investigation on the Forming Limit of 2024-O Aluminum Alloy Sheet at Cryogenic Temperatures. *Met. Mater. Int.* **2021**, *27*, 5199–5211. [CrossRef]
430. Fiedler, T.; Al-Sahlani, K.; Linul, P.A.; Linul, E. Mechanical Properties of A356 and ZA27 Metallic Syntactic Foams at Cryogenic Temperature. *J. Alloys Compd.* **2020**, *813*, 152181. [CrossRef]
431. Sagar, S.R.; Srikanth, K.M.; Jayasimha, R. Effect of Cryogenic Treatment and Heat Treatment on Mechanical and Tribological Properties of A356 Reinforced with SiC. *Mater. Today Proc.* **2021**, *45*, 184–190. [CrossRef]
432. Zhao, Z.; Hong, S.Y. Cooling Strategies for Cryogenic Machining from a Materials Viewpoint. *J. Mater. Eng. Perform.* **1992**, *1*, 669–678. [CrossRef]
433. Jovičević-Klug, M.; Jovičević-Klug, P.; Sever, T.; Feizpour, D.; Podgornik, B. Extraordinary Nanocrystalline Pb Whisker Growth from Bi-Mg-Pb Pools in Aluminum Alloy 6026 Moderated through Oriented Attachment. *Nanomaterials* **2021**, *11*, 1842. [CrossRef] [PubMed]
434. Jovičević-Klug, M.; Tegg, L.; Jovičević-Klug, P.; Dražić, G.; Almásy, L.; Lim, B.; Cairney, J.M.; Podgornik, B. Multiscale Modification of Aluminum Alloys with Deep Cryogenic Treatment for Advanced Properties. *J. Mater. Res. Technol.* **2022**, *21*, 3062–3073. [CrossRef]
435. Jovičević-Klug, M.; Verbovšek, T.; Jovičević-Klug, P.; Batič, B.Š.; Ambrožič, B.; Dražić, G.; Podgornik, B. Revealing the Pb Whisker Growth Mechanism from Al-Alloy Surface and Morphological Dependency on Material Stress and Growth Environment. *Materials* **2022**, *15*, 2574. [CrossRef] [PubMed]
436. Wang, X.; Fan, X.; Chen, X.; Yuan, S. Forming Limit of 6061 Aluminum Alloy Tube at Cryogenic Temperatures. *J. Mater. Process. Technol.* **2022**, *306*, 117649. [CrossRef]
437. Wang, X.; Fan, X.; Chen, X.; Yuan, S. Cryogenic Deformation Behavior of 6061 Aluminum Alloy Tube under Biaxial Tension Condition. *J. Mater. Process. Technol.* **2022**, *303*, 117532. [CrossRef]
438. Bouzada, F.; Cabeza, M.; Merino, P.; Trillo, S. Effect of Deep Cryogenic Treatment on the Microstructure of an Aerospace Aluminum Alloy. *Adv. Mat. Res.* **2012**, *445*, 965–970. [CrossRef]
439. Mohan, K.; Suresh, J.A.; Ramu, P.; Jayaganthan, R. Microstructure and Mechanical Behavior of Al 7075-T6 Subjected to Shallow Cryogenic Treatment. *J. Mater. Eng. Perform.* **2016**, *25*, 2185–2194. [CrossRef]
440. Siyi, M.; Su, R.; Wang, K.; Yang, Y.; Qu, Y.; Li, R. Effect of Deep Cryogenic Treatment on Wear and Corrosion Resistance of an Al–Zn–Mg–Cu Alloy. *Russ. J. Non-Ferr. Met.* **2021**, *62*, 89–96. [CrossRef]
441. Wei, L.; Wang, D.; Li, H.; Xie, D.; Ye, F.; Song, R.; Zheng, G.; Wu, S. Effects of Cryogenic Treatment on the Microstructure and Residual Stress of 7075 Aluminum Alloy. *Metals* **2018**, *8*, 273. [CrossRef]
442. Zhang, P.; Liu, Z.; Liu, J.; Yu, J.; Mai, Q.; Yue, X. Effect of Aging plus Cryogenic Treatment on the Machinability of 7075 Aluminum Alloy. *Vacuum* **2023**, *208*, 111692. [CrossRef]
443. Deshpande, Y.V.; Andhare, A.B.; Padole, P.M. How Cryogenic Techniques Help in Machining of Nickel Alloys? A Review. *Mach. Sci. Technol.* **2018**, *22*, 543–584. [CrossRef]
444. Baig, A.; Jaffery, S.H.I.; Khan, M.A.; Alruqi, M. Statistical Analysis of Surface Roughness, Burr Formation and Tool Wear in High Speed Micro Milling of Inconel 600 Alloy under Cryogenic, Wet and Dry Conditions. *Micromachines* **2022**, *14*, 13. [CrossRef] [PubMed]
445. Satyanarayana, K.; Krishna, B.R.; Bhargavi, M.; Vasuki, R.E.; Kiran, K.R. Taguchi Optimization in Machining Inconel 600 with WEDM Process Using Cryogenically Treated Brass Wire. *E3S Web Conf.* **2021**, *309*, 01110. [CrossRef]
446. Mandal, P.K.; Michael Saji, A.; Kurian Lalu, A.; Krishnan, A.; Nair, A.S.; Jacob, M.M. Microstructural Study and Mechanical Properties of TIG Welded Inconel 617 Superalloy. *Mater. Today Proc.* **2022**, *62*, 3561–3568. [CrossRef]
447. Akgün, M.; Demir, H. Optimization of Cutting Parameters Affecting Surface Roughness in Turning of Inconel 625 Superalloy by Cryogenically Treated Tungsten Carbide Inserts. *SN Appl. Sci.* **2021**, *3*, 277. [CrossRef]
448. Yıldırım, Ç.V. Experimental Comparison of the Performance of Nanofluids, Cryogenic and Hybrid Cooling in Turning of Inconel 625. *Tribol. Int.* **2019**, *137*, 366–378. [CrossRef]
449. Anburaj, R.; Pradeep Kumar, M. Experimental Studies on Cryogenic CO2 Face Milling of Inconel 625 Superalloy. *Mater. Manuf. Process.* **2020**, *36*, 814–826. [CrossRef]
450. Makhesana, M.A.; Patel, K.M.; Khanna, N. Analysis of Vegetable Oil-Based Nano-Lubricant Technique for Improving Machinability of Inconel 690. *J. Manuf. Process.* **2022**, *77*, 708–721. [CrossRef]
451. Jovičevič Klug, P.; Jovičević Klug, M.; Podgornik, B. Potential in Deep Cryogenic Treatment of Non-Ferrous Alloys. In Proceedings of the European Cryogenics Days 2021, Virtual Conference, 4 November 2021; Cryogenics Society of Europe: Darmstadt, Germany, 2021.

452. Palanisamy, A.; Jeyaprakash, N.; Sivabharathi, V.; Sivasankaran, S. Effects of Dry Turning Parameters of Incoloy 800H Superalloy Using Taguchi-Based Grey Relational Analysis and Modeling by Response Surface Methodology. *Proc. Inst. Mech. Eng. Part C J. Mech. Eng. Sci.* **2022**, *236*, 607–623. [CrossRef]
453. Palanisamy, A.; Selvaraj, T.; Sivasankaran, S. Optimization of Turning Parameters of Machining Incoloy 800H Superalloy Using Cryogenically Treated Multilayer CVD-Coated Tool. *Arab. J. Sci. Eng.* **2018**, *43*, 4977–4990. [CrossRef]
454. Palanisamy, A.; Selvaraj, T. Optimization of Turning Parameters for Surface Integrity Properties on Incoloy 800 h Superalloy Using Cryogenically Treated Multi-Layer Cvd Coated Tool. *Surf. Rev. Lett.* **2019**, *26*, 1850139. [CrossRef]
455. Dhananchezian, M. Study the Machinability Characteristics of Nicked Based Hastelloy C-276 under Cryogenic Cooling. *Measurement* **2019**, *136*, 694–702. [CrossRef]
456. Nas, E.; Kara, F. Optimization of EDM Machinability of Hastelloy C22 Super Alloys. *Machines* **2022**, *10*, 1131. [CrossRef]
457. Akincioğlu, S.; Gökkaya, H.; Akincioğlu, G.; Karataş, M.A. Taguchi Optimization of Surface Roughness in the Turning of Hastelloy C22 Super Alloy Using Cryogenically Treated Ceramic Inserts. *Proc. Inst. Mech. Eng. Part C J. Mech. Eng. Sci.* **2020**, *234*, 3826–3836. [CrossRef]
458. Ekambaram, P. Study of Mechanical and Metallurgical Properties of Hastelloy X at Cryogenic Condition. *J. Mater. Res. Technol.* **2019**, *8*, 6413–6419. [CrossRef]
459. Wu, Y.; Yuan, X.; Wen, X.; Jiao, M. Body-Centered Cubic High-Entropy Alloys. In *Materials Horizons: From Nature to Nanomaterials*; Springer: Singapore, 2022; pp. 3–34. [CrossRef]
460. Hu, W.; Dong, Z.; Wang, H.; Ahamad, T.; Ma, Z. Microstructure Refinement and Mechanical Properties Improvement in the W-Y$_2$O$_3$ Alloys via Optimized Freeze-Drying. *Int. J. Refract. Met. Hard Mater.* **2021**, *95*, 105453. [CrossRef]
461. Yildiz, Y.; Sundaram, M.M.; Rajurkar, K.P.; Nalbant, M. The Effects of Cold and Cryogenic Treatments on the Machinability of Beryllium-Copper Alloy in Electro Discharge Machinability. In Proceedings of the 44th CIRP Conference on Manufacturing Systems, Madison, WI, USA, 1–3 June 2011.
462. Bagherzadeh, A.; Kuram, E.; Budak, E. Experimental Evaluation of Eco-Friendly Hybrid Cooling Methods in Slot Milling of Titanium Alloy. *J. Clean. Prod.* **2021**, *289*, 125817. [CrossRef]
463. Uygur, I.; Gerengi, H.; Arslan, Y.; Kurtay, M. The Effects of Cryogenic Treatment on the Corrosion of AISI D3 Steel. *Mater. Res.* **2015**, *18*, 569–574. [CrossRef]
464. Shinde, T.; Pruncu, C.; Dhokey, N.B.; Parau, A.C.; Vladescu, A. Effect of Deep Cryogenic Treatment on Corrosion Behavior of AISI H13 Die Steel. *Materials* **2021**, *14*, 7863. [CrossRef] [PubMed]
465. Wang, Y.M.; Liang, Y.; Zhai, Y.D.; Zhang, Y.S.; Sun, H.; Liu, Z.G.; Su, G.Q. Study on the Role of Cryogenic Treatment on Corrosion and Wear Behaviors of High Manganese Austenitic Steel. *J. Mater. Res. Technol.* **2023**, *24*, 5271–5285. [CrossRef]
466. Akhbarizadeh, A.; Amini, K.; Javadpour, S. Effects of Applying an External Magnetic Field during the Deep Cryogenic Heat Treatment on the Corrosion Resistance and Wear Behavior of 1.2080 Tool Steel. *Mater. Des.* **2012**, *41*, 114–123. [CrossRef]
467. Wang, L.; Dong, C.; Cao, Y.; Liang, J.; Xiao, K.; Li, X. Co-Enhancing the Mechanical Property and Corrosion Resistance of Selective Laser Melted High-Strength Stainless Steel via Cryogenic Treatment. *J. Mater. Eng. Perform.* **2020**, *29*, 7052–7062. [CrossRef]
468. Baldissera, P.; Delprete, C. Deep Cryogenic Treatment of AISI 302 Stainless Steel: Part II—Fatigue and Corrosion. *Mater. Des.* **2010**, *31*, 4731–4737. [CrossRef]
469. Cai, Y.; Luo, Z.; Zeng, Y. Influence of Deep Cryogenic Treatment on the Microstructure and Properties of AISI304 Austenitic Stainless Steel A-TIG Weld. *Sci. Technol. Weld. Join.* **2016**, *22*, 236–243. [CrossRef]
470. He, X.; Lü, X.-Y.; Wu, Z.-W.; Li, S.-H.; Yong, Q.-L.; Liang, J.-X.; Su, J.; Zhou, L.-X.; Li, J. M23C6 Precipitation and Si Segregation Promoted by Deep Cryogenic Treatment Aggravating Pitting Corrosion of Supermartensitic Stainless Steel. *J. Iron Steel Res. Int.* **2021**, *28*, 629–640. [CrossRef]
471. Cabeza, M.; Feijoo, I.; Merino, P.; Trillo, S. Effect of the Deep Cryogenic Treatment on the Stress Corrosion Cracking Behaviour of AA 2017-T4 Aluminium Alloy. *Mater. Corros.* **2016**, *67*, 504–512. [CrossRef]
472. Diekman, F. Cold and Cryogenic Treatment of Steel. In *Steel Heat Treating Fundamentals and Processes*; ASM International: Almere, The Netherlands, 2013.

Disclaimer/Publisher's Note: The statements, opinions and data contained in all publications are solely those of the individual author(s) and contributor(s) and not of MDPI and/or the editor(s). MDPI and/or the editor(s) disclaim responsibility for any injury to people or property resulting from any ideas, methods, instructions or products referred to in the content.

MDPI
St. Alban-Anlage 66
4052 Basel
Switzerland
www.mdpi.com

Coatings Editorial Office
E-mail: coatings@mdpi.com
www.mdpi.com/journal/coatings

Disclaimer/Publisher's Note: The statements, opinions and data contained in all publications are solely those of the individual author(s) and contributor(s) and not of MDPI and/or the editor(s). MDPI and/or the editor(s) disclaim responsibility for any injury to people or property resulting from any ideas, methods, instructions or products referred to in the content.